Stochastic Modeling and Optimization

With Applications in Queues, Finance, and Supply Chains

Springer
New York
Berlin
Heidelberg
Hong Kong
London
Milan
Paris
Tokyo

David D. Yao Hanqin Zhang
Xun Yu Zhou
Editors

Stochastic Modeling and Optimization

With Applications in Queues, Finance, and Supply Chains

With 30 Illustrations

Springer

David D. Yao
Department of Operations Research
and Industrial Engineering
Columbia University
New York, NY 10027
USA
yao@fermi.ieor.columbia.edu

Hanqin Zhang
Academy of Mathematics and
System Sciences
Chinese Academy of Science
Beijing
China
hzhang@amath3.amt.ac.cn

Xun Yu Zhou
Department of Systems Engineering
and Engineering Management
Chinese University of Hong Kong
Shatin, Hong Kong
China
xyzhou@se.cuhk.edu.hk

Library of Congress Cataloging-in-Publication Data
Stochastic modeling and optimization : with applications in queues, finance, and supply chains / editors, David D. Yao, Hanzin [i.e. Hanqin] Zhang, Xun Yu Zhao.
p. cm.
Parttially based on papers presented at conferences held in Beijing on May 23-25, 2001 and in Hong Kong on May 28-29, 2001.
Includes bibliographical references and index.
ISBN 0-387-95582-8 (alk. paper)
1. Economics—Mathematical models—Congresses. 2. Finance—Mathematical models—Congresses. 3. Stochastic analysis—Congresses. I. Yao, David D., 1950– II. Zhang, Hanqin. III. Zhou, Xun Yu.
HB135.S736 2003
330′.01′519232—dc21 2002030567

ISBN 0-387-95582-8 Printed on acid-free paper.

Printed in the United States of America.

9 8 7 6 5 4 3 2 1 SPIN 10892637

www.springer-ny.com

Springer-Verlag New York Berlin Heidelberg
A member of BertelsmannSpringer Science+Business Media GmbH

Preface

The objective of this volume is to highlight through a collection of chapters some of the recent research works in applied probability, specifically stochastic modeling and optimization.

The volume is organized loosely into four parts. The first part is a collection of several basic methodologies: singularly perturbed Markov chains (Chapter 1), and related applications in stochastic optimal control (Chapter 2); stochastic approximation, emphasizing convergence properties (Chapter 3); a performance-potential based approach to Markov decision programming (Chapter 4); and interior-point techniques (homogeneous self-dual embedding and central path following) applied to stochastic programming (Chapter 5).

The three chapters in the second part are concerned with queueing theory. Chapters 6 and 7 both study processing networks – a general class of queueing networks – focusing, respectively, on limit theorems in the form of strong approximation, and the issue of stability via connections to related fluid models. The subject of Chapter 8 is performance asymptotics via large deviations theory, when the input process to a queueing system exhibits long-range dependence, modeled as fractional Brownian motion.

The third part features several studies in finance engineering: continuous-time mean-variance portfolio selection based on newly developed stochastic linear-quadratic control approaches, emphasizing the difference from single-period models (Chapter 9); a general approach to minimizing variance in dynamic portfolio optimization, using convexification and separation techniques to overcome analytical and computational difficulties (Chapter 10); and a Markov-chain approach to option pricing, including a GARCH based pricing model that handles stochastic volatility (Chapter 11).

The last part illustrates three different applications that address the general issue of coordinating supply and demand: managing patient flows to improve the quality of service in a hospital system modeled as a queueing network (Chapter 12); revenue management in terms of optimal booking control in airlines, formulated and solved as a stochastic dynamic programming problem (Chapter 13); and modeling information dynamics and analyzing their impact on inventory-service performance in a supply chain (Chapter 14).

Every effort has been expended to make each chapter self-contained, logically organized, and written in a style that is similar to lecture notes, suitable for classroom or seminar studies at the graduate level. Wherever possible, key ideas are presented and elaborated in simple albeit less general forms, with all main results proved in full, and supplemented by outlines of possible generalizations and extensions. Each chapter is concluded with a section of bibliographical notes, pointing to further readings. Researchers and practitioners, as well as graduate students, should find the volume a useful and informative study guide to the topics covered.

The planning on the volume originated from a workshop, *Stochastic Models and Optimization,* jointly organized by the Academy of Mathematics and Systems Science of the Chinese Academy of Science (CAS), and the Department of Systems Engineering and Engineering Management of the Chinese University of Hong Kong (CUHK). The workshop started in Beijing, May 23-25, 2001; and continued in Hong Kong, May 28-29, 2001. A total of about 60 presentations were delivered at the workshop. Most of the contributors of this volume were participants at the workshop. The chapters collected here, however, are not necessarily based on their presentations.

We thank the two institutions, CAS and CUHK, for their support and sponsorship. Additional acknowledgments go to Columbia University and its Center for Applied Probability (David Yao), the National Science Foundation of China (Hanqin Zhang), and the Research Grant Council of Hong Kong (David Yao and Xun Yu Zhou). Several colleagues, in addition to many of the contributors, served as reviewers of the chapters. For their generous assistance, we thank Frank Chen (CUHK), Shijie Deng (Georgia Tech), Andrew Lim (Columbia/Berkeley), John Moore (Australian National University), Xiaoqing Wang (CUHK), Yinyu Ye (Iowa/Stanford), Raymond Yeung (CUHK), and Gongyun Zhao (National University of Singapore).

DDY, HZ, XYZ (June, 2002)

Contents

1

Discrete-time Singularly Perturbed Markov Chains

G. Yin and Q. Zhang

This chapter provides a survey on recent developments for the study of singularly perturbed Markov chains in discrete time. We illustrate the use of singular perturbation techniques to treat large-scale discrete-time hybrid stochastic dynamic systems involving discrete event interventions. To reduce the complexity of the systems, time-scale separation and singularly perturbed Markov chains are used in the modeling and analysis. Asymptotic expansions of probability vectors and the structural properties of these Markov chains are provided.

The chapter is organized as follows. In Section 1.1, we give motivation and the precise formulation of the singularly perturbed Markov chains in discrete time. Section 1.2 proceeds with the construction of asymptotic expansions of the corresponding probability vectors and transition matrices. In Section 1.3, we define sequences of unscaled and scaled occupation measures and obtain limit properties of such sequences. For ease of presentation, proofs of results are given for the case that the transition matrices satisfy weak irreducibility conditions (precise definitions will be given shortly). Results for transition matrices with more complex structures are presented although the verbatim proofs are omitted. Furthermore, to make it be more accessible, we concentrate on the cases that the transition matrices are time independent (i.e., the corresponding Markov chains are time homogeneous, also known as having stationary transition probabilities) and the state space is finite. In Section 1.4, we extend the results in Sections 1.2 and 1.3 to more complex situations in which multi-block weakly irreducible matrices are considered. We also look at an application of the results obtained to nearly optimal controls of a nonlinear system. Further remarks are collected in Section 1.5, including extensions to nonhomogeneous cases, some of the recent developments on singularly perturbed Markov chains, and applications to control and optimization problems.

1.1 Singularly Perturbed Markov Chains

1.1.1 Motivation

Originating from a wide range of applications in optimization and control of large-scale systems such as wireless communications, queueing networks, and manufacturing systems, hybrid systems have drawn much attention lately. Perhaps, one of the reasons is that such a formulation allows us to treat systems that are subject to not only the usual dynamics modeled by differential or difference equations, but also certain discrete event interventions. These discrete events evolve randomly and influence the underlying dynamic systems by providing a set of configurations, or frames, or regimes. Across such regimes, the dynamic behavior of the systems can be markedly different. This is especially the case when the underlying random processes are not observable and act as hidden processes. Naturally, a Markovian formulation provides an appropriate model. Owing to the computational requirement by using digital computers, finite state space models are particularly appealing. The rapid progress of science and computing technology offers us with unprecedented opportunities. At the same time, it also poses much challenges because the underlying systems tend to be of large scale. For example, the modulating hidden Markov chain to have a large state space. When the state space of the Markov chain becomes fairly large, the computational task for obtaining optimal control policies in various applications can be rather difficult due to the curse of dimensionality. Straight forward numerical schemes may be infeasible. Alternative approaches are thus necessary. With such a primary concern, this chapter presents an approach by exploiting the structure of the underlying systems and hierarchical architecture. Using such a hierarchical approach, the problem is reformulated by noting the high contrast of the fast and slow motions.

For a Markov chain with a large state space, a decomposition approach is often attractive. Ideally, we would like to split the underlying problem into subproblems that can be solved completely independently; we can then paste together the solutions of the subproblems to obtain the solution of the entire problem. Consider a homogeneous Markov chain. If, for example, the transition matrix is decomposable into several sub-transition matrices (in a diagonal block form), then the problem can be solved easily by the aforementioned decomposition methods. Unfortunately, the real world is not an ideal one. One normally does not have the leisure to deal with complete decomposability. Nevertheless, more often than not, one encounters nearly completely decomposable cases. In this chapter, we treat such a model via the formulation of discrete-time singularly perturbed Markov chains, and provide a survey on some of the recent developments. The modeling task is conveniently done by introducing a small parameter $\varepsilon > 0$ into the transition probability matrix. The transition matrix then becomes one with a small perturbation by a generator. In the subsequent sections, we concentrate on the asymptotic study of such systems. We derive properties associated with the probability distribution vectors and transition probability matrices. We then further our understanding about the unscaled and scaled occupation measures. The combined use of analytic and probabilistic methods allows us to have a comprehensive understanding of the structures of the Markovian models. To solve many problems arising in various applications, a clear understanding of the properties of the Markov chains is of crucial importance. This chapter is devoted to such a study, whereas the reduction of complexity and applications of nearly controls can be found in [61].

1.1.2 Preliminary

The definition of a discrete-time Markov chain can be found in a standard textbook in stochastic processes. (see [24, 25]). Before proceeding further, we recall the definition of weak irreducibility and quasi-stationary distribution.

Definition 1.1. Suppose that $\{\alpha_n : n \geq 0\}$ is a discrete-time Markov chain with a finite state space $\mathcal{M} = \{1, \ldots, m\}$ and transition probability matrix P (that could be time dependent). The matrix P or the corresponding Markov chain is said to be weakly irreducible if the system of equations

$$\begin{aligned} \nu P &= \nu, \\ \nu \mathbb{1}_m &= 1 \end{aligned} \tag{1.1}$$

has a unique solution $\nu = (\nu^1, \ldots, \nu^m) \in \mathbb{R}^{1\times m}$ and ν is nonnegative. Here nonnegativity means that $\nu^i \geq 0$, for each $i \in \mathcal{M}$, where $\mathbb{1}_m$ denotes an m-dimensional column vector with all component being equal to 1. The unique nonnegative solution given in (1.1) is termed a quasi-stationary distribution. If the transition matrix is time dependent, i.e., $P = P_n$, then weak irreducibility means P_n to be weak irreducible for each n and $\nu = \nu_n$. The matrix P or the corresponding Markov chain is said to be irreducible if the system of equations (1.1) has a unique solution and ν is positive (i.e., $\nu^i > 0$ for each $i \in \mathcal{M}$).

Remark 1.2. The usual notion of irreducibility given in the standard textbook of stochastic processes is usually derived from communication properties. Suppose that p_n^{ij} denotes the ijth entry of the n-step transition probability matrix. State j is accessible from state i if for some $n > 0$, $p_n^{ij} > 0$. Two states i and j are accessible to each other are said to be communicate. It is well-known that communication is an equivalence relation; see [24, 39]. For any states i and j, define f_n^{ij} to be the probability that, starting in i, the first transition to j happens at time n, i.e.,

$$\begin{aligned} f_0^{ij} &= 0, \\ f_n^{ij} &= \mathsf{P}(\alpha_n = j, \alpha_k \neq j, k = 1, \ldots, n-1 | \alpha_0 = i). \end{aligned}$$

Define also

$$f^{ij} = \sum_{n=1}^{\infty} f_n^{ij},$$

which is the probability of a transition into state j starting from i. According to the Kolmogorov's classification of states, state j is recurrent if $f^{ij} = 1$ and is transient otherwise. According to the traditional definition, a Markov chain is said to be irreducible if there is only one class. That is, all states communicate with each other. It is also known that in a finite-state Markov chain, there is at least one recurrent state (i.e., not all states are transient). Thus it contains either all recurrent states, or it also includes some transient states in addition to recurrent states.

The traditional definition of irreducibility, in fact, is equivalent to Definition 1.1. To see this, let μ^{ii} be the expected number of transitions needed to return to state i. According to [39, Theorem 4.3.1, p. 108] (see also [24, p. 86]), if i communicates with j, then

$$\lim_{n\to\infty} \frac{1}{n} \sum_{k=1}^{n} p_k^{ji} = 1/\mu^{ii} \text{ and } \lim_{n\to\infty} p_{nd(i)}^{ji} = d(i)/\mu^{ii}, \tag{1.2}$$

where $d(i)$ denotes the periodicity of the state i. In particular if the chain is aperiodic (i.e., $d(i) = 1$), the above limit is just $1/\mu^{ii}$. Note that the limit is independent of the initial state j. This is a property enjoyed by Markov chains. Roughly, after a large number of transitions, the n-step transition probability matrix tends to stabilize around its limit distribution in the sense that it will have identical rows when $n \to \infty$. Suppose that a Markov chain α_k is irreducible according to the definition of the standard textbook. If the chain is aperiodic, then (1.1) has a unique positive solution and the solution $\nu^i = \lim_n p_n^{ii}$. In the periodic case (see [39, p. 111], (1.1) still has a unique positive solution such that ν^i is the long-run proportion of time that the chain is in state i. Therefore, $\nu^i = 1/\mu^{ii}$, and the limit probability of going from i to i in $nd(i)$ steps, is $\lim_{n\to\infty} p_{nd(i)}^{ii} = d(i)/\mu^{ii} = d(i)\nu^i$.

Conversely, if a Markov chain is irreducible in accordance with Definition 1.1, we demonstrate that it is also irreducible in the traditional sense. Since in a finite-state Markov chain, there is at least one state that is positive recurrent, there exists at least one positive recurrent class. Through relabeling (elementary row and column operations) if necessary, we may assume that the states $\{1, \ldots, r\}$ form a positive recurrent class. Thus the matrix P has the form $P = \begin{pmatrix} P^1 & 0 \\ B & C \end{pmatrix}$, where P^1 is an $r \times r$ matrix with

$r \le m$; see [25, p. 3]. We argue that it must be $r = m$. For suppose not $r < m$. By the basic theorem of Markov chain [24, Theorem 1.3 of Chapter 3], P^1 has a stationary distribution $\widetilde{\nu} = (\widetilde{\nu}^1, \dots, \widetilde{\nu}^r) \in \mathbb{R}^{1\times r}$. That is, $\widetilde{\nu}P^1 = \widetilde{\nu}$ and $\widetilde{\nu}\mathbb{1}_r = 1$. Denote $\nu = (\widetilde{\nu}^1, \dots, \widetilde{\nu}^r, 0, \dots, 0) \in \mathbb{R}^{1\times m}$. Then ν is a solution of (1.1), but it is not strictly positive. This is a contradiction.

The definitions of weak irreducibility and quasi-stationary distribution are generalizations of the usual notions of irreducibility and stationary distribution, respectively. First, time dependence and nonhomogeneity are allowed. Second, even if the transition matrix is independent of n, the set of weakly irreducible matrices contains the collection of irreducible matrices as a subset. It is easily seen that the usual notion of irreducibility implies the weak irreducibility. That is, if a matrix is irreducible then it is weakly irreducible. In fact, if it is irreducible, all states belong to the same class. Then the stationary distribution is also the quasi-stationary distribution. However, the reverse implication does not hold. For example,

$$P = \begin{pmatrix} 1/2 & 1/4 & 1/4 \\ 1/4 & 1/2 & 1/4 \\ 0 & 0 & 1 \end{pmatrix} \tag{1.3}$$

is weakly irreducible, but it is not irreducible. The state space of the corresponding Markov chain is $\mathcal{M} = \{1, 2, 3\}$, where 3 is an absorbing state and the other two states are transient. The state space is partitioned to $\mathcal{M} = \{1, 2\} \cup \{3\}$. As far as the quasi-stationary distributions are concerned, first we allow them to be time dependent. Secondly, there is no positive recurrence assumption as in the usual notion of stationary distribution. Thus some of the components of the quasi-stationary distributions may well be 0. For example, the quasi-stationary distribution corresponding to the matrix given in (1.3) is $\nu = (0, 0, 1)$.

For a finite-state Markov chain, we know more about its transition matrix P. In fact ([25, pp. 4-9]), the multiplicity of the eigenvalue 1 is equal to the number of recurrent classes associated with P. In addition, any eigenvalues of modulus 1 is a root of unity. The dth roots of unit are eigenvalues of P if and only if P has a recurrent class with period d. The multiplicity of each dth root of unity is just the number of recurrent classes of period d.

Finally, as far as the subsequent developments are concerned, the most important fact we need is the existence of the unique solution of (1.1). Thus we adopt the alternative definition henceforth.

1.1.3 Singularly Perturbed Models

Let $\varepsilon > 0$ be a small parameter and k be the index of the discrete time with $k = 0, 1 \dots, \lfloor T/\varepsilon \rfloor$, for some $T > 0$, where $\lfloor b \rfloor$ denotes the largest integer that is less than or equal to b. For notational simplicity, we shall abandon the notation $\lfloor T/\varepsilon \rfloor$ henceforth and write it simply as T/ε with the understanding that we are using its integer part. Let $\alpha_{\varepsilon,k}$ be a discrete-time Markov chain with state space $\mathcal{M} = \{1, \dots, m\}$ and transition probability matrix

$$P_\varepsilon = P + \varepsilon Q, \tag{1.4}$$

where $P \in \mathbb{R}^{m\times m}$ is a transition matrix of an appropriate Markov chain and $Q \in \mathbb{R}^{m\times m}$ is a generator of another continuous-time Markov chain, i.e., $Q = (q^{ij})$ satisfies $q^{ii} \le 0$ and $\sum_{j=1}^m q^{ij} = 0$ for each i. The last equality can also be written as $Q\mathbb{1}_m = 0$. For simplicity, throughout this chapter, detailed proofs are given only for the cases where the Markov chain has stationary transition probabilities. Nonstationary cases are much more complex and more difficult to deal with, we state the corresponding results in Section 1.4, and omit the proofs with appropriate references noted.

To illustrate the idea of time-scale separation, we consider the following example. Suppose that we have a discrete-time Markov chain with a transition matrix given by

$$P = \begin{pmatrix} 0.9 & 0.05 & 0.05 & 0 \\ 0 & 0.69 & 0.01 & 0.3 \\ 0 & 0.7 & 0.05 & 0.25 \\ 0.25 & 0.05 & 0.7 & 0 \end{pmatrix}.$$

We demonstrate in what follows that such a matrix can be converted to the form as in (1.4). In accordance with the magnitude of the entries, we can split it as two matrices

$$P = \begin{pmatrix} 0.9 & 0 & 0 & 0 \\ 0 & 0.69 & 0 & 0.3 \\ 0 & 0.70 & 0 & 0.25 \\ 0.25 & 0.70 & 0 & 0 \end{pmatrix} + \begin{pmatrix} 0 & 0.05 & 0.05 & 0 \\ 0 & 0 & 0.01 & 0 \\ 0 & 0 & 0.05 & 0 \\ 0 & 0.05 & 0 & 0 \end{pmatrix}.$$

Interchanging their rows and columns to make the first matrix in block diagonal form leads to

$$\begin{pmatrix} 0.90 & 0 & 0 & \\ 0.25 & 0.7 & 0 & 0 \\ 0 & 0 & 0.7 & 0.25 \\ 0 & 0 & 0.69 & 0.3 \end{pmatrix} + \begin{pmatrix} 0 & 0.05 & 0.05 & 0 \\ 0 & 0 & 0.05 & 0 \\ 0 & 0.05 & 0 & 0 \\ 0 & 0.01 & 0 & 0 \end{pmatrix}.$$

By adding and subtracting appropriate amount from the first and second matrices, we obtain

$$\begin{pmatrix} 1 & 0 & 0 & 0 \\ 0.25 & 0.75 & 0 & 0 \\ 0 & 0 & 0.75 & 0.25 \\ 0 & 0 & 0.69 & 0.31 \end{pmatrix} + \varepsilon \begin{pmatrix} -1 & 0.5 & 0.5 & 0 \\ 0 & -0.5 & 0.5 & 0 \\ 0 & 0.5 & -0.5 & 0 \\ 0 & 0.1 & 0 & -0.1 \end{pmatrix},$$

where $\varepsilon = 0.1$. This example shows that how a matrix with inherent two-time scales can be put in a canonical form.

Singularly perturbed Markovian models arise frequently from many applications in control and optimization of dynamic systems, and in Markov decision processes. The rationale for considering (1.4) has been mentioned in the introduction. Ideally, we would like to reduce the state space of the underlying Markov chain. For example, if we consider a control problem involving a singularly perturbed Markov chain. If P is weakly irreducible, as can be seen in what follows, we can find a deterministic limit system that is averaged out with respect to the quasi-stationary measures. If P can be partitioned into several submatrices, although we cannot have a complete separability, we can obtain nearly completely decomposability.

In view of (1.4), apparently, the matrix P in (1.4) is the dominating force. Thus to model the decomposability and the partition, we begin with this matrix. It is well known that any finite state Markov chain must have at least one recurrent state. To model a Markov chain with a large state space $\mathcal{M}$, if the states belong to l weakly irreducible classes, the matrix P can be partitioned as

$$P = \text{diag}(P^1, \dots, P^l) = \begin{pmatrix} P^1 & & \\ & \ddots & \\ & & P^l \end{pmatrix}, \tag{1.5}$$

where each P^i for $i = 1, 2, \dots, l$ is a transition matrix of appropriate dimension. The corresponding state space is then split in accordance with

$$\mathcal{M} = \mathcal{M}_1 \cup \mathcal{M}_2 \cdots \cup \mathcal{M}_l,$$

where $\mathcal{M}_i$ represents a subspace with transition matrix P^i.

To some extent, each of the weakly irreducible classes can be thought of as a "recurrent class." The reason is as follows. If all the states in this class are recurrent, then the term recurrent class is clear, and the weak irreducibility implies that all the states are in the same "ergodic" class. If in a weakly irreducible class, there is one absorbing state and the rest of states are transient, the transient states will wonder around, hover about the states, and eventually be absorbed into this absorbing state. In either case, once a state is in the class, it will be remain there.

Suppose that in addition to the l weakly irreducible classes, the Markov chain also has another class of transient states, then the transition matrix P defined in (1.4) consisting of l weakly irreducible classes and a collection of transient states, takes the form

$$P = \begin{pmatrix} P^1 & & & & \\ & P^2 & & & \\ & & \ddots & & \\ & & & P^l & \\ P^{*,1} & P^{*,2} & \cdots & P^{*,l} & P^* \end{pmatrix}, \tag{1.6}$$

where each P^i is a transition matrix within the ith weakly irreducible class for $i \le l$, and the last row $(P^{*,1}, \cdots P^{*,l}, P^*)$ corresponds to the transient states. Thus, we have the following decomposition

$$\mathcal{M} = \mathcal{M}_1 \cup \mathcal{M}_2 \cup \cdots \cup \mathcal{M}_l \cup \mathcal{M}_*,$$

where for each $i = 1, \ldots, l$, $\mathcal{M}_i$ denote the subspace corresponding to the states associated with the transition matrix P^i and $\mathcal{M}_*$ denotes that of the collection of transient states. The matrices given in (1.5) and (1.6) take care of all practical concerns (see [23]).

To achieve the goal of reduction of complexity, we can lump all the states within a weakly irreducible class (or subspace) $\mathcal{M}_i$ into a single aggregated state. As a result, the work required to analyze or design controllers for such systems will be greatly reduced. For ease of presentation and as a building block preparation, the theoretical development presented here focuses on the case in which P is *weakly irreducible.*

Remark 1.3. Discrete-time vs. Continuous-time Models. There is a close connection between continuous-time singularly perturbed Markov chains and the discrete-time counter part. The discussion below, inspired by the ideas exploited in [30] and motivated by our previous work [51], reveals the natural connection.

Motivated by many applications in manufacturing systems [41, 51], consider a continuous time Markov chain $\alpha_\varepsilon(t)$ with state space $\mathcal{M} = \{1, \ldots, m\}$ and generator $\widetilde{Q}/\varepsilon + \widehat{Q}$, where $\widetilde{Q}$ and $\widehat{Q}$ are generators of some Markov chains with stationary transition probabilities. Then the probability vector $p_\varepsilon(t) = (\mathsf{P}(\alpha_\varepsilon(t) = 1), \ldots, \mathsf{P}(\alpha_\varepsilon(t) = m))$ satisfies the forward equation

$$\dot{x}_\varepsilon(t) = x_\varepsilon(t)\left((\widetilde{Q}/\varepsilon) + \widehat{Q}\right), \; x(0) = x_0.$$

Introduce a new variable $\gamma = t/\varepsilon$. Then the probability vector under the fast time scale can be viewed as a function of γ. That is, $x_\varepsilon(t) = x(\gamma)$, which satisfies the rescaled forward equation

$$\dot{x}(\gamma) = (d/d\gamma)x(\gamma) = x(\gamma)(\widetilde{Q} + \varepsilon\widehat{Q}).$$

Denote

$$Q_\varepsilon = \widetilde{Q} + \varepsilon\widehat{Q} = (q_\varepsilon^{ij}) \text{ and}$$
$$q_0 = \max_i(-q_\varepsilon^{ii}) = \max_i \sum_{j \ne i} q_\varepsilon^{ij},$$

and fix $q \ge q_0$. Define

$$\widetilde{P}_\varepsilon = (I + (1/q)(\widetilde{Q} + \varepsilon\widehat{Q})).$$

Then all entries of $\widetilde{P}_\varepsilon$ are nonnegative and $\widetilde{P}_\varepsilon \mathbb{1}_m = 1$. Therefore, $\widetilde{P}_\varepsilon$ is a transition probability matrix, and

$$\dot{x}(\gamma) = x(\gamma)[q(\widetilde{P}_\varepsilon - I)].$$

The formal solution of the above forward equation is

$$x(\gamma) = x_0 \exp(\gamma q(\widetilde{P}_\varepsilon - I)) = x_0 \exp(-\gamma q) \sum_{j=0}^{\infty} ((\gamma q)^j / j!)(\widetilde{P}_\varepsilon)^j.$$

Consider a discrete-time Markov chain $\widetilde{\alpha}_{\varepsilon,k}$ having transition matrix $\widetilde{P}_\varepsilon$. Then the corresponding probability vector

$$\widetilde{x}(k) = (P(\widetilde{\alpha}_{\varepsilon,k} = 1), \ldots, P(\widetilde{\alpha}_{\varepsilon,k} = m))$$

with $\widetilde{x}(0) = x_0$ satisfies $\widetilde{x}(k) = x_0(\widetilde{P}_\varepsilon)^k$. This together with the solution of the forward equation yields

$$x(\gamma) = \sum_{j=0}^{\infty} \exp(-\gamma q)((q\gamma)^j / j!)\widetilde{x}(j).$$

Despite the connections between discrete- and continuous-time Markov chains, the corresponding singular perturbations are different. First, the two-time scale is naturally interpreted as fast vs. slow in a continuous-time setting, whereas in a discrete-time model, it becomes normal vs. slow in a longtime horizon. Second, the time-space separation method in discrete-time singular perturbation outlined in [35] does not seem to work in our formulation. Nevertheless, the idea of time-scale separation can still be used for asymptotic expansion in discrete-time Markov chains thanks to the work of Hoppensteadt and Miranker [20].

1.1.4 Motivating Examples

In this section, we give several examples, including a discrete optimization problem, a discrete-time manufacturing system, and a discrete-time Markov chains arising from discretization.

Example 1.4. Let us consider a discrete optimization problem. Given an objective function $f : \mathcal{M} \to \mathbb{R}$, our objective is to find

$$\imath_* = \arg\min_{\imath \in \mathcal{M}} f(\imath),$$

where $\mathcal{M} = \{1, \ldots, m\}$ is the search space and each element of $\mathcal{M}$ is called a configuration or design). In the problem of interest, there is no analytic expression of $f(\imath)$ available or the formula is complicated and difficult to evaluate. Thus we have to resolve the problem by using appropriate observational data. Let $h(\imath)$ be the estimate of $f(\imath)$. we need to use a form of Monte Carlo algorithm to resolve the issue. Let α_k denote the estimate of $\imath_*$ at the end of the kth iteration of the algorithm and suppose that

$\alpha_k = \imath_k \in \mathcal{M}$. At the $(k+1)$th iteration, another candidate $\widetilde{\alpha}_{k+1}$ is randomly chosen from $\mathcal{M} - \{\imath_k\}$. We assume that there is some $\eta > 0$ such that

$$\mathsf{P}(\widetilde{\alpha}_{k+1} = \jmath_k | \alpha_k = \imath_k) \geq \eta, \tag{1.7}$$

for all k and all $\jmath_k \in \mathcal{M} - \{\imath_k\}$. For instance, $\widetilde{\alpha}_{k+1}$ may be randomly chosen as uniformly distributed in $\mathcal{M} - \{\imath_k\}$. Next, M_k independent and identically distributed (i.i.d.) samples are then generated for both $h(\imath_k)$ and $h(\jmath_k)$. Let $\zeta_k = h(\imath_k) - h(\jmath_k)$, and let $h^l(\cdot)$ be the lth sample of $h(\cdot)$. Then we have the i.i.d. samples $\zeta_k^l = h^l(\imath_k) - h^l(\jmath_k)$, $l = 1, \ldots, M_k$. To update the estimate of $\imath_*$, use the test hypotheses

$$H_0 : \; g(\imath_k) - g(\jmath_k) < 0,$$
$$H_1 : \; g(\imath_k) - g(\jmath_k) > 0$$

based on ζ_k^l, $l = 1, \ldots, M_k$, and set $\alpha_{k+1} = \widetilde{\alpha}_{k+1}$ if H_1 is accepted; $\alpha_{k+1} = \alpha_k$, otherwise. Testing is accomplished by defining a function $t_k : R^{M_k} \to R$ and a test statistic $T_k = t_k(\zeta_k^1, \ldots, \zeta_k^{M_k})$, and by using the decision rule:

$$\begin{aligned} \alpha_{k+1} &= \imath_k \;\; \text{if } \; T_k < \tau_k; \\ &= \jmath_k \;\; \text{otherwise.} \end{aligned}$$

The threshold τ_k could be different for different k's. The iterations proceed until a stopping criterion is met. The values of the threshold sequence $\{\tau_k\}$ and the sample-size sequence $\{M_k\}$, together with the forms of the test statistics $\{T_k\}$ and the distributions of the samples determine the convergence properties of the algorithm. It is clear that we would like the probability of detection (i.e., the probability that $\alpha_{k+1} = \jmath_k$ when H_1 is true) to be large and the probability of false alarm (i.e., the probability that $\alpha_{k+1} = \jmath_k$ when H_0 is true) to be small. But they are conflicting goals, and must be balanced in order to have a convergent algorithm.

The algorithm so constructed was shown to be convergent in the sense of w.p.1 (with probability one) in [16]. In what follows, we demonstrate that the rate of convergence issues can also be studied with the help of the singularly perturbed Markov chain formulation. To be more specific, it is an interesting application of the singularly perturbed Markov chains under weak irreducibility.

To begin, we partition the state space $\mathcal{M}$ as

$$\mathcal{M} = \{1, 2, \ldots, m-1\} \cup \{m\} = \mathcal{M}_{tr} \cup \mathcal{M}_a,$$

where $\mathcal{M}_{tr}$ contains all of the transient states and $\mathcal{M}_a$ includes the absorbing state $m = \imath_*$. If the Markov chain initially belongs to $\mathcal{M}_{tr}$, it will eventually be absorbed into $\mathcal{M}_a$. Nevertheless, the current situation differs from the classical notion of absorbing state in Markov chains. This is due to the presence of (1.7), which prevents the last row of the transition matrix having the form $(0, 0, \ldots, 1)$. That is, the state $\{m\}$ can be best of "approximately absorbing." To accommodate this scenario, we introduce a slight perturbation in the transition matrix so that none of the last row of the transition matrix will be 0 and the last element (corresponding to the absorbing state) is close to 1. In addition, it is possible that the transition matrix be time dependent. Generally, the rate of convergence study of the underlying algorithm is a difficult task. We resolve the problem from a different perspective by using the techniques of singularly perturbed Markov chains.

Introducing a small parameter $\varepsilon > 0$, we let the transition matrix of the Markov chain $\alpha_k = \alpha_{\varepsilon,k}$ be

$$P_{\varepsilon,k} = P + \varepsilon Q, \quad k = 0, 1, \ldots, \widetilde{T}/\varepsilon, \quad \text{for some } \widetilde{T} > 0, \tag{1.8}$$

where P is a transition matrix of the form

$$P = \begin{pmatrix} p_{11} & \cdots & p_{1,m-1} & p_{1m} \\ \cdots & \cdots & \cdots & \cdots \\ p_{m-1,1} & \cdots & p_{m-1,m-1} & p_{m-1,m} \\ 0 & 0 & 0 & 1 \end{pmatrix}, \tag{1.9}$$

Q for each $t \in [0, \widetilde{T}]$ is a generator of a continuous-time Markov chain (i.e., $q_{ii} \leq 0$ and $\sum_{i=1}^{s} q_{ij}(t) = 0$ for each $i \in \mathcal{M}$). With such a formulation, the rate of convergence can be obtained. In fact, we have

$$p_{\varepsilon,k} = \nu + O(\varepsilon + \lambda^k), \quad \text{for some } \ 0 < \lambda < 1,$$

where $p_{\varepsilon,k}$ denotes the probability distribution vector at time k. The small parameter $\varepsilon > 0$ has the meaning being the ratio of the probability of detection and false alarm.

Example 1.5 Consider a manufacturing system in discrete time. Suppose that the system is given by

$$y_{k+1} = y_k + \varepsilon(u_k - z_k), \quad y_0 \ \text{ is given}, \tag{1.10}$$

where y_k represents the surplus, u_k is the rate of production, and z_k is the rate of demand at time k. There is a constraint $0 \leq u_k \leq \alpha_{\varepsilon,k}$ where $\alpha_{\varepsilon,k}$, being a discrete-time singularly perturbed Markov chain with state space $\mathcal{M} = \{1, \ldots, m\}$, represents the capacity of the machine. Our objective is to choose the control subject to (1.10) and the capacity constraint so that the cost function

$$J_\varepsilon(u, \alpha) = \mathsf{E} \sum_{j=0}^{T/\varepsilon} \varepsilon g(y_j, u_j, \alpha_{\varepsilon,j})$$

is minimized. Using a continuous-time flow approximation, continuous-time counterparts of such models have been studied extensively in [41] among others. Since in daily operation, the production decision is often made at discrete time, it is important to study a discrete-time formulation given above as well. What is the rationale for the use of the small parameter $\varepsilon > 0$ in $\alpha_{\varepsilon,k}$? Suppose that we have a manufacturing system consisting of two machines that are subject to breakdown and repair. Suppose that machine one is less reliable than machine two. Then a small parameter ε can be introduced to distinguish the different jump probabilities. For instance, assuming each of the machines has two states up (denoted by 1) and down (denoted by 0). Then the system has four states $\{(1,1),(0,1),(1,0),(0,0)\}$. Suppose that the transition probability matrix is

$$\widehat{P} = \begin{pmatrix} 1/4 & 1/2 & 1/4 & 0 \\ 1/2 & 1/4 & 0 & 1/4 \\ 1/4 & 0 & 1/4 & 1/2 \\ 0 & 1/4 & 1/2 & 1/4 \end{pmatrix}.$$

Note that at this point, there is no small parameter involved. To distinguish the different levels of reliability, we introduce a small ε. Using $\varepsilon = 1/4$, we obtain

$$\widehat{P} = P_\varepsilon = \begin{pmatrix} 1/2 & 1/2 & 0 & 0 \\ 1/2 & 1/2 & 0 & 0 \\ 0 & 0 & 1/2 & 1/2 \\ 0 & 0 & 1/2 & 1/2 \end{pmatrix} + \varepsilon \begin{pmatrix} -1 & 0 & 1 & 0 \\ 0 & -1 & 0 & 1 \\ 1 & 0 & -1 & 0 \\ 0 & 1 & 0 & -1 \end{pmatrix}.$$

1.2 Asymptotic Expansions

This section is devoted to the construction of asymptotic expansions of the probability vectors and transition matrices. It appears to be more instructive to work out less complicated situations. We thus provide the detailed proofs for the case that P is a weakly irreducible. The results of more complex situation are then stated without proofs.

Define

$$p_{\varepsilon,k} = (\mathsf{P}(\alpha_{\varepsilon,k} = 1), \ldots, \mathsf{P}(\alpha_{\varepsilon,k} = m)) \in \mathbb{R}^{1\times m}.$$

Then $p_{\varepsilon,k}$ satisfies the equation

$$p_{\varepsilon,k+1} = p_{\varepsilon,k} P_\varepsilon, \quad p_{\varepsilon,0} = p_0. \tag{1.11}$$

Note that p_0 is a probability vector that satisfies $p_0^i \geq 0$ and $\sum_i p_0^i = 1$. We will show in what follows that $p_{\varepsilon,k}$ converges to $\nu = (\nu^1, \ldots, \nu^m)$, the quasi-stationary distribution. In what follows we first analyze the asymptotic properties of $p_{\varepsilon,k}$. Note that the convergence takes place as $\varepsilon \to 0$, $k \to \infty$, but εk remains to be bounded. For future use, for an appropriate function $f : \mathbb{R}^{1\times m} \mapsto \mathbb{R}^{1\times m}$, define an operator $\mathcal{L}^\varepsilon$ as

$$\mathcal{L}^\varepsilon f(k) = f(k+1) - f(k) P_{\varepsilon,k}. \tag{1.12}$$

Theorem 1.6. Suppose that P in (1.4) is weakly irreducible and aperiodic.

(a) Then for an $n > 0$, for some $T > 0$, and for any $0 \le k \le T/\varepsilon$, as $\varepsilon \to 0$ and $k \to \infty$, $p_{\varepsilon,k} \to \nu$, where ν is the stationary distribution corresponding to P. Moreover, there exist two sequences $\{\varphi_i(t)\}_{i=0}^n$, $0 \le t \le T$, and $\{\psi_i(k)\}_{i=0}^n$ such that $|\psi_i(k)| \le K\lambda_0^k$ for some $0 < \lambda_0 < 1$, that $\varphi_i(\cdot)$ for $i = 0, \dots, n$ are sufficiently smooth, and that

$$\sup_{0 \le k \le T/\varepsilon} \left| p_{\varepsilon,k} - \sum_{i=0}^n \varepsilon^i \varphi_i(\varepsilon k) - \sum_{i=0}^n \varepsilon^i \psi_i(k) \right| = O(\varepsilon^{n+1}).$$

(b) Denote the k-step transition matrix associated with P_ε by $P_\varepsilon(k)$. Then

$$P_\varepsilon(k) = \overline{P} + \Psi_0(k) + \varepsilon \Phi_1(t) + \varepsilon \Psi_1(k) + O\left(\varepsilon^2\right) \tag{1.13}$$

such that $\overline{P} = \mathbb{1}_m \nu$, $\Phi_1(t) = \mathbb{1}_m \varphi_1(t)$, and $|\Psi_i(k)| \le K\lambda_0^k$ for some $K > 0$ and $i = 0, 1$, where $0 < \lambda_0 < 1$.

Remark 1.7. Note that λ_0 given above is corresponding to the absolute value of the "largest" eigenvalues that are non-unity. Note that any transition matrix always has an eigenvalue 1. The aperiodic implies that no eigenvalues other than 1 are on the unit circle. In view of (b) above, $\varphi_1(t)$ and $\Phi_1(t)$ are independent of the initial conditions and $\Phi_1(t)$ has identical rows. In the subsequent study, for example, in the study of the functional invariance theorem, we need the representation of the initial layer term $\Psi_0(k)$. For future use, we note that

$$\begin{aligned} \Psi_0(k) &= \Psi_0(k-1)P, \\ \Psi_0(0) &= I - \overline{P}. \end{aligned} \tag{1.14}$$

Proof of Theorem 1.6. The proof is divided into several steps. The first step is to construct the formal expansions, the second step derives certain desired properties, and the third step validates the asymptotic expansions by providing error estimates.

Step 1. Formal expansions. Recall that now $P_{\varepsilon,k} = P + \varepsilon Q$. There are two-time scales, fast time scale $k \le T/\varepsilon$ and slow time scale εk. Define a new variable $t = \varepsilon k$.

Our task to follow is to construct approximation of solution of (1.11). We first seek outer expansion of the form

$$U_\varepsilon(t) = \sum_{i=0}^{n+1} \varepsilon^i \varphi_i(t).$$

Using a Taylor expansion of $\varphi_i(t+\varepsilon)$ about t, we obtain

$$\begin{aligned}\varphi_i(t+\varepsilon)-\varphi_i(t) &= \sum_{j=1}^{n+1-i}\frac{\varepsilon^j}{j!}\frac{d^j\varphi_i(t)}{dt^j}+O(\varepsilon^{n+2-i}),\ 0\le i\le n,\\ \varphi_{n+1}(t+\varepsilon)-\varphi_{n+1}(t) &= \varepsilon\frac{d\varphi_{n+1}(t)}{dt}+O(\varepsilon^2).\end{aligned} \tag{1.15}$$

Substituting $U_\varepsilon(t)$ into $\mathcal{L}^\varepsilon U_\varepsilon(t)=0$ leads to

$$\begin{aligned}\varphi_0(t)(P-I) &= 0,\\ \varphi_i(t)(P-I) &= \widetilde{\varphi}_{i-1}(t)\\ &\stackrel{\text{def}}{=}\sum_{j=1}^{i}\frac{1}{j!}\frac{d^j\varphi_{n+1-j}(t)}{dt^j}-\varphi_{i-1}(t)Q,\quad\text{for }\ 1\le i\le n+1.\end{aligned} \tag{1.16}$$

Define an augmented matrix $Q_c=((P-I)\vdots\mathbb{1}_m)$. The system of homogeneous equations

$$\begin{cases}\varphi_0(t)(P-I)=0,\\ \varphi_0(t)\mathbb{1}_m=1\end{cases}$$

can be rewritten as

$$\varphi_0(t)Q_c=(0_m\vdots 1). \tag{1.17}$$

The weak irreducibility implies that Q_cQ_c' has full rank, i.e., $\text{rank}[Q_cQ_c']=m$. The unique solution of (1.17) is given by

$$\varphi_0(t)=(0_m\vdots 1)Q_c'[Q_cQ_c']^{-1}. \tag{1.18}$$

Using exactly the same technique, we obtain the solution of the rest of systems of nonhomogeneous equations as

$$\varphi_i(t)=\widetilde{\varphi}_{i-1}(t)Q_c'[Q_cQ_c']^{-1},\quad\text{for }\ 1\le i\le n+1. \tag{1.19}$$

Note that these $\varphi_i(t)$'s are obtained in a consecutive manner. Thus, when we solve the equation for $\varphi_i(t)$, $\varphi_j(t)$ for $j\le i-1$ have been found and $\widetilde{\varphi}_{i-1}(t)$ is a known function.

Note that since P is a constant matrix, the k-step transition matrix depends only on the time difference k. Note that $\overline{P}$ is a matrix having identical rows with each row being the quasi-stationary distribution.

The outer expansions provide a good approximation for t to be away from 0. However, it does not satisfy the initial condition. To compensate, we need to use the initial layer corrections. To do so, substituting

$$V_\varepsilon(t)=\sum_{i=0}^{n+1}\varepsilon^i\psi_i(k)$$

into (1.11) yields

$$\begin{aligned} \psi_0(k+1) &= \psi_0(k)P, \\ \psi_\imath(k+1) &= \psi_\imath(k)P + \psi_{\imath-1}(k)Q, \quad 0 < \imath \le n+1. \end{aligned} \tag{1.20}$$

To ensure the match of the outer expansions and the initial layer corrections, we choose the initial conditions to be

$$\begin{aligned} \varphi_0(0) + \psi_0(0) &= p_0, \\ \varphi_\imath(0) + \psi_\imath(0) &= 0. \end{aligned}$$

Then it is easily seen that

$$\begin{aligned} \psi_0(k+1) &= (p_0 - \varphi_0(0))(P)^{k+1}, \\ \psi_\imath(k+1) &= -\varphi_\imath(0)(P)^{k+1} + \sum_{j=0}^{k} \psi_{\imath-1}(k)Q(P)^{k-j}, \ 0 < \imath \le n+1. \end{aligned}$$

Since P is weakly irreducible and aperiodic, all of its non-unity eigenvalues are inside the unit circle. Noting that $[p_0 - \varphi_0(0)]\overline{P} = 0$, it can be shown that

$$|\psi_0(k)| = |(p_0 - \varphi_0(0))((P)^k - \overline{P})| \le K\lambda_0^k,$$

for some $K > 0$ and some $0 < \lambda_0 < 1$. Similarly, it can be verified that $|\psi_\imath(k)| \le K\lambda_0^k$. Thus all the outer expansions and initial layer correction terms have been found and the initial layer terms decay geometrically fast.

As for the proof of (b), we seek expansions of the form

$$\Phi_0(t) + \varepsilon\Phi_1(t) + \Psi_0(k) + \varepsilon\Psi_1(k) + \widetilde{e}_\varepsilon(k),$$

with $t = \varepsilon k$, and $\widetilde{e}_\varepsilon(k)$ represents the error. To illustrate, consider $\Phi_0(t)$ and $\Psi_0(k)$. Then it is easily seen that $\Phi_0(t)(P-I) = 0$. Use $\Phi_0^i(t)$ to denote the ith row in $\Phi_0(t)$. Then

$$\Phi_0^i(t)(P - I) = 0, \text{ and } \Phi_0^i(t)\mathbb{1}_m = 1.$$

The system above has a unique solution $\Phi_0^i(t) = \nu$ since P is irreducible and aperiodic. Thus $\Phi_0^i(t) = \nu$ is independent of t for each $i \in \mathcal{M}$. As for the initial layer correction term, we need only verify the geometric decay property. Similar to part (a), in view of (1.14), since $(I - \overline{P})\overline{P} = 0$,

$$\begin{aligned} |\Psi_0(k)| &= |(I - \overline{P})(P)^k| = |(I - \overline{P})((P)^k - \overline{P})| \\ &\le |I - \overline{P}||(P)^k - \overline{P}| \le K\lambda^k, \end{aligned}$$

for some $K > 0$ and some $0 < \lambda < 1$.

Similar calculations can be obtained for the first order expansion terms $\Phi_1(t)$ and $\Psi_1(k)$, respectively. Moreover, the techniques can be used to obtain a full asymptotic expansion of the transition matrix.

Step 2: Verification of the asymptotic expansion. As alluded to in the end of last part, the proof for treating Part (b) is similar to that of the probability vector, we thus only concentrate on the verification of the asymptotic expansions for the probability vectors in this step. Define

$$e_\imath(k) = \sum_{j=0}^{\imath} \varepsilon^j \varphi_j(t) + \sum_{j=0}^{\imath} \varepsilon^j \psi_j(k) - p_{\varepsilon,k}. \tag{1.21}$$

Lemma 1.8 For each $\imath = 0, \ldots, n+1$, $\mathcal{L}^\varepsilon e_\imath(k) = O(\varepsilon^{\imath+1})$, where $\mathcal{L}^\varepsilon$ is defined in (1.12).

Proof. Since $\mathcal{L}^\varepsilon p_{\varepsilon,k} = 0$, we have for each $\imath = 0, \ldots, n+1$,

$$\mathcal{L}^\varepsilon_\varepsilon e_\imath(k) = \mathcal{L}^\varepsilon_\varepsilon \left(\sum_{j=0}^{\imath} \varepsilon^j \varphi_j(\varepsilon k) + \sum_{j=0}^{\imath} \varepsilon^j \psi_j(k) \right).$$

Then

$$\begin{aligned} &\mathcal{L}^\varepsilon \left(\sum_{j=0}^{\imath} \varepsilon^j \varphi_j(\varepsilon k) + \sum_{j=0}^{\imath} \varepsilon^j \psi_j(k) \right) \\ &= \sum_{j=0}^{\imath} \varepsilon^j [\varphi_j(\varepsilon(k+1)) - \varphi_j(\varepsilon k)(P + \varepsilon Q)] \\ &\quad + \sum_{j=0}^{\imath} \varepsilon^j [\psi_j(k+1) - \psi_j(k)(P + \varepsilon Q)]. \end{aligned}$$

Furthermore,

$$\sum_{j=0}^{\imath} \varepsilon^j [\psi_j(k+1) - \psi_j(k)(P + \varepsilon Q)] = -\varepsilon^{\imath+1} \psi_\imath(k) Q,$$

and that

$$\begin{aligned} &\sum_{j=0}^{\imath} \varepsilon^j [\varphi_j(\varepsilon(k+1)) - \varphi_j(\varepsilon k)(P + \varepsilon Q)] \\ &= \sum_{j=0}^{\imath} \varepsilon^j [\varphi^j(\varepsilon(k+1)) - \varphi_j(\varepsilon k)] - \sum_{j=0}^{\imath} \varepsilon^j \varphi_j(\varepsilon k)((P - I) + \varepsilon Q) \\ &= \sum_{j=0}^{\imath} \sum_{\iota=1}^{\imath-j} \frac{\varepsilon^\iota}{\iota!} \frac{d^\iota \varphi_j(\varepsilon k)}{dt^\iota} + O(\varepsilon^{\imath+1}) - \sum_{j=0}^{\imath} \varepsilon^j \varphi_\imath(\varepsilon k)((P - I) + \varepsilon Q) \\ &= O(\varepsilon^{\imath+1}). \end{aligned}$$

The desired lemma then follows. □

Lemma 1.9 Suppose that for some $\imath > 0$, $\{e(k) : 0 \le k \le T/\varepsilon\}$ is a solution of

$$\mathcal{L}^{\varepsilon} e(k) = O(\varepsilon^{\imath+1}), \ e(0) = 0,$$

where $\mathcal{L}^{\varepsilon}$ is defined in (1.12). Then the solution satisfies

$$\sup_{0 \le k \le T/\varepsilon} |e(k+1)| = O(\varepsilon^{\imath}).$$

Proof. Since $e(k)$ is the solution of

$$\begin{aligned} &e(k+1) = e(k)P_{\varepsilon} + O(\varepsilon^{\imath+1}), \\ &e(0) = 0, \end{aligned}$$

it can be written

$$e(k+1) = \sum_{j=0}^{k} O(\varepsilon^{\imath+1})(P_{\varepsilon})^{j}, \quad \text{for } k = 0, \ldots, \frac{T}{\varepsilon}, \tag{1.22}$$

and as a result

$$\sup_{0 \le k \le T/\varepsilon} |e(k+1)| \le O(\varepsilon^{\imath+1})(T/\varepsilon) = O(\varepsilon^{\imath})$$

as desired. □

Lemma 1.10.

$$\sup_{0 \le k \le T/\varepsilon} |e_n(k)| = O(\varepsilon^{n+1}).$$

Proof. By virtue of Lemma 1.8, $\mathcal{L}^{\varepsilon} e_{n+1}(k) = O(\varepsilon^{n+2})$. Applying Lemma 1.9 to $e_{n+1}(k)$ leads to

$$\sup_{0 \le k \le T/\varepsilon} |e_{n+1}(k+1)| = O(\varepsilon^{n+1}).$$

Since

$$e_{n+1}(k+1) = e_n(k+1) + \varepsilon^{n+1}\varphi_{n+1}(\varepsilon(k+1)) + \varepsilon^{n+1}\psi^{n+1}(k+1),$$

the boundedness of $\varphi_{n+1}(\varepsilon k)$ and $\psi_{n+1}(k)$ then implies

$$e_{n+1}(k+1) = e_n(k+1) + O(\varepsilon^{n+1}),$$

and hence

$$\begin{aligned} e_n(k+1) &= e_{n+1}(k+1) + O(\varepsilon^{n+1}) \\ &= O(\varepsilon^{n+1}) + O(\varepsilon^{n+1}) \\ &= O(\varepsilon^{n+1}). \end{aligned}$$

The proof of the theorem is thus concluded. □

1.3 Occupation Measures

To further characterize the asymptotic behavior, we show that the time that the Markov chain spends in a state i can be "approximated" by ν^i in a suitable sense. Moreover, a scaled sequence of such approximations lead to a limit diffusion process. To be more specific, for $k \le T/\varepsilon$, define a sequence of occupation measures by

$$\begin{aligned} w^i_{\varepsilon,k} &= \varepsilon \sum_{j=0}^{k-1} [I_{\{\alpha_{\varepsilon,j}=i\}} - \nu^i], \\ w_{\varepsilon,k} &= (w^1_{\varepsilon,k}, \ldots, w^m_{\varepsilon,k}) \in \mathbb{R}^{1\times m}, \end{aligned} \tag{1.23}$$

Note that $w^i_{\varepsilon,k}$ is (a centered sequence of) the amount of time the Markov chain spends in the state i and hence it is termed occupation measure. The terminology appears to be self explanatory. First we obtain the following lemma.

Lemma 1.11. For each $i \in \mathcal{M}$,

$$\sup_{0\le k\le T/\varepsilon} \mathsf{E}(w^i_{\varepsilon,k})^2 = O(\varepsilon). \tag{1.24}$$

Proof. Note that

$$\begin{aligned} \mathsf{E}(w^i_{\varepsilon,k})^2 &= \varepsilon^2 \sum_{k_1=0}^{k-1} \sum_{k_2=0}^{k-1} \mathsf{E}[I_{\{\alpha_{\varepsilon,k_1}=i\}} - \nu^i][I_{\{\alpha_{\varepsilon,k_2}=i\}} - \nu^i] \\ &= \varepsilon^2 \sum_{k_1=0}^{k-1} \sum_{k_2=0}^{k-1} \Big[\mathsf{P}(\alpha_{\varepsilon,k_1}=i, \alpha_{\varepsilon,k_2}=i) - \nu^i \mathsf{P}(\alpha_{\varepsilon,k_1}=i) \\ &\qquad -\nu^i \mathsf{P}(\alpha_{\varepsilon,k_2}=i) + (\nu^i)^2\Big] \\ &\le K\varepsilon^2 \sum_{k_1=0}^{k-1} \sum_{k_2\ge k_1}^{k-1} \Big[\mathsf{P}(\alpha_{\varepsilon,k_1}=i)\mathsf{P}(\alpha_{\varepsilon,k_2}=i|\alpha_{\varepsilon,k_1}=i) \\ &\qquad -\nu^i \mathsf{P}(\alpha_{\varepsilon,k_1}=i) - \nu^i \mathsf{P}(\alpha_{\varepsilon,k_2}=i) + (\nu^i)^2\Big] \\ &\le K\varepsilon^2 \sum_{k_1=0}^{k-1} \sum_{k_2\ge k_1}^{k-1} \Big[(\nu^i + O(\varepsilon+\lambda^{k_1}))(\nu^i + O(\varepsilon\lambda^{k_2-k_1})) \\ &\qquad -(\nu^i)^2 + O(\varepsilon + \lambda^{k_1} + \lambda^{k_2})\Big]. \end{aligned} \tag{1.25}$$

In the above, we have used the asymptotic expansions (see Theorem 1.6) to obtain

$$\begin{aligned} &\mathsf{P}(\alpha_{\varepsilon,k_2}=i|\alpha_{\varepsilon,k_1}) = \nu^i + O(\varepsilon+\lambda^{k_2-k_1}), \quad \text{for } \ k_2 \ge k_1, \\ &\mathsf{P}(\alpha_{\varepsilon,j}=i) = \nu^i + O(\varepsilon+\lambda^j), \quad \text{for } \ j=k_1 \ \text{ and } \ j=k_2. \end{aligned}$$

Using (1.25) and noting

$$\sum_{j=0}^{k-1} \lambda^j = \frac{1-\lambda^k}{1-\lambda} \le \frac{1}{1-\lambda} = O(1),$$

and $k \le T/\varepsilon$, we obtain (1.24). □

Now define

$$w^i_\varepsilon(t) = w^i_{\varepsilon,k} \ \text{ on } \ [\varepsilon k, \varepsilon k + \varepsilon).$$

Note that

$$w^i_\varepsilon(t) = \varepsilon \sum_{j=0}^{t/\varepsilon - 1} [I_{\{\alpha_{\varepsilon,j}=i\}} - \nu^i], \ \text{ for } \ i \in \mathcal{M}.$$

Using similar estimates as above, we obtain the following corollary.

Corollary 1.12. For each $i \in \mathcal{M}$,

$$\sup_{0\le t\le T} \mathsf{E}(w^i_\varepsilon(t))^2 = O(\varepsilon).$$

To proceed, define a sequence of the scaled occupation measures as

$$\begin{aligned} n^i_{\varepsilon,k} &= \sqrt{\varepsilon} \sum_{j=0}^{k} [I_{\{\alpha_{\varepsilon,j}=i\}} - \nu^i], \ i \in \mathcal{M}, \\ n_{\varepsilon,k} &= (n^1_{\varepsilon,k}, \ldots, n^m_{\varepsilon,k}) \in \mathbb{R}^{1\times m}. \end{aligned} \tag{1.26}$$

Define a sequence of piecewise constant functions by

$$n_\varepsilon(t) = n_{\varepsilon,k} \ \text{ for } \ t \in [\varepsilon k, \varepsilon k + \varepsilon).$$

Note that $n_\varepsilon(t) = (w^1_\varepsilon(t)/\sqrt{\varepsilon}, \ldots, w^m_\varepsilon(t)/\sqrt{\varepsilon})$. We will figure out the limit property of $n_\varepsilon(\cdot)$.

Theorem 1.13. Under the conditions of Theorem 1.6, $n_\varepsilon(\cdot)$ converges weakly to $n(\cdot)$, a Brownian motion whose covariance is given by Σt, where $\Sigma = (\sigma^{ij}) \in \mathbb{R}^{m\times m}$ with

$$\sigma^{ij} = \nu^i \sum_{\iota=0}^{\infty} \psi^{ij}_0(\iota) + \nu^j \sum_{\iota=1}^{\infty} \psi^{ji}_0(\iota), \tag{1.27}$$

and $\Psi_0(\iota) = (\psi^{ij}_0(\iota))$ is given by (1.14).

Proof. We divide the proof into several steps.

Step 1: Calculate the limit mean of $n_\varepsilon(\cdot)$. In view of Theorem 1.6, we have

$$\begin{aligned}\mathsf{E}n^i_\varepsilon(t) &= \sqrt{\varepsilon}\sum_{\iota=0}^{t/\varepsilon-1}[\mathsf{P}(\alpha_{\varepsilon,\iota}=i)-\nu^i]\\ &= \sqrt{\varepsilon}\sum_{\iota=0}^{t/\varepsilon-1}\lambda^\iota+\sum_{\iota=0}^{t/\varepsilon-1}O(\varepsilon^{3/2})\\ &= O(\sqrt{\varepsilon}),\quad \text{for each } i\in\mathcal{M}.\end{aligned}$$

Thus we conclude that

$$\mathsf{E}n_\varepsilon(t)\to 0 \text{ as } \varepsilon\to 0.$$

Step 2: Compute the limit covariance of $n(\cdot)$. We begin with the examination of $\mathsf{E}n^i_\varepsilon(t)n^j_\varepsilon(t)$ for $i,j\in\mathcal{M}$. Using the asymptotic expansion Theorem 1.6, detailed calculation reveals that

$$\begin{aligned}&\mathsf{E}n^i_\varepsilon(t)n^j_\varepsilon(t)\\ &=\varepsilon\sum_{k_2=0}^{t/\varepsilon-1}\sum_{k_1=0}^{k_2}\mathsf{P}(\alpha_{\varepsilon,k_1}=i)[\mathsf{P}(\alpha_{\varepsilon,k_2}=j|\alpha_{\varepsilon,k_1}=i)-\nu^j]\\ &+\varepsilon\sum_{k_1=0}^{t/\varepsilon-1}\sum_{k_2=0}^{k_1-1}\mathsf{P}(\alpha_{\varepsilon,k_2}=j)[\mathsf{P}(\alpha_{\varepsilon,k_1}=i|\alpha_{\varepsilon,k_2}=j)-\nu^i]+o(1),\end{aligned}\tag{1.28}$$

where $o(1)\to 0$ as $\varepsilon\to 0$ uniformly in $t\in[0,T]$. Again, by virtue of Theorem 1.6, making substitution $\iota=k_2-k_1$, interchanging the orders of summations, and using the Dirichlet summation formula,

$$\begin{aligned}&\varepsilon\sum_{k_2=0}^{t/\varepsilon-1}\sum_{k_1=0}^{k_2}\mathsf{P}(\alpha_{\varepsilon,k_1}=i)[\mathsf{P}(\alpha_{\varepsilon,k_2}=j|\alpha_{\varepsilon,k_1}=i)-\nu^j]\\ &\quad=\varepsilon\nu^i\sum_{k_2=0}^{t/\varepsilon-1}\sum_{\iota=0}^{k_2}\psi_0^{ij}(\iota)+o(1)\\ &\quad=\varepsilon\nu^i\sum_{\iota=0}^{t/\varepsilon-1}\sum_{k_2=\iota}^{t/\varepsilon-1}\psi_0^{ij}(\iota)+o(1)\\ &\quad=\varepsilon\nu^i\sum_{\iota=0}^{t/\varepsilon-1}(t/\varepsilon-\iota)\psi_0^{ij}(\iota)+o(1),\end{aligned}\tag{1.29}$$

where $o(1)\to 0$ as $\varepsilon\to 0$ uniformly in $t\in[0,T]$. Likewise, we also obtain

$$\begin{aligned}&\varepsilon\sum_{k_1=0}^{t/\varepsilon-1}\sum_{k_2=0}^{k_1-1}\mathsf{P}(\alpha_{\varepsilon,k_2}=j)[\mathsf{P}(\alpha_{\varepsilon,k_1}=i|\alpha_{\varepsilon,k_2}=j)-\nu^i]\\ &\quad=\varepsilon\nu^j\sum_{\iota=1}^{t/\varepsilon-1}(t/\varepsilon-\iota)\psi_0^{ji}(\iota)+o(1),\end{aligned}\tag{1.30}$$

where $o(1) \to 0$ as $\varepsilon \to 0$ uniformly in $t \in [0,T]$. Therefore, we need only figure out the limits on the last line of (1.29) and that of (1.30).

Letting $\varepsilon k_1 = t$ and $\varepsilon k_2 = s$, we have

$$\begin{aligned}
\nu^i \mathsf{P}(\alpha_{\varepsilon,k_2} = j) &= \nu^i\nu^j + \varepsilon\nu^i\varphi_1^j(s) + \nu^i\psi_0^j(k_2) + \varepsilon\nu^i\psi_1^j(k_2) + O(\varepsilon^2),\\
\nu^j \mathsf{P}(\alpha_{\varepsilon,k_1} = i) &= \nu^i\nu^j + \varepsilon\nu^j\varphi_1^i(t) + \nu^j\psi_0^i(k_1) + \varepsilon\nu^j\psi_1^i(k_1) + O(\varepsilon^2).
\end{aligned}$$

Note that if $k_2 \ge k_1$,

$$\begin{aligned}
&\mathsf{P}(\alpha_{\varepsilon,k_1} = i, \alpha_{\varepsilon,k_2} = j)\\
&\quad = \mathsf{P}(\alpha_{\varepsilon,k_1} = i)\mathsf{P}(\alpha_{\varepsilon,k_2} = j | \alpha_{\varepsilon,k_1} = i)\\
&\quad = (\nu^i + \varepsilon\varphi_1^i(t) + \psi_0^i(k_1) + \varepsilon\psi_1^i(k_1))\\
&\qquad\qquad \times(\nu^j + \varepsilon\varphi_1^j(s) + \psi_0^{ij}(k_2-k_1) + \varepsilon\psi_1^{ij}(k_2-k_1) + O(\varepsilon^2))\\
&\quad = \nu^i\nu^j + \varepsilon\nu^i\varphi_1^j(s) + \nu^i\psi_0^{ij}(k_2-k_1) + \varepsilon\nu^i\psi_1^{ij}(k_2-k_1)\\
&\qquad +\varepsilon\nu^j\varphi_1^i(t) + \varepsilon^2\varphi_1^i(t)\varphi_1^j(s) + \varepsilon\varphi_1^i(t)\psi_0^{ij}(k_2-k_1) + \varepsilon^2\varphi_1^i(t)\psi_1^{ij}(k_2-k_1)\\
&\qquad +\nu^j\psi_0^i(k_1) + \varepsilon\varphi_1^j(s)\psi_0^i(k_1) + \psi_0^i(k_1)\psi_0^{ij}(k_2-k_1) + \varepsilon\psi_0^i(k_1)\psi_1^{ij}(k_2-k_1)\\
&\qquad +\varepsilon\nu^j\psi_1^i(k_1) + \varepsilon^2\psi_1^i(k_1)\varphi_1^j(s) + \varepsilon\psi_1^i(k_1)\psi_0^{ij}(k_2-k_1)\\
&\qquad +\varepsilon^2\psi_1^i(k_1)\psi_1^{ij}(k_2-k_1) + O(\varepsilon^2).
\end{aligned}$$

Using the asymptotic expansions, we have

$$\begin{aligned}
&\varepsilon \sum_{k_1=0}^{t/\varepsilon-1}\sum_{k_2\ge k_1} \psi_0^{ij}(k_2-k_1) = O(1),\\
&\sum_{k_1=0}^{t/\varepsilon-1}\sum_{k_2\ge k_1} \psi_0^i(k_1)\psi_1^{ij}(k_2-k_1) = O(1),\\
&\varepsilon \sum_{k_1=0}^{t/\varepsilon-1}\sum_{k_2\ge k_1} \psi_0^i(k_1) = O(1).
\end{aligned}$$

These bounds hold uniformly in $t \in [0,T]$. Note that

$$\varepsilon \sum_{\iota=0}^{t/\varepsilon-1} (t/\varepsilon - \iota)\psi_0^{ij}(\iota) \to t\sum_{\iota=0}^{\infty}\psi_0^{ij}(\iota) \quad \text{as } \varepsilon \to 0.$$

In addition, using the geometric decay property of $\Psi_0(\iota)$ and the error bounds,

$$\begin{aligned}
\varepsilon\left|\sum_{\iota=0}^{t/\varepsilon-1} \iota\psi_0^{ij}(\iota)\right| &\le \varepsilon\sum_{\iota=0}^{t/\varepsilon-1}\iota\lambda^\iota = \lambda\varepsilon\sum_{\iota=1}^{t/\varepsilon-1}\iota\lambda^{\iota-1}\\
&\le \lambda\varepsilon\frac{d}{d\lambda}\left(\sum_{\iota=0}^{\infty}\lambda^\iota\right) \to 0 \quad \text{as } \varepsilon \to 0.
\end{aligned}$$

Putting the above two estimates together, we have

$$\varepsilon \sum_{\iota=0}^{t/\varepsilon-1} (t/\varepsilon - \iota)\psi_0^{ij}(\iota) \to \sum_{\iota=0}^{\infty} \psi_0^{ij}(\iota). \tag{1.31}$$

Similarly, we have that as $\varepsilon \to 0$,

$$\varepsilon \sum_{\iota=0}^{t/\varepsilon-1} (t/\varepsilon - \iota)\psi_0^{ji}(\iota) \to \sum_{\iota=0}^{\infty} \psi_0^{ji}(\iota). \tag{1.32}$$

The desired covariance then follows from (1.31) and (1.32).

Step 3: Establish the tightness of $\{n_\varepsilon(\cdot)\}$ in $D^m[0,T]$, the space of $\mathbb{R}^m$-valued functions that are right continuous and have left-hand limits, endowed with the Skorohod topology [4]. The well-known result due to Cramér and Wold states that problems involving finite-dimensional random vectors can be reduced to problems involving scalar random variables (see [4, p. 48]). Similar to the previous step, for $t \in [0,T]$, $s > 0$, and $t+s \in [0,T]$, we have for each $i \in \mathcal{M}$,

$$\begin{aligned}
&\mathsf{E}\left|n_\varepsilon^i(t+s) - n_\varepsilon^i(t)\right|^2 \\
&= \mathsf{E}\left|\sqrt{\varepsilon} \sum_{k=t/\varepsilon}^{(t+s)/\varepsilon-1} [I_{\{\alpha_{\varepsilon,k}=i\}} - \nu^i]\right|^2 \\
&= \varepsilon \mathsf{E} \sum_{k=t/\varepsilon}^{(t+s)/\varepsilon-1} \sum_{j=t/\varepsilon}^{(t+s)/\varepsilon-1} [I_{\{\alpha_{\varepsilon,k}=i\}} - \nu^i][I_{\{\alpha_{\varepsilon,j}=i\}} - \nu^i] \\
&\le K\varepsilon \sum_{k=t/\varepsilon}^{(t+s)/\varepsilon-1} \sum_{j=t/\varepsilon}^{(t+s)/\varepsilon-1} [O(\varepsilon) + \lambda^j + \lambda^k] \\
&\le O(s).
\end{aligned} \tag{1.33}$$

It can also be shown that for each $i \in \mathcal{M}$,

$$\mathsf{E}\left|n_\varepsilon^i(t+s) - n_\varepsilon^i(t)\right|^4 \le O(s^2). \tag{1.34}$$

Thus (1.33), (1.34), and the Cramér-Wold device mentioned above imply that $\{n_\varepsilon(\cdot)\}$ is tight in $D^m[0,T]$, and that all limits have continuous sample paths with probability one.

We remark that it can be shown that, $n_\varepsilon(\cdot)$ is ϕ-mixing with geometric mixing rate, and

$$\mathsf{E}|n_\varepsilon(t) - n_\varepsilon(s)|^{2\iota} = O(|t-s|^\iota), \quad \text{for} \quad \iota = 1, 2.$$

Step 4: As a consequence of Step 3, the limit $n(\cdot)$ has continuous paths w.p.1. Using characteristic functions, we can show $n(\cdot)$ has independent increments. Thus it is necessarily a Gaussian process with desired mean and covariance as calculated in Steps 1 and 2. □

Remark 1.14. We should like to remark that the limit covariance matrix is not positive definite but only positive semidefinite or nonnegative definite. Thus the corresponding limit diffusion is "degenerate". This can be easily seen by noting that the initial layer correction terms $\Psi_0(k)$ is orthogonal to $\mathbb{1}_m$.

1.4 Nonstationary Markov Chains and Applications

In the previous two sections, we have demonstrated how to obtain the asymptotic expansions of the probability vectors and transition matrices of singularly perturbed Markov chains and how to derive further properties of sequences of scaled and unscaled occupation measures. A crucial condition used there is the weak irreducibility of the matrix P given in (1.4). In addition, P and Q were also assumed to be time invariant constant matrices in (1.4). In this section, we relax the weak irreducibility, deal with transition matrix P having multiple classes; we also remove the time homogeneity condition. As an applications, we examine a nonlinear control problem under relaxed control formulation. Relaxed control is a convenient way for getting limit result. Its essence is to use a variational approach, instead of looking at the actual control, examine certain associated measure-valued processes; see [32, 51]. This section is divided into three parts. The first part is concerned with smooth transition functions, the second part examines bounded and measurable transition functions, and the third part considers a nearly optimal control problem under relaxed control formulation. Since the verbatim proofs are rather complex and technically demanding, we will omit the details and state only the main results with appropriate references noted. It seems to be more instructive to present the main ideas than getting into all the technical details in this tutorial.

1.4.1 Asymptotic Properties for Smooth Transition Matrices

The results of the last two sections can be generalized to the case of time-varying transition matrices and inclusion of several weakly irreducible classes and a collection of additional transient states. In what follows, we set up the problem and present the desired results. The proofs of the asymptotic expansions can be found in [52] and the asymptotic properties of the scaled and unscaled sequences of occupation measures can be found in [56].

Generally, time-varying transition functions are difficult to deal with. Here, we extend our results to a class of discrete-time Markov chains with nonstationary transition probabilities. Let $\varepsilon > 0$ be a small parameter and $\alpha_{\varepsilon,k}$ be a singularly perturbed Markov chain with time-dependent transition matrix $P_\varepsilon(k)$ of the form

$$P_\varepsilon(k) = P(\varepsilon k) + \varepsilon Q(\varepsilon k), \tag{1.35}$$

where for each k, $P(\varepsilon k)$ is a transition matrix of a time-inhomogeneous Markov chain, and $Q(\varepsilon k)$ is a time-dependent generator. As far as the transition matrix $P(\varepsilon k)$ is concerned, we assume that it is either of the form

$$P(\varepsilon k) = \mathrm{diag}(P^1(\varepsilon k), \ldots, P^l(\varepsilon k)), \tag{1.36}$$

or

$$P(\varepsilon k) = \begin{pmatrix} P^1(\varepsilon k) & & & & \\ & P^2(\varepsilon k) & & & \\ & & \ddots & & \\ & & & P^l(\varepsilon k) & \\ P^{*,1}(\varepsilon k) & P^{*,2}(\varepsilon k) & \cdots & P^{*,l}(\varepsilon k) & P^*(\varepsilon k) \end{pmatrix}. \tag{1.37}$$

The primary motivation for considering such transition matrices stems from, for example, the discretization of systems in continuous time. Consider a singularly perturbed, time-inhomogeneous, continuous-time Markov chain. Write down its forward equation

$$\dot{p}_\varepsilon(t) = p_\varepsilon(t)\left(\frac{\widetilde{Q}(t)}{\varepsilon} + \widehat{Q}(t)\right).$$

Making change of variable $\gamma = t/\varepsilon$ and treating $p_\varepsilon(t)$ as a function of the new variable lead to

$$\frac{dp(\gamma)}{d\gamma} = p(\gamma)(\widetilde{Q}(\varepsilon\gamma) + \varepsilon\widehat{Q}(\varepsilon\gamma)).$$

Discretizing this equation with a step size $h > 0$ yields the following difference equation

$$p_{k+1} = p_k[I + h\widetilde{Q}(\varepsilon k)] + \varepsilon p_k h\widehat{Q}(\varepsilon k).$$

Note that $h > 0$ can be properly chosen so that $I + h\widetilde{Q}(\varepsilon k) \stackrel{\text{def}}{=} P(\varepsilon k)$ becomes a transition probability matrix. It is also clear that $h\widehat{Q}(\varepsilon k) \stackrel{\text{def}}{=} Q(\varepsilon k)$ is a generator. To proceed, we pose the following conditions.

(A1) For each k, $P_\varepsilon(k)$ and $P(\varepsilon k)$ are transition probability matrices such that $P(\varepsilon k)$ is given either by (1.36) or (1.37) satisfying for each $i \le l$, $P^i(0)$ is aperiodic, $P^i(\varepsilon k)$ is irreducible, and $P^*(\varepsilon k)$ is a matrix having all of its eigenvalues inside the unit circle. The matrix-valued function $P(t)$ is $(n+1)$-times continuously differentiable with the $(n+1)$st derivatives being bounded and $Q(t)$ is Lipschitz continuous and bounded, where $t = \varepsilon k$.

Although the constructions of the asymptotic expansions are similar in spirit to the time-homogeneous and weakly irreducible cases, the calculations are much more involved. First quantities such as quasi-stationary distributions $\nu^i(t)$ ($i \in \mathcal{M}$) become time dependent. More importantly, the outer expansion and initial layer corrections cannot be constructed separately. In fact, both the outer expansions and inner expansions require the solutions of algebric-differential equations. These terms are intertwined. One cannot expect the procedure to be "short and sweet". Rather the outer and inner expansions must be obtained jointly. We refer to the reader to [52] for the details, and state only the final result in what follows.

Theorem 1.15. Assume (A1) holds. Then two sequences $\{\varphi_i(\varepsilon k)\}_{i=0}^n$ and $\{\psi_i(k)\}_{i=0}^n$ can be constructed such that for each $i = 0, \dots, n$, $\varphi_i(\cdot)$ is sufficiently smooth and $|\psi_i(k)| \le K\lambda^k$ for some $0 < \lambda < 1$. Furthermore,

$$\sup_{0 \le k \le T/\varepsilon} \left| p_{\varepsilon,k} - \sum_{j=0}^{n} \varepsilon^j \varphi_j(\varepsilon k) - \sum_{j=0}^{n} \varepsilon^j \psi_j(k) \right| = O(\varepsilon^{n+1}), \tag{1.38}$$

where $p_{\varepsilon,k}$ is the solution of (1.11).

In addition to the asymptotic expansions, we can also obtain further asymptotic properties of scaled and unscaled sequences of occupation measures. If $P_\varepsilon(\varepsilon k)$ has the form (1.36), the state space $\mathcal{M}$ is decomposed as

$$\begin{aligned} \mathcal{M} &= \{s_{11}, \dots, s_{1m_1}\} \cup \{s_{21}, \dots, s_{2m_2}\} \cup \dots \cup \{s_{l1}, \dots, s_{lm_l}\} \\ &= \mathcal{M}_1 \cup \mathcal{M}_2 \cup \dots \cup \mathcal{M}_l, \end{aligned} \tag{1.39}$$

with $m = m_1 + m_2 + \dots + m_l$. The subspace $\mathcal{M}_i$ consists of weakly irreducible states corresponding to the transition matrix $P^i(\varepsilon k)$. In this case, define an aggregated process $\overline{\alpha}_{\varepsilon,k}$ by

$$\overline{\alpha}_{\varepsilon,k} = i \;\text{ if }\; \alpha_{\varepsilon,k} \in \mathcal{M}_i, \quad \text{for } i = 1, \dots, l, \tag{1.40}$$

and continuous-time interpolations (for $t \in [0, T]$) with the interpolation interval ε as

$$\alpha_\varepsilon(t) = \alpha_{\varepsilon,k}, \;\text{ and }\; \overline{\alpha}_\varepsilon(t) = \overline{\alpha}_{\varepsilon,k} \;\text{ if } t \in [\varepsilon k, \varepsilon(k+1)). \tag{1.41}$$

Define for each $i = 1, \dots, l$, $j = 1, \dots, m_i$,

$$\begin{aligned} w_{ij}(\varepsilon k, \alpha) &= I_{\{\alpha = s_{ij}\}} - \nu^i_j(\varepsilon k) I_{\{\alpha \in \mathcal{M}_i\}}, \\ o^{ij}_{\varepsilon,k} &= \varepsilon \sum_{r=0}^{k-1} w^{ij}(\varepsilon r, \alpha_{\varepsilon,r}), \\ n^{ij}_{\varepsilon,k} &= \sqrt{\varepsilon} \sum_{r=0}^{k-1} w^{ij}(\varepsilon r, \alpha_{\varepsilon,r}), \end{aligned}$$

where $\nu^i_j(\varepsilon k)$ denotes the jth component of the quasi-stationary distribution $\nu^i(\varepsilon k)$. Define also

$$\begin{aligned} W(\varepsilon k,\alpha) &= (w^{11}(\varepsilon k,\alpha),\ldots,w^{lm_l}(\varepsilon k,\alpha)),\\ O_{\varepsilon,k} &= (o^{11}(\varepsilon k,\alpha),\ldots,o^{lm_l}(\varepsilon k,\alpha)),\\ n_{\varepsilon,k} &= (n^{11}(\varepsilon k,\alpha),\ldots,n^{lm_l}(\varepsilon k,\alpha)), \end{aligned}$$

$O_{\varepsilon,k}$, $n_{\varepsilon,k}$, and their continuous-time interpolations similar as before.

Theorem 1.16. Assume (A1) with $n = 1$. Then

(1) the estimates of Lemma 1.11 and Corollary 1.12 continue to hold;

(2) $\overline{\alpha}_\varepsilon(\cdot)$ converges weakly to $\overline{\alpha}(\cdot)$, a Markov chain generated by

$$\overline{Q}(t) = \widetilde{\nu}(t)Q(t)\widetilde{\mathbb{1}}, \tag{1.42}$$

where

$$\widetilde{\nu}(t) = \mathrm{diag}(\nu^1(t),\ldots,\nu^l(t)),\ \widetilde{\mathbb{1}} = \mathrm{diag}(\mathbb{1}_{m_1},\ldots,\mathbb{1}_{m_l});$$

(3) $n_\varepsilon(\cdot)$ converges weakly to $n(\cdot)$, a diffusion with the limit operator given by

$$\mathcal{L}f(x,i) = \frac{1}{2}\sum_{j_1=1}^{m_i}\sum_{j_2=1}^{m_i} a_{j_1j_2}(t,i)\frac{\partial^2 f(x,i)}{\partial x_{ij_1}\partial x_{ij_2}} + \overline{Q}(t)f(x,\cdot)(i);$$

(4) For each $i = 1,\ldots,l$, the covariance of the limit process is given by $S(t,i) = \Sigma(t,i)\Sigma'(t,i)$, where

$$\begin{aligned} &\Sigma(t,i)\\ &\quad = \mathrm{diag}(0_{m_1\times m_1},\ldots,0_{m_{i-1}\times m_{i-1}},\sigma(t,i),0_{m_{i+1}\times m_{i+1}},\ldots,0_{m_l\times m_l}), \end{aligned} \tag{1.43}$$

$\sigma(t,i) \in \mathbb{R}^{m_i\times m_i}$ satisfying $\sigma(t,i)\sigma'(t,i) = A(t,i) = (a_{j_1j_2}(t,i))$, and

$$A(t,i) = \nu^i_{\mathrm{diag}}(t)\Psi(0,i) + \nu^i_{\mathrm{diag}}(t)\sum_{k=1}^{\infty}\Psi(k,i) + \sum_{k=1}^{\infty}\Psi'(k,i)\nu^i_{\mathrm{diag}}(t), \tag{1.44}$$

with $\nu^i_{\mathrm{diag}}(t) = \mathrm{diag}(\nu^i_1(t),\ldots,\nu^i_{m_i}(t)) \in \mathbb{R}^{m_i\times m_i}$.

If $P(\varepsilon k)$ has the form (1.37), the state space $\mathcal{M}$ is decomposed to

$$\mathcal{M} = \mathcal{M}_1 \cup \mathcal{M}_2 \cup \cdots \cup \mathcal{M}_l \cup \mathcal{M}_*, \tag{1.45}$$

where $\mathcal{M}_*$ is the collection of transient states that are not in any of the weakly irreducible classes. In this case, assume:

(A2) For each fixed k, with t denoting εk, there exists an $m_* \times m_*$ nonsingular matrix $B(t)$ and constant matrices P^* and $P^{*,i}$ satisfying $P^*(t) - I = B(t)(P^* - I)$, and $P^{*,i}(t) = B(t)P^{*,i}$, for $i = 1, \ldots, l$. All the eigenvalues of $P^*(t)$ are inside the unit circle.

Partition the matrix $Q(\varepsilon k)$ as

$$Q(\varepsilon k) = \begin{pmatrix} Q^{11}(\varepsilon k) & Q^{12}(\varepsilon k) \\ Q^{21}(\varepsilon k) & Q^{22}(\varepsilon k) \end{pmatrix}, \tag{1.46}$$

where

$$Q^{11}(\varepsilon k) \in \mathbb{R}^{(m-m_*)\times(m-m_*)}, \; Q^{12}(\varepsilon k) \in \mathbb{R}^{(m-m_*)\times m_*},$$
$$Q^{21}(\varepsilon k) \in \mathbb{R}^{m_*\times(m-m_*)}, \text{ and } Q^{22}(\varepsilon k) \in \mathbb{R}^{m_*\times m_*}.$$

Write

$$\overline{Q}_*(\varepsilon k) = \operatorname{diag}(\nu^1(\varepsilon k), \ldots, \nu^l(\varepsilon k))(Q^{11}(\varepsilon k)\widetilde{\mathbb{1}} + Q^{12}(\varepsilon k)A_*(\varepsilon k)), \tag{1.47}$$

where

$$\begin{aligned} A_*(\varepsilon k) &= (a_1(\varepsilon k), \ldots, a_l(\varepsilon k)) \in \mathbb{R}^{m_*\times l} \in \mathbb{R}^{m_*\times l}, \quad \text{with} \\ a_i(\varepsilon k) &= -(P^*(\varepsilon k) - I)^{-1}P^{*,i}(\varepsilon k)\mathbb{1}_{m_i}, \text{ for } i = 1, \ldots, l. \end{aligned} \tag{1.48}$$

It is readily seen that

$$a_i(\varepsilon k) = a_i = (P^* - I)^{-1}P^{*,i}\mathbb{1}_{m_i}$$

is independent of k so $A_*(\varepsilon k) = A_*$ is a constant matrix. Note that $a_i = (a_{i,1}, \ldots, a_{i,m_*})' \in \mathbb{R}^{m_*\times 1}$ and that $a_{i,j}$ is the probability of entering the ith weakly irreducible class starting from transient state s_{*j}.

Define

$$\widetilde{\mathbb{1}}_* = \begin{pmatrix} \widetilde{\mathbb{1}} & 0_{(m-m_*)\times m_*} \\ A_* & 0_{m_*\times m_*} \end{pmatrix},$$

$$\nu_*(\varepsilon k) = \operatorname{diag}(\nu^1(\varepsilon k), \ldots, \nu^l(\varepsilon k), 0_{m_*\times m_*}),$$

define sequences of centered occupation measures

$$o^{ij}_{\varepsilon,k} = \begin{cases} \varepsilon \displaystyle\sum_{r=0}^{k-1} (I_{\{\alpha_{\varepsilon,r}=s_{ij}\}} - \nu^i_j(\varepsilon r)I_{\{\overline{\alpha}_{\varepsilon,r}=i\}}), & \text{for } i = 1, \ldots, l, \\ \varepsilon \displaystyle\sum_{r=0}^{k-1} I_{\{\alpha_{\varepsilon,r}=s_{*j}\}}, & \text{for } i = *, \end{cases} \tag{1.49}$$

and define the vector $O_{\varepsilon,k}$, and the continuous-time interpolation of $O_{\varepsilon,k}$. Consider an aggregated process given by

$$\overline{\alpha}_{\varepsilon,k} = \begin{cases} i, & \text{if } \alpha_{\varepsilon,k} \in \mathcal{M}_i, \\ U_j, & \text{if } \alpha_{\varepsilon,k} = s_{*j}, \end{cases} \tag{1.50}$$

where U_j is given by

$$U_j = I_{\{0 \le U \le a_{1,j}\}} + 2I_{\{a_{1,j} < U \le a_{1,j}+a_{2,j}\}} + \dots + lI_{\{a_{1,j}+\dots+a_{l-1,j} < U \le 1\}},$$

and U is a random variable uniformly distributed on $[0,1]$, independent of $\alpha_{\varepsilon,k}$.

Theorem 1.17. Under the conditions of Theorem 1.16 and (A2), the following assertions hold.

(1) The probability vector $p_{\varepsilon,k}$ admits the following asymptotic expansion:

$$p_{\varepsilon,k} = (\theta(\varepsilon k)\mathrm{diag}(\nu^1(\varepsilon k), \dots, \nu^l(\varepsilon k)), 0_{m_*}) + O(\varepsilon + \lambda^k),$$

where $0_{m_*} \in \mathbb{R}^{1\times m_*}$ and $\theta(t) = (\theta_1(t), \dots, \theta_l(t)) \in \mathbb{R}^{1\times l}$ satisfies

$$\begin{aligned} \frac{d\theta(t)}{dt} &= \theta(t)\overline{Q}_*(t), \\ \theta_i(0) &= p^{\{i\}}(0)\mathbb{1}_{m_i} - p^{\{*\}}(0)a_i, \end{aligned}$$

where we have used $p = (p^{\{1\}}, \dots, p^{\{l\}}, p^{\{*\}})$ and $p^{\{i\}} \in \mathbb{R}^{1\times m_i}$. The transition matrix $P_\varepsilon(\varepsilon k_0, \varepsilon k)$ satisfies

$$P_\varepsilon(\varepsilon k_0, \varepsilon k) = \Phi(t_0, t) + \Psi(k_0, k) + \varepsilon\widehat{\Phi}(t_0, t) + \varepsilon\widehat{\Psi}(k_0, k) + O(\varepsilon^2),$$

for some λ with $0 < \lambda < 1$, where

$$\Phi(t_0, t) = \widetilde{\mathbb{1}}_*\Theta_*(t_0, t)\nu_*(t) \tag{1.51}$$

with

$$\Theta_*(t_0, t) = \mathrm{diag}(\Theta(t_0, t), I_{m_*\times m_*}),$$

where $\Theta(t_0, t) = (\theta_{ij}(t_0, t))$ satisfies the differential equation

$$\begin{aligned} \frac{\partial\Theta(t_0, t)}{\partial t} &= \Theta(t_0, t)\overline{Q}_*(t), \\ \Theta(t_0, t_0) &= I, \end{aligned} \tag{1.52}$$

where $\overline{Q}_*(t)$ is given by (1.47);

(2) For each $j = 1, \dots, m_i$,

$$\sup_{t\in[0,T]} \mathsf{E}|o_\varepsilon^{ij}(t)|^2 = \begin{cases} O(\varepsilon), & \text{for } i = 1, \dots, l, \\ O(\varepsilon^2), & \text{for } i = *; \end{cases}$$

(3) $\overline{\alpha}_\varepsilon(\cdot)$ converges weakly to $\overline{\alpha}(\cdot)$, a Markov chain generated by $\overline{Q}_*(t)$;

(4) the assertions of Theorem 1.16 part (3) continue to hold with the limit operator given by

$$\mathcal{L}f(x,i) = \frac{1}{2}\sum_{j_1=1}^{m_i}\sum_{j_2=1}^{m_i} a_{j_1j_2}(t,i)\frac{\partial^2 f(x,i)}{\partial x_{ij_1}\partial x_{ij_2}} + \overline{Q}_*(t)f(x,\cdot)(i).$$

1.4.2 Bounded and Measurable Transition Matrices

For some $T > 0$, we work with $0 \le k \le T/\varepsilon$. Suppose that the Markov chain $\alpha_{\varepsilon,k}$ has a transition matrix given by (1.35), where $P(\varepsilon k)$ is either of the form (1.36) or (1.37). If we have assumed the smoothness conditions as in Section 5.1, we would have obtained the asymptotic expansions and derived the asymptotic normality of the scaled occupation measures. Nevertheless, in certain applications, one may not have the smoothness conditions fulfilled. Consequently, the asymptotics have to be examined from a different angle, namely, certain weak formulation. Let us begin with the following conditions.

(**A3**) For both $P(\varepsilon k)$ given by (1.36) and (1.37), for each k, $P_\varepsilon(\varepsilon k)$, and $P(\varepsilon k)$ are transition probability matrices. The matrix-valued functions $P(\varepsilon k)$ and $Q(\varepsilon k)$ are Borel measurable and $Q(\cdot)$ is uniformly bounded for $0 \le k \le T/\varepsilon$. For each $i \le l$, $P^i(\varepsilon k)$ is aperiodic and weakly irreducible for each fixed k.

In addition, for $P(\varepsilon k)$ given by (1.37), there exists an $m_* \times m_*$ nonsingular matrix Borel measurable and bounded matrix-valued function $B(\cdot)$ and constant matrices P^* and $P^{*,i}$ such that for each k, $P^*(\varepsilon k) - I = B(\varepsilon k)(P^* - I)$, and $P^{*,i}(\varepsilon k) = B(\varepsilon k)P^{*,i}$, for $i = 1, \ldots, l$. All the eigenvalues of $P^*(\varepsilon k)$ are inside the unit circle.

Since we do not assume the smoothness, no asymptotic expansion can be obtained (unless one works with weak derivatives). Nevertheless, we can obtain convergence of the probability vectors in the weak topology. We can also obtain certain asymptotic results regarding aggregation. The next two theorems provide such results. The first is concerned with the case of $P(\varepsilon k)$ containing l weakly irreducible classes, whereas the second one takes care of the case of inclusion of transient states in addition to the l weakly irreducible classes.

Theorem 1.18. Assume (A3) and $P(\varepsilon k)$ being of the form (1.36). Then the following assertions hold:

(a) The sequence $p_\varepsilon(\cdot)$ converges under the weak topology of $L^2[0,T]$ (denoted by $p_\varepsilon(\cdot)\xrightarrow{w}\theta(\cdot)\widetilde{\nu}(\cdot)$), i.e.,

$$\int_0^T p_\varepsilon(t)f(t)dt \to \int_0^T \theta(t)\widetilde{\nu}(t)f(t)dt,$$

for any square integrable function $f(\cdot)$ on $[0,T]$ ($f(t)$ has compatible dimension with $p_\varepsilon(t)$), where

$$\begin{aligned} &\dot{\theta}(t) = \theta(t)\overline{Q}(t), \quad \theta(0) = p_0\mathbb{1}, \quad \text{and} \\ &\overline{Q}(t) = \widetilde{\nu}(t)Q(t)\widetilde{\mathbb{1}}. \end{aligned} \tag{1.53}$$

(b) Suppose that $P_\varepsilon(t_0,t)$ is the interpolation of the transition matrix with $t_0 = \varepsilon k_0$ and $t = \varepsilon k$. Then $P_\varepsilon(t_0,\cdot) \xrightarrow{w} P(t_0,\cdot) = \mathbb{1}\Theta(t_0,\cdot)\widetilde{\nu}(\cdot)$ such that

$$\widetilde{\mathbb{1}}\Theta(t_0,t)\widetilde{\nu}(t) = \begin{pmatrix} \mathbb{1}_{m_1}\nu^1(t)\theta_{11}(t_0,t) & \dots & \mathbb{1}_{m_1}\nu^l(t)\theta_{1l}(t_0,t) \\ \dots & \dots & \dots \\ \mathbb{1}_{m_l}\nu^1(t)\theta_{l1}(t_0,t) & \dots & \mathbb{1}_{m_l}\nu^l(t)\theta_{ll}(t_0,t) \end{pmatrix},$$

and $\Theta(t_0,t) = (\theta_{ij}(t_0,t)) \in \mathbb{R}^{l\times l}$ satisfies the differential equation

$$\frac{\partial \Theta(t_0,t)}{\partial t} = \Theta(t_0,t)\overline{Q}(t),$$
$$\Theta(t_0,t_0) = I.$$

(c) For each $i = 1,\dots,l$ and $j = 1,\dots,m_i$, as $\varepsilon \to 0$,

$$\mathsf{E}\left|\varepsilon \sum_{k_1=0}^{T/\varepsilon} [I_{\{\alpha_{\varepsilon,k_1}=s_{ij}\}} - \nu_j^i(\varepsilon k_1) I_{\{\overline{\alpha}_{\varepsilon,k_1}=i\}}]\right|^2 \to 0. \tag{1.54}$$

(d) $\overline{\alpha}_\varepsilon(\cdot)$ converges weakly to a continuous-time Markov chain $\overline{\alpha}(\cdot)$ generated by $\overline{Q}(\cdot)$ as $\varepsilon \to 0$.

Remark 1.19 The derivations involve detailed estimations, we refer the reader to [2] for the verbatim proofs. However, to get a flavor of the proofs of such results, we take a look at the convergence of the probability vector (i.e., part (a) in the above theorem).

To prove part (a), note that $\{p_\varepsilon(\cdot)\}$ is uniformly bounded in $L^2[0,T]$. It is well-known that $L^2[0,T]$ is a Hilbert space and hence weakly compact (i.e., any bounded sequence has a convergent subsequence in the sense of convergence in the weak topology). Extract a convergent subsequence and still index it by ε (without loss of generality) with limit denoted by $p(\cdot)$. We will identify the limit. Since $P(\cdot) - I \in L^2[0,T]$, for $0 \le s \le t \le T$,

$$\begin{aligned}
&\int_s^t p(s)(P(s)-I)ds \\
&= \lim_{\varepsilon\to 0} \int_s^t p_\varepsilon(s)(P(r)-I)dr \\
&= \lim_{\varepsilon\to 0} \left(\int_s^t p_\varepsilon(r+\varepsilon)dr - \int_s^t p_\varepsilon(r)dr - \varepsilon \int_s^t p_\varepsilon(r) Q(\varepsilon\lfloor r/\varepsilon\rfloor)dr \right) \\
&= 0.
\end{aligned}$$

The arbitrariness of t and s then implies that

$$p(t)(P(t)-I) = 0.$$

By the irreducibility of $P^k(t)$ for $k = 1, \ldots, l$, we obtain $p(t) = \theta(t)\widetilde{\nu}(t)$, for some function $\theta(t)$, which will be specified next. Note that

$$p_{\varepsilon,k+1} = p_{\varepsilon,k} + p_{\varepsilon,k}(P_\varepsilon(\varepsilon k) - I).$$

Iterating on the above recursion and using the interpolation,

$$p_\varepsilon(t) = p_0 + \sum_{j=0}^{t/\varepsilon-1} p_{\varepsilon,j}[P(\varepsilon j) - I)] + \varepsilon \sum_{j=0}^{t/\varepsilon-1} p_{\varepsilon,j} Q(\varepsilon j).$$

Post-multiplying the last equation by $\mathbb{1}$ and noting $(P(\varepsilon j) - I)\mathbb{1} = 0$ yield

$$\begin{aligned} p_\varepsilon(t)\mathbb{1} &= p_0\mathbb{1} + \varepsilon \sum_{j=0}^{t/\varepsilon-1} p_{\varepsilon,j} Q(\varepsilon j)\mathbb{1} \\ &= p_0\mathbb{1} + \int_0^t p_\varepsilon(s) Q(s)\mathbb{1} ds. \end{aligned}$$

The weak convergence of $p_\varepsilon(\cdot)$ to $p(\cdot) = \theta(\cdot)\widetilde{\nu}(\cdot)$ then implies the limit in $L^2[0,T]$ is

$$\begin{aligned} \theta(t) &= p_0\mathbb{1} + \int_0^t \theta(s)\widetilde{\nu}(s)Q(s)\mathbb{1} ds \\ &= p_0\mathbb{1} + \int_0^t \theta(s)\overline{Q}(s) ds. \end{aligned}$$

Remark 1.20. If additional transient states are included, that is assuming (A3) with $P(\varepsilon k)$ being of the form (1.37). Then the assertions will be changed to:

(a) The sequence $p_\varepsilon(\cdot)$ converges under the weak topology of $L^2[0,T]$, i.e.,

$$p_\varepsilon(\cdot) \xrightarrow{w} (\theta(\cdot)\tilde{\nu}(\cdot), 0_{m_*}),$$

where

$$\begin{aligned} \dot{\theta}(t) &= \theta(t)\overline{Q}(t), \\ \theta(0) &= (p_0^1 \mathbb{1}_{m_1} + p_0^* a_1, \ldots, p_0^l \mathbb{1}_{m_l} + p_0^* a_l). \end{aligned} \tag{1.55}$$

(b) Suppose that $P_\varepsilon(t_0, t)$ is the interpolation of the transition matrix with $t_0 = \varepsilon k_0$ and $t = \varepsilon k$. Then

$$P_\varepsilon(t_0, \cdot) \xrightarrow{w} \mathbb{1}_* \Theta_*(t_0, \cdot)\nu_*(\cdot),$$

with $\Theta_*(t_0, t) = \mathrm{diag}(\Theta(t_0, t), I_{m_* \times m_*})$, where the matrix-valued function $\Theta(t_0, t)$ satisfies the differential equation

$$\begin{aligned} \frac{\partial \Theta(t_0, t)}{\partial t} &= \Theta(t_0, t)\overline{Q}(t), \\ \Theta(t_0, t_0) &= I. \end{aligned}$$

(c) For each $j = 1, \ldots, m_i$, as $\varepsilon \to 0$,

$$\begin{aligned}
&\mathsf{E}\left|\varepsilon \sum_{k_1=0}^{T/\varepsilon} \left[I_{\{\alpha_{\varepsilon,k_1}=s_{ij}\}} - \nu_j^i(\varepsilon k_1) I_{\{\overline{\alpha}_{\varepsilon,k_1}=i\}}\right]\right|^2 \to 0, \text{ for } i = 1, \ldots, l \\
&\mathsf{E}\left|\varepsilon \sum_{k_1=0}^{T/\varepsilon} I_{\{\alpha_{\varepsilon,k_1}=s_{*j}\}}\right|^2 \to 0, \quad \text{for } \ i = *.
\end{aligned} \tag{1.56}$$

(d) If we assume furthermore that $B(t)$ is Lipschitz continuous on $t \in [0, T]$. Then, for all $0 < k \leq \lfloor T/\varepsilon \rfloor$, $\mathsf{P}(\alpha_{\varepsilon,k} = s_{*j}) \to 0$ as $\varepsilon \to 0$, for all $j = 1, \ldots, m_*$.

(e) Under conditions of Part (d) above, $\overline{\alpha}_\varepsilon(\cdot)$ converges weakly to $\overline{\alpha}(\cdot)$, a Markov chain generated by $\overline{Q}_*(\cdot)$, as $\varepsilon \to 0$.

1.4.3 Applications to Nearly Optimal Controls

This section is concerned with nearly optimal controls of large-scale discrete-time nonlinear dynamic systems driven by singularly perturbed Markov chains. It presents an application of the two-time scale in modeling and controls of hybrid systems. The features of the system include time-varying and possibly nonsmooth transition probabilities. Our main goal is to reduce the complexity of the underlying systems; the motivation comes from a wide range of applications arising in stochastic networks, wireless communication, queueing theory, and manufacturing systems, in which the underlying systems are subject to random noise fluctuations of pure jump type.

The optimal or suboptimal controls sought are often not explicitly representable in a closed form. Due to the large dimensionality of the state space of the Markov chain, a straightforward implementation of a numerical scheme can be infeasible. Thus one has to resolve the problem by alternative means. Based on the idea of hierarchical decomposition as in [42], we formulate the problem as an optimal control problem involving singularly perturbed Markov chains. The techniques we use include time scale separation and asymptotic analysis for singularly perturbed Markov chains. To highlight the time-scale separation, we introduce a small parameter $\varepsilon > 0$, and let the discrete-time Markov chain be $\alpha_{\varepsilon,k}$. The detailed proofs of this section can be found in [2].

Let

$$\begin{aligned}
f(\cdot,\cdot,\cdot) &: \mathbb{R}^n \times \mathbb{R}^{n_1} \times \mathcal{M} \mapsto \mathbb{R}^n, \\
c(\cdot,\cdot,\cdot) &: \mathbb{R}^n \times \mathbb{R}^{n_1} \times \mathcal{M} \mapsto \mathbb{R}.
\end{aligned}$$

Consider a stochastic dynamic system with the state $x_{\varepsilon,k} \in \mathbb{R}^n$ and control

$u_k \in \Gamma \subset \mathbb{R}^{n_1}$. The problem of interest is:

$$\begin{aligned}&\text{minimize: } J_\varepsilon(u) = \varepsilon \mathsf{E} \sum_{k=0}^{T/\varepsilon} c(x_{\varepsilon,k}, u_k, \alpha_{\varepsilon,k}),\\ &\text{subject to: } x_{\varepsilon,k+1} = x_{\varepsilon,k} + \varepsilon f(x_{\varepsilon,k}, u_k, \alpha_{\varepsilon,k}),\ x_{\varepsilon,0} = x.\end{aligned} \tag{1.57}$$

where $\mathsf{E} = \mathsf{E}_{x,\alpha}$ is the expectation given $\alpha_{\varepsilon,0} = \alpha$ and $x_{\varepsilon,0} = x$. Different from the usual stochastic control problems for controlled diffusions, there is no diffusion here. The sample paths of the disturbance is pure jump type. Thus, instead of one real-valued cost function, we have m ($m = |\mathcal{M}|$, the cardinality of $\mathcal{M}$) real-valued cost functions. In many stochastic networks, m is inevitably large, as a consequence, one must deal with a large number of value functions, so any attempt on reduction of complexity will be helpful.

(A4) (a) The control space Γ is a compact subset in $\mathbb{R}^{n_1}$.

(b) For each γ and each α, $f(\cdot, \gamma, \alpha)$ is continuously differentiable; for each α, $f(\cdot, \cdot, \alpha)$ is a continuous function; for each (x, γ, α), $|f(x, \gamma, \alpha)| \le K(1 + |x|)$;

(c) for each α, $c(\cdot, \cdot, \alpha)$ is continuous, for each (γ, α), $c(\cdot, \gamma, \alpha)$ is Lipschitz continuous, and $|c(x, \gamma, \alpha)| \le K(1 + |x|^{k_0})$ for some positive integer k_0 and for all triples (x, γ, α).

To solve the problem (1.57), we use the relaxed control formulation; see [13] and [32, pp. 47-54].

The problem in terms of the relaxed control is given by

$$\mathcal{P}_\varepsilon : \begin{cases} \text{minimize: } J_\varepsilon(m_\varepsilon) = \varepsilon \mathsf{E} \displaystyle\sum_{k=0}^{T/\varepsilon} \int c(x_{\varepsilon,k}, \gamma, \alpha_{\varepsilon,k}) m_{\varepsilon,k}(d\gamma), \\ \text{subject to: } x_{\varepsilon,k+1} = x_{\varepsilon,k} + \varepsilon \displaystyle\int f(x_{\varepsilon,k}, \gamma, \alpha_{\varepsilon,k}) m_{\varepsilon,k}(d\gamma), \\ x_{\varepsilon,0} = x, \\ \alpha_{\varepsilon,0} = \alpha, \end{cases} \tag{1.58}$$

where $\{m_{\varepsilon,k}(\cdot)\}$ is a sequence of discrete-parameter relaxed controls such that $m_{\varepsilon,k}(\cdot)$ is $\mathcal{F}_{\varepsilon,k}$-measurable. We will use $\Re(\Gamma \times [0,T])$ to denote the collection of relaxed controls on $\Gamma \times [0, T]$.

To proceed, let $\mathcal{B}(\Gamma \times [0,T])$ and $\mathcal{B}(\Gamma)$ be the σ-algebras of Borel subsets of $\Gamma \times [0,T]$ and Borel subsets of Γ, respectively. We use weak topology on the space of measures defined on $\mathcal{B}(\Gamma \times [0,T])$. That is, $m_k(\cdot) \to m(\cdot)$ in this topology if and only if for a smooth function $g(\cdot)$,

$$\int_0^T \int g(\gamma, t) m_k(d\gamma dt) \to \int_0^T \int g(\gamma, t) m(d\gamma dt).$$

Thus the space is compact under this topology, which in turn, allows us conveniently to extract convergent subsequences. For any $A \in \mathcal{B}(\Gamma)$ define

$$m_\varepsilon(A \times [0,t]) = \varepsilon \sum_{j=0}^{\lfloor t/\varepsilon \rfloor -1} m_{\varepsilon,j}(A) + (t - \varepsilon\lfloor t/\varepsilon \rfloor) m_{\varepsilon,\lfloor t/\varepsilon \rfloor}(A). \tag{1.59}$$

For notational convenience, we will write $m_\varepsilon(A \times [0,t])$ simply as $m_\varepsilon(A,t)$. Since for each i, $m_{\varepsilon,i}(\Gamma) = 1$, it follows

$$\begin{aligned} m_\varepsilon(\Gamma,t) &= \varepsilon \sum_{j=0}^{\lfloor t/\varepsilon \rfloor -1} m_{\varepsilon,j}(\Gamma) + (t - \varepsilon\lfloor t/\varepsilon \rfloor) m_{\varepsilon,\lfloor t/\varepsilon \rfloor}(\Gamma) \\ &= \varepsilon\lfloor t/\varepsilon \rfloor + t - \varepsilon\lfloor t/\varepsilon \rfloor = t. \end{aligned}$$

Therefore, $m_\varepsilon(\cdot)$ is a continuous-parameter relaxed control (see [32]). In addition, for any $A \in \Gamma$, $m_\varepsilon(A,\cdot)$ is nondecreasing. Hence the derivative exists almost everywhere, and is given by

$$m_{\varepsilon,t}(A) = \lim_{\delta \to 0} \frac{1}{\delta}[m_\varepsilon(A,t) - m_\varepsilon(A,t-\delta)]. \tag{1.60}$$

The definition above ensures the nonanticipativity of the process.

Note that for $t \in [\varepsilon k, \varepsilon k + \varepsilon)$,

$$m_\varepsilon(A,t) = \varepsilon \sum_{j=0}^{k-1} m_{\varepsilon,j}(A) + (t - \varepsilon k) m_{\varepsilon,k}(A). \tag{1.61}$$

Using (1.61), it is easily seen that for $t \in (\varepsilon k, \varepsilon k + \varepsilon)$, $m_{\varepsilon,t}(A) = m_{\varepsilon,k}(A)$. This also holds for $t = \varepsilon k$. To ensure the measurability of $m_{\varepsilon,t}(\cdot)$ w.r.t. $\mathcal{F}_{\varepsilon,t}$, we consider the following one-sided derivative modification. Using (1.61), for sufficiently small $\delta > 0$, consider the neighborhood $\varepsilon k \le t - \delta < t < \varepsilon k + \varepsilon$, we obtain $\lfloor t/\varepsilon \rfloor = k$ and $\lfloor (t-\delta)/\varepsilon \rfloor = k$ which leads to

$$\begin{aligned} m_{\varepsilon,t}(A) &= \lim_{\delta \to 0+} \frac{1}{\delta}\left[m_\varepsilon(A,t) - m_\varepsilon(A,t-\delta)\right] \\ &= \lim_{\delta \to 0+} \frac{1}{\delta}\left[\varepsilon \sum_{j=0}^{k-1} m_{\varepsilon,j}(A) + \delta m_{\varepsilon,k}(A) - \varepsilon \sum_{j=0}^{k-1} m_{\varepsilon,j}(A)\right] \\ &= m_{\varepsilon,k}(A). \end{aligned}$$

Therefore,

$$m_{\varepsilon,t}(A) = m_{\varepsilon,k}(A), \text{ if } t \in [\varepsilon k, \varepsilon k + \varepsilon). \tag{1.62}$$

Thus, $m_{\varepsilon,k}(A)$ is essentially the "derivative" of the relaxed control. To proceed, define a continuous-time interpolation of $x_{\varepsilon,k}$ by

$$x_\varepsilon(t) = x_{\varepsilon,k}, \text{ if } t \in [\varepsilon k, \varepsilon k + \varepsilon). \tag{1.63}$$

Let $\mathcal{R}_\varepsilon$, a subset of $\Re(\Gamma \times [0,T])$, be the set of admissible relaxed controls for $\mathcal{P}_\varepsilon$, i.e.,

$$\begin{aligned}\mathcal{R}_\varepsilon &= \{m_\varepsilon(\cdot) : m_\varepsilon(\cdot) \text{ is a measure on } \mathcal{B}(\Gamma \times [0,T]) \\ &\qquad \text{such that } m_\varepsilon(\Gamma, t) = t \text{ for all } t \geq 0; \\ &\qquad \text{and } m_\varepsilon(\cdot) \text{ is } \mathcal{F}_{\varepsilon,t}\text{-adapted } \}.\end{aligned} \tag{1.64}$$

Note that $\mathcal{R}_\varepsilon$ is a subset of $\Re(\Gamma \times [0,T])$. In what follows, we only consider admissible controls in $\mathcal{R}_\varepsilon$.

Although (1.58) is a discrete-time problem, its associated limit is a continuous-time problem given by

$$\mathcal{P}_0 : \begin{cases} \text{minimize: } J(m) = \mathsf{E}\displaystyle\int_0^T \int \overline{c}(x(s), \gamma, \overline{\alpha}(s)) m_s(d\gamma) ds, \\ \text{subject to: } x(t) = x + \displaystyle\int_0^t \int \overline{f}(x(s), \gamma, \overline{\alpha}(s)) m_s(d\gamma) ds, \\ x(0) = x, \\ \overline{\alpha}(0) = \alpha \mathbb{1}, \end{cases} \tag{1.65}$$

where $\overline{c}(\cdot)$, and $\overline{f}(\cdot)$ are defined as

$$\begin{aligned}\overline{c}(x, \gamma, i) &= \sum_{j=1}^{m_i} \nu_j^i(t) c(x, \gamma, s_{ij}), \\ \overline{f}(x, \gamma, i) &= \sum_{j=1}^{m_i} \nu_j^i(t) f(x, \gamma, s_{ij}),\end{aligned} \tag{1.66}$$

respectively. Denote by $\mathcal{R}_0$ the set of admissible controls for the limit problem, i.e.,

$$\begin{aligned}\mathcal{R}_0 &= \{m(\cdot) : m(\cdot) \text{ is a measure on } \mathcal{B}(\Gamma \times [0,T]) \\ &\qquad \text{such that } m(\Gamma, t) = t \text{ for all } t \geq 0; \\ &\qquad \text{and } m(\cdot) \text{ is } \mathcal{F}_t\text{-adapted}\},\end{aligned} \tag{1.67}$$

where $\{\mathcal{F}_t\}$ is a collection of increasing σ-algebras such that $\mathcal{F}_t$ measures at least $\{\overline{\alpha}(s); s \leq t\}$. To proceed, we first obtain the *a priori* bounds on the trajectories given by $\mathcal{P}_\varepsilon$ and $\mathcal{P}_0$, respectively. In fact, we can show that the solution $x_{\varepsilon,k}$ of the equation in $\mathcal{P}_\varepsilon$, its interpolation $x_\varepsilon(t)$, and the solution $x(t)$ in $\mathcal{P}_0$, satisfy $\sup_{0 \leq k \leq T/\varepsilon} |x_{\varepsilon,k}| \leq K(1+|x|)$, $\sup_{0 \leq t \leq T} |x_\varepsilon(t)| \leq K(1+|x|)$, $\sup_{0 \leq t \leq T} |x(t)| \leq K(1+|x|)$. The following theorem is stated with $P(\varepsilon k)$ given by (1.37). The assertions corresponding to $P(\varepsilon k)$ given by (1.36) are similar. The reader may consult the paper [2].

Theorem 1.21. Suppose that all the conditions in Theorem 1.20 and (A4) hold, and that $m_\varepsilon(\cdot)$ is a δ_ε an admissible relaxed control for $\mathcal{P}_\varepsilon$. Then

(a) $\{x_\varepsilon(\cdot), m_\varepsilon(\cdot)\}$ is tight.

(b) Suppose $(x(\cdot), m(\cdot))$ is the limit of a weakly convergent subsequence of $(x_\varepsilon(\cdot), m_\varepsilon(\cdot))$. Then $m(\cdot) \in \mathcal{R}_0$ and $(x(\cdot), m(\cdot))$ is a solution to $\mathcal{P}_0$. Moreover, $J_\varepsilon(m_\varepsilon) \to J(m)$ as $\varepsilon \to 0$.

Theorem 1.22. Suppose that the conditions of Theorem 1.21 are fulfilled and that the differential equation in the limit problem $\mathcal{P}_0$ has a unique solution for each initial condition. Then

$$\limsup_{\varepsilon \to 0} |J_\varepsilon(x, \iota, u^{\varepsilon,\delta}) - v_\varepsilon(x, \iota)| \le \delta, \tag{1.68}$$

for each $\iota \in \mathcal{M}$.

1.5 Notes and Remarks

1.5.1 Notes on the Literature

As was alluded to in the introduction, the main motivation of our study stems from the effort of reduction of complexity for large-scale hybrid dynamic systems. The approach that we are taking is along the line of hierarchical approach. Perhaps, Simon and Ando were among one of the first to recognize the nearly completely decomposable models [42]. As a continuation of their work, Courtois [6] studied the near decomposability in queueing and computer systems applications.

The subject matter discussed in this chapter is at the verge of the theory of stochastic processes and singular perturbation theory. There have been well-documented references on these areas. The discrete-time Markov chains were brought into being by A.A. Morkov in 1907 with the intention to generalize classical theory of sum of independent and identically distributed random variables. The development of the theory began in 1930 by A.N. Kolmogorov's systematic treatment. For general reading and background materials on stochastic processes, see [4, 7, 12, 23, 24, 25, 39, 40]. The well-written book by Chung [7] provides a clear exposition for homogeneous Markov chains; another useful source is [24, 25]. The recent work of Davis [8] gives a new interpretation in the piecewise deterministic process setup; Ethier and Kurtz' book presents an illuminating discussion on Markov processes and their generators among others. Introduction of stochastic control can be found in [3, 14, 30, 31] among others; introduction to Markov decision processes are included in [17, 45].

The field of singular perturbation contains a vast literature, which may be traced back to the study of Prandtl at the beginning of the 20th century, which established the foundation of boundary layer theory. Singular perturbation methods have been extensively used in, for example, various branches of physics including statistical mechanics, solid state physics, chemical physics, molecular biophysics. For the related applications in control theory and optimization, we refer the reader to Phillips and Kokotovic [38] for a continuous-time problem; see also Pan and Başar [36] and Pervozvanskii and Gaitsgori [37]. Discrete-time singularly perturbed Markov chains have been considered by Abbad, Filar, Bielecki [1], Blankenship [5], Delebecque and Quadrat [9] among others. The idea of two-time-scale expansion has also found emerging applications in communication theory (see Tse, Gallager, and Tsitsiklis [44] among others). This opens up new lines of applications.

Parallel to the advances in the deterministic theory, there is a stochastic version of the averaging methods. It began with the work of a seminal paper of R.Z. Khasminskii in 1966, continued by the large deviations approach of M. Friedlin and A. Wentzell, and the martingale averaging methods of Kushner (e.g., [32] and the references therein).

Much of our studies were studied with the motivation of treating manufacturing systems, in which the systems under consideration are of hybrid type and jump processes are included. The recent book of Sethi and Zhang contains a wide range of applications of production planning and optimal control problems. Our systematic study began with [28, 29]; our subsequent effort has been devoted to such singularly perturbed systems and related control problems; see [51, 52, 53, 54, 55, 57, 58, 59, 60, 62, 63, 65], and the references therein. Related work in singularly perturbed diffusions and singularly perturbed switching diffusions can be found in [21, 22, 27, 47].

1.5.2 Possible Future Research Topics

As further research topics, a number of issues deserve to be considered in the future study and investigation. We mention them below for an interested reader.

- There have been resurgent interests in using Markovian formulations to treat various economics problems; see [18, 19, 26, 49]. The reduction of complexity in such hybrid systems are worthwhile.

- Very recently attempts on using Markovian hybrid geometric Brownian motion models in finance have drawn much attention. Further investigation along this line will be an important effort.

- To study related singularly perturbed Markovian models with countable state spaces is particularly important for applications arising in queueing applications.

- Finally partially observed systems and filtering problems also deserve further study and examination.

Acknowledgments

The work of G. Yin was supported in part by the National Science Foundation under grant DMS-9877090. The work of Q. Zhang was supported in part by the USAF Grant F30602-99-2-0548 and ONR Grant N00014-96-1-0263.

1.6 References

[1] M. Abbad, J. A. Filar, and T. R. Bielecki, Algorithms for singularly perturbed limiting average Markov control problems, *IEEE Trans. Automat. Control*, **AC-37** (1992), 1421-1425.

[2] G. Badowski, G. Yin, and Q. Zhang, Nearly optimal controls of discrete-time dynamic systems driven by singularly perturbed Markov chains, preprint, 2001.

[3] D. Bertsekas, *Dynamic Programming and Optimal Control*, Vol. I & II, Athena Scientific, Belmont, MA, 1995.

[4] P. Billingsley, *Convergence of Probability Measures*, J. Wiley, New York, 1968.

[5] G. Blankenship, Singularly perturbed difference equations in optimal control problems, *IEEE Trans. Automat. Control*, **T-AC 26** (1981), 911-917.

[6] P. J. Courtois, *Decomposability: Queueing and Computer System Applications*, Academic Press, New York, 1977.

[7] K. L. Chung, *Markov Chains with Stationary Transition Probabilities*, Second Edition, Springer-Verlag, New York, 1967.

[8] M. H. A. Davis, *Markov Models and Optimization*, Chapman & Hall, London, 1993.

[9] F. Delebecque and J. Quadrat, Optimal control for Markov chains admitting strong and weak interactions, *Automatica*, **17** (1981), 281-296.

[10] A. Dembo and O. Zeitouni, *Large Deviations Techniques and Applications*, Springer-Verlag, New York, 1998.

[11] G. B. Di Masi and Yu. M. Kabanov, A first order approximation for the convergence of distributions of the Cox processes with fast Markov switchings, *Stochastics Stochastics Rep.*, **54** (1995), 211-219.

[12] S. N. Ethier and T. G. Kurtz, *Markov Processes: Characterization and Convergence*, J. Wiley, New York, 1986.

[13] W. H. FLEMING, Generalized solution in optimal stochastic control, *Proc. URI Conf. on Control*, 1982, 147-165.

[14] W. H. FLEMING AND R. W. RISHEL, *Deterministic and Stochastic Optimal Control*, Springer-Verlag, New York, 1975.

[15] S. B. GERSHWIN, *Manufacturing Systems Engineering*, Prentice Hall, Englewood Cliffs, 1994.

[16] W.-B. GONG, P. KELLY, AND W. ZHAI, Comparison schemes for discrete optimization with estimation algorithms, *Proc. IEEE Conf. Decision Control*, 1993, 2211-2216.

[17] F. S. HILLIER AND G. J. LIEBERMAN, *Introduction to Operations Research*, McGraw-Hill, New York, 6th Ed., 1995.

[18] J. D. HAMILTON AND R. SUSMEL. Autoregressive conditional heteroskedasticity and changes in regime, *J. Econometrics*, **64** (1994), 307-333.

[19] B. E. HANSEN, The likelihood ratio test under nonstandard conditions: Testing Markov trend model of GNP, *J. Appl. Economics*, **7** (1992), S61-S82.

[20] F. C. HOPPENSTEADT AND W. L. MIRANKER, Multitime methods for systems of difference equations, *Studies Appl. Math.*, **56** (1977), 273-289.

[21] A. IL'IN, R.Z. KHASMINSKII, AND G. YIN, Singularly perturbed switching diffusions: Rapid switchings and fast diffusions, *J. Optim. Theory Appl.*, **102** (1999), 555-591.

[22] A. IL'IN, R.Z. KHASMINSKII, AND G. YIN, Asymptotic expansions of solutions of integro-differential equations for transition densities of singularly perturbed switching diffusions: Rapid switchings, *J. Math. Anal. Appl.*, **238** (1999), 516-539.

[23] M. IOSIFESCU, *Finite Markov Processes and Their Applications*, Wiley, Chichester, 1980.

[24] S. KARLIN AND H. M. TAYLOR, *A First Course in Stochastic Processes*, 2nd Ed., Academic Press, New York, 1975.

[25] S. KARLIN AND H. M. TAYLOR, *A Second Course in Stochastic Processes*, Academic Press, New York, 1981.

[26] D. KENDRICK, On the Leontief dynamic inverse, *Quart. J. Economics*, **86** (1972), 693-696.

[27] R. Z. Khasminskii and G. Yin, On transition densities of singularly perturbed diffusions with fast and slow components, *SIAM J. Appl. Math.*, **56** (1996), 1794-1819.

[28] R. Z. Khasminskii, G. Yin and Q. Zhang, Asymptotic expansions of singularly perturbed systems involving rapidly fluctuating Markov chains, *SIAM J. Appl. Math.*, **56** (1996), 277-293.

[29] R. Z. Khasminskii, G. Yin and Q. Zhang, Constructing asymptotic series for probability distribution of Markov chains with weak and strong interactions, *Quart. Appl. Math.*, **LV** (1997), 177-200.

[30] P. R. Kumar and P. Varaiya, *Stochastic Systems: Estimation, Identification and Adaptive Control*, Prentice-Hall, Englewood Cliffs, N.J., 1984.

[31] H. J. Kushner, *Introduction to Stochastic Control Theory*, Holt, Rinehart and Winston, New York, 1972.

[32] H. J. Kushner, *Weak Convergence Methods and Singularly Perturbed Stochastic Control and Filtering Problems*, Birkhäuser, Boston, 1990.

[33] H. J. Kushner and G. Yin, *Stochastic Approximation Algorithms and Applications*, Springer-Verlag, New York, 1997.

[34] R. H. Liu, Q. Zhang, and G. Yin, Nearly optimal control of singularly perturbed Markov decision processes in discrete time, *Appl. Math. Optim.*, **44** (2001), 105-129.

[35] D. S. Naidu, *Singular Perturbation Methodology in Control Systems*, Peter Peregrinus Ltd., Stevenage Herts, UK, 1988.

[36] Z. G. Pan and T. Başar, H^∞-control of Markovian jump linear systems and solutions to associated piecewise-deterministic differential games, in *New Trends in Dynamic Games and Applications*, G. J. Olsder (Ed.), 61-94, Birkhäuser, Boston, 1995.

[37] A. A. Pervozvanskii and V. G. Gaitsgori, *Theory of Suboptimal Decisions: Decomposition and Aggregation*, Kluwer, Dordrecht, 1988.

[38] R. G. Phillips and P. V. Kokotovic, A singular perturbation approach to modelling and control of Markov chains, *IEEE Trans. Automat. Control*, **26** (1981), 1087-1094.

[39] S. Ross, *Stochastic Processes,* J. Wiley & Sons, New York, 1983.

[40] E. Seneta, *Non-negative Matrices and Markov Chains*, Springer-Verlag, New York, 1981.

[41] S. P. SETHI AND Q. ZHANG, *Hierarchical Decision Making in Stochastic Manufacturing Systems*, Birkhäuser, Boston, 1994.

[42] H. A. SIMON AND A. ANDO, Aggregation of variables in dynamic systems, *Econometrica*, **29** (1961), 111-138.

[43] W. A. THOMPSON, JR., *Point Process Models with Applications to Safety and Reliability*, Chapman and Hall, New York, 1988.

[44] D. N. C. TSE, R. G. GALLAGER, AND J. N. TSITSIKLIS, Statistical multiplexing of multiple time-scale Markov streams, *IEEE J. Selected Areas Comm.*, **13** (1995), 1028-1038.

[45] D. J. WHITE, *Markov Decision Processes*, Wiley, New York, 1992.

[46] H. YANG, G. YIN, K. YIN, AND Q. ZHANG, Control of singularly perturbed Markov chains: A numerical study, to appear in *J. Australian Math. Soc. Ser. B: Appl. Math.*

[47] G. YIN, On limit results for a class of singularly perturbed switching diffusions, *J. Theoretical Probab.*, **14** (2001), 673-697.

[48] G. YIN AND M. KNIAZEVA, Singularly perturbed multidimensional switching diffusions with fast and slow switchings, *J. Math. Anal. Appl.*, **229** (1999), 605-630.

[49] G. YIN AND J.F. ZHANG, Hybrid singular systems of differential equations, to appear in *Scientia Sinica*.

[50] G. YIN AND Q. ZHANG, Control of dynamic systems under the influence of singularly perturbed Markov chains, *J. Math. Anal. Appl.*, **216** (1997), 343-367.

[51] G. YIN AND Q. ZHANG, *Continuous-time Markov Chains and Applications: A Singular Perturbation Approach,* Springer-Verlag, New York, 1998.

[52] G. YIN AND Q. ZHANG, Singularly perturbed discrete-time Markov chains, *SIAM J. Appl. Math.*, **61** (2000), 834-854.

[53] G. YIN, Q. ZHANG, AND G. BADOWSKI, Asymptotic properties of a singularly perturbed Markov chain with inclusion of transient states, *Ann. Appl. Probab.*, **10** (2000), 549-572.

[54] G. YIN, Q. ZHANG, AND G. BADOWSKI, Singularly perturbed Markov chains: Convergence and aggregation, *J. Multivariate Anal.*, **72** (2000), 208-229.

[55] G. YIN, Q. ZHANG, AND G. BADOWSKI, Occupation measures of singularly perturbed Markov chains with absorbing states, *Acta Math. Sinica*, **16** (2000), 161-180.

[56] G. YIN, Q. ZHANG, AND G. BADOWSKI, Decomposition and aggregation of large-dimensional Markov chains in discrete time, *Proc. 40th Conf. Decision Control*, 2001.

[57] G. YIN, Q. ZHANG, AND Q.G. LIU, Error bounds for occupation measure of singularly perturbed Markov chains including transient states, *Probab. Eng. Informational Sci.*, **14** (2000), 511-531.

[58] G. YIN, Q. ZHANG, H. YANG, AND K. YIN, Discrete-time dynamic systems arising from singularly perturbed Markov chains, *Nonlinear Anal., Theory, Methods Appl.*, **47** (2001), 4763-4774.

[59] Q. ZHANG, Risk sensitive production planning of stochastic manufacturing systems: A singular perturbation approach, *SIAM J. Control Optim.*, **33** (1995), 498-527.

[60] Q. ZHANG, Finite state Markovian decision processes with weak and strong interactions, *Stochastics Stochastics Rep.*, **59** (1996), 283-304.

[61] Q. ZHANG, R.H. LIU, AND G. YIN, Nearly optimal controls of Markovian systems, in *Stochastic Models and Optimization*, D. Yao, H.Q. Zhang, and X.Y. Zhou eds., Springer-Verlag.

[62] Q. ZHANG AND G. YIN, A central limit theorem for singularly perturbed nonstationary finite state Markov chains, *Ann. Appl. Probab.*, **6** (1996), 650-670.

[63] Q. ZHANG AND G. YIN, Structural properties of Markov chains with weak and strong interactions, *Stochastic Process Appl.*, **70** (1997), 181-197.

[64] Q. ZHANG, G. YIN, AND E. K. BOUKAS, Controlled Markov chains with weak and strong interactions: Asymptotic optimality and application in manufacturing, *J. Optim. Theory Appl.*, **94** (1997), 169-194.

[65] Q. ZHANG AND G. YIN, On nearly optimal controls of hybrid LQG problems, *IEEE Trans. Automat. Control*, **44** (1999), 2271-2282.

2

Nearly Optimal Controls of Markovian Systems

Q. Zhang, R.H. Liu, and G. Yin

This chapter is concerned with asymptotically optimal controls of Markovian systems with weak and strong interactions in discrete time. The primary motivation of our study stems from many applications in operations research such as resource allocation, queueing networks, machine replacement, and command control. Due to the complexity of the real-world problems, one often has to deal with large dimensional systems under uncertainty. Frequently, the underlying Markov chain displays jump behavior having different rates of jumps (some of them vary fast, whereas others change slowly). It is natural to divide the states of such Markov chain into several groups so that transitions among the states within each group occur much more frequently than the transitions among the states belonging to different groups. The basic idea of our approach is to aggregate the states in each group as a single state and to derive a limit problem of smaller dimension. Using an optimal control of the limit problem, we construct controls of the original problem, leading to asymptotic optimality. Our hierarchical approach provides an efficient tool for treating large and complex systems and for finding approximately optimal solutions.

We treat discrete-time models. The motivation for studying discrete-time models comes from the following considerations. First, various system measurements are frequently recorded in discrete time. In addition, much of the success of obtaining optimal controls of various continuous-time systems depends on efficient computation procedures that in turn depend on the corresponding discretized dynamic systems. One of the interesting aspects is that the limit problem derived is a continuous-time control problem although the original one is in discrete time. We explore the interface between continuous-time and discrete-time problems in a natural way.

The chapter consists of two parts. The first part focuses on singularly perturbed Markov decision processes (MDPs) and the second part studies a hybrid linear quadratic (LQ) control problem. Throughout the chapter, we provide proofs in simple cases to demonstrate the main ideas and techniques without undue complex notation.

2.1 Singularly Perturbed MDP

A common practice in dealing with MDPs is the method of dynamic programming (DP). Using such an approach, to find the optimal solutions, one needs to solve a set of associated DP equations. However, such an approach is computationally feasible only when the dimension of the underlying system is not very large. For large-scale systems, one has to resort to approximately optimal schemes. In this section, we consider Markov decision processes with weak and strong interactions. Our main results are based upon recent development in singularly perturbed Markov chains. We first study a discounted problem. Then we examine a long-run-average cost problem.

We begin with an irreducible MDP. Results on MDPs with more general transition matrices are summarized with the proofs omitted.

Consider a discrete-time Markov chain $\alpha^\varepsilon_\cdot = \{\alpha^\varepsilon_n,\ n = 0, 1, \ldots\}$ with finite state space $\mathcal{M} = \{1, \ldots, m\}$, where $\varepsilon > 0$ is a small parameter. Let the control (or policy) space Γ be a compact subset of an Euclidean space. We consider feedback control $u_n = u(\alpha^\varepsilon_n)$ such that $u_n \in \Gamma$, $n = 0, 1, \ldots$.

Definition 2.1. Let $u(\cdot) = \{u(i) : \ i \in \mathcal{M}\}$ be a function such that $u(i) \in \Gamma$ for all states $i \in \mathcal{M}$. $u(\cdot)$ is called an admissible control and the collection of all such functions is denoted by $\mathcal{A}_f$.

Let $P^\varepsilon(u) = (p^\varepsilon_{ij}(u))_{m\times m}$ denote the probability transition matrix by

$$P^\varepsilon(u) = P(u) + \varepsilon Q(u), \quad \text{for } u \in \Gamma, \tag{2.1}$$

where $P(u) = (p_{ij}(u))_{m\times m}$ is a probability transition matrix, i.e., $0 \le p_{ij}(u) \le 1$, $\sum_j p_{ij}(u) = 1$, and $Q(u) = (q_{ij}(u))_{m\times m}$ is a generator of a continuous-time Markov chain, i.e., $q_{ij}(u) \ge 0$ for $j \ne i$ and $\sum_j q_{ij}(u) = 0$.

Consider the following cost functional $J^\varepsilon_d(i, u(\cdot))$ defined on $\mathcal{M} \times \mathcal{A}_f$:

$$J^\varepsilon_d(i, u(\cdot)) = \mathsf{E}\left(\varepsilon \sum_{n=0}^{\infty} (1 - \beta\varepsilon)^n g(\alpha^\varepsilon_n, u(\alpha^\varepsilon_n))\right), \tag{2.2}$$

where $i = \alpha^\varepsilon_0$ is the initial state of the chain, $g(i, u)$ is the cost-to-go function, and $\beta > 0$ is a given constant. Here the subscript d identifies the criterion function with the discounted cost. We consider the case when ε is small enough such that $1 - \beta\varepsilon > 0$. The objective is to find a function $u(\cdot) \in \mathcal{A}_f$ that minimizes $J^\varepsilon_d(i, u(\cdot))$.

The original MDP problem termed $\mathcal{P}_d^\varepsilon$ takes the form

$$\mathcal{P}_d^\varepsilon : \begin{cases} \text{minimize: } J_d^\varepsilon(i, u(\cdot)) = \mathsf{E}\left(\varepsilon \displaystyle\sum_{n=0}^{\infty} (1-\beta\varepsilon)^n g(\alpha_n^\varepsilon, u(\alpha_n^\varepsilon))\right), \\ \text{subject to: } \alpha_n^\varepsilon \sim P^\varepsilon(u(\alpha_n^\varepsilon)),\ n = 0, 1, \ldots,\ u(\cdot) \in \mathcal{A}_f, \\ \text{value function: } v_d^\varepsilon(i) = \displaystyle\inf_{u(\cdot)\in\mathcal{A}_f} J_d^\varepsilon(i, u(\cdot)), \end{cases}$$

where by $\alpha_n^\varepsilon \sim P^\varepsilon(u(\alpha_n^\varepsilon))$ we mean that α_n^ε is a discrete-time Markov chain whose probability transition matrix is $P^\varepsilon(u(\alpha_n^\varepsilon))$.

We can use the DP approach to solve the problem $\mathcal{P}_d^\varepsilon$. For each $i \in \mathcal{M}$, the associated discrete-time DP equation is

$$v_d^\varepsilon(i) = \min_{u\in\Gamma}\left\{\varepsilon g(i,u) + (1-\beta\varepsilon)\sum_j p_{ij}^\varepsilon(u) v_d^\varepsilon(j)\right\}. \tag{2.3}$$

Subtracting $(1-\beta\varepsilon)v_d^\varepsilon(i)$ from both sides leads to

$$\beta\varepsilon v_d^\varepsilon(i) = \min_{u\in\Gamma}\left\{\varepsilon g(i,u) + (1-\beta\varepsilon)\sum_j (p_{ij}^\varepsilon(u) - \delta_{ij}) v_d^\varepsilon(j)\right\},$$

where $\delta_{ij} = \begin{cases} 1, & \text{if } i = j, \\ 0, & \text{if } i \neq j. \end{cases}$ Dividing both sides by ε yields

$$\beta v_d^\varepsilon(i) = \min_{u\in\Gamma}\left\{g(i,u) + (1-\beta\varepsilon)\sum_j b_{ij}^\varepsilon(u) v_d^\varepsilon(j)\right\}, \tag{2.4}$$

where

$$(b_{ij}^\varepsilon(u))_{m\times m} = \frac{1}{\varepsilon}(P(u) - I) + Q(u). \tag{2.5}$$

Here and hereafter I always denotes the identity matrix of appropriate dimensions, (e.g., in (2.5), I is an $m \times m$ identity matrix).

It can be shown that $v_d^\varepsilon(i)$ is the unique solution to the DP equation (2.4); see Bertsekas [4]. Moreover, for each $i \in \mathcal{M}$, let $u_{o,\varepsilon}(i)$ be the minimizer of

$$g(i,u) + (1-\beta\varepsilon)\sum_j b_{ij}^\varepsilon(u) v_d^\varepsilon(j).$$

Then $u_{o,\varepsilon}(\cdot) = \{u_{o,\varepsilon}(\alpha_n^\varepsilon)\}$ is optimal. That is, $J_d^\varepsilon(i, u_{o,\varepsilon}(\cdot)) = v_d^\varepsilon(i)$.

To obtain the optimal solution of $\mathcal{P}_d^\varepsilon$, one has to solve m DP equations of the form (2.4). If the dimension m is very large, this approach is not computationally feasible in practice. To resolve this problem, we use a hierarchical control approach to be discussed as in the subsequent sections.

2.1.1 Irreducible MDP under Discounted Cost

To reduce the dimensionality of the problem under consideration, we will show that as $\varepsilon \to 0$, the original problem can be approximated by a limit problem. This section is devoted to the derivation of the corresponding limit control problem.

Next, we define the control set for the limit problem as in [40, Chap. 8]. Define

$$\overline{\Gamma} = \{U := (u^1, \dots, u^m) : u^i \in \Gamma, \ i = 1, 2, \dots, m\}.$$

For each $U \in \overline{\Gamma}$, let

$$P(U) = (p_{ij}(u^i))_{m \times m},$$

$$Q(U) = (q_{ij}(u^i))_{m \times m},$$

i.e., to replace the variable u in the ith row of $P(u)$ and $Q(u)$ by u^i. We define

$$\overline{P}^{\varepsilon}(U) = P(U) + \varepsilon Q(U).$$

For each $u(\cdot) \in \mathcal{A}_f$, it is easy to show that the Markov chain determined by $P^{\varepsilon}(u(\alpha_n^{\varepsilon}))$ and the one determined by $\overline{P}^{\varepsilon}(U)$, with $u(i) = u^i$, have the same probability distribution. The following assumptions are made on the transition matrix $P^{\varepsilon}(u)$ and the cost-to-go function $g(i, u)$.

(A1) For each $\varepsilon > 0$, $P^{\varepsilon}(u)$ is a continuous function of u. Moreover, for each $U \in \overline{\Gamma}$, $P(U)$ is irreducible and aperiodic.

(A2) For each $i \in \mathcal{M}$, $g(i, \cdot)$ is a continuous function on Γ.

The original problem $\mathcal{P}_d^{\varepsilon}$ depends on $\varepsilon > 0$. We aim to show that as $\varepsilon \to 0$, there is a limit problem in an appropriate sense. Our task to follow is to characterize the limit and to prove the convergence.

For each $U \in \overline{\Gamma}$, let $\mu(U) = (\mu_1(U), \dots, \mu_m(U))$ be the stationary distribution of $P(U)$, i.e., $\mu(U)$ is the unique solution of the following system of equations

$$\mu(U)P(U) = \mu(U), \qquad \sum_{i=1}^{m} \mu_i(U) = 1.$$

Using $\mu(U)$, we define an averaged cost function

$$\overline{g}(U) = \sum_{i=1}^{m} \mu_i(U) g(i, u^i). \tag{2.6}$$

Now we have an auxiliary optimization problem:

$$\mathcal{P}_d^0 : \begin{cases} \text{minimize:} \ \ J_d^0(U) = \displaystyle\int_0^{\infty} e^{-\beta t} \overline{g}(U) dt = \frac{1}{\beta} \overline{g}(U), \\ \text{value function:} \ \ v_d = \displaystyle\inf_{U \in \overline{\Gamma}} J_d^0(U) = \min_{U \in \overline{\Gamma}} \frac{1}{\beta} \overline{g}(U). \end{cases}$$

Clearly, this is merely an optimization problem with no dynamics given, which is much easier to solve. In general, when the transition matrix has more than one class of recurrent states, such problem corresponds to a smaller dimension MDP. Our task to follow is to show that $\mathcal{P}_d^0$ is the limit problem in an appropriate sense.

Asymptotic Optimality. In this section, we study the convergence of the sequence of value functions $v_d^\varepsilon(\cdot)$ to $v_d(\cdot)$ and the construction of asymptotic optimal controls for $\mathcal{P}_d^\varepsilon$.

Theorem 2.2. Assume (A1) and (A2). For each $i \in \mathcal{M}$,

$$\lim_{\varepsilon\to 0} v_d^\varepsilon(i) = v_d = \frac{1}{\beta}\min_{U\in\overline{\Gamma}} \overline{g}(U). \tag{2.7}$$

Proof. Note that $|g| \le K$ for some constant K. This implies that $|v_d^\varepsilon(i)| \le K\varepsilon\sum_{n=0}^\infty (1-\beta\varepsilon)^n = K/\beta < \infty$, i.e., $v_d^\varepsilon(i)$ is uniformly bounded. Thus, for each subsequence of $\varepsilon \to 0$, there exists a further subsequence (for notational simplicity, still index both subsequences by ε) and a $v_d(i)$, such that $v_d^\varepsilon(i) \to v_d(i)$. In the following, we show that such a limit $v_d(i)$ is independent of i and is equal to $(1/\beta)\min_{U\in\overline{\Gamma}}\overline{g}(U)$.

Let $\alpha_0^\varepsilon = i \in \mathcal{M}$. For $U \in \overline{\Gamma}$, the DP equation (2.4) implies

$$\beta v_d^\varepsilon(i) \le g(i,u^i) + (1-\beta\varepsilon)\sum_j b_{i,j}^\varepsilon(u^i)v_d^\varepsilon(j). \tag{2.8}$$

Using the hypothesis $v_d^\varepsilon(i) \to v_d^0(i)$ and multiplying both sides of (2.8) by ε and sending $\varepsilon \to 0$ lead to

$$0 \le \sum_{j=1}^m (p_{ij}(u^i) - \delta_{ij})v_d^0(j).$$

Now, the irreducibility of $P(U)$ and Lemma A.39 in [40] imply that $v_d^0(1) = \cdots = v_d^0(m) := v_d^0$. In addition, it follows from (2.8) that

$$\beta\sum_{i=1}^m \mu_i(U)v_d^\varepsilon(i) \le \sum_{i=1}^m \mu_i(U)g(i,u^i) + (1-\beta\varepsilon)\sum_{i=1}^m \mu_i(U)\sum_j b_{i,j}^\varepsilon(u^i)v_d^\varepsilon(j).$$

Recall the definition of stationary distribution $\mu(U)$. It follows that the second term goes to 0 as $\varepsilon \to 0$. Letting $\varepsilon \to 0$, we have $\beta v_d \le \overline{g}(U)$. Since this inequality holds for every $U \in \overline{\Gamma}$, we have

$$\beta v_d \le \min_{U\in\overline{\Gamma}} \overline{g}(U).$$

To derive the reverse inequality, let $U_{o,\varepsilon} \in \overline{\Gamma}$ denote a minimizer of the right-hand side of (2.4). Then

$$\beta v_d^\varepsilon(i) = g(i, u_{o,\varepsilon}^i) + (1 - \beta\varepsilon) \sum_j b_{i,j}^\varepsilon(u_{o,\varepsilon}^i) v_d^\varepsilon(j).$$

Thus, we have

$$\beta \sum_{i=1}^m \mu_i(U_{o,\varepsilon}) v_d^\varepsilon(i) = \sum_{i=1}^m \mu_i(U_{o,\varepsilon}) g(i, u_{o,\varepsilon}^i) + (1 - \beta\varepsilon) \sum_{i=1}^m \mu_i(U_{o,\varepsilon}) \sum_j b_{i,j}^\varepsilon(u_{o,\varepsilon}^i) v_d^\varepsilon(j).$$

Since $\overline{\Gamma}$ is a bounded set, there exists a further subsequence of $\varepsilon \to 0$ (still indexed by ε) such that $U_{o,\varepsilon} \to U_o \in \overline{\Gamma}$. Hence, letting $\varepsilon \to 0$ leads to

$$\beta v_d = \overline{g}(U_o) \geq \min_{U \in \overline{\Gamma}} \{\overline{g}(U)\}.$$

This completes the proof. □

Next we construct an asymptotic optimal control policy for the original problem $\mathcal{P}_d^\varepsilon$ from the optimal decision of the limit problem $\mathcal{P}_d^0$. Let $U_o \in \overline{\Gamma}$ be an optimal control for the limit problem $\mathcal{P}_d^0$. For this $U_o \in \overline{\Gamma}$, define a control $u(\cdot) = \{u(\alpha) : \alpha \in \mathcal{M}\}$ for the original problem $\mathcal{P}_d^\varepsilon$:

$$u(\alpha) = \sum_{i=1}^m I_{\{\alpha = i\}} u_o^i, \tag{2.9}$$

where I_A is the indicator function of a set A. It is clear that $u(\cdot) \in \mathcal{A}_f$. In what follows, we will show that the constructed control $u(\cdot)$ is asymptotically optimal. The following lemma gives an error bound for the corresponding probability vector expansion in terms of ε.

Lemma 2.3. Given $U \in \overline{\Gamma}$, let $\alpha^\varepsilon(\cdot)$ denote the discrete-time Markov chain with probability transition matrix $\overline{P}^\varepsilon(U)$. Then there exist positive constants K and $0 < \lambda < 1$ (both independent of ε and n) such that for $j = 1, \ldots, m_k$, $k = 1, \ldots, l$,

$$|\mathsf{P}(\alpha_n^\varepsilon = i) - \mu_i(U)| \leq K(n\varepsilon^2 + \varepsilon + \lambda^n). \tag{2.10}$$

Remark 2.4. This lemma is a modification of the corresponding results in [39]. In that part, the asymptotic expansion is obtained for some finite $T > 0$ and for $n \leq T/\varepsilon$. Here, we do not have such a restriction on n. As a result, there is an extra term $\varepsilon^2 n$ in the error bound (2.10). Owing to the discount factor in the cost function, the desired convergence properties still hold. In fact, what is needed is the property that $n \to \infty$ but $n\varepsilon^2 \to 0$. For example, if we choose $n\varepsilon = t$ for a given $t > 0$, then the discount rate at time n is $(1 - \beta\varepsilon)^n = (1 - (t\beta)/n)^n \sim \exp(-\beta t)$, which corresponds to a discount rate in a continuous-time problem.

Theorem 2.5. The control $u_\varepsilon(\cdot) = \{u(\alpha^\varepsilon.)\}$ constructed in (2.9) is asymptotically optimal in that

$$\lim_{\varepsilon \to 0} |J_d^\varepsilon(i, u_\varepsilon(\cdot)) - v_d^\varepsilon(i)| = 0, \text{ for } i \in \mathcal{M}.$$

Proof. In view of the convergence of $v_d^\varepsilon(i)$, it suffices to show that for $\alpha^\varepsilon(0) = i \in \mathcal{M}$,

$$\lim_{\varepsilon \to 0} J_d^\varepsilon(i, u_\varepsilon(\cdot)) = v_d. \tag{2.11}$$

Let $\alpha^\varepsilon(\cdot)$ denote the discrete-time Markov chain with transition matrix $P^\varepsilon(u_\varepsilon(\alpha_n^\varepsilon))$, $n = 0, 1, \ldots$, with $u_\varepsilon(\alpha)$ given in (2.9). Then $P^\varepsilon(u_\varepsilon(\alpha_n^\varepsilon))$ and $\overline{P}^\varepsilon(U_o)$ both determine the chains with identical probability distribution.

Using the definition of $u_\varepsilon(\alpha)$, we have, for any $i \in \mathcal{M}$,

$$\begin{aligned} J_d^\varepsilon(i, u_\varepsilon(\cdot)) &= \mathsf{E}\left(\varepsilon \sum_{n=0}^{\infty} (1 - \beta\varepsilon)^n g(\alpha_n^\varepsilon, u(\alpha_n^\varepsilon))\right) \\ &= \mathsf{E}\left(\sum_{i=1}^{m} \sum_{n=0}^{\infty} \varepsilon(1 - \beta\varepsilon)^n g(i, u_o^i) I_{\{\alpha_n^\varepsilon = i\}}\right) \\ &= \sum_{i=1}^{m} g(i, u_o^i) \left(\sum_{n=0}^{\infty} \varepsilon(1 - \beta\varepsilon)^n \mathsf{P}(\alpha_n^\varepsilon = i)\right). \end{aligned}$$

Now, we consider the limit cost $J_d^0(U_o)$. In view of the definition of $\overline{g}(U)$ by (2.6), we write

$$J_d^0(U_o) = \int_0^\infty e^{-\beta t} \sum_{i=1}^{m} \mu_i(U_o) g(i, u_o^i) dt.$$

Then we have

$$\begin{aligned} &|J_d^\varepsilon(i, u_\varepsilon(\cdot)) - v_d| \\ &= \left|J_d^\varepsilon(i, u_\varepsilon(\cdot)) - J_d^0(U_o)\right| \\ &\leq \sum_{i=1}^{m} |g(i, u_o^i)| \left|\sum_{n=0}^{\infty} \varepsilon(1 - \beta\varepsilon)^n \mathsf{P}(\alpha_n^\varepsilon = i) - \int_0^\infty e^{-\beta t} \mu_i(U_o) dt\right|. \end{aligned} \tag{2.12}$$

Recall that $|g(i, u)| \leq C$ for some constant C and for $i \in \mathcal{M}$, $u \in \Gamma$.

Note that

$$\begin{aligned}
&\left|\sum_{n=0}^{\infty}\varepsilon(1-\beta\varepsilon)^n \mathsf{P}(\alpha_n^\varepsilon = i) - \int_0^\infty e^{-\beta t}\mu_i(U_o)dt\right| \\
&\quad \le \sum_{n=0}^{\infty}\varepsilon(1-\beta\varepsilon)^n \left|\mathsf{P}(\alpha_n^\varepsilon = i) - \mu_i(U_o)\right| \\
&\quad\quad + \mu_i(U_o)\left|\sum_{n=0}^{\infty}\varepsilon(1-\beta\varepsilon)^n - \sum_{n=0}^{\infty}\varepsilon e^{-\beta\varepsilon n}\right| \\
&\quad\quad + \mu_i(U_o)\left|\sum_{n=0}^{\infty}\varepsilon e^{-\beta\varepsilon n} - \int_0^\infty e^{-\beta t}dt\right|.
\end{aligned} \tag{2.13}$$

In view of Lemma 2.3, we have

$$\begin{aligned}
&\sum_{n=0}^{\infty}\varepsilon(1-\beta\varepsilon)^n \left|\mathsf{P}(\alpha_n^\varepsilon = i) - \mu_i(U_o)\right| \\
&\quad \le \sum_{n=0}^{\infty}\varepsilon(1-\beta\varepsilon)^n K(n\varepsilon^2 + \varepsilon + \lambda^n) \le K_1\varepsilon.
\end{aligned} \tag{2.14}$$

Since $e^{-\beta\varepsilon} \ge 1 - \beta\varepsilon$, it is easy to show that

$$\begin{aligned}
&\left|\sum_{n=0}^{\infty}\varepsilon(1-\beta\varepsilon)^n - \sum_{n=0}^{\infty}\varepsilon e^{-\beta\varepsilon n}\right| \\
&\quad \le \sum_{n=0}^{\infty}\varepsilon(e^{-\beta\varepsilon n} - (1-\beta\varepsilon)^n) \le K_1\varepsilon.
\end{aligned} \tag{2.15}$$

Finally, it is clear that the last line of (2.13) goes to 0 as $\varepsilon \to 0$. Then we obtain the convergence is of the order ε. In fact,

$$\begin{aligned}
\left|\sum_{n=0}^{\infty}\varepsilon e^{-\beta\varepsilon n} - \int_0^\infty e^{-\beta t}dt\right| &\le \sum_{n=0}^{\infty}\int_{\varepsilon n}^{\varepsilon(n+1)} \left|e^{-\beta\varepsilon n} - e^{-\beta t}\right| dt \\
&= \sum_{n=0}^{\infty}\int_{\varepsilon n}^{\varepsilon(n+1)} (e^{-\beta\varepsilon n} - e^{-\beta t})dt \\
&\le K_1\varepsilon.
\end{aligned} \tag{2.16}$$

Combining (2.12)–(2.16), we have

$$|J_d^\varepsilon(i, u_\varepsilon(\cdot)) - v_d| = O(\varepsilon), \text{ for } i \in \mathcal{M}. \tag{2.17}$$

This completes the proof. □

Convergence Rates and Error Bounds. It is interesting from a computational point of view to estimate the convergence rate of $v_d^\varepsilon(\cdot)$ to v_d and to obtain the error bound of the control $u_\varepsilon(\cdot)$ constructed in (2.9). The following theorem demonstrates that the convergence rate and error bound are of the order ε.

Theorem 2.6. Assume that (A1) and (A2) hold and that the control set Γ contains finitely many elements. Then for all $i \in \mathcal{M}$,

$$v_d^\varepsilon(i) - v_d = O(\varepsilon), \text{ and}$$
$$J_d^\varepsilon(i, u_\varepsilon(\cdot)) - v_d^\varepsilon(i) = O(\varepsilon).$$

Proof. Using the triangle inequalities, we have

$$|J_d^\varepsilon(i, u_\varepsilon(\cdot)) - v_d^\varepsilon(i)| \le |J_d^\varepsilon(i, u_\varepsilon(\cdot)) - v_d| + |v_d^\varepsilon(i) - v_d|, \; i \in \mathcal{M}.$$

In view of (2.17), it suffices to show that $v_d^\varepsilon(i) - v_d = O(\varepsilon)$, $i \in \mathcal{M}$. Note that the inequality (2.17) implies that

$$v_d^\varepsilon(i) - v_d \le J_d^\varepsilon(i, u_\varepsilon(\cdot)) - v_d \le O(\varepsilon), \text{ for } i \in \mathcal{M},$$

To derive the reverse inequalities, let $u_{o,\varepsilon}$ denote an optimal control for $\mathcal{P}_d^\varepsilon$. Using the same arguments as in the proof of (2.17), one gets

$$\left|J_d^\varepsilon(i, u_{o,\varepsilon}(\cdot)) - J_d^0(U_{o,\varepsilon})\right| \le O(\varepsilon), \text{ for } i \in \mathcal{M}. \tag{2.18}$$

The control set Γ contains finitely many elements by the hypothesis, so is $\overline{\Gamma}$. Suppose $\overline{\Gamma} = \{\gamma_1, \ldots, \gamma_L\}$ for some positive integer L. Define

$$\mathcal{E}_j = \{\varepsilon : \varepsilon \in (0,1),\, U_{o,\varepsilon} = \gamma_j\}, \text{ for } j = 1, \ldots, L.$$

Then $\{\mathcal{E}_j\}$ consists of finitely many sets such that $(0,1) = \mathcal{E}_1 \cup \ldots \cup \mathcal{E}_L$. For fixed j and any $\varepsilon \in \mathcal{E}_j$, consider $\alpha^\varepsilon(\cdot)$ with transition matrix $\overline{P}^\varepsilon(U_{o,\varepsilon})$. In view of (2.18) and the optimality of $u_{o,\varepsilon}(\cdot)$, we have

$$v_d^\varepsilon(i) = J_d^\varepsilon(i, u_{o,\varepsilon}(\cdot)) = J_d^0(\gamma_j) + O(\varepsilon) \ge v_d + O(\varepsilon), \text{ for } i \in \mathcal{M}.$$

Thus for $0 < \varepsilon < 1$ and for $i \in \mathcal{M}$, $v_d^\varepsilon(i) - v_d \ge O(\varepsilon)$. This completes the proof. □

2.1.2 Irreducible MDP under Long-Run Average Cost

Up to now, we have discussed singularly perturbed discrete-time MDPs having discounted cost criteria. In this section we consider MDPs with a long-run average cost. The results in this section are similar in spirit to that of [5] and [6]. The major differences are that we consider classical feedback control policies and obtain asymptotic optimality of these policies as $\varepsilon \to 0$ while in [5] and [6] relaxed control is considered and only nearly optimal policies are obtained.

In what follows, we need to revise the assumptions (A1) and (A2), and to require a stronger version of irreducibility.

Definition 2.7. Let $P(U)$ be a transition matrix of a discrete-time Markov chain. If it is irreducible and the corresponding stationary distribution $\mu(U) = (\mu_1(U), \dots, \mu_m(U))$, satisfies $\mu_i(U) > 0$, $i = 1, \dots, m$, then $P(U)$ is said to be irreducible.

(B1) For each $U \in \overline{\Gamma}$, both $P(U)$ and $\overline{P}^\varepsilon(U)$ are irreducible.

(B2) Γ contains finitely many elements.

Since we are dealing with long-run average costs, aperiodicity is not needed. Consider the following problem:

$$\mathcal{P}^\varepsilon_{\text{av}} : \begin{cases} \text{minimize: } J^\varepsilon_d(u(\cdot)) = \limsup_{N\to\infty} \frac{1}{N} \mathsf{E} \sum_{n=0}^{N} g(\alpha^\varepsilon_n, u(\alpha^\varepsilon_n)), \\ \text{subject to: } \alpha^\varepsilon_n \sim P^\varepsilon(u(\alpha^\varepsilon_n)),\ n = 0, 1, \dots,\ u(\cdot) \in \mathcal{A}_f, \\ \text{value function: } \lambda^\varepsilon_d = \inf_{u(\cdot)\in\mathcal{A}_f} J^\varepsilon_d(u(\cdot)). \end{cases}$$

Note that for any given $U \in \overline{\Gamma}$, $\overline{P}^\varepsilon(U)$ is irreducible. Thus the associated Markov chain $\alpha^\varepsilon(\cdot)$ has a stationary distribution. Consequently, the average cost function $J^\varepsilon_d(u(\cdot))$ does not depend on the initial state $\alpha^\varepsilon(0) = i$, nor does the value function λ^ε_d.

Using DP approach, one can write the associated DP equation of $\mathcal{P}^\varepsilon_{\text{av}}$ as follows:

$$\lambda^\varepsilon_d + h^\varepsilon(i) = \min_{u\in\overline{\Gamma}} \left\{ g(i,u) + \sum_{j=1}^{m} p^\varepsilon_{ij}(u) h^\varepsilon(j) \right\}, \tag{2.19}$$

where $h^\varepsilon(i)$ is a function to be determined later. Subtracting $h^\varepsilon(i)$ from both sides of (2.19) leads to

$$\lambda^\varepsilon_d = \min_{u\in\Gamma} \left\{ g(i,u) + \sum_{j=1}^{m} (p^\varepsilon_{ij}(u) - \delta_{ij}) h^\varepsilon(j) \right\}. \tag{2.20}$$

Note that the irreducibility of $\overline{P}^\varepsilon(U)$ implies that $\overline{P}^\varepsilon(U) - I$ is a irreducible generator for a continuous-time Markov chain. Using a standard proof as in Ross [36], one gets the following theorem that reveals the optimality of problem $\mathcal{P}^\varepsilon_{\text{av}}$.

Theorem 2.8. Under (B1) and (B2), the following assertions hold:

(a) For each fixed $\varepsilon > 0$, there exists a pair $(\lambda^\varepsilon_d, h^\varepsilon(\cdot))$ that satisfies the DP equation (2.20).

(b) The DP equation (2.20) has a unique solution in the sense that if $(\tilde{\lambda}^\varepsilon_d, \tilde{h}^\varepsilon(\cdot))$ is another solution to (2.20), then $\tilde{\lambda}^\varepsilon_d = \lambda^\varepsilon_d$ and for some constant K_0, $\tilde{h}^\varepsilon(i) = h^\varepsilon(i) + K_0$, for $i \in \mathcal{M}$.

(c) Let $u_{o,\varepsilon}(\cdot) = \{u_{o,\varepsilon}(i) \in \Gamma,\ i \in \mathcal{M}\}$ denote a minimizer of the right-hand side of (2.20). Then $u_{o,\varepsilon}(\cdot) \in \mathcal{A}_f$ is optimal and

$$J_d^\varepsilon(u_{o,\varepsilon}(\cdot)) = \lambda_d^\varepsilon.$$

Let $\mathcal{A}^0$ denote a class of functions $U(\cdot) = \{U(k) :\ k \in \overline{\mathcal{M}}\}$ such that $U(k) \in \Gamma_k$ for $k = 1, \ldots, l$. For convenience, call $U = (U(1), \ldots, U(l)) \in \mathcal{A}^0$ an admissible control for the limit problem.

Analogous to the discounted cost case, the limit problem with the long-run average cost is given as follows. Again we obtain a limit optimization problem under long-run average cost.

$$\mathcal{P}_{\text{av}}^0 : \begin{cases} \text{minimize: } J_d^0(U) = \limsup\limits_{T\to\infty} \dfrac{1}{T} \mathsf{E} \displaystyle\int_0^T \overline{g}(U)dt = \overline{g}(U), \\ \text{value function: } \lambda_d^0 = \inf\limits_{U(\cdot)\in\mathcal{A}^0} J_d^0(U(\cdot)) = \min\limits_{U\in\overline{\Gamma}} \overline{g}(U). \end{cases}$$

The following lemma concerns with an approximation to the stationary probability corresponding to $\overline{P}^\varepsilon(U)$.

Lemma 2.9. For any $U \in \overline{\Gamma}$, let $\mu_\varepsilon(U)$ denote the stationary distribution of $\overline{P}^\varepsilon(U)$. Then $\mu_\varepsilon(U) = \mu(U) + O(\varepsilon)$, where $\mu(U)$ is the stationary distribution of $P(U)$.

This lemma can be proved using the irreducibility condition of $\overline{P}^\varepsilon$. Next, we present the main theorem of this section.

Theorem 2.10. Let $U_o \in \overline{\Gamma}$ be an optimal solution for $\mathcal{P}_{\text{av}}^0$ and construct $u_\varepsilon(\cdot)$ as in (2.9). Then $u_\varepsilon(\cdot)$ is asymptotically optimal with an ε-order bound, i.e.,

$$J_d^\varepsilon(u_\varepsilon(\cdot)) - \lambda_d^\varepsilon = O(\varepsilon).$$

Proof. Let α_n^ε denote the discrete-time Markov chain with transition probability matrix $\overline{P}^\varepsilon(U_o)$. In view of the irreducibility of $\overline{P}^\varepsilon(U_o)$, we have

$$\lim_{n\to\infty} \mathsf{P}(\alpha_n^\varepsilon = i) = \mu_\varepsilon^i(U_o).$$

Therefore,

$$\begin{aligned} J_d^\varepsilon(u_\varepsilon(\cdot)) &= \limsup_{N\to\infty} \frac{1}{N} \mathsf{E} \sum_{n=0}^N g(\alpha_n^\varepsilon, u(\alpha_n^\varepsilon)) \\ &= \limsup_{N\to\infty} \frac{1}{N} \sum_{n=0}^N \sum_{i=1}^m g(i, u_o^i) \mathsf{P}(\alpha_n^\varepsilon = i) \qquad (2.21) \\ &= \sum_{i=1}^m g(i, u_o^i) \mu_\varepsilon^i(U_o), \end{aligned}$$

and

$$\lambda_d^0 = J_d^0(U_o) = \limsup_{T\to\infty} \frac{1}{T}\int_0^T \overline{g}(U_o))dt = \overline{g}(U_o) = \sum_{i=1}^m \mu_i(U_o) g(i, u_o^i). \tag{2.22}$$

By virtue of Lemma 2.9, we have

$$\left|\mu_\varepsilon^i(U_o) - \mu_i(U_o)\right| = O(\varepsilon).$$

It follows that

$$J_d^\varepsilon(u_\varepsilon(\cdot)) - \lambda_d^0 = O(\varepsilon). \tag{2.23}$$

As a result,

$$\lambda_d^\varepsilon \le J_d^\varepsilon(u_\varepsilon(\cdot)) \le \lambda_d^0 + O(\varepsilon). \tag{2.24}$$

Next, let $u_{o,\varepsilon}(\cdot) = \{u_{o,\varepsilon}(\alpha) : \alpha \in \mathcal{M}\} \in \mathcal{A}_f$ be an optimal control for $\mathcal{P}_{av}^\varepsilon$.
Using the same arguments as in the proof of (2.23), we obtain

$$J_d^\varepsilon(u_{o,\varepsilon}(\cdot)) - J_d^0(U_{o,\varepsilon}) = O(\varepsilon). \tag{2.25}$$

Since the control set Γ contains finitely many elements, so does $\overline{\Gamma}$, the control set for the limit problem $\mathcal{P}_{av}^0$. As in the proof of Theorem 2.6, suppose $\overline{\Gamma} = \{\gamma_1, \ldots, \gamma_L\}$ for some positive integer L. Define

$$\mathcal{E}_j = \{\varepsilon : \varepsilon \in (0,1),\ U_{o,\varepsilon} = \gamma_j\}, \text{ for } j = 1, \ldots, L.$$

Then $\{\mathcal{E}_j\}$ consists of finitely many sets such that $(0,1) = \mathcal{E}_1 \cup \ldots \cup \mathcal{E}_L$. For fixed j and any $\varepsilon \in \mathcal{E}_j$, in view of the optimality of $u_{o,\varepsilon}(\cdot)$ and (2.25), we have

$$\lambda_d^\varepsilon = J_d^\varepsilon(u_{o,\varepsilon}(\cdot)) = J_d^0(U_{o,\varepsilon}) + O(\varepsilon) \ge \lambda_d^0 + O(\varepsilon). \tag{2.26}$$

Thus for $0 < \varepsilon < 1$,

$$\lambda_d^\varepsilon \ge \lambda_d^0 + O(\varepsilon). \tag{2.27}$$

Combining (2.24) and (2.27), we get

$$\lambda_d^\varepsilon = \lambda_d^0 + O(\varepsilon). \tag{2.28}$$

Finally, in view of (2.23) and (2.28), we obtain

$$J_d^\varepsilon(u_\varepsilon(\cdot)) - \lambda_d^\varepsilon = (J_d^\varepsilon(u_\varepsilon(\cdot)) - \lambda_d^0) + (\lambda_d^0 - \lambda_d^\varepsilon) = O(\varepsilon).$$

This completes the proof. □

2.1.3 MDP with General Transition Matrices

In this section, we consider the MDP problems with more general transition matrices. Proofs of results in this section will be omitted and can be found in [27]. Note that a Markov decision process corresponding to $P(u)$ can either consist of a number of recurrent (ergodic) classes of states or also include transient states in addition to the ergodic classes.

Suppose that the Markov chain with transition matrix $P(u)$ consists of only a number of recurrent classes, then we write it as

$$P(u) = \begin{pmatrix} P_1(u) & & \\ & \ddots & \\ & & P_l(u) \end{pmatrix}. \tag{2.29}$$

For $k = 1, \ldots, l$, let $\mathcal{M}_k = \{s_{k1}, \ldots, s_{km_k}\}$ denote the recurrent sub-state space of $\alpha^\varepsilon(\cdot)$ corresponding to the block $P_k(u)$. The entire state space of $\alpha^\varepsilon(\cdot)$ can be decomposed to

$$\mathcal{M} = \mathcal{M}_1 \cup \ldots \cup \mathcal{M}_l = \{s_{11}, \ldots, s_{1m_1}, \ldots, s_{l1}, \ldots, s_{lm_l}\}$$

with $m_1 + \cdots + m_l = m$.

If the Markov chain with transition matrix $P(u)$ also includes transient states, we write

$$P(u) = \begin{pmatrix} P_1(u) & & & \\ & \ddots & & \\ & & P_l(u) & \\ P_{*,1}(u) & \cdots & P_{*,l}(u) & P_*(u) \end{pmatrix}. \tag{2.30}$$

In this case, denote the subspace of transient states by $\mathcal{M}_* = \{s_{*1}, \ldots, s_{*m_*}\}$. Then the state space can be written as

$$\mathcal{M} = \mathcal{M}_1 \cup \ldots \cup \mathcal{M}_l \cup \mathcal{M}_* = \{s_{11}, \ldots, s_{1m_1}, \ldots, s_{l1}, \ldots, s_{lm_l}, s_{*1}, \ldots, s_{*m_*}\}$$

with $m_1 + \cdots + m_l + m_* = m$.

Since the transition matrix with recurrent states is a special case of the model with transient states, we only consider the latter. Let $p^k_{ij}(u)$ denote the ijth component of $P_k(u)$ i.e., $P_k(u) = (p^k_{ij}(u))_{m_k \times m_k}$. Similarly, define $p_{*ij}(u)$ and $p^k_{*ij}(u)$ such that $P_*(u) = (p_{*ij}(u))_{m_* \times m_*}$ and $P_{*,k}(u) = (p^k_{*ij}(u))_{m_* \times m_k}$, respectively. Next, we define the control set for the limit problem. For each $k = 1, \ldots, l, *$, define

$$\begin{aligned} \Gamma_k &= \{U^k := (u^{k1}, \ldots, u^{km_k}) : \; u^{kj} \in \Gamma, \; j = 1, \ldots, m_k\}, \\ \overline{\Gamma} &= \Gamma_1 \times \cdots \times \Gamma_l = \{U = (U^1, \ldots, U^l) : \; U^k \in \Gamma_k, \; k = 1, \ldots, l\}, \\ \widetilde{\Gamma} &= \overline{\Gamma} \times \Gamma_* = \{\widetilde{U} = (U, U^*) : \; U = (U^1, \ldots, U^l) \in \overline{\Gamma}, \; U^* \in \Gamma_*\}. \end{aligned}$$

For each $\widetilde{U} = (U, U^*) = (U^1, \ldots, U^l, U^*) \in \widetilde{\Gamma}$, let

$$\begin{aligned}
P_k(U^k) &= (p_{ij}^k(u^{ki}))_{m_k \times m_k}, \text{ for } k = 1, \ldots, l, \\
P_*(U^*) &= (p_{*ij}(u^{*i}))_{m_* \times m_*}, \\
P_{*,k}(U^*) &= (p_{*ij}^k(u^{*i}))_{m_* \times m_k}, \text{ for } k = 1, \ldots, l,
\end{aligned}$$

and $Q(\widetilde{U}) = (q_{ij}(\widetilde{U}))_{m \times m}$, where

$$q_{ij}(\widetilde{U}) = \begin{cases} q_{ij}(u^{1i}), & \text{if } 1 \le i \le m_1, \\ q_{ij}(u^{2(i-m_1)}), & \text{if } m_1 < i \le m_1 + m_2, \\ \cdots & \cdots \\ q_{ij}(u^{l(i-m+m_l+m_*)}), & \text{if } m - m_l - m_* < i \le m - m_*, \\ q_{ij}(u^{*(i-m+m_*)}), & \text{if } m - m_* < i \le m, \end{cases}$$

and

$$\overline{P}^\varepsilon(\widetilde{U}) = \begin{pmatrix} P_1(U^1) & & & \\ & \ddots & & \\ & & P_l(U^l) & \\ P_{*,1}(U^*) & \cdots & P_{*,l}(U^*) & P_*(U^*) \end{pmatrix} + \varepsilon Q(\widetilde{U}). \quad (2.31)$$

Here the matrices P and Q may have different block decomposition structure. Note that if we denote $\widetilde{U} \in \widetilde{\Gamma}$ as an m-vector $\widetilde{U} = (u^1, \ldots, u^m)$, then all these newly defined matrices above are obtained from $P^\varepsilon(u)$ (defined in 2.1) by replacing the control variable u in the ith row with u^i. Such a definition reveals the dependence of the matrices on the controls.

For each $u(\cdot) \in \mathcal{A}_f$, if we define

$$\widetilde{U} = (u^{11}, \ldots, u^{1m_1}, \ldots, u^{l1}, \ldots, u^{lm_l}, u^{*1}, \ldots, u^{*m_*}), \quad \text{with } u^{ij} = u(s_{ij}),$$

then the Markov chain determined by $P^\varepsilon(u(\alpha_n^\varepsilon))$ and the one determined by $\overline{P}^\varepsilon(\widetilde{U})$ have the same probability distribution.

General MDP under Discounted Cost. For each $k = 1, \ldots, l$, let $a_k = (a_{k,1}, \ldots, a_{k,m_*})' \in \mathbb{R}^{m_* \times 1}$. Next, for each $U^* \in \Gamma_*$, if $(P_*(U^*) - I)$ is invertible, we define

$$a_k(U^*) = -(P_*(U^*) - I)^{-1} P_{*,k}(U^*) \mathbb{1}_{m_k}, \text{ for } k = 1, \ldots, l, \quad (2.32)$$

where $\mathbb{1}_{m_k} = (1, \ldots, 1)' \in \mathbb{R}^{m_k \times 1}$. For each $U^k \in \Gamma_k$, $k = 1, \ldots, l$, let $\mu^k(U^k) = (\mu_1^k(U^k), \ldots, \mu_{m_k}^k(U^k))$ be the stationary distribution of $P_k(U^k)$.

To proceed, we impose the following additional conditions on the matrices $P_*(U^*)$ and $P_{*,k}(U^*)$, $k = 1, \ldots, l$.

(A1') For each $\varepsilon > 0$, $P^\varepsilon(u)$ is a continuous function of u. Moreover, for each $U^k \in \Gamma_k$, $k = 1, \ldots, l$, $P_k(U^k)$ is irreducible and aperiodic; for each $U^* \in \Gamma_*, P_*(U^*)$ has all of its eigenvalues inside the unit circle. Moreover, for each $k = 1, \ldots, l$, $P_*(U^*) - I = B(U^*)Q_*$, and $P_{*,k}(U^*) = B(U^*)Q_{*,k}$, where $B(U^*) \in \mathbb{R}^{m_* \times m_*}$ is a Lipschitz continuous function, $Q_{*,k} \in \mathbb{R}^{m_* \times m_k}$ and $Q_* \in \mathbb{R}^{m_* \times m_*}$ are constant matrices.

Remark 2.11. Under (A1'), it is readily seen that $B(U^*)$ is nonsingular for each $U^* \in \Gamma_*$. Moreover, $a_k(U^*) = a_k$ (i.e., they are independent of the control U^*). Furthermore, it can be verified that $a_{k,j} \geq 0$ and $\sum_{k=1}^{l} a_{k,j} = 1$ for each $j = 1, \ldots, m_*$. That is, $(a_{1,j}, \ldots, a_{l,j})$ is a probability row vector.

Remark 2.12. One of the main focuses of this part is the limit behavior of the system as $\varepsilon \to 0$. Intuitively, as ε gets smaller and smaller, from any transient state s_{*j}, the Markov chain will jump to the set of recurrent states with smaller and smaller probability of return. The role of control in the set of transient states is not as important as that of the recurrent states.

Partition $Q(\widetilde{U})$ as

$$Q(\widetilde{U}) = \begin{pmatrix} Q^{11}(U) & Q^{12}(U) \\ Q^{21}(U^*) & Q^{22}(U^*) \end{pmatrix},$$

where $Q^{11}(U) \in \mathbb{R}^{(m-m_*)\times(m-m_*)}$, $Q^{12}(U) \in \mathbb{R}^{(m-m_*)\times m_*}$, $Q^{21}(U^*) \in \mathbb{R}^{m_* \times (m-m_*)}$, and $Q^{22}(U^*) \in \mathbb{R}^{m_* \times m_*}$. For $U = (U^1, \ldots, U^l) \in \overline{\Gamma}$, let

$$\overline{Q}(U) = \operatorname{diag}(\mu^1(U^1), \ldots, \mu^l(U^l)) \left(Q^{11}(U)\tilde{\mathbb{1}} + Q^{12}(U)(a_1, \ldots, a_l)\right), \tag{2.33}$$

where

$$\tilde{\mathbb{1}} = \operatorname{diag}(\mathbb{1}_{m_1}, \ldots, \mathbb{1}_{m_l}) \in \mathbb{R}^{(m-m_*)\times l},$$

and $\operatorname{diag}(A_1, \ldots, A_l)$ denotes a block diagonal matrix with matrix entries $A_1, \ldots, A_l$.

It is easy to check that $\overline{Q}(U)$ is a generator. Let $\tilde{x}(\cdot) = \{\tilde{x}(t) : t \geq 0\}$ be the continuous-time Markov chain generated by $\overline{Q}(U)$. The state space of $\tilde{x}(\cdot)$ is denoted by $\overline{\mathcal{M}} = \{1, \ldots, l\}$. Note that for $k = 1, \ldots, l$, the kth row of $\overline{Q}(U)$ depends only on U^k. i.e., $\overline{Q}(U) = (\bar{q}_{kk'}(U^k))_{l\times l}$. For a function $f(\cdot)$ defined on $\overline{\mathcal{M}}$, with a slight abuse of notation, write $\overline{Q}(U^k)f(\cdot)(k)$ instead of $\overline{Q}(U)f(\cdot)(k)$, where

$$\overline{Q}(U^k)f(\cdot)(k) = \sum_{k' \neq k} \bar{q}_{kk'}(U^k)(f(k') - f(k)).$$

Thus, the process $\tilde{x}(\cdot)$ generated by $\overline{Q}(U)$ can be viewed as a Markov chain generated by $\overline{Q}(U(\tilde{x}(t)))$, $t \geq 0$, with the understanding that $U(\tilde{x}(t)) = U^k$ if $\tilde{x}(t) = k$.

To proceed, define

$$\overline{g}(k, U^k) = \sum_{j=1}^{m_k} \mu_j^k(U^k) g(s_{kj}, u^{kj}), \ k = 1, \ldots, l.$$

Now we have an auxiliary Markov decision problem:

$$\mathcal{P}_d^0 : \begin{cases} \text{minimize:} \ \ J_d^0(k, U) = \mathsf{E} \displaystyle\int_0^\infty e^{-\beta t} \overline{g}(\tilde{x}(t), U(\tilde{x}(t)))dt, \\ \text{subject to:} \ \ \tilde{x}(t) \sim \overline{Q}(U(\tilde{x}(t))), \ t \geq 0, \ \tilde{x}(0) = k, \ U \in \mathcal{A}^0, \\ \text{value function:} \ \ v_d(k) = \displaystyle\inf_{U \in \mathcal{A}^0} J_d^0(k, U). \end{cases}$$

Our task to follow is to show that $\mathcal{P}_d^0$ is the limit problem in an appropriate sense. The DP equation for the limit problem $\mathcal{P}_d^0$ is :

$$\beta v_d(k) = \min_{U^k \in \Gamma_k} \left\{ \overline{g}(k, U^k) + \overline{Q}(U^k) v_d(\cdot)(k) \right\}, \quad \text{for } k = 1, \ldots, l. \tag{2.34}$$

Let $U_o = (U_o^1, \ldots, U_o^l) \in \overline{\Gamma}$ denote a minimizer of the right-hand side of (2.34). Then $U_o \in \mathcal{A}^0$ is optimal for $\mathcal{P}_d^0$.

Remark 2.13. Note that the limit problem is a continuous-time Markov decision process although the original problem $\mathcal{P}_d^\varepsilon$ is a discrete-time Markov decision process. Effectively, we are using solutions of continuous-time MDPs to approximate that of discrete-time MDPs.

Theorem 2.14. Assume (A1), (A1') and (A2). For each $i \in \mathcal{M}_k$, $k = 1, \ldots, l$,

$$\lim_{\varepsilon \to 0} v_d^\varepsilon(i) = v_d(k),$$

where $v_d(k)$ is the value function of the limit problem $\mathcal{P}_d^0$; for each $i = s_{*j} \in \mathcal{M}_*$, $j = 1, \ldots, m_*$,

$$\lim_{\varepsilon \to 0} v_d^\varepsilon(s_{*j}) = v_d(*j), \tag{2.35}$$

where $v_d(*j) = a_{1,j} v_d(1) + \cdots + a_{l,j} v_d(l)$.

Next we construct an asymptotic optimal control policy for the original problem $\mathcal{P}_d^\varepsilon$ from the optimal decision of the limit problem $\mathcal{P}_d^0$. Let $U_o = (U_o^1, \ldots, U_o^l) \in \mathcal{A}^0$ be an optimal control for the limit problem $\mathcal{P}_d^0$, which is obtained by minimizing the right-hand side of (2.34). Pick out any vector

$U^* = (u^{*1}, \ldots, u^{*m_*})$ in Γ_*. For this $\widetilde{U}_o = (U_o, U^*) \in \widetilde{\Gamma}$, define a control $u_\varepsilon(\cdot) = \{u_\varepsilon(\alpha) : \alpha \in \mathcal{M}\}$ for the original problem $\mathcal{P}_d^\varepsilon$:

$$u(\alpha) = \sum_{k=1}^{l} \sum_{j=1}^{m_k} I_{\{\alpha = s_{kj}\}} u_o^{kj} + \sum_{j=1}^{m_*} I_{\{\alpha = s_{*j}\}} u^{*j}. \tag{2.36}$$

It is clear that $u_\varepsilon(\cdot) = u(\alpha^\varepsilon.) \in \mathcal{A}_f$.

Remark 2.15. Note that in the above construction, the control corresponding to the transient state $i \in \mathcal{M}_*$ can be taken to be any value. This indicates that for the purpose of establishing nearly optimal policies, the control on the transition probabilities of transient states is unimportant.

Theorem 2.16. Assume (A1), (A1') and (A2). The control $u_\varepsilon(\cdot) = u(\alpha^\varepsilon.)$ constructed in (2.36) is asymptotically optimal in that

$$\lim_{\varepsilon \to 0} |J_d^\varepsilon(i, u_\varepsilon(\cdot)) - v_d^\varepsilon(i)| = 0, \quad \text{for } i \in \mathcal{M}.$$

Theorem 2.17. Assume that (A1), (A1') and (A2) hold and that the control set Γ contains finitely many elements. Then for all $i \in \mathcal{M}_k$, $k = 1, \ldots, l$, $v_d^\varepsilon(i) - v_d(k) = O(\varepsilon)$ and $J_d^\varepsilon(i, u_\varepsilon(\cdot)) - v_d^\varepsilon(i) = O(\varepsilon)$, for all $i = s_{*j} \in \mathcal{M}_*$, $j = 1, \ldots, m_*$,

$$v_d^\varepsilon(s_{*j}) - v_d(*j) = O(\varepsilon) \text{ and } J_d^\varepsilon(s_{*j}, u_\varepsilon(\cdot)) - v_d^\varepsilon(s_{*j}) = O(\varepsilon).$$

Long-Run Average Cost. Analogous to the discounted cost case, the limit problem with the long-run average cost is given as follows. Note that again we obtain a continuous-time MDP limit problem.

$$\mathcal{P}_{\text{av}}^0 : \begin{cases} \text{minimize: } J_d^0(U) = \limsup_{T \to \infty} \dfrac{1}{T} \mathsf{E} \displaystyle\int_0^T \overline{g}(\tilde{x}(t), U(\tilde{x}(t)))dt, \\ \text{subject to: } \tilde{x}(t) \sim \overline{Q}(U(\tilde{x}(t))),\ t \geq 0,\ \tilde{x}(0) = k,\ U \in \mathcal{A}^0, \\ \text{value function: } \lambda_d^0 = \displaystyle\inf_{U \in \mathcal{A}^0} J_d^0(U). \end{cases}$$

Remark 2.18. To relate the original singularly perturbed problem and the limit problem, it is essential that there exists a stationary distribution of $\tilde{x}(\cdot)$. A sufficient condition that guarantees the existence of this stationary distribution is the irreducibility of the generator $\overline{Q}(u)$. Nevertheless, the irreducibility of $\overline{P}^\varepsilon(U)$ does not imply that of $\overline{Q}(U)$. To illustrate, consider the Markov chain with transition matrix given by

$$\overline{P}^{\varepsilon} = \begin{pmatrix} 1 & 0 & 0 & 0 & 0 \\ 0 & 1 & 0 & 0 & 0 \\ 0 & 0 & 1 & 0 & 0 \\ \frac{1}{2} & 0 & 0 & \frac{1}{2} & 0 \\ 0 & \frac{1}{2} & 0 & 0 & \frac{1}{2} \end{pmatrix} + \varepsilon \begin{pmatrix} -1 & 0 & 0 & 1 & 0 \\ 0 & -1 & 0 & 0 & 1 \\ 0 & 1 & -1 & 0 & 0 \\ 1 & 1 & 1 & -4 & 1 \\ 1 & 1 & 1 & 1 & -4 \end{pmatrix},$$

where $\varepsilon < 1/8$. In this example, we have

$$P_1 = P_2 = P_3 = 1, \; P_* = \begin{pmatrix} \frac{1}{2} & 0 \\ 0 & \frac{1}{2} \end{pmatrix}.$$

It follows that

$$\mu^1 = \mu^2 = \mu^3 = 1, \; a_1 = (1,0)', \; a_2 = (0,1)', \; a_3 = (0,0)'.$$

Thus by (2.33) one can find that

$$\overline{Q} = \mathrm{diag}(\mu^1, \mu^2, \mu^3)\left(Q^{11}\tilde{\mathbb{1}} + Q^{12}(a_1, a_2, a_3)\right) = \begin{pmatrix} 0 & 0 & 0 \\ 0 & 0 & 0 \\ 0 & 1 & -1 \end{pmatrix},$$

which is not irreducible since the rank of $\overline{Q}$ is only 1. On the other hand, it is easy to see that $\overline{P}^{\varepsilon}$ is irreducible for sufficiently small $\varepsilon > 0$. In fact, the unique positive stationary distribution is given by

$$\left(\frac{2(5\varepsilon+1)}{5(10\varepsilon+1)}, \frac{20\varepsilon+3}{5(10\varepsilon+1)}, \frac{10\varepsilon}{5(10\varepsilon+1)}, \frac{4\varepsilon}{5(10\varepsilon+1)}, \frac{6\varepsilon}{5(10\varepsilon+1)}\right).$$

To guarantee the desired irreducibility, we need a condition, which is given as (B1') below, on the vectors a_k for $k = 1, \ldots, l$. Then we establish the irreducibility of the generator $\overline{Q}(U)$.

(B1') for $k = 1, \ldots, l$, $j = 1, \ldots, m_*$, $0 < a_{k,j} < 1$.

Lemma 2.19. Assume (B1), (B1') and (B2). The generator $\overline{Q}(U)$ is irreducible for each $U \in \overline{\Gamma}$.

To proceed, let us consider the DP equation for the limit problem $\mathcal{P}^0_{\mathrm{av}}$,

$$\lambda_d^0 = \min_{U^k \in \Gamma_k} \left\{ \overline{g}(i, U^k) + \overline{Q}(U^k) h^0(\cdot)(k) \right\}, \tag{2.37}$$

for some function $h^0(\cdot)$. The verification theorem on the limit problem is given below; its proof is standard (see Ross [36]).

Theorem 2.20. Assume (B1), (B1') and (B2). Then the following assertions hold:

(a) There exists a pair $(\lambda_d^0, h^0(\cdot))$ that satisfies the DP equation (2.37).

(b) The DP equation (2.37) has a unique solution in the sense that if $(\tilde{\lambda}_d^0, \tilde{h}^0(\cdot))$ is another solution to (2.37), then $\tilde{\lambda}_d^0 = \lambda_d^0$ and for some constant K_0, $\tilde{h}^0(k) = h^0(k) + K_0$, for $k = 1, \ldots, l$.

(c) Let $U_o = (U_o^1, \ldots, U_o^l) \in \overline{\Gamma}$ be a minimizer of the right-hand side of (2.37). Then $U_o \in \mathcal{A}^0$ is optimal and $J_d^0(U_o) = \lambda_d^0$.

Theorem 2.21. Assume (B1), (B1') and (B2). Let $U_o \in \mathcal{A}_0$ be an optimal control for $\mathcal{P}_{\text{av}}^0$ and construct $u_\varepsilon(\cdot)$ as in (2.36). Then $u_\varepsilon(\cdot)$ is asymptotically optimal with an ε-order bound, i.e.,

$$J_d^\varepsilon(u_\varepsilon(\cdot)) - \lambda_d^\varepsilon = O(\varepsilon).$$

2.1.4 Historical Notes

Singularly perturbed MDPs have been studied by many researchers in the past decades. In [17], a class of DP equations arising from continuous-time discounted MDPs was considered. The underlying Markov chain is assumed to have weak and strong interactions. Under such a probabilistic structure, an asymptotic expansion of solutions to these DP equations was derived. Then the results in [17] were extended in [16] and [34] to incorporate Markov chains with multiple-time scales. The corresponding probabilistic interpretation of the reduction of perturbed Markov chains was discussed in depth in [16]. The main focus in these papers is on asymptotic expansions of DP equations rather than construction of nearly optimal policies via limit control problems.

In [1] and [5], a discrete-time singularly perturbed MDP with average cost was considered. The basic idea is to let the time go to infinity and then study the dependence of the system on the small parameter ε. As $\varepsilon \to 0$, a limit problem is obtained. Using this limit problem, a nearly optimal control policy was derived. In [1], it is shown that there exists a deterministic control policy that is uniformly optimal for ε sufficiently small, while the emphasize of [5] was on the relaxed control policies. Then these results were extended to general state space in [6]. Additional results along this line can be found in [2], which contains several algorithms for solving the corresponding limit problems. For more detailed discussion on singularly perturbed Markov systems and review of the literature, we refer the reader to the book [33].

Classical treatments of discrete-time-MDP models can be found in [18], [36], and [37] among others. As for related literature on discrete-time singularly perturbed systems, see, for example, [9], [20], [31], and references therein.

In connection with sensitivity analysis of these Markov systems, Cao and his colleagues developed a matrix analysis approach of Poisson equations and the corresponding potentials. We refer the reader to their recent papers [11, 12, 13, 14, 15].

2.2 Hybrid LQG Control

This part aims at finding nearly optimal controls for a class of hybrid discrete-time linear quadratic (LQ) regulator problems. The classical design of linear feedback controls is based on a plant with fixed-deterministic parameters. The advantage of using a LQ model is that the resulting control laws are linear with respect to the state variable and are easy to compute. The disadvantage is, however, it fails to capture system structural changes due to uncertainty. In recent years, much effort has been directed to the design of the so-called "robust" controls that can adapt varying system environment. Hybrid systems, in which the usual dynamic systems are coupled with discrete-event interventions, have been developed. The research activities are motivated by various applications in process control, speech and/or pattern recognition, signal processing, telecommunications, and manufacturing, among others. In this connection, hybrid LQG systems whose dynamics are governed by a number of linear subsystems coupled by a jump process (often assumed to be a Markov chain) dictating structural changes, have received much attention.

We focus on a hybrid linear system consisting of a large number of configurations modulated by a finite-state Markov chain. At any given instance, the system takes one of the possible system configurations, in which the coefficients depend on the state of the underlying Markov chain. Clearly, this model has greater capability to account for various types of random disturbances in reality. The inclusion of additional variables makes the system large and complex, however. For the control problem of the hybrid system, using a commonly utilized machinery–dynamic programming (DP) principle, one obtains a system of Riccati equations that leads to optimal control laws. Since the state space of the Markov chain is frequently very large in various applications, a large number of Riccati equations have to be solved using the DP approach. To overcome the difficulty, using time-scale separation, we introduce a small parameter in the underlying Markov chains to reflect the different rates of changes among different states resulting in two-time scale (a fast-time scale and a slowly varying one). This leads to singularly perturbed Markovian models with weak and strong interactions.

In this part, we first decompose the state space of the underlying Markov chain into a number of recurrent classes and a group of transient states according to the different jump rates. We then aggregate the states and replace the original system with its "average" and show that under suitable scaling, a limit control system that has fewer number of Riccati equations to be solved, is obtained. One interesting aspect is that the corresponding limit problem is a continuous-time one. Using the optimal control law of the limit system, we construct controls for the original system, which leads to a feasible approximation scheme. We demonstrate that controls so constructed are asymptotically optimal. Since the LQ problem is completely determined by the solutions of Riccati equations, by decomposition and aggregation, we can substantially reduce the complexity of the problem via reduction of dimensionality.

Let $\{\alpha_n^\varepsilon : n \geq 0\}$ be a finite-state Markov chain with state space $\mathcal{M} = \{1, 2, \ldots, m\}$ and transition matrix $P^\varepsilon = (p_{ij}^\varepsilon)_{m \times m}$. For a given $0 < T < \infty$, the discrete-time control system is governed by

$$\begin{cases} x_{n+1} = x_n + \varepsilon[A(\alpha_n^\varepsilon)x_n + B(\alpha_n^\varepsilon)u_n] + \sqrt{\varepsilon}\xi_n, \\ x_0 = x, \ \alpha_0^\varepsilon = \alpha, \ 0 \leq n \leq \lfloor T/\varepsilon \rfloor, \end{cases} \tag{2.38}$$

where $x_n \in \mathbb{R}^{n_1}$ represents the state (column) vector, $\xi_n \in \mathbb{R}^{n_1}$ denotes the system disturbance, $u_n \in \mathbb{R}^{n_2}$ is the control, $A(i) \in \mathbb{R}^{n_1 \times n_1}$ and $B(i) \in \mathbb{R}^{n_1 \times n_2}$, and for any $z \in \mathbb{R}$, $\lfloor z \rfloor$ denotes the integer part of z. Let $u(\cdot) = \{u_0, u_1, \ldots, u_{\lfloor T/\varepsilon \rfloor - 1}\}$ denote the control sequence. We consider the quadratic cost function

$$J^\varepsilon(x, \alpha, u(\cdot)) = \mathsf{E}\left\{\varepsilon \sum_{n=0}^{\lfloor T/\varepsilon \rfloor - 1} [x_n' M(\alpha_n^\varepsilon)x_n + u_n' N(\alpha_n^\varepsilon)u_n] + x'_{\lfloor T/\varepsilon \rfloor} D x_{\lfloor T/\varepsilon \rfloor}\right\}, \tag{2.39}$$

where for each $i \in \mathcal{M}$, $M(i) \in \mathbb{R}^{n_1 \times n_1}$, $N(i) \in \mathbb{R}^{n_2 \times n_2}$, and $D \in \mathbb{R}^{n_1 \times n_1}$, respectively.

Remark 2.22. The control problem (2.38) and (2.39) may be obtained from a continuous-time problem via discretization. To see this, consider a continuous-time hybrid control problem:

minimize

$$J^\varepsilon = \mathsf{E}\left\{\int_0^T (x'(t)M(\alpha^\varepsilon(t))x(t) + u'(t)N(\alpha^\varepsilon(t))u(t))\, dt + x'(T)Dx(T)\right\}$$

$$\text{subject to } dx(t) = (A(\alpha^\varepsilon(t))x(t) + B(\alpha^\varepsilon(t))u(t))dt + dw(t), \tag{2.40}$$

where $w(\cdot)$ is a standard Brownian motion and $\alpha^\varepsilon(t)$ is a continuous-time Markov chain generated by $\widetilde{Q}/\varepsilon + \widehat{Q}$, where both $\widetilde{Q}$ and $\widehat{Q}$ are generators and $\widetilde{Q}$ has block-diagonal form (see [40, Chapter 9] or [45] for more detail). In (2.40), making a time change $\tau = t/\varepsilon$ and then discretizing the resulting system using a step size ε lead to a control problem of the form given by (2.38) and (2.39). It should also be mentioned that discrete-time control problems also come from many applications directly because in many cases observations and state data can often only be collected in discrete time. We make the following assumptions.

(C1) For system (2.38),

(a) for each $i \in \mathcal{M}$, $M(i)$ is symmetric and nonnegative definite, and $N(i)$ and D are symmetric positive definite.

(b) $\{\xi_n\}$ is a sequence of independent and identically distributed (i.i.d) Gaussian random variables with mean 0 and variance Σ.

(c) $\{\alpha_n^\varepsilon\}$ and $\{\xi_n\}$ are independent.

The problem of interest is to choose $u(\cdot)$ to minimize J^ε. We use the DP approach to solve the problem (see Bertsekas [4]). Let

$$\lambda_n^\varepsilon(x_n, \alpha_n) = \min_{u_n,\ldots,u_{\lfloor T/\varepsilon\rfloor -1}} \mathsf{E}\left\{ \varepsilon \sum_{k=n}^{\lfloor T/\varepsilon\rfloor -1} [x_k' M(\alpha_k^\varepsilon) x_k + u_k' N(\alpha_k^\varepsilon) u_k] \right. \\ \left. + x'_{\lfloor T/\varepsilon\rfloor} D x_{\lfloor T/\varepsilon\rfloor} \right\}$$

with $\alpha_n^\varepsilon = \alpha_n \in \mathcal{M}$. Then

$$\lambda_0^\varepsilon(x_0, \alpha_0) = \min_{u(\cdot)} J^\varepsilon(x, \alpha, u(\cdot)) \text{ with } x_0 = x,\ \alpha_0 = \alpha.$$

The DP equations are given by

$$\begin{cases} \lambda^\varepsilon_{\lfloor T/\varepsilon\rfloor}(x_{\lfloor T/\varepsilon\rfloor}, \alpha_{\lfloor T/\varepsilon\rfloor}) = x'_{\lfloor T/\varepsilon\rfloor} D x_{\lfloor T/\varepsilon\rfloor}, \\ \lambda_n^\varepsilon(x_n, \alpha_n) = \min_{u_n} \mathsf{E}\left\{ \varepsilon [x_n' M(\alpha_n) x_n + u_n' N(\alpha_n) u_n] \right. \\ \qquad \left. + \lambda^\varepsilon_{n+1}(x_{n+1}, \alpha_{n+1}) \right\},\ 0 \le n < \lfloor T/\varepsilon\rfloor. \end{cases} \tag{2.41}$$

Thus, given x_n, for $\alpha_n = i \in \mathcal{M}$, in view of the independence of $\{\alpha_n^\varepsilon\}$ and $\{\xi_n\}$, we have

$$\lambda_n^\varepsilon(x_n, i) = \min_{u_n} \left\{ \varepsilon [x_n' M(i) x_n + u_n' N(i) u_n] + \sum_{j=1}^m p_{ij}^\varepsilon \lambda^\varepsilon_{n+1}(x_{n+1}, j) \right\}. \tag{2.42}$$

To solve (2.42), we consider $\lambda_n^\varepsilon(x, i)$ having the following quadratic form

$$\lambda_n^\varepsilon(x, i) = x' K_n^\varepsilon(i) x + q_n^\varepsilon(i), \tag{2.43}$$

where $K_n^\varepsilon(i) \in \mathbb{R}^{n_1 \times n_1}$ and $q_n^\varepsilon(i) \in \mathbb{R}^1$ are functions to be determined. We will show in what follows that both $K_n^\varepsilon(i)$ and $q_n^\varepsilon(i)$ exist uniquely and $\lambda_n^\varepsilon(x,i)$ is indeed a solution to (2.42). Then following the verification theorem approach in [19], it is easy to show that $\lambda_n^\varepsilon(x,i)$ is the only solution to (2.42) and the corresponding minimizer u_n^* of the right-hand side of (2.42) is an optimal control.

To proceed, we use (2.38), (2.42) and (2.43) to obtain

$$\begin{aligned}\lambda_n^\varepsilon(x_n,i) = \ & \min_{u_n}\Bigg\{ \varepsilon[x_n' M(i)x_n + u_n' N(i)u_n] \\ & + \sum_{j=1}^m p_{ij}^\varepsilon \Big[x_n'(I+\varepsilon A(i))' K_{n+1}^\varepsilon(j)(I+\varepsilon A(i))x_n \\ & + \varepsilon^2 u_n' B'(i) K_{n+1}^\varepsilon(j) B(i) u_n \\ & + \varepsilon \operatorname{tr}\left(K_{n+1}^\varepsilon(j)\Sigma\right) + 2\varepsilon x_n'(I+\varepsilon A(i))' K_{n+1}^\varepsilon(j) B(i) u_n \\ & + q_{n+1}^\varepsilon(j)\Big]\Bigg\}.\end{aligned} \tag{2.44}$$

For any given function $f(\cdot)$ and for $i \in \mathcal{M}$, define

$$\widehat{f}(i) = \sum_{j=1}^m p_{ij}^\varepsilon f(j).$$

Then $\widehat{K}_{n+1}^\varepsilon(i) = \sum_{j=1}^m p_{ij}^\varepsilon K_{n+1}^\varepsilon(j)$. The DP equation (2.44) can be rewritten as

$$\begin{aligned}\lambda_n^\varepsilon(x_n,i) = \ & \min_{u_n}\Big\{ \varepsilon x_n' M(i)x_n + \varepsilon u_n' N(i)u_n \\ & + x_n'(I+\varepsilon A(i))' \widehat{K}_{n+1}^\varepsilon(i)(I+\varepsilon A(i))x_n \\ & + \varepsilon^2 u_n' B'(i) \widehat{K}_{n+1}^\varepsilon(i) B(i) u_n \\ & + \varepsilon \operatorname{tr}\left(\widehat{K}_{n+1}^\varepsilon(i)\Sigma\right) + 2\varepsilon x_n'(I+\varepsilon A(i))' \widehat{K}_{n+1}^\varepsilon(i) B(i) u_n \\ & + \widehat{q}_{n+1}^\varepsilon(i)\Big\}.\end{aligned} \tag{2.45}$$

Differentiating the right-hand side of (2.45) w.r.t. u_n and sending the derivative to be 0, we obtain the corresponding minimizer

$$\begin{aligned}u_n^*(i,x_n) &= -\left(N(i) + \varepsilon B'(i)\widehat{K}_{n+1}^\varepsilon(i)B(i)\right)^{-1} B'(i)\widehat{K}_{n+1}^\varepsilon(i)(I+\varepsilon A(i))x_n \\ &\overset{\text{def}}{=} -\Phi_n(i)x_n.\end{aligned} \tag{2.46}$$

Using (2.43)–(2.46), we obtain a system of Riccati equations satisfied by

$K_n^\varepsilon(i)$,

$$\begin{cases} K_n^\varepsilon(i) & = \varepsilon M(i) + (I+\varepsilon A(i))'\widehat{K}_{n+1}^\varepsilon(i)(I+\varepsilon A(i)) \\ & -\varepsilon(I+\varepsilon A(i))'\widehat{K}_{n+1}^\varepsilon(i)B(i)\left(N(i)+\varepsilon B'(i)\widehat{K}_{n+1}^\varepsilon(i)B(i)\right)^{-1} \\ & \times B'(i)\widehat{K}_{n+1}^\varepsilon(i)(I+\varepsilon A(i)), \\ K_{\lfloor T/\varepsilon \rfloor}^\varepsilon(i) & = D,\ i=1,\ldots,m, \end{cases} \tag{2.47}$$

and a system of equations satisfied by $q_n^\varepsilon(\cdot)$,

$$\begin{cases} q_n^\varepsilon(i) & = \varepsilon \mathrm{tr}\left(\widehat{K}_{n+1}^\varepsilon(i)\Sigma\right) + \widehat{q}_{n+1}^\varepsilon(i), \\ q_{\lfloor T/\varepsilon \rfloor}^\varepsilon(i) & = 0,\ i=1,\ldots,m. \end{cases} \tag{2.48}$$

Note that (2.47) and (2.48) both consist of m equations coupled by P^ε. To find the optimal control sequence $\{u_n^*\}$, we have to solve these equations. However, if the state space of α_n^ε is large, the computation becomes difficult. One has to resort to approximation schemes. In the next section, we present an approach based on the aggregation of the underlying Markov chain.

2.2.1 Aggregation and Approximation

In this section, we show that there exists a limit problem as $\varepsilon \to 0$, which is a continuous-time LQ control problem. The limit problem is simple to solve. We then use the optimal controls of the limit problem to construct controls for the original problem, which are asymptotically optimal.

Suppose $P^\varepsilon = (p_{ij}^\varepsilon)_{m\times m}$, the transition probability matrix of the Markov chain α_n^ε has the form

$$P^\varepsilon = P + \varepsilon Q, \tag{2.49}$$

where $P = (p_{ij})_{m\times m}$ is a probability transition matrix (i.e., $p_{ij} \ge 0$ and, for each i, $\sum_j p_{ij} = 1$), and $Q = (q_{ij})_{m\times m}$ is a generator of a continuous-time Markov chain (i.e., $q_{ij} \ge 0$ for $i \ne j$ and, for each $i \in \mathcal{M}$, $\sum_j q_{ij} = 0$). For simplicity, we first consider:

(C2) P is irreducible and aperiodic.

Generalizations will be given in Section 2.2.3. We begin with equations (2.47) and (2.48). First rewrite (2.47) as:

$$\begin{aligned} K_n^\varepsilon(i) = \ & \widehat{K}_{n+1}^\varepsilon(i) + \varepsilon A'(i)\widehat{K}_{n+1}^\varepsilon(i) + \varepsilon \widehat{K}_{n+1}^\varepsilon(i)A(i) + \varepsilon M(i) \\ & -\varepsilon \widehat{K}_{n+1}^\varepsilon(i)B(i)\left(N(i)+\varepsilon B'(i)\widehat{K}_{n+1}^\varepsilon(i)B(i)\right)^{-1} B'(i)\widehat{K}_{n+1}^\varepsilon(i) \\ & +O(\varepsilon^2). \end{aligned} \tag{2.50}$$

Define

$$Q^\varepsilon = \frac{1}{\varepsilon}(P - I) + Q = (q^\varepsilon_{ij}) = \left(\frac{1}{\varepsilon}(p^\varepsilon_{ij} - \delta_{ij}) + q_{ij}\right). \tag{2.51}$$

Using q^ε_{ij} instead of p^ε_{ij}, we have

$$\widehat{K}^\varepsilon_{n+1}(i) = K^\varepsilon_{n+1}(i) + \varepsilon \sum_{j=1}^{m} q^\varepsilon_{ij} K^\varepsilon_{n+1}(j). \tag{2.52}$$

Here δ_{ij} is the usual Kronecker's δ, i.e., $\delta_{ij} = 1$ if $i = j$, $\delta_{ij} = 0$ otherwise. Using (2.52), rewrite (2.50) as

$$\begin{aligned} K^\varepsilon_{n+1}(i) &= K^\varepsilon_n(i) - \varepsilon A'(i)\widehat{K}^\varepsilon_{n+1}(i) - \varepsilon \widehat{K}^\varepsilon_{n+1}(i)A(i) - \varepsilon M(i) \\ &\quad + \varepsilon \widehat{K}^\varepsilon_{n+1}(i)B(i)\left(N(i) + \varepsilon B'(i)\widehat{K}^\varepsilon_{n+1}(i)B(i)\right)^{-1} B'(i)\widehat{K}^\varepsilon_{n+1}(i) \\ &\quad - \varepsilon \sum_{j=1}^{m} q^\varepsilon_{ij} K^\varepsilon_{n+1}(j) + O(\varepsilon^2). \end{aligned} \tag{2.53}$$

To proceed, for each $i \in \mathcal{M}$, define the piecewise constant interpolations $K^\varepsilon(\cdot, i)$ and $q^\varepsilon(\cdot, i)$ as follows

$$K^\varepsilon(t,i) = K^\varepsilon_n(i),\ q^\varepsilon(t,i) = q^\varepsilon_n(i), \quad \text{for} \ \ t \in [n\varepsilon,\ n\varepsilon + \varepsilon). \tag{2.54}$$

We now concentrate on the convergence of $K^\varepsilon(\cdot, i)$ and $q^\varepsilon(\cdot, i)$. With a modification of the argument of the proofs of Lemmas 1 and 2 in [45], we obtain:

Lemma 2.23. Under Assumption (C1), the following assertions hold

(a) $K^\varepsilon_n(i)$ and $K^\varepsilon(t,i)$ are positive definite for each $n \le \lfloor T/\varepsilon \rfloor$ and each $t \in [0, T]$, respectively.

(b) There exists a constant C_T depending only on T such that for each $i \in \mathcal{M}$,

$$|K^\varepsilon_n(i)| \le C_T,\ |K^\varepsilon(t,i)| \le C_T,\ |q^\varepsilon_n(i)| \le C_T,\ |q^\varepsilon(t,i)| \le C_T.$$

For given matrices $H(i)$, $i \in \mathcal{M}$, define

$$\overline{H} = \sum_{i=1}^{m} \mu_i H(i),$$

where $(\mu_1, \ldots, \mu_n)$ is the stationary distribution of P.

Theorem 2.24. Under Assumptions (C1). As $\varepsilon \to 0$, for each $i \in \mathcal{M}$,

$$K^\varepsilon(t,i) \to \overline{K}(t)$$

and

$$q^\varepsilon(t,i) \to \overline{q}(t).$$

and $\overline{K}(t)$ and $\overline{q}(t)$ are the unique solutions to the following differential equations

$$\begin{cases} \dot{\overline{K}}(t) = -\overline{K}(t)\overline{A} - \overline{A'}\,\overline{K}(t) - \overline{M} + \overline{K}(t)\overline{BN^{-1}B'}\,\overline{K}(t) \\ \overline{K}(T) = D, \end{cases} \tag{2.55}$$

and

$$\begin{cases} \dot{\overline{q}}(t) = -\mathrm{tr}\left(\overline{K}(t)\Sigma\right). \\ \overline{q}(T) = 0, \end{cases} \tag{2.56}$$

respectively.

Proof. For any $t, s \in [0,T]$ satisfying $t+s \le T$, for simplicity, we often use $(t+s)/\varepsilon$ and t/ε in lieu of $\lfloor (t+s)/\varepsilon \rfloor$ and $\lfloor t/\varepsilon \rfloor$.

Define the piecewise constant interpolation $v^\varepsilon(\cdot,x,i)$ as

$$v^\varepsilon(t,x,i) = \lambda_n^\varepsilon(x,i) \quad \text{for} \quad t \in [n\varepsilon,\ n\varepsilon+\varepsilon). \tag{2.57}$$

Using (2.46) and $\Phi_n(\cdot)$ in (2.46) and following the proof of Lemma 3 in [45], we have

$$\begin{aligned} &v^\varepsilon(t+s,x,i) - v^\varepsilon(t,x,i) = \\ &\quad -\varepsilon \mathsf{E} \sum_{k=t/s}^{(t+s)/\varepsilon-1} \Big(x_k' M(\alpha_k^\varepsilon) x_k + x_k' \Phi_k'(\alpha_k^\varepsilon) N(\alpha_k^\varepsilon) \Phi_k(\alpha_k^\varepsilon) x_k \Big). \end{aligned}$$

Moreover, it can be shown that $\mathsf{E}|x_k|^2 \le K < \infty$. Using the boundedness of $M(\cdot)$, $N(\cdot)$, and $K_n^\varepsilon(\cdot)$, it follows that for any $\eta > 0$, there is a $\Delta > 0$ such that

$$\limsup_{\varepsilon\to 0} \sup_{\substack{0\le s\le \Delta \\ 0\le t+s\le T}} |v^\varepsilon(t+s,x,i) - v^\varepsilon(t,x,i)| \le \eta.$$

Thus, $v^\varepsilon(\cdot)$ is equicontinuous in the extended sense (see [26, p. 73] for a definition). Note that

$$v^\varepsilon(t,x,i) = x' K^\varepsilon(t,i) x + q^\varepsilon(t,i), \tag{2.58}$$

taking $x = 0$ in the above yields the equicontinuity of $q^\varepsilon(\cdot)$ in the extended sense. Since $x'K^\varepsilon(t,i)x = v^\varepsilon(t,x,i) - q^\varepsilon(t,i)$, the quadratic form is equicontinuous in the extended sense. By repeatedly choosing appropriate vectors x's, we can show all the entries of $K^\varepsilon(t,i)$ are equicontinuous in the extended sense.

Take $i \in \mathcal{M}$. Since $\{K^\varepsilon(\cdot, i)\}$ is equicontinuous in the extended sense and is uniformly bounded, [26, Theorem 4.2.2] implies that it has a subsequence that converges uniformly on $[0,T]$ to a continuous limit $K^0(\cdot, i)$. We next characterize the limit. Using (2.52), we write

$$\varepsilon A'(i)\widehat{K}^\varepsilon_{n+1}(i) = \varepsilon A'(i)K^\varepsilon_{n+1}(i) + g^\varepsilon_n,$$

where

$$g^\varepsilon_n = \varepsilon^2 A'(i)\sum_{j=1}^{m} q^\varepsilon_{ij}K^\varepsilon_{n+1}(j).$$

Similarly, we can treat the other terms involving $K^\varepsilon_{n+1}(i)$. Note that $Q^\varepsilon = (q^\varepsilon_{ij})$ is a generator of a singularly perturbed Markov chain in continuous time with $Q^\varepsilon = \widetilde{Q}/\varepsilon + Q$ and $\widetilde{Q} = P - I$. Now we write (2.53) as

$$\begin{aligned} K^\varepsilon_{n+1}(i) = {} & K^\varepsilon_n(i) - \varepsilon A'(i)K^\varepsilon_{n+1}(i) - \varepsilon K^\varepsilon_{n+1}(i)A(i) - \varepsilon M(i) \\ & + \varepsilon K^\varepsilon_{n+1}(i)B(i)\left(N(i) + O(\varepsilon)\right)^{-1}B'(i)K^\varepsilon_{n+1}(i) \\ & - \varepsilon Q^\varepsilon K^\varepsilon_{n+1}(\cdot)(i) + \overline{g}^\varepsilon_n, \end{aligned}$$

where $\overline{g}^\varepsilon_n$ denotes a collection of the terms involves g^ε_n. Consequently,

$$\begin{aligned} & K^\varepsilon(t+s,i) - K^\varepsilon(t,i) \\ & \quad = -\varepsilon \sum_{k=t/\varepsilon}^{(t+s)/\varepsilon-1} \left[A'(i)K^\varepsilon_{k+1}(i) + K^\varepsilon_{k+1}(i)A(i) + M(i)\right] \\ & \qquad + \varepsilon \sum_{k=t/\varepsilon}^{(t+s)/\varepsilon-1} K^\varepsilon_{k+1}(i)B(i)(N(i)+O(\varepsilon))^{-1}B'(i)K^\varepsilon_{k+1}(i) \\ & \qquad - \varepsilon \sum_{k=t/\varepsilon}^{(t+s)/\varepsilon-1} Q^\varepsilon K^\varepsilon_{k+1}(\cdot)(i) + G^\varepsilon(t), \end{aligned} \tag{2.59}$$

where $G^\varepsilon(t) = \sum_{k=t/\varepsilon}^{(t+s)/\varepsilon-1} \overline{g}^\varepsilon_k$.

Replacing

$$\int_t^{t+s} \widetilde{Q}K^\varepsilon(r,\cdot)(i)dr$$

by

$$\varepsilon \sum_{k=t/\varepsilon}^{(t+s)/\varepsilon-1} (P-I)K^\varepsilon_{k+1}(\cdot)(i),$$

and following the proof of [45, Theorem 1], we have $G^\varepsilon(t) \to 0$, $K^\varepsilon(t,i) \to \overline{K}(t)$ as $\varepsilon \to 0$ such that $\overline{K}(t)$ satisfies (2.55). Similarly, we can show

$$q^{\varepsilon}(t,i) \to \overline{q}(t) \text{ as } \varepsilon \to 0.$$

□

As expected, note that the limit Riccati equation in (2.55) is identical to the limit Riccati equation corresponding to the continuous-time LQG problem; see [45].

Corollary 2.25. As $\varepsilon \to 0$, $v^{\varepsilon}(t,x,i) \to v(t,x)$ for $i = 1, \dots, m$, where

$$v(t,x) = x'\overline{K}(t)x + \overline{q}(t). \tag{2.60}$$

Proof. Following from (2.54), (2.57), and (2.58), the convergence of $K^{\varepsilon}(\cdot,i)$ and $q^{\varepsilon}(\cdot,i)$, we obtain the convergence of the interpolated value functions $v^{\varepsilon}(\cdot,x,i)$. □

Next we introduce a limit control problem whose corresponding value function is $v(t,x)$. First, let $\mathcal{U}$ denote the control set for the limit control system

$$\mathcal{U} = \{U = (u^1, \dots, u^m) : u^i \in \mathbb{R}^{n_2}\}. \tag{2.61}$$

Define

$$f(x,U) = \overline{A}x + \sum_{i=1}^{m} \mu_i B(i) u^i,$$
$$\widetilde{N}(U) = \sum_{i=1}^{m} \mu_i \left(u^{i,\prime} N(i) u^i\right).$$

It can be shown that $v(t,x)$ satisfies the following HJB equation

$$\begin{aligned} 0 = \frac{\partial v(t,x)}{\partial t} + \min_{U \in \mathcal{U}} \Bigg\{ & f(x,U)\frac{\partial v(t,x)}{\partial x} + x'\overline{M}x \\ & + \widetilde{N}(U) + \frac{1}{2}\operatorname{tr}\left(\frac{\partial^2 v(t,x)}{\partial x^2}\Sigma\right)\Bigg\} \end{aligned} \tag{2.62}$$

with $v(T,x) = x'Dx$. The corresponding limit control problem is

$$\begin{cases} \text{minimize: } J(s,x,U(\cdot)) \\ \qquad = \mathsf{E}\left\{\displaystyle\int_s^T \left[\overline{x}'(t)\overline{M}\,\overline{x}(t) + \widetilde{N}(U(t))\right] dt + \overline{x}'(T)D\overline{x}(T)\right\}, \\ \text{subject to: } d\overline{x}(t) = f(\overline{x}(t), U(t))dt + \sigma dw(t), \quad \overline{x}(s) = x, \\ \text{value function: } v(s,x) = \displaystyle\min_{U(\cdot)} J(s,x,U(\cdot)), \end{cases}$$

where $\sigma\sigma' = \Sigma$ and $\omega(t)$ is a standard Brownian motion. Note that the limit problem is a continuous-time quadratic regulator problem. Using the approach as in [45], we can find the optimal control for this limit control problem, which is given by

$$U^o(t,x) = \left(u^{1,o}(t,x), \ldots, u^{m,o}(t,x)\right), \tag{2.63}$$

where

$$u^{i,o}(t,x) = -N^{-1}(i)B'(i)\overline{K}(t)x, \; i = 1, \ldots, m.$$

Now we construct a control for the original discrete-time control system as

$$u_n^\varepsilon = u_n^\varepsilon(\alpha, x_n) = u^\varepsilon(n\varepsilon, x_n, \alpha), \tag{2.64}$$

where

$$u^\varepsilon(t,x,\alpha) = \sum_{i=1}^m I_{\{\alpha=i\}} u^{i,o}(t,x).$$

Equivalently,

$$u_n^\varepsilon(i, x_n) = -N^{-1}(i)B'(i)\overline{K}(n\varepsilon)x_n. \tag{2.65}$$

We will show in the following section that such a control is asymptotically optimal.

2.2.2 Asymptotic Optimality

In this section, we consider the asymptotic optimality of our approximation scheme.

Theorem 2.26. Under Assumptions (C1) and (C2), the control $u^\varepsilon(\cdot)$ defined in (2.65) is asymptotically optimal for the original control system (2.38) and (2.39) in the sense that

$$\lim_{\varepsilon\to 0} |J^\varepsilon(x, \alpha, u^\varepsilon(\cdot)) - \lambda_0^\varepsilon(x,\alpha)| = 0. \tag{2.66}$$

Proof. First, we need to derive some convergence results. Let

$$F(t,i) = A(i) - B(i)N^{-1}(i)B'(i)\overline{K}(t). \tag{2.67}$$

Then under the optimal control $U^o(t,x)$ in (2.63), the limit control system equation can be written as

$$d\overline{x}(t) = \sum_{i=1}^m \mu_i F(t,i)\overline{x}(t)dt + \sigma dw(t), \; \overline{x}(0) = x. \tag{2.68}$$

Let

$$\widetilde{F}(t) = \sum_{i=1}^m \mu_i F(t,i).$$

We write (2.68) as

$$d\overline{x}(t) = \widetilde{F}(t)\overline{x}(t)dt + \sigma dw(t). \tag{2.69}$$

Under the feedback control (2.65), the system equation of the original discrete-time problem (2.38) becomes

$$x_{n+1} = x_n + \varepsilon F(n\varepsilon, \alpha_n^\varepsilon)x_n + \sqrt{\varepsilon}\xi_n, \quad x_0 = x. \tag{2.70}$$

Next, we define the piecewise constant interpolated processes $x^\varepsilon(\cdot)$ and $\alpha^\varepsilon(\cdot)$ as

$$x^\varepsilon(t) = x_n, \quad \alpha^\varepsilon(t) = \alpha_n, \quad \text{for } t \in [n\varepsilon,\ n\varepsilon + \varepsilon). \tag{2.71}$$

We will show that $\mathsf{E}\,|x^\varepsilon(t) - \overline{x}(t)| \to 0$ as $\varepsilon \to 0$.

In view of the system (2.70) and the limit system (2.69), we obtain

$$\sup_{0\le t\le T} \mathsf{E}|x^\varepsilon(t)|^4 < \infty, \quad \sup_{0\le t\le T} \mathsf{E}|\overline{x}(t)|^4 < \infty. \tag{2.72}$$

Next, let us establish an estimate of $\mathsf{E}\,|x^\varepsilon(t) - \overline{x}(t)|$. For $0 \le t \le T$, let $n_t^\varepsilon = \lfloor \frac{t}{\varepsilon} \rfloor$. Then, in view of (2.71), we have

$$\begin{aligned} x^\varepsilon(t) &= x + \sum_{i=1}^m \int_0^t I_{\{\alpha^\varepsilon(s)=i\}} F(s,i)x^\varepsilon(s)ds \\ &\quad - \sum_{i=1}^m \int_{n_t^\varepsilon \varepsilon}^t I_{\{\alpha^\varepsilon(s)=i\}} F(s,i)x^\varepsilon(s)ds \\ &\quad + \sum_{i=1}^m \sum_{n=0}^{n_t^\varepsilon - 1} \int_{n\varepsilon}^{(n+1)\varepsilon} I_{\{\alpha_n^\varepsilon = i\}} \Big[F(n\varepsilon, i) - F(s,i)\Big] x_n ds + \sqrt{\varepsilon} \sum_{n=0}^{n_t^\varepsilon - 1} \xi_n. \end{aligned}$$

It is easy to see from definition (2.67) and Lemma 2.23 that $|F(t,i)| \le C$. We have

$$\begin{aligned} \mathsf{E}\,|x^\varepsilon(t) - \overline{x}(t)| &\le C \int_0^t \mathsf{E}\,|x^\varepsilon(s) - \overline{x}(s)|\, ds \\ &\quad + C \sum_{i=1}^m \mathsf{E} \left| \int_0^t \left(I_{\{\alpha^\varepsilon(s)=i\}} - \mu_i\right) \overline{x}(s) ds \right| \\ &\quad + C \int_{n_t^\varepsilon \varepsilon}^t \mathsf{E}\,|x^\varepsilon(s)|\, ds \\ &\quad + C \sum_{i=1}^m \sum_{n=0}^{n_t^\varepsilon - 1} \int_{n\varepsilon}^{(n+1)\varepsilon} |F(n\varepsilon, i) - F(s,i)|\, \mathsf{E}\,|x_n|\, ds \\ &\quad + \mathsf{E} \left| \sqrt{\varepsilon} \sum_{n=0}^{n_t^\varepsilon - 1} \xi_n - \sigma \int_0^t dw(s) \right|. \end{aligned} \tag{2.73}$$

Integrating the term in the second line above by parts yields

$$\begin{aligned}\mathsf{E}\left|\int_0^t \left(I_{\{\alpha^\varepsilon(s)=i\}}-\mu_i\right)\overline{x}(s)ds\right| &\le \mathsf{E}\left|\overline{x}(t)\int_0^t \left(I_{\{\alpha^\varepsilon(s)=i\}}-\mu_i\right)ds\right| \\ &+\mathsf{E}\left|\int_0^t\left(\int_0^s \left(I_{\{\alpha^\varepsilon(\tau)=i\}}-\mu_i\right)d\tau\right)d\overline{x}(s)\right|.\end{aligned}$$

By virtue of the Cauchy-Schwarz inequality and (2.72), as $\varepsilon \to 0$,

$$\begin{aligned}&\mathsf{E}\left|\overline{x}(t)\int_0^t \left(I_{\{\alpha^\varepsilon(s)=i\}}-\mu_i\right)ds\right| \\ &\qquad \le \mathsf{E}^{1/2}|\overline{x}(t)|^2\mathsf{E}^{1/2}\left(\int_0^t \left(I_{\{\alpha^\varepsilon(s)=i\}}-\mu_i\right)ds\right)^2 \to 0.\end{aligned}$$

Similarly, as $\varepsilon \to 0$,

$$\begin{aligned}&\mathsf{E}\left|\int_0^t\left(\int_0^s \left(I_{\{\alpha^\varepsilon(\tau)=i\}}-\mu_i\right)d\tau\right)d\overline{x}(s)\right| \\ &\qquad \le C\int_0^t \mathsf{E}\left|\left(\int_0^s \left(I_{\{\alpha^\varepsilon(\tau)=i\}}-\mu_i\right)d\tau\right)\overline{x}(s)\right|ds \\ &\qquad\quad +\sigma\int_0^t \mathsf{E}\left|\left(\int_0^s \left(I_{\{\alpha^\varepsilon(\tau)=i\}}-\mu_i\right)d\tau\right)\right|dw(s) \to 0.\end{aligned}$$

Thus, we obtain

$$\mathsf{E}\left|\int_0^t \left(I_{\{\alpha^\varepsilon(s)=i\}}-\mu_i\right)\overline{x}(s)ds\right| \to 0 \text{ as } \varepsilon \to 0. \tag{2.74}$$

We claim that the rest terms in (2.73) all go to zero as $\varepsilon \to 0$. In fact, for line 3, in view of (2.72),

$$\int_{n_t^\varepsilon \varepsilon}^t \mathsf{E}\left|x^\varepsilon(s)\right|ds \le C(t-n_t^\varepsilon\varepsilon) \le C\varepsilon \to 0.$$

For line 4, in view of (2.55) and (2.67), it is readily seen that $F(t,i)$ is Lipschitz in t, i.e., $|F(t,i)-F(s,i)| \le C\,|t-s|$. Then

$$\begin{aligned}&\sum_{n=0}^{n_t^\varepsilon-1}\int_{n\varepsilon}^{(n+1)\varepsilon} |F(n\varepsilon,i)-F(s,i)|\,\mathsf{E}\,|x_n|\,ds \\ &\qquad \le C\sum_{n=0}^{n_t^\varepsilon-1}\int_{n\varepsilon}^{(n+1)\varepsilon}(s-n\varepsilon)ds \le C\sum_{n=0}^{n_t^\varepsilon-1}\varepsilon^2 \le Cn_t^\varepsilon\varepsilon^2 \to 0.\end{aligned}$$

The last line of (2.73) goes to zero due to the convergence of

$$\sqrt{\varepsilon}\sum_{n=0}^{n_t^\varepsilon-1}\xi_n \to \sigma\int_0^t dw(s).$$

Now, applying the Gronwall's inequality to (2.73), we obtain

$$\mathsf{E}\,|x^{\varepsilon}(t) - \overline{x}(t)| \to 0 \text{ as } \varepsilon \to 0. \tag{2.75}$$

We are ready to complete the proof of the theorem. Note that $\lambda_0^{\varepsilon}(x,\alpha) = v^{\varepsilon}(0,x,\alpha)$. For recurrent state $\alpha = i \in \mathcal{M}$,

$$\begin{aligned} &|J^{\varepsilon}(x,\alpha,u^{\varepsilon}(\cdot)) - \lambda_0^{\varepsilon}(x,\alpha)| \\ &\quad \le |J^{\varepsilon}(x,\alpha,u^{\varepsilon}(\cdot)) - v(0,x)| + |v^{\varepsilon}(0,x,\alpha) - v(0,x)|. \end{aligned}$$

In view of Corollary 2.25, we have that as $\varepsilon \to 0$,

$$|v^{\varepsilon}(0,x,\alpha) - v(0,x)| \to 0 \quad \text{for} \quad \alpha \in \mathcal{M}.$$

Thus, to prove (2.66), it suffices to show

$$|J^{\varepsilon}(x,\alpha,u^{\varepsilon}(\cdot)) - v(0,x)| \to 0 \quad \text{for} \quad \alpha \in \mathcal{M}.$$

Let

$$S(t,i) = M(i) + \overline{K}(t)B(i)N^{-1}(i)B'(i)\overline{K}(t), \text{ for } i = i. \tag{2.76}$$

Then applying the control (2.65) in (2.39), for any $\alpha \in \mathcal{M}$, we have

$$\begin{aligned} &J^{\varepsilon}(x,\alpha,u^{\varepsilon}(\cdot)) = \\ &\quad \sum_{i=1}^{m} \mathsf{E}\left(\sum_{n=0}^{\lfloor T/\varepsilon \rfloor -1} \int_{n\varepsilon}^{(n+1)\varepsilon} I_{\{\alpha^{\varepsilon}(t)=i\}} x^{\varepsilon,\prime}(t) S(t,i) x^{\varepsilon}(t)\,dt \right) \\ &\quad + \sum_{i=1}^{m} \mathsf{E}\left(\sum_{n=0}^{\lfloor T/\varepsilon \rfloor -1} \int_{n\varepsilon}^{(n+1)\varepsilon} I_{\{\alpha_n^{\varepsilon}=i\}} x_n' \Big[S(n\varepsilon,i) - S(t,i) \Big] x_n\,dt \right) \\ &\quad + \mathsf{E}(x'_{\lfloor T/\varepsilon \rfloor} D x_{\lfloor T/\varepsilon \rfloor}). \end{aligned}$$

Since $U^o(\cdot)$ is the optimal control of the limit system, we have

$$\begin{aligned} v(0,x) &= J(0,x,U^o(\cdot)) \\ &= \mathsf{E}\left\{ \int_0^T \Big[\overline{x}'(t)\overline{M}\,\overline{x}(t) + \widetilde{N}(U^o(t)) \Big] dt + \overline{x}'(T)D\overline{x}(T) \right\} \\ &= \sum_{i=1}^{m} \mathsf{E}\left\{ \int_0^T \mu_i \overline{x}'(t) S(t,i) \overline{x}(t)\,dt \right\} + \mathsf{E}\left(\overline{x}'(T)D\overline{x}(T)\right). \end{aligned}$$

It follows that

$$\begin{aligned}
&|J^\varepsilon(x,\alpha,u^\varepsilon(\cdot)) - v(0,x)| \\
&\le \left| \mathsf{E}\left(x'_{\lfloor T/\varepsilon \rfloor} D x_{\lfloor T/\varepsilon \rfloor} - \overline{x}'(T) D \overline{x}(T) \right) \right| \\
&\quad + \sum_{i=1}^{m} \sum_{n=0}^{\lfloor T/\varepsilon \rfloor -1} \int_{n\varepsilon}^{(n+1)\varepsilon} \mathsf{E} \left| I_{\{\alpha_n^\varepsilon = i\}} x_n' \Big[S(n\varepsilon,i) - S(t,i) \Big] x_n \right| dt \\
&\quad + \sum_{i=1}^{m} \left| \mathsf{E} \left(\int_0^{\lfloor T/\varepsilon \rfloor \varepsilon} I_{\{\alpha^\varepsilon(t)=i\}} x^\varepsilon(t)' S(t,i) x^\varepsilon(t) dt - \int_0^T \mu_i \overline{x}'(t) S(t,i) \overline{x}(t) dt \right) \right| .
\end{aligned} \tag{2.77}$$

Note that for $i \in \mathcal{M}$, $|S(t,i)| \le C$, $|S(t,i) - S(s,i)| \le C\,|t-s|$. In view of (2.72), it is easy to show that as $\varepsilon \to 0$,

$$\begin{aligned}
&\left| \mathsf{E}\left(x'_{\lfloor T/\varepsilon \rfloor} D x_{\lfloor T/\varepsilon \rfloor} - \overline{x}'(T) D \overline{x}(T) \right) \right| \to 0, \\
&\sum_{n=0}^{\lfloor T/\varepsilon \rfloor -1} \int_{n\varepsilon}^{(n+1)\varepsilon} \mathsf{E} \left| I_{\{\alpha_n^\varepsilon = i\}} x_n' \Big[S(n\varepsilon,i) - S(t,i) \Big] x_n \right| dt \to 0.
\end{aligned}$$

For the last line of (2.77), using the triangle inequality, we obtain

$$\begin{aligned}
&\left| \mathsf{E} \left(\int_0^{\lfloor T/\varepsilon \rfloor \varepsilon} I_{\{\alpha^\varepsilon(t)=i\}} x^\varepsilon(t)' S(t,i) x^\varepsilon(t) dt - \int_0^T \mu_i \overline{x}'(t) S(t,i) \overline{x}(t) dt \right) \right| \\
&\qquad \le \left| \mathsf{E} \int_0^{\lfloor T/\varepsilon \rfloor \varepsilon} I_{\{\alpha^\varepsilon(t)=i\}} \left[x^\varepsilon(t)' S(t,i) x^\varepsilon(t) - \overline{x}(t)' S(t,i) \overline{x}(t) \right] dt \right| \\
&\qquad\quad + \left| \mathsf{E} \int_0^{\lfloor T/\varepsilon \rfloor \varepsilon} \left[I_{\{\alpha^\varepsilon(t)=i\}} - \mu_i \right] \overline{x}(t)' S(t,i) \overline{x}(t) dt \right| \\
&\qquad\quad + \left| \mathsf{E} \int_{\lfloor T/\varepsilon \rfloor \varepsilon}^{T} \mu_i \overline{x}'(t) S(t,i) \overline{x}(t) dt \right| .
\end{aligned} \tag{2.78}$$

In view of (2.72), we can show that the second and the fourth line goes to zero as $\varepsilon \to 0$. It can be proved (via integration by parts) as in the proof of (2.74) that the third line also converges to zero. As a result,

$$|J^\varepsilon(x,\alpha,u^\varepsilon(\cdot)) - v(0,x)| \to 0 \text{ for any } \alpha \in \mathcal{M}. \tag{2.79}$$

This completes the proof. □

2.2.3 Hybrid LQG with General Transition Matrices

In this section, we present results with general transition matrices. Proofs of these results can be found in [28]. Suppose that P has partitioned block form

$$P = \begin{pmatrix} P_1 & & & \\ & \ddots & & \\ & & P_l & \\ P_{*,1} & \cdots & P_{*,l} & P_* \end{pmatrix},$$

where the matrices are such that $P_\iota = (p^\iota_{ij}) \in \mathbb{R}^{m_\iota \times m_\iota}$, $P_{*,\iota} = (p^\iota_{*ij}) \in \mathbb{R}^{m_* \times m_\iota}$, for $\iota = 1, \ldots, l$, $P_* = (p_{*ij}) \in \mathbb{R}^{m_* \times m_*}$. As will be seen that P_ι corresponds to recurrent states and P_* and $P_{*,\iota}$ correspond to transient states. Let $\mathcal{M}_\iota = \{s_{\iota 1}, \ldots, s_{\iota m_\iota}\}$ denote the recurrent sub-state space of α^ε_n corresponding to the block P_ι, for $\iota = 1, \ldots, l$, and let $\mathcal{M}_* = \{s_{*1}, \ldots, s_{*m_*}\}$ denote the subspace of transient states. Then the state space of α^ε_n can be written as

$$\begin{aligned} \mathcal{M} &= \mathcal{M}_1 \cup \ldots \cup \mathcal{M}_l \cup \mathcal{M}_* \\ &= \{s_{11}, \ldots, s_{1m_1}\} \cup \cdots \cup \{s_{l1}, \ldots, s_{lm_l}\} \cup \{s_{*1}, \ldots, s_{*m_*}\} \end{aligned}$$

with $m_1 + \cdots + m_l + m_* = m$. For the Markov chain α^ε_n, assume:

(C2') The transition probability matrices satisfy

(a) P_ι is a transition probability matrix and is irreducible and aperiodic, for each $\iota = 1, \ldots, l$;

(b) P_* has all of its eigenvalues inside the unit circle.

As for the MDP part, for each $i = 1, \ldots, l$, $P_{*,i}$ represents the transition probabilities from the transient states to the ith recurrent class, and that P_* is the transition probabilities going into the transient states. Thus as $n \to \infty$, $P^n_* \to 0$. The natural condition for this is the eigenvalues all being inside the unit circle.

Under assumption (C2'), there exists stationary distribution of the sub-Markov chain corresponding to P_ι for each $\iota = 1, \ldots, l$. Let $\mu^\iota = (\mu^\iota_1, \ldots, \mu^\iota_{m_\iota})$ denote such stationary distribution.

Since all the eigenvalues of P_* are inside the unit circle, $(P_* - I)$ is invertible. Define

$$a_\iota = (a_{\iota,1}, \ldots, a_{\iota,m_*})' = -(P_* - I)^{-1} P_{*,\iota} \mathbb{1}_{m_\iota}, \text{ for } \iota = 1, \ldots, l, \tag{2.80}$$

where $\mathbb{1}_\iota = (1, \ldots, 1)' \in \mathbb{R}^{\iota \times 1}$. Similarly as in Section 4, partition Q as

$$Q = \begin{pmatrix} Q^{11} & Q^{12} \\ Q^{21} & Q^{22} \end{pmatrix}.$$

Let

$$\overline{Q} = \text{diag}(\mu^1, \dots, \mu^l) \left(Q^{11}\tilde{\mathbb{1}} + Q^{12}(a_1, \dots, a_l) \right).$$

Let $\overline{\alpha}(\cdot) = \{\overline{\alpha}(t) : t \geq 0\}$ be a continuous-time Markov chain generated by $\overline{Q}$. The state space of $\overline{\alpha}(\cdot)$ is $\overline{\mathcal{M}} = \{1, \dots, l\}$.

Let ζ be a random variable uniformly distributed on $[0, 1]$ that is independent of $\{\alpha_n^\varepsilon\}$. For each $j = 1, \dots, m_*$, define an integer-valued random variable ζ_j by

$$\zeta_j = I_{\{0 \leq \zeta \leq a_{1,j}\}} + 2I_{\{a_{1,j} < \zeta \leq a_{1,j} + a_{2,j}\}} + \cdots + lI_{\{a_{1,j} + \cdots + a_{l-1,j} < \zeta \leq 1\}}.$$

Define an aggregated process $\overline{\alpha}_n^\varepsilon$ by

$$\overline{\alpha}_n^\varepsilon = \begin{cases} \iota, & \text{if } \alpha_n^\varepsilon \in \mathcal{M}_\iota, \\ \zeta_j, & \text{if } \alpha_n^\varepsilon = s_{*j} \in \mathcal{M}_*. \end{cases}$$

Define also the continuous-time interpolations $\alpha^\varepsilon(\cdot)$ and $\overline{\alpha}^\varepsilon(\cdot)$ as

$$\alpha^\varepsilon(t) = \alpha_n^\varepsilon \text{ and } \overline{\alpha}^\varepsilon(t) = \overline{\alpha}_n^\varepsilon, \quad \text{for } t \in [n\varepsilon, n\varepsilon + \varepsilon).$$

Note that the state space of $\overline{\alpha}^\varepsilon(t)$ (resp. $\overline{\alpha}_n^\varepsilon$) is $\overline{\mathcal{M}} = \{1, \dots, l\}$. In addition, $P(\overline{\alpha}_n^\varepsilon = \iota | \alpha_n^\varepsilon = s_{*j}) = a_{\iota,j}$. As shown in [42], the aggregated process $\overline{\alpha}^\varepsilon(\cdot)$ converges weakly to $\overline{\alpha}(\cdot)$ generated by $\overline{Q}$. Moreover, as $\varepsilon \to 0$, for $\iota = 1, \dots, l$,

$$\mathsf{E} \left(\varepsilon \sum_{n=1}^{\lfloor T/\varepsilon \rfloor - 1} \left(I_{\{\alpha_n^\varepsilon = s_{\iota j}\}} - \mu_j^\iota I_{\{\overline{\alpha}_n^\varepsilon \in \mathcal{M}_\iota\}} \right) \right)^2 \to 0, \quad \text{for } j = 1, \dots, m_\iota,$$

$$\mathsf{E} \left(\varepsilon \sum_{n=1}^{\lfloor T/\varepsilon \rfloor - 1} I_{\{\alpha_n^\varepsilon = s_{*j}\}} \right)^2 \to 0, \quad \text{for } j = 1, \dots, m_*.$$

For given matrices $H(s_{\iota j})$, $s_{\iota j} \in \mathcal{M}$, define

$$\overline{H}(\iota) = \sum_{j=1}^{m_\iota} \mu_j^\iota H(s_{\iota j}).$$

Theorem 2.27. Under Assumptions (C1) and (C2'). As $\varepsilon \to 0$, we have $K^\varepsilon(t, s_{\iota j}) \to \overline{K}(t, \iota)$ and $q^\varepsilon(t, s_{\iota j}) \to \overline{q}(t, \iota)$ for $\iota = 1, \dots, l$ and $j = 1, \dots, m_\iota$; $K^\varepsilon(t, s_{*j}) \to \overline{K}(t, *j)$ and $q^\varepsilon(t, s_{*j}) \to \overline{q}(t, *j)$ for $j = 1, \dots, m_*$, uniformly on $[0, T]$ where

$$\begin{aligned} \overline{K}(t, *j) &= a_{1,j}\overline{K}(t, 1) + \cdots + a_{l,j}\overline{K}(t, l), \\ \overline{q}(t, *j) &= a_{1,j}\overline{q}(t, 1) + \cdots + a_{l,j}\overline{q}(t, l), \end{aligned}$$

and for $\iota = 1,\ldots,l$, $\overline{K}(t,\iota)$ and $\overline{q}(t,\iota)$ are the unique solutions to the following differential equations

$$\left\{\begin{array}{rcl}\dot{\overline{K}}(t,\iota) &=& -\overline{K}(t,\iota)\overline{A}(\iota) - \overline{A'}(\iota)\overline{K}(t,\iota) - \overline{M}(\iota) \\ && +\overline{K}(t,\iota)\overline{BN^{-1}B'}(\iota)\overline{K}(t,\iota) - \overline{QK}(t,\cdot)(\iota), \\ \overline{K}(T,\iota) &=& D,\end{array}\right. \tag{2.81}$$

and

$$\left\{\begin{array}{rcl}\dot{\overline{q}}(t,\iota) &=& -\mathrm{tr}\left(\overline{K}(t,\iota)\Sigma\right) - \overline{Q}\overline{q}(t,\cdot)(\iota), \\ \overline{q}(T,\iota) &=& 0,\end{array}\right.$$

respectively.

As expected, note that the limit Riccati equation in (2.81) is identical to the limit Riccati equation corresponding to the continuous-time LQG problem; see [45].

Corollary 2.28. As $\varepsilon \to 0$, $v^{\varepsilon}(t,x,s_{\iota j}) \to v(t,x,\iota)$ for $\iota = 1,\ldots,l$, $j = 1,\ldots,m_{\iota}$, and $v^{\varepsilon}(t,x,s_{*j}) \to v(t,x,*j)$ for $j = 1,\ldots,m_*$, where

$$\begin{aligned}v(t,x,*j) &= a_{1,j}v(t,x,1) + \cdots + a_{l,j}v(t,x,l), \\ v(t,x,\iota) &= x'\overline{K}(t,\iota)x + \overline{q}(t,\iota), \quad \text{for } \iota = 1,\ldots,l.\end{aligned}$$

Next, let $\mathcal{U}$ denote the control set for the limit control system

$$\mathcal{U} = \left\{U = (U^1,\ldots,U^l) : U^{\iota} = (u^{\iota 1},\ldots,u^{\iota m_{\iota}}),\ u^{\iota j} \in \mathbb{R}^{n_2}\right\}.$$

Define

$$\begin{aligned}f(\iota,x,U) &= \overline{A}(\iota)x + \sum_{j=1}^{m_{\iota}} \mu_j^{\iota} B(s_{\iota j})u^{\iota j}, \\ \widetilde{N}(\iota,U) &= \sum_{j=1}^{m_{\iota}} \mu_j^{\iota}\left(u^{\iota j,\prime}N(s_{\iota j})u^{\iota j}\right),\end{aligned}$$

for $\iota = 1,\ldots,l$. It can be shown that $v(t,x,\iota)$ satisfies the following HJB equation

$$\begin{aligned}0 = \frac{\partial v(t,x,\iota)}{\partial t} + \min_{U\in\mathcal{U}}\Bigg\{ & f(\iota,x,U)\frac{\partial v(t,x,\iota)}{\partial x} + x'\overline{M}(\iota)x \\ & +\widetilde{N}(\iota,U) + \frac{1}{2}\mathrm{tr}\left(\frac{\partial^2 v(t,x,\iota)}{\partial x^2}\Sigma\right) + \overline{Q}v(t,x,\cdot)(\iota)\Bigg\}\end{aligned}$$

with $v(T,x,\iota)=x'Dx$. The corresponding limit control problem is

$$\begin{cases} \text{minimize: } J(s,x,\iota,U(\cdot)) \\ \qquad = \mathsf{E}\left\{\int_s^T \left[\overline{x}'(t)\overline{M}(\overline{\alpha}(t))\overline{x}(t)+\widetilde{N}(\overline{\alpha}(t),U(t))\right]dt+\overline{x}'(T)D\overline{x}(T)\right\}, \\ \text{subject to: } d\overline{x}(t)=f(\overline{\alpha}(t),\overline{x}(t),U(t))dt+\sigma d\omega(t), \quad \overline{x}(s)=x, \\ \text{value function: } v(s,x,\iota)=\min_{U(\cdot)} J(s,x,\iota,U(\cdot)), \end{cases}$$

where $\sigma\sigma'=\Sigma$ and $\omega(t)$ is a standard Brownian motion. Note that the limit problem is a continuous-time quadratic regulator problem. Using the approach as in [45], we can find the optimal control for this limit control problem, which is given by

$$U^o(t,x)=\left(U^{1,o}(t,x),\ldots,U^{l,o}(t,x)\right),$$

where for $1\le\iota\le l$, $U^{\iota,o}(t,x)=\left(u^{\iota 1,o}(t,x),\ldots,u^{\iota m_\iota,o}(t,x)\right)$ and

$$u^{\iota j,o}(t,x)=-N^{-1}(s_{\iota j})B'(s_{\iota j})\overline{K}(t,\iota)x,\ j=1,\ldots,m_\iota.$$

Now we construct a control for the original discrete-time control system as

$$u_n^\varepsilon=u_n^\varepsilon(\alpha,x_n)=u^\varepsilon(n\varepsilon,x_n,\alpha),$$

where

$$u^\varepsilon(t,x,\alpha)=\sum_{\iota=1}^{l}\sum_{j=1}^{m_\iota}I_{\{\alpha=s_{\iota j}\}}u^{\iota j,o}(t,x)+\sum_{j=1}^{m_*}I_{\{\alpha=s_{*j}\}}u^{*j,o}(t,x),$$

where $u^{\iota j,o}(t,x)$ is given above and

$$u^{*j,o}(t,x)=-N^{-1}(s_{*j})B'(s_{*j})\overline{K}(t,*j)x.$$

Equivalently,

$$u_n^\varepsilon(\alpha,x_n)=\begin{cases} -N^{-1}(s_{\iota j})B'(s_{\iota j})\overline{K}(n\varepsilon,\iota)x_n, & \text{if } \alpha=s_{\iota j}\in\mathcal{M}_\iota, \\ -N^{-1}(s_{*j})B'(s_{*j})\overline{K}(n\varepsilon,*j)x_n, & \text{if } \alpha=s_{*j}\in\mathcal{M}_*. \end{cases} \tag{2.82}$$

Theorem 2.29. Under Assumptions (C1) and (C2'), the control $u^\varepsilon(\cdot)$ defined in (2.82) is asymptotically optimal for the original control system (2.38) and (2.39) in the sense that

$$\lim_{\varepsilon\to 0}|J^\varepsilon(x,\alpha,u^\varepsilon(\cdot))-\lambda_0^\varepsilon(x,\alpha)|=0.$$

Remark 2.30. Some of the conditions in (C1) and (C2) might be relaxed at the expense of more complex notation and technical difficulty. Since the main purpose of this part is to demonstrate the near optimality of our scheme, rather than finding the weakest conditions possible, we are contented with the current setting.

It would be interesting to come up with an error-bound estimate of (2.66) in terms of ε. However, such bound is usually difficult to obtain since the near optimality proof is based on weak convergence ideas yielding no error estimates. Nevertheless, in certain cases, error bounds can be obtained, for example, in the MDP part. In addition, the numerical example in the next section suggests the error bound is of the order $\sqrt{\varepsilon}$.

2.2.4 A Numerical Example

In this section, we present a numerical example. Consider a Markov chain $\alpha_n^\varepsilon \in \mathcal{M} = \{1,2,3\}$ with transition matrix P^ε

$$P^\varepsilon = \begin{pmatrix} 0.8 & 0.2 & 0 \\ 0.1 & 0.9 & 0 \\ 0.3 & 0.2 & 0.5 \end{pmatrix} + \varepsilon \begin{pmatrix} -0.5 & 0.3 & 0.2 \\ 0.2 & -0.5 & 0.3 \\ 0.3 & 0.2 & -0.5 \end{pmatrix}.$$

Note that the state space $\mathcal{M}$ consists of a pair of recurrent states (namely, $\alpha = 1, 2$) and a transient state ($\alpha = 3$). As a result, to obtain the optimal control law, one needs to solve a system of three Riccati equations (2.47). Nevertheless, one need only solve a single scalar equation (2.55) for the limit problem to get the nearly optimal control policy. This demonstrates considerable savings in computational effort.

Consider the following one-dimensional discrete-time dynamic system

$$x_{n+1} = x_n + \varepsilon[A(\alpha_n^\varepsilon)x_n + B(\alpha_n^\varepsilon)u_n] + \sqrt{\varepsilon}\xi_n,$$

and the objective function (2.39) with the following specifications:

$$\begin{aligned} &A(1) = 0.5,\ A(2) = -0.3,\ A(3) = 0.2,\ B(1) = 1,\ B(2) = 2,\ B(3) = 1.5, \\ &M(1) = M(2) = M(3) = N(1) = N(2) = N(3) = D = 1. \end{aligned}$$

Take $T = 10,\ x_0 = x = 0,\ \alpha_0^\varepsilon = \alpha = 1$. Let $\lambda^\varepsilon = \lambda_0^\varepsilon(0,1)$, $v = v(0,0,1)$, and $J^\varepsilon = J^\varepsilon(0,1,u^\varepsilon(\cdot))$. Define the norms

$$|K^\varepsilon - \overline{K}| = \frac{1}{\lfloor T/\varepsilon \rfloor} \sum_{n=0}^{\lfloor T/\varepsilon \rfloor - 1} \Big(|K_n^\varepsilon(1) - \overline{K}(n\varepsilon)| + |K_n^\varepsilon(2) - \overline{K}(n\varepsilon)| + |K_n^\varepsilon(3) - \overline{K}(n\varepsilon)| \Big),$$

$$|x^\varepsilon - \overline{x}| = \frac{1}{\lfloor T/\varepsilon \rfloor} \sum_{n=1}^{\lfloor T/\varepsilon \rfloor} |x_n - \overline{x}_n|,$$

where $K_n^\varepsilon(i)$, $i = 1,2,3$, are the solutions of (2.47), $\overline{K}(t)$ denotes the solution of the limit Riccati equation (2.55), x_n is the optimal trajectory under the optimal control law (2.46), and $\overline{x}_n$ is the near optimal trajectory under the asymptotic control (2.65). Table 1 gives error bounds for various ε values, which are based on computations using 100 sample paths.

ε	$\lvert K^\varepsilon - \overline{K}\rvert$	$\lvert x^\varepsilon - \overline{x}\rvert$	$\lvert\lambda^\varepsilon - v\rvert$	$\lvert J^\varepsilon - \lambda^\varepsilon\rvert$
0.1	7.54ε	0.42ε	18.78ε	$4.99\sqrt{\varepsilon}$
0.01	11.63ε	0.10ε	23.85ε	$9.52\sqrt{\varepsilon}$
0.001	12.20ε	0.069ε	24.58ε	$15.30\sqrt{\varepsilon}$
0.0001	12.26ε	0.11ε	26.59ε	$26.07\sqrt{\varepsilon}$

Table 1. Error Bounds for Various Values of ε

Sample paths of various trajectories of α_n^ε, x_n, $\overline{x}_n$, $K_n^\varepsilon(1)$, $K_n^\varepsilon(2)$, $K_n^\varepsilon(3)$, $\overline{K}(t)$ are given in Fig. 1 for $\varepsilon = 0.01$. Note that with the total iterations fixed at 10000, the average CPU time for solving the original Riccati equations is 42.1 seconds, whereas the average CPU time for solving the limit Riccati equation is only 15.7 seconds. Thus compared with the solution of the original system, only a little more than one third of the computational effort is used to find the nearly optimal control policy. Meanwhile, our scheme performs well. The numerical bounds provided in Table 1 together with Fig. 1 gives an illuminating illustration.

For simplicity, we treated a scalar dynamic system (with the Markov chain having $m = 3$ states) for demonstration purpose. More numerical experiments and multidimensional problems of singularly perturbed Markovian systems were carried out and reported in [38]. The improvement and reduction of complexity will be more pronounced for large-dimensional systems.

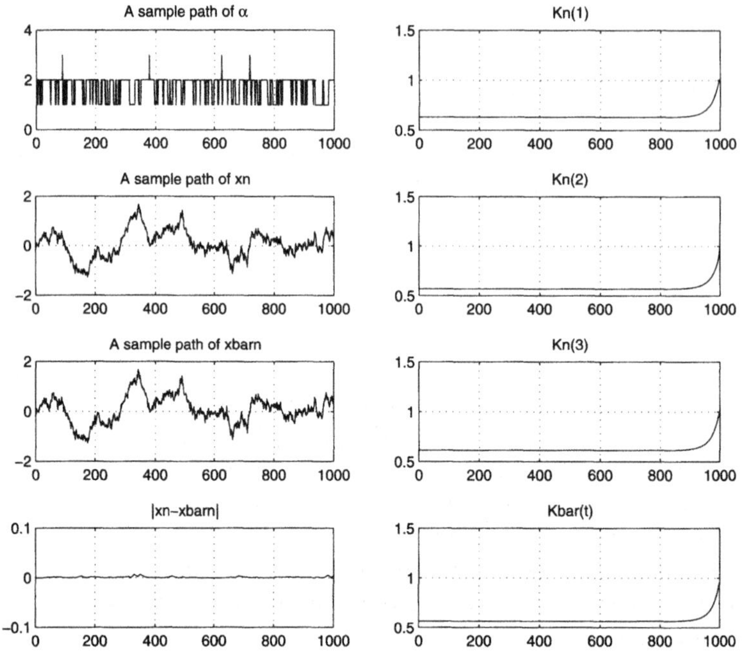

FIGURE 2.1. Various sample paths with $\varepsilon = 0.01$

Remark 2.31. Note that in order to implement this procedure in application, the value of ε does not have to be too small. For example, if all the coefficients are of order 1, then $\varepsilon = 0.1$ should yield a decent solution.

2.2.5 Historical Notes

For LQ control problems, see Fleming and Rishel [19] and Bertsekas [4], among others.

For results on optimal control and related issues such as system stability, we refer the reader to Blair and Sworder [8], Caines and Chen [10], Mariton [30], Rishel [35], Ji and Chizeck [22], [23], Zhang and Yin [45]; see also the related work of Altman and Gaitsgory [3] and the references therein.

For singularly perturbed Markovian models with weak and strong interactions. see for example, Abbad, Filar, and Bielecki [2], Pan and Basar [32], Pervozvanskii and Gaitsgory [33] among others.

Results of continuous-time Markov chains are obtained in [24], [25], [40], and [41]. A continuous-time hybrid LQG problem with singularly perturbed Markov chain is considered in [45] in which a limit control problem is obtained and nearly optimal control policy is constructed using the solution of the limit problem. Additional applications to control and filtering can be found in [29] and [44].

2.3 Conclusions

This chapter deals with the singularly perturbed Markov decision processes and hybrid LQG problems with weak and strong interactions. We have developed a systematic approach that gives asymptotically optimal strategies and that addresses practical concerns of dynamic control problems with finite-state Markov chains. The hierarchical control approach developed in this chapter is useful for reduction of dimensionality of a wide variety of stochastic systems. For example, in the LQG part, to find the approximate control, one need only solve l Riccati equations. The computational load is reduced considerably compared with the optimal solution to the original problem that requires solving m Riccati equations. For example, if $m = 120$, $l = 10$, $m_1 = m_2 = \cdots = m_{10} = 10$, and $m_* = 20$, then instead of solving 120 Riccati equations jointly, one only has to deal with 10 Riccati equations. This is the most attractive feature of our approach. Furthermore, the asymptotic optimality ensures that such an approximation is almost as good as the optimal one for sufficiently small ε.

In [14], Markov chains whose transition matrices have similar structure as considered in this chapter are treated. However, a different approach, namely time aggregation is used. It is interesting to see how these two approaches are related and how they differ. Further study and exploration may be directed in this direction.

Acknowledgments

The work of Q. Zhang and R.H. Liu was supported in part by the USAF Grant F30602-99-2-0548 and ONR Grant N00014-96-1-0263. The work of G. Yin was supported in part by the National Science Foundation under grant DMS-9877090.

2.4 References

[1] M. Abbad and J.A. Filar, Perturbation and stability theory for Markov control problems. *IEEE Trans. Automat. Control*, **AC-37** (1992), 1415-1420.

[2] M. Abbad, J.A. Filar, and T.R. Bielecki, Algorithms for singularly perturbed limiting average Markov control problems, *IEEE Trans. Automat. Control*, **AC-37** (1992), 1421-1425.

[3] E. Altman and V. Gaitsgory, Control of a hybrid stochastic system, *Systems Control Lett.*, **20** (1993), 307-314.

[4] D. Bertsekas, *Dynamic Programming: Deterministic and Stochastic Models,* Prentice-Hall, Englewood Cliffs, NJ, 1987.

[5] T.R. Bielecki and J.A. Filar, Singularly perturbed Markov control problem: Limiting average cost. *Ann. Oper. Res.*, **28** (1991), 153-168.

[6] T.R. Bielecki and L. Stettner, Ergodic control of a singularly perturbed Markov process in discrete time with general state and compact action spaces, *Appl. Math. Optim.*, **38** (1998), 261-281.

[7] P. Billingsley, *Convergence of Probability Measures*, 2nd Ed., J. Wiley, New York, 1999.

[8] W.P. Blair and D.D. Sworder, Feedback control of a class of linear discrete systems with jump parameters and quadratic cost criteria, *Internat. J. Control*, **21** (1986), 833-841.

[9] G. Blankenship, Singularly perturbed difference equations in optimal control problems. *IEEE Trans. Automat. Control*, **AC-26** (1981), 911-917.

[10] P.E. Caines and H.F. Chen, Optimal adaptive LQG control for systems with finite state process parameters, *IEEE Trans. Automat. Control*, **30** (1985), 185-189.

[11] X.R. Cao, The relation among potentials, perturbation analysis, and Markov decision processes, *J. Discrete Event Dynamic Systems*, **8** (1998), 71-87.

[12] X.R. Cao, A unified approach to Markov decision problems and performance sensitivity analysis, *Automatica*, **36** (2000), 771-774.

[13] X.R. Cao, The Maclaurin series for performance functions of Markov chains, *Adv. Applied Probability*, **30** (2000), 676-692.

[14] X.R. Cao and Z.Y. Ren, S. Bhatnagar, M. Fu, and S. Marcus, A time aggregation approach to Markov decision processes, *Automatica*, in press.

[15] X.R. Cao, Semi-Markov decision problems and performance sensitivity analysis, Working paper.

[16] F. Delebecque, A reduction process for perturbed Markov chains. *SIAM J. Appl. Math.*, **48** (1983), 325-350.

[17] F. Delebecque and J.P. Quadrat, Optimal control for Markov chains admitting strong and weak interactions. *Automatica*, **17** (1981), 281-296.

[18] C. Derman, Finite State Markovian Decision Processes. Academic Press, New York, 1970.

[19] W.H. FLEMING AND R.W. RISHEL, *Deterministic and Stochastic Optimal Control,* Springer-Verlag, New York, 1975.

[20] F.C. HOPPENSTEADT AND W.L. MIRANKER, Multitime methods for systems of difference equations. *Studies Appl Math.*, **56** (1977), 273-289.

[21] M. IOSIFESCU, *Finite Markov Processes and Their Applications*, Wiley, Chichester, 1980.

[22] Y. JI AND H.J. CHIZECK, Controllability, stabilizability, and continuous-time Markovian jump linear quadratic control, *IEEE Trans. Automat. Control*, **35** (1990), 777-788.

[23] Y. JI AND H.J. CHIZECK, Jump linear quadratic Gaussian control in continuous-time, *IEEE Trans. Automat. Control*, **37** (1992), 1884-1892.

[24] R.Z. KHASMINSKII, G. YIN, AND Q. ZHANG, Asymptotic expansions of singularly perturbed systems involving rapidly fluctuating Markov chains, *SIAM J. Appl. Math.*, **56** (1996), 277-293.

[25] R.Z. KHASMINSKII, G. YIN, AND Q. ZHANG, Constructing asymptotic series for probability distribution of Markov chains with weak and strong interactions, *Quart. Appl. Math.*, **LV** (1997), 177-200.

[26] H. J. KUSHNER AND G. YIN, *Stochastic Approximation Algorithms and Applications*, Springer-Verlag, New York, 1997.

[27] R.H. LIU, G. YIN, AND Q. ZHANG, Nearly optimal control of singularly perturbed Markov decision processes in discrete time, *Appl. Math. Optim.*, **44** (2001), 105-129.

[28] R.H. LIU, Q. ZHANG, AND G. YIN, Asymptotically optimal controls of hybrid linear quadratic regulators in discrete time, preprint.

[29] R.H. LIU, Q. ZHANG AND G. YIN, Nearly optimal control of nonlinear Markovian systems subject to weak and strong interactions, *Stoch. Anal. Appl.*, **19** (2001), 361-386.

[30] M. MARITON, Robust jump linear quadratic control: A mode stabilizing solution, *IEEE Trans. Automat. Control*, **30** (1985), 1145-1147.

[31] D.S. NAIDU, Singular Perturbation Methodology in Control Systems. Peter Peregrinus Ltd, Stevenage Herts, UK, 1988.

[32] Z. PAN AND T. BASAR, H^∞ control of large-scale jump linear systems via averaging and aggregation, *Internat. J. Control*, **72** (1999), 866-881.

[33] A.A. Pervozvanskii and V.G. Gaitsgory, *Theory of Suboptimal Decisions: Decomposition and Aggregation*, Kluwer, Dordrecht, 1988.

[34] J.P. Quadrat, Optimal control of perturbed Markov chains: The multitime scale case. Singular Perturbations in Systems and Control (edited by MD Aradema), CISM Courses and Lectures No. 280, Springer 216-239, 1983.

[35] R. Rishel, Control of systems with jump Markov disturbances, *IEEE Trans. Automat. Control*, **20** (1975), 241-244.

[36] S. Ross, Introduction to Stochastic Dynamic Programming, Academic Press, New York, 1983.

[37] D.J. White, Markov Decision Processes, Wiley, New York, 1992.

[38] H. Yang, G. Yin, K. Yin, and Q. Zhang, Control of singularly perturbed Markov chains: A numerical study, to appear in *ANZIAM J.*, (formerly known as *J. Australian Math. Society, Series B: Appl. Math.*

[39] G. Yin and Q. Zhang, Singularly perturbed discrete-time Markov chains, *SIAM J. Appl. Math.*, **61** (2000), 834-854.

[40] G. Yin and Q. Zhang, Continuous-time Markov Chains and Applications: A Singular Perturbation Approach, Springer-Verlag, New York, 1998.

[41] G. Yin , Q. Zhang and G. Badowski, Asymptotic properties of a singularly perturbed Markov chain with inclusion of transient states, *Ann. Appl. Probab.*, **10** (2000), 549-572.

[42] G. Yin, Q. Zhang, and G. Badowski, Asymptotic properties of singularly perturbed Markov chains in discrete time, preprint.

[43] Q. Zhang, Finite state Markovian decision processes with weak and strong interactions, *Stochastics Stochastics Rep.*, **59** (1996), 283-304.

[44] Q. Zhang, Hybrid filtering for linear systems with non-Gaussian disturbances, *IEEE Trans. Automat. Control*, **45** (2000), 50-61.

[45] Q. Zhang and G. Yin, On nearly optimal controls of hybrid LQG problems, *IEEE Trans. Automat. Control*, **44** (1999), 2271-2282.

3

Stochastic Approximation, with Applications

Han-Fu Chen

Optimization is ubiquitous in various research and application fields. Many theoretical and practical problems are often reduced to optimizing some function $L(\cdot)$, i.e., finding its minimum (or maximum). If $L(\cdot)$ is differentiable, then the optimization problem reduces to finding the roots of $f(x)$, where $f(x) = \frac{dL(x)}{dx}$, the derivative of $L(\cdot)$. In the case where $L(\cdot)$ or $f(\cdot)$ can be observed without errors, in optimization theory there are many numerical methods for solving the problem. In contrast to this, stochastic approximation (SA) is devoted to finding zeros of an unknown function $f(\cdot)$ which can be observed, but the observations are corrupted by errors. SA is now widely applied in identification, adaptive control, signal processing and other areas.

This chapter aims at proposing a well-behaving SA algorithm, establishing convergence theorems, and demonstrating some applications of the proposed algorithm.

In Section 3.1 the classical Robbins-Monro (RM) algorithm is introduced, its restrictions are pointed out, and the SA algorithm with expanding truncations is defined. Its convergence is proved in Section 3.2 under a general noise condition with no growth restriction on $f(\cdot)$. In Section 3.3 a state-independent noise condition is given for convergence of the algorithm. Some applications of the proposed SA algorithm are demonstrated in Section 3.4 showing that conditions used in the existing results can greatly be weakened by applying theorems given in the present chapter. A brief review on the path-wise convergence analysis of SA algorithms is given in Section 3.5.

3.1 SA Algorithms

Let the measurement be of the form

$$y_{k+1} = f(x_k) + \epsilon_{k+1}, \tag{3.1}$$

where $x_k \in I\!R^l$ denotes the estimate for the sought-for zero of $f(\cdot)$ at time k, while ϵ_{k+1} is the measurement noise.

The well-known RM algorithm suggests to recursively estimate the root of $f(\cdot)$ by

$$x_{k+1} = x_k + a_k y_{k+1}, \tag{3.2}$$

where a_k is the step size.

From (3.1) it is seen that the measurement of $f(\cdot)$ is carried out at x_k. In the classical theory, the noise ϵ_{k+1} is normally assumed to be independent of x_k. In this case $\{\epsilon_k\}$ is sometimes said to be state-independent, because x_k resembles the state of a system. However, in many problems of practical importance, the noise ϵ_{k+1} does depend on x_k. For example, in the case where $f(\cdot)$ is a function of temperature, and the noise intensity may depend on the temperature itself. Furthermore, any parameter estimation problem based on the observation data $\{y_k\}$ can formally be transformed to a root-seeking problem, since y_{k+1} can be viewed as an observation on $f(x_k)$ for an arbitrary function $f(\cdot)$, i.e., y_{k+1} can be written in the form of (3.1) with error expressed as

$$\epsilon_{k+1} = y_{k+1} - f(x_k). \tag{3.3}$$

In this case the error ϵ_{k+1} depends not only on the random observation noise, but also on the structural error reflecting the difference between the selected $f(\cdot)$ and the true one. It is worth noting that the structural error normally depends on x_k, and is hardly to have a certain kind of probabilistic properties such as independence, stationarity or martingale property.

Concerning function $f(\cdot)$, the classical theory of SA either requires its growth rate not faster than linear or directly assumes the boundedness of $\{x_k\}$ generated by (3.1) and (3.2). These conditions are restrictive, but, in a certain sense, are also unavoidable for convergence of $\{x_k\}$ given by (3.1) and (3.2). To see this, let us consider the following example, where the growth rate of $f(\cdot)$ is faster than linear:

$$f(x) = -(x-10)^3, \quad \epsilon_k \equiv 0, \quad a_k = \frac{1}{k+1}.$$

Simple calculation shows that x_k given by the RM algorithm rapidly diverges:

$$x_0 = 0, \quad x_1 = 1000, \quad x_2 = -485148500, \quad x_3 \approx 3.8 \times 10^{25}.$$

From this one might conclude that the growth rate restriction would be necessary. However, if we take the initial value x_0 with $|x_0 - 10| < 1$, then x_k given by the RM algorithm converges to the root $x^0 = 10$. To reduce initial value x_0, in a certain sense, it is equivalent to use step size not from a_0 but from a_{k_0} for some k_0. The difficulty consists in that from which a_{k_0} we should start the algorithm. This is one of the motivations to use the expanding truncation version of (3.1) and (3.2).

To guarantee the boundedness of $\{x_k\}$ one might use algorithms truncated at fixed bounds, but it may happen that the sought-for roots are outside the truncation region and then x_k will never converge to the roots. This is another motivation to consider algorithms with expanding truncations.

Let $f(\cdot)$ be an $\mathbb{R}^l \to \mathbb{R}^l$ function with root set $J = \{x \in \mathbb{R} : f(x) = 0\}$. Let $\{M_k\}$ be a sequence of positive numbers increasingly diverging to infinity, and let x^* be a fixed point in $\mathbb{R}^l$. Fix an arbitrary initial value x_0, and denote by x_k the estimate at time k serving as the kth approximation to J. Define x_k by the following algorithm with expanding truncations:

$$x_{k+1} = (x_k + a_k y_{k+1}) I_{[\|x_k + a_k y_{k+1}\| \le M_{\sigma_k}]} + x^* I_{[\|x_k + a_k y_{k+1}\| > M_{\sigma_k}]}, \tag{3.4}$$

$$\sigma_k = \sum_{i=1}^{k-1} I_{[\|x_i + a_i y_{i+1}\| > M_{\sigma_i}]}, \quad \sigma_0 = 0, \tag{3.5}$$

$$y_{k+1} = f(x_k) + \epsilon_{k+1}, \tag{3.6}$$

where $I_{[inequality]}$ is an indicator function meaning that it equals 1 if the inequality indicated in the bracket is fulfilled, and 0 if the inequality does not hold.

We explain the algorithm. σ_k is the number of truncations up-to time k. M_{σ_k} serves as the truncation bound when the $(k+1)th$ estimate is generated. From (3.4) it is seen that if the estimate at time $k+1$ calculated by the RM algorithm remains in the truncation region, i.e., if $\|x_k + a_k y_{k+1}\| \le M_{\sigma_k}$, then the algorithm evolves as the RM algorithm. If $(x_k + a_k y_{k+1})$ exits from the sphere with radius M_{σ_k}, i.e., if $\|x_k + a_k y_{k+1}\| > M_{\sigma_k}$, then the estimate at time $k+1$ is pulled back to the pre-specified point x^*, and the truncation bound is enlarged from M_{σ_k} to $M_{\sigma_{k+1}}$. Consequently, if it can be shown that the number of truncations is finite, or equivalently, $\{x_k\}$ generated by (3.4)–(3.6) is bounded, then the algorithm (3.4)–(3.6) turns to be the one without truncations, i.e., turns to the RM algorithm in a finite number of steps. This actually is the key step when we prove convergence of (3.4)–(3.6).

Before establishing convergence of x_k, it is even unknown if x_k is bounded or not. So, in the case where ϵ_{k+1} depends on $x_j, j \le k$ (state-dependent noise), it is difficult to analyze properties of $\{\epsilon_{k+1}\}$. The method used in the chapter proposes to carry out the analysis at a fixed trajectory (sample path) for convergent subsequences of $\{x_k\}$. This is why we call the method of analysis used here as TS method. As a rule, the behavior of $\{\epsilon_{k+1}\}$ is easier to be analyzed along the convergent subsequences $\{x_{n_k}\}$ in comparison with along the whole sequence $\{x_k\}$.

3.2 General Convergence Theorems by TS Method

In this section we establish convergence of the RM algorithm with expanding truncations defined by (3.4)–(3.6) under general conditions by TS method. Let us first list conditions to be used.

A1 Let $\{a_k\}$ be a sequence of real number such that $a_k > 0, \quad a_k \to 0,$ and $\sum_{k=1}^{\infty} a_k = \infty$.

A2 There is a continuously differentiable function (not necessarily being non-negative) $v(\cdot) : I\!R^l \to R$ such that

$$\sup_{\delta \le d(x,J) \le \Delta} f^T(x) v_x(x) < 0, \tag{3.7}$$

for any $\Delta > \delta > 0$, and $v(J) \triangleq \{v(x) : x \in J\}$ is nowhere dense, where J is the zero set of $f(\cdot)$, i.e., $f(x) = 0, \forall x \in J, d(x, J) = \inf_y\{\|x - y\| : y \in J\}$ and $v_x(\cdot)$ denotes the gradient of $v(\cdot)$. Further, x^* used in (4) is such that $v(x^*) < \inf_{\|x\|=c_0} v(x)$ for some $c_0 > 0$ and $\|x^*\| < c_0$.

For introducing the condition on noise let us denote by $(\Omega, \mathcal{F}, \Pr)$ the probability space. Let $\epsilon_{k+1}(\cdot,\cdot) = (I\!R^l \times \Omega, \mathcal{B}^l \times \mathcal{F}) \to (I\!R^l \times \mathcal{B}^l)$ be a measurable function defined on the product space. Fixing an $\omega \in \Omega$ means that a sample path is under consideration. Let the noise ϵ_{k+1} be given by

$$\epsilon_{k+1} = \epsilon_{k+1}(x_k, \omega), \quad \omega \in \Omega.$$

Thus, the noise is state-dependent, and for a fixed x, $\epsilon_{k+1}(x, \omega)$ may be random.

A3 For the sample path ω under consideration for any sufficiently large integer $N (\ge N_0)$ and any $T_k \in [0, T]$

$$\lim_{T \to 0} \limsup_{k \to \infty} \frac{1}{T} \| \sum_{i=n_k}^{m(n_k, T_k)} a_i \epsilon_{i+1}(x_i(\omega), \omega) I_{[\|x_i(\omega)\| \le N]} \| = 0, \tag{3.8}$$

for any $\{n_k\}$ such that $\{x_{n_k}(\omega)\}$ converges, where $x_i(\omega)$ denotes x_i given by (3.4)–(3.6) and valued at the sample path ω and

$$m(k, T) = max\{m : \sum_{i=k}^{m} a_i \le T\}. \tag{3.9}$$

A4 $f(\cdot)$ is measurable and locally bounded.

In the sequel, the algorithm (3.4)–(3.6) is considered for a fixed ω for which A3 holds, and ω in $x_i(\omega)$ is often suppressed if no confusion will be caused. It is worth noting that $a_{n_k}\epsilon_{n_k+1} \to 0$ as $k \to \infty$ if $\{x_{n_k}\}$ converges. To see this it suffices to take $T_k = a_{n_k}$ in (3.8).

Theorem 3.1 Let $\{x_k\}$ be given by (3.4)–(3.6) for a given initial value x_0. Assume A1–A4 hold. Then, $d(x_k, J^*) \to 0$ for the sample path ω for which (3.8) holds, where J^* is a connected subset contained in $\bar{J}$, the closure of J.

Proof. The proof is completed by seven steps by considering convergent subsequences $\{x_{n_k}\}$ at the fixed trajectory.

Step 1. We show that there are constants $M > 0$, $T > 0$ such that for any $t \in [0, T]$ there exists $k_t > 0$ such that for any $k > k_t$,

$$\|\sum_{i=n_k}^{m+1} a_i y_{i+1}\| \le M, \quad \forall m : n_k - 1 \le m \le m(n_k, t), \tag{3.10}$$

if $\{x_{n_k}\}$ is a convergent subsequence of $\{x_k\}$: $x_{n_k} \to \bar{x}$ as $k \to \infty$, where M is independent of t and k.

Since $a_{n_k} f(x_{n_k}) \to 0$ and $a_{n_k}\epsilon_{n_{k+1}} \to 0$ as $k \to \infty$, we need only to prove (3.10) for $m : n_k \le m \le m(n_k, t)$.

If the number of truncations in (3.4)–(3.6) is finite, then there is an N such that $\sigma_k = \sigma_N, \forall k \ge N$, i.e., there is no more truncation for $k \ge N$. Hence, $\|\sum_{i=n_k}^{m+1} a_i y_{i+1}\| = \|x_{m+2} - x_{n_k}\| \le 2M_{\sigma_N}$, whenever $n_k \ge N$. In this case, we may take $M = 2M_{\sigma_N}$ in (3.10).

We now prove (3.10) for the case where $\sigma_k \to \infty$ as $k \to \infty$.

Assume the converse that (3.10) is not true. Take $c > \|\bar{x}\|$. There is k_c such that

$$\|x_{n_k}\| \le (c + \|\bar{x}\|)/2, \quad \forall k \ge k_c \tag{3.11}$$

Take a sequence of positive real numbers $t_j > 0$ and $t_j \to 0$ as $j \to \infty$. Since (3.10) is not true, for $j = 1$ there are $k_1 > k_c$ and $m_1 : n_{k_1} \le m_1 \le m(n_{k_1}, t_1)$ such that

$$\|\sum_{i=n_{k_j}}^{m_j+1} a_i y_{i+1}\| > (c - \|\bar{x}\|)/2,$$

and for any $j > 1$ there are $k_j > k_{j-1}$ and $m_j : n_{k_j} \le m_j \le m(n_{k_j}, t_j)$ such that

$$\|\sum_{i=n_{k_j}}^{m_j+1} a_i y_{i+1}\| > (c - \|\bar{x}\|)/2. \tag{3.12}$$

Without loss of generality we may assume

$$m_j = \inf\{m : \|\sum_{i=k_j}^{m+1} a_i y_{i+1}\| > (c - \|\bar{x}\|)/2\}. \tag{3.13}$$

Then for any $m : n_{k_j} \le m \le m_j$, from (3.11) and (3.13) it follows that

$$\|x_{n_{k_j}} + \sum_{i=n_{k_j}}^{m} a_i y_{i+1}\| < c. \tag{3.14}$$

Since $\sigma \to \infty$, there is j_0 such that $M_{\sigma_{n_j}} > c$, $\forall j \ge j_0$. Then from (3.14) it follows that

$$x_{m+1} = x_m + a_m y_{m+1}, \quad \forall m : n_{k_j} \le m \le m_j, \tag{3.15}$$

and by (3.11),(3.14) and (3.15)

$$\|x_m\| \le c, \quad \text{and hence} \quad \|f(x_m)\| \le c, \quad \forall m : n_{k_j} \le m \le m_j + 1, \tag{3.16}$$

where c_1 is a constant.

Let $N > \max\{c, N_0\}$, where N_0 is specified in A3. Then from A3 for any $T_j : 0 \le T_j < T$,

$$\limsup_{j \to \infty} \|\sum_{i=n_{k_j}}^{m(n_{k_j}, T_j)} a_i \epsilon_{i+1} I_{[\|x_i\| \le N]}\| = 0. \tag{3.17}$$

For any fixed $T > 0$, if j is large enough, then $t_j < T$ and $m_j + 1 < m(n_{k_j}, T)$, and by (3.17)

$$\limsup_{j \to \infty} \|\sum_{i=n_{k_j}}^{m_j+1} a_i \epsilon_{i+1} I_{[\|x_i\| \le N]}\| = 0. \tag{3.18}$$

Since $\|x_m\| \le c < N$, $\forall m : n_{k_j} \le m \le m_j + 1$, from (3.18) it follows that

$$\limsup_{j \to \infty} \|\sum_{i=n_{k_j}}^{m_j+1} a_i \epsilon_{i+1}\| = 0. \tag{3.19}$$

Taking $T_j = \sum_{i=n_{k_j}}^{m_j} a_i$ and $T_j = \sum_{i=n_{k_j}}^{m_j+1} a_i$ respectively in (3.17) and noticing from (3.16) $\|x_i\| \le c < N$, $\forall i : n_{k_j} \le i \le m_j + 1$, we then have

$$\limsup_{j \to \infty} \|\sum_{i=n_{k_j}}^{m_j} a_i \epsilon_{i+1}\| = 0 \quad \text{and} \quad \limsup_{j \to \infty} \|\sum_{i=n_k}^{m_j+1} a_i \epsilon_{i+1}\| = 0,$$

and hence

$$\lim_{j \to \infty} a_{m_j+1} \epsilon_{m_j+2} = 0. \tag{3.20}$$

From (3.15) it follows that

$$\|x_{m_j+1} - x_{n_{k_j}}\| \le \sum_{i=n_{k_j}}^{m_j} a_i \|f(x_i)\| + \|\sum_{i=n_{k_j}}^{m_j} a_i \epsilon_{i+1}\| \to 0 \quad \text{as} \quad j \to \infty, \tag{3.21}$$

where the second term on the right-hand side of the inequality tends to zero by (3.19) and (3.20), while the first term tends to zero because

$$\sum_{i=n_{k_j}}^{m_j} a_i \|f(x_i)\| \le c_1 \sum_{i=n_{k_j}}^{m_j} a_i \le c_1 t_j \to 0 \quad \text{as} \quad j \to \infty.$$

Noticing that $a_{m+j+1} y_{m_j+2} = a_{m_j+1}(f(x_{m_j+1}) + \epsilon_{m_j+2}) \to 0$ as $j \to \infty$ by (3.16) and (3.20), by (3.21) we then have

$$\begin{aligned}&\|x_{m_j+1} - x_{n_{k_j}} + a_{m_j+1} y_{m_j+2}\| \le \|x_{m_j+1} - x_{n_{k_j}}\| + \|a_{m_j+1} y_{m_j+2}\| \\ &\to 0 \quad \text{as} \quad j \to \infty.\end{aligned}$$

On the other hand, by (3.13) we have

$$\|x_{m_j+1} - x_{n_{k_j}} + a_{m_j+1} y_{m_j+2}\| = \|\sum_{i=n_{k_j}}^{m_j+1} a_i y_{i+1}\| > (c - \|\bar{x}\|)/2, \quad \forall j.$$

The contradiction proves (3.10).

Step 2. We now show that for all large enough k

$$\|x_{m+1} - x_{n_k}\| \le 2c_2 t, \quad \forall m : n_k \le m \le m(n_k, t), \quad \forall t \in [0, T], \tag{3.22}$$

where c_2 is a constant.

If the number of truncations in (3.4)–(3.6) is finite, then $\{x_k\}$ is bounded and hence $\{f(x_k)\}$ is also bounded. Then for large enough k there is no truncation, and by (3.8) for $\forall m : n_k \le m \le m(n_k, t)$

$$\begin{aligned}\|x_{m+1} - x_{n_k}\| &\le \|\sum_{i=n_k}^{m} a_i f(x_i)\| + \|\sum_{i=n_k}^{m} a_i \epsilon_{i+1}\| \\ &\le \sup_i \|f(x_i)\| \sum_{i=1}^{m} a_i + o(t) \le c_3 t,\end{aligned} \tag{3.23}$$

where for the last inequality the boundedness of $\{x_k\}$ is invoked and c_3 is a constant. Thus, it suffices to prove (3.22) for the case where $\sigma_k \to \infty$.

From (3.10) it follows that for any $t \in [0, T]$

$$\|x_{n_k} + \sum_{i=n_k}^{m} a_i y_{i+1}\| \le M + \|\bar{x}\| + 1 \le M_{\sigma_k}, \quad \forall m : n_k \le m \le m(n_k, t), \tag{3.24}$$

if k is large enough.

This implies that for $\forall m : n_k \le m \le m(n_k, t)$

$$x_{m+1} = x_m + a_m y_{m+1}, \quad \|x_{m+1}\| \le M + 1 + \|\bar{x}\|, \quad \|f(x_m)\| \le c_2, \tag{3.25}$$

where c_2 is a constant. The last inequality in (3.25) yields

$$\|\sum_{i=n_k}^{m} a_i f(x_i)\| \le c_2 t. \tag{3.26}$$

With $N \ge \max\{(M + 1 + \|\bar{x}\|), N_0\}$ in A3, from (3.8) we have

$$\|\sum_{i=n_k}^{m(n_k,t)} a_i \epsilon_{i+1} I_{[\|x_i\| \le N]}\| = \|\sum_{i=n_k}^{m(n_k,t)} a_i \epsilon_{i+1}\| \le c_2 t, \tag{3.27}$$

for large enough k and small enough t.

Combining (3.25),(3.26) and (3.27) leads to

$$\|x_{m+1} - x_{n_k}\| \le 2c_2 t, \quad \forall m : n_k \le m \le m(n_k, t),$$

for all large enough k. This together with (3.23) verifies (3.22).

Step 3. We now show the following assertion.

For any interval $[\delta_1, \delta_2]$ with $\delta_1 < \delta_2$ and $d([\delta_1, \delta_2], v(J)) > 0$, the sequence $\{v(x_k)\}$ cannot cross $[\delta_1, \delta_2]$ infinitely many times with $\{\|x_{n_k}\|\}$ bounded, where by "crossing $[\delta_1, \delta_2]$ by $v(x_{n_k}, \cdots, v(x_{m_k})$" we mean that $v(x_{n_k}) \le \delta_2, v(x_{m_k}) \ge \delta_2$, and $\delta_1 < v(x_i) < \delta_2, \forall i : n_k < i < m_k$.

Assume the converse: infinitely many crossings $v(x_{n_k}), \cdots, v(x_{m_k}), k = 1, 2, \cdots$, occur and $\{\|x_{n_k}\|\}$ is bounded. By boundedness of $\{\|x_{n_k}\|\}$, without loss of generality, we may assume $x_{n_k} \to \overline{x}$ as $k \to \infty$.

By setting $t = a_{n_k}$ in (3.22), we have

$$\|x_{n_k+1} - x_{n_k}\| \le 2c_2 a_{n_k} \to 0 \quad \text{as} \quad k \to \infty.$$

But by definition $v(x_{n_k+1}) > \delta_1 \ge v(x_{n_k})$, so we have

$$v(x_{n_k}) \to \delta_1 = v(\bar{x}) \quad \text{as} \quad k \to \infty, \quad \text{and} \quad d(\bar{x}, J) \triangleq \delta > 0. \tag{3.28}$$

From (3.22) we see that if we take t sufficiently small, then

$$d(x_m, J) \ge \frac{\delta}{2}, \quad \forall m : n_k \le m \le m(n_k, t), \tag{3.29}$$

for sufficiently large k.

By (3.25) and (3.22), for large k we then have

$$v(x_{m(n_k,t)+1}) - v(x_{n_k}) = \sum_{i=n_k}^{m(n_k,t)} a_i y_{i+1}^T v_x(\bar{x}) + o(t)$$

$$= \sum_{i=n_k}^{m(n_k,t)} a_i f^T(x_i) v_x(x_i) + \sum_{i=n_k}^{m(n_k,t)} a_i f^T(x_i)(v_x(\bar{x}) - v_x(x_i))$$

$$+ \sum_{i=n_k}^{m(n_k,t)} a_i v_x^T(\bar{x}) \epsilon_{i+1} + o(t), \tag{3.30}$$

where $v_x(\cdot)$ denotes the gradient of $v(\cdot)$ and $o(t) \to 0$ as $t \to 0$.

For $N \geq M + 1 + \max\{\|\bar{x}\|, N_0\}$, condition (3.8) implies that

$$\lim_{t\to 0} \limsup_{k\to\infty} \frac{1}{t} \|v_x^T(\bar{x}) \sum_{i=n_k}^{m(n_k,t)} a_i \epsilon_{i+1}\| = 0. \tag{3.31}$$

By (3.22), (3.25) it follows that

$$\| \sum_{i=n_k}^{m(n_k,t)} a_i f^T(x_i)(v_x(\bar{x}) - v_x(x_i))\| = o(t). \tag{3.32}$$

Then, by (3.29), (3.7) and (3.30)–(3.32) it follows that there are $\alpha > 0$ and $t > 0$ such that

$$v(x_{m(n_k,t)+1}) - v(x_{n_k}) \leq -\alpha t, \tag{3.33}$$

for all sufficiently large k.

Noticing (3.28), from (3.33) we derive

$$\limsup_{k\to\infty} v(x_{m(n_k,t)+1}) \leq \delta_1 - \alpha t. \tag{3.34}$$

However, by (3.22) we have

$$\lim_{t\to 0} \max_{n_k \leq m \leq m(n_k,t)} |v(x_{m+1}) - v(x_{n_k}))| = 0,$$

which implies that $m(n_k, t) + 1 < m_k$ for small enough t.

This means that $v(x_{m(n_k,t)+1}) \in [\delta_1, \delta_2]$, which contradicts (3.34).

Step 4. We now show that the number of truncations is bounded.

By A2, $v(J)$ is nowhere dense and hence a nonempty interval $[\delta_1, \delta_2]$ exists such that $[\delta_1, \delta_2] \subset (v(x^*), \inf_{\|x\|=c_0} v(x))$ and $d([\delta_1, \delta_2], v(J)) > 0$. If $\sigma_k \to \infty$, then x_k, starting from x^*, will cross the sphere $\{x : \|x\| = c_0\}$ infinitely many times. Consequently, $v(x_k)$ will cross $[\delta_1, \delta_2]$ infinitely often with $\{x_{n_k}\}$ bounded. In Step 3, we have shown this is impossible . Therefore, starting from some k_0, the algorithm (3.4)–(3.6) will have no truncations and $\{x_k\}$ is bounded.

This means that the algorithm (3.4)–(3.6) turns to be the conventional RM algorithm for $k \geq k_0$ and a stronger than (3.8) condition is satisfied:

$$\lim_{T\to 0} \limsup_{k\to\infty} \frac{1}{T} \| \sum_{i=n_k}^{m(n_k,T_k)} a_i \epsilon_{i+1}\| = 0, \quad \forall T_k \in [0, T], \tag{3.35}$$

for any $\{n_k\}$ such that $\{x_{n_k}\}$ converges.

Step 5. We now show that $v(x_k)$ converges. Let

$$v_1 \triangleq \liminf_{k\to\infty} v(x_k) \leq \limsup_{k\to\infty} v(x_k) \triangleq v_2.$$

We have to show $v_1 = v_2$.

If $v_1 < v_2$ and one of v_1 and v_2 does not belong to $v(J)$, then $[\delta_1, \delta_2] \subset [v_1, v_2]$ exists such that $d([\delta_1, \delta_2], v(J)) > 0$ and $\delta_2 > \delta_1$. By Step 3 this is impossible. So, both v_1 and v_2 belong to $v(J)$ and

$$\lim_{k\to\infty} d(v(x_k), v(J)) = 0. \tag{3.36}$$

If we can show that $\{v(x_k)\}$ is dense in $[v_1, v_2]$, then from (3.36) it will follow that $v(J)$ is dense in $[v_1, v_2]$, which contradicts to the assumption that $v(J)$ is nowhere dense. This will prove the convergence of $v(x_k)$.

To show that $\{v(x_k)\}$ is dense in $[v_1, v_2]$, it suffices to show that $x_{k+1} - x_k \to 0$. Assume the converse: there is a subsequence

$$\lim_{k\to\infty} \|x_{l_k+1} - x_{l_k}\| \triangleq \beta > 0. \tag{3.37}$$

Without loss of generality, we may assume x_{l_k} converges. Otherwise, a convergent subsequence can be extracted, which is possible because $\{x_k\}$ is bounded. However, if we take $t = a_{l_k}$ in (3.22), we have

$$\|x_{l_k+1} - x_{l_k}\| \leq 2c_2 a_{l_k} \to 0 \quad \text{as} \quad k \to \infty,$$

which contradicts (3.37). Thus $v_1 = v_2$ and $v(x_k)$ converges.

Step 6. For proving $d(x_k, J) \to 0$, it suffices to show that any limit points of $\{x_k\}$ belong to J.

Assume the converse: $x_{n_k} \to \bar{x} \notin J$, $d(\bar{x}, J) \triangleq \delta > 0$. By (3.22) we have

$$d(x_m, J) > \frac{\delta}{2}, \quad \forall m : n_k \leq m \leq m(n_k, t),$$

for all large k if t is small enough. By (3.7) it follows that

$$v_x^T(x_m) f(x_m) < -b < 0, \quad \forall m : n_k \leq m \leq m(n_k, t),$$

and from (3.30)

$$v(x_{m(n_k,t)+1}) - v(x_{n_k}) \leq -\frac{bt}{2}, \tag{3.38}$$

for small enough t. This leads to a contradiction because $v(x_k)$ converges and the left hand side of (3.38) tends to zero as $k \to \infty$. Thus, we conclude $d(x_k, J) \to 0$.

Step 7 Denote by J^* the set of limit points of $\{x_k\}$. Assume the converse: i.e., J^* is disconnected. In other words, closed sets J_1^* and J_2^* exist such that$J^* = J_1^* \cup J_2^*$ and $d(J_1^*, J_2^*) > 0$.

Define

$$\rho \triangleq \frac{1}{3} d(J_1^*, J_2^*).$$

Since $d(x_k, J^*) \to 0$, a k_0 exists such that

$$x_k \in B(J_1^*, \rho) \cup B(J_2^*, \rho), \quad \forall k \geq k_0,$$

where $B(A, \rho)$ denotes the ρ-neighborhood of set A.

Define

$$n_0 = \inf\{k > k_0, d(x_k, J_1^*) < \rho\},$$
$$m_l = \inf\{k > n_l, d(x_k, J_2^*) < \rho\},$$
$$n_{l+1} = \inf\{k > m_l, d(x_k, J_1^*) < \rho\}.$$

It is clear that $m_l < \infty, n_l < \infty, \forall l$, and

$$x_{n_l} \in B(J_1^*, \rho), \qquad x_{n_l - 1} \in B(J_2^*, \rho).$$

Consequently, by $d(J_1^*, J_2^*) = 3\rho$ we have

$$\|x_{n_l} - x_{n_l - 1}\| \geq \rho. \tag{3.39}$$

By boundedness of $\{x_{n_l - 1}\}$, we may assume that $x_{n_l - 1}$ converges. Then, by taking $t = a_{n_l - 1}$ in (3.22), we derive

$$\|x_{n_l} - x_{n_l - 1}\| \leq 2c_2 a_{n_l - 1} \to 0 \quad \text{as} \quad l \to \infty,$$

which contradicts (3.39) and proves the theorem. □

Remark 3.2 In (3.4)–(3.6) the spheres with expanding radiuses $\{M_k\}$ are used for truncations. Obviously, the spheres can be replaced by other expanding sets. At a first glance, the point x^* in (3.4) may be arbitrarily chosen, but the condition is actually imposed on the existence of c_0 such that $v(x^*) < inf_{\|x\|=c_0} v(x)$. This condition is obviously satisfied if $v(x) \to \infty$ as $\|x\| \to \infty$ because the availability of c_0 is not required.

Remark 3.3 If J is not dense in any connected set, then under the conditions of Theorem 2.2.1, $\{x_k\}$ given by (3.4)–(3.6) converges to a point in $\bar{J}$. This is because in the present case any connected set in $\bar{J}$ consists of a single point.

Example 3.4 Consider again the example given in Introduction:

$$f(x) = -(x - 10)^3, \quad J = \{10\}, \quad x_0 = 0, \quad a_k = \frac{1}{k+1}.$$

It was shown in Introduction that the RM algorithm rapidly diverges to $\pm\infty$ even in the noise-free case.

We now assume the observations are noise-corrupted:

$$y_{k+1} = f(x_k) + \epsilon_{k+1},$$

where $\{\epsilon_k\}$ is an ARMA process driven by the independent identically distributed normal random variables $w_k \in N(0, \sigma)$,

$$\epsilon_{k+1} + a_1 \epsilon_k = w_{k+1} + c_1 w_k,$$

where $a_1 = -0.9, \quad c_1 = 0.5, \quad \sigma = 0.1.$

We use the algorithm (3.4)–(3.6) with $M_k = 2^{k+1}, \quad x^* = 0.5$. The computation shows

$$\begin{aligned} x_{10} &= 0.5, x_{20} = 43.4, x_{30} = -23.9, x_{40} = 21.9, x_{50} = 7.76, x_{60} = 8.74, \\ x_{70} &= 8.97, x_{100} = 9.26, x_{200} = 9.46, x_{300} = 9.52, x_{400} = 9.61, \end{aligned}$$

which tends to the sought-for root 10.

Example 3.5 Let $f(x) = \sin x, \quad x^* = 3, \quad M_k = k, \quad a_k = \frac{1}{k}$. Then $J = \{k\pi, k = 0, \pm 1, \pm 2, \cdots\}$.

Clearly, A1 and A4 hold. Concerning A2, we may take $\cos x$ to serve as $v(x)$. Since $f(x) v_x(x) = -(\sin x)^2$, (3.7) is satisfied. The existence of c_0 required in A2 is obvious, for example, $c_0 = 5\pi/3$: $\cos 3 < 0 < \cos(\pm 5\pi/3)$.

Finally, $v(J) = \{-1, 1\}$ is nowhere dense. So A2 also holds.

Now assume the noise is such that

$$\sum_{k=1}^{\infty} \frac{1}{k} \epsilon_{k+1} < \infty.$$

Then A3 is satisfied too. By Remark 2, $\{x_k\}$ given by (3.4)–(3.6) converges to a point $\in J$.

If for the conventional (untruncated) RM algorithm

$$x_{k+1} = x_k + a_k y_{k+1}, \quad y_{k+1} = f(x_k) + \epsilon_{k+1}, \tag{3.40}$$

it is *a priori* known that $\{x_k\}$ is bounded, then we have the following theorem.

Theorem 3.6 Assume A1–A4 hold, where in A2 "Further, x^* used in (3.4) is such that $v(x^*) < \inf_{\|x\|=c_0} v(x)$ for some $c_0 > 0$ and $\|x^*\| < c_0$" is removed. If $\{x_k\}$ produced by (3.40) is bounded, then $d(x_k, J^*) \to 0$ for the sample path ω for which A3 holds, where J^* is a connected subset of $\bar{J}$.

Proof. By the boundedness of $\{x_k\}$, (3.10) and (3.22) follow immediately. Steps 3, 5, 6 and 7 in the proof of Theorem 3.1 remain unchanged, while Step 4 in no longer needed. □

Theorems 3.1 and 3.6 concern with the time-invariant function $f(\cdot)$, but the results can easily be extended to the time-varying functions, i.e., the measurements are carried out for $f_k(x_k)$:

$$y_{k+1} = f_k(x_k) + \epsilon_{k+1},$$

where $f_k(\cdot)$ depends on time k.

Conditions A2 and A4 are respectively replaced by the following conditions:

A5 There is a continuously differentiable function $v(\cdot) : \mathbb{R}^l \to R$ such that

$$\sup_k \sup_{\delta \le d(x,J) \le \Delta} f_k^T(x) v_x(x) < 0,$$

for any $\Delta > \delta > 0$ and $v(J) \triangleq \{v(x) : x \in J\}$ is nowhere dense, where $J = \bigcup_{j=1}^{\infty} \bigcap_{k=1}^{\infty} J_{k+j}$ and $J_k = \{x : f_k(x) = 0\}$, and $v_x(\cdot)$ denotes the gradient of $v(\cdot)$. Further, x^* used in (4) is such that $v(x^*) < \inf_{\|x_0\|=c_0} v(x)$ for some $c_0 > 0$ and $\|x^*\| < c_0$.

A6 $f_k(\cdot), \forall k$ are measurable and uniformly locally bounded, i.e., for any constant c

$$\sup_k \sup_{\|x\|<c} \|f_k(x)\| < \infty.$$

Theorem 3.7 Let $\{x_k\}$ be given by (3.4)–(3.6) for a given initial value x_0. Assume A1, A5 and A6 hold. Then $d(x_k, J^*) \to 0$ for the sample path ω for which A3 holds, where J^* is a connected subset of $\bar{J}$.

Proof. It suffices to replace $f(x_k)$ by $f_k(x_k)$ everywhere in the proof of Theorem 3.1. □

3.3 Convergence Theorems Under State-Independent Conditions

We have established convergence theorems under general conditions. These theorems take a sample-path-based form: under A1, A2 and A4, $\{x_k\}$ converges at those trajectories for which A3 holds. Condition A3 looks rather complicated, but it is so weak that it is necessary for a large class of $f(\cdot)$. However, the formulation of condition A3 is state-dependent in the sense that the condition itself depends on the behavior of $\{x_k\}$. This makes it not always possible to verify the condition beforehand. We now give convergence theorems under conditions with no state $\{x_k\}$ involved. For this we have to reformulate Theorem 3.1.

As defined in the previous section $\epsilon_{k+1} = \epsilon_{k+1}(x_k, \omega)$, where $\epsilon_{k+1}(\cdot, \cdot)$ is a measurable function. In lieu of A3 we introduce the following condition.

A7 For any sufficiently large integer $N(\geq N_0)$ there is an $\omega-$set Ω_N with $\Pr\Omega_N = 1$ such that for any $\omega \in \Omega_N$ and any $T_k \in [0,T)$

$$\lim_{T\to 0}\limsup_{k\to\infty}\frac{1}{T}\|\sum_{i=n_k}^{m(n_k,T_k)} a_i\epsilon_{i+1}(x_i(\omega),\omega)I_{[\|x_i(\omega)\|\leq N]}\| = 0, \tag{3.41}$$

for any $\{n_k\}$ such that $\{x_{n_k}\}$ converges.

Theorem 3.8 Assume A1, A2, A4 and A7 hold. Then $d(x_k, J^*) \to 0$ a.s. for $\{x_k\}$ generated by (3.4)–(3.6) with a given initial value x_0, where J^* is a connected subset contained in $\bar{J}$, the closure of J.

Proof. Let $\Omega' \triangleq \bigcap_{N=N_0}^{\infty}\Omega_N$. It is clear that $\Pr\Omega' = 1 - \Pr(\Omega')^c = 1 - \Pr\bigcup_{N=N_0}^{\infty}\Omega_N^c \geq 1 - \sum_{N=N_0}^{\infty}\Pr\Omega_N^c = 1$, *i.e.*, $\Pr\Omega' = 1$. Then for any $\omega \in \Omega'$, A3 is fulfilled with N_0 possibly depending on ω, and the conclusion of the theorem follows from Theorem 3.1. □

We now introduce a state-independent condition on noise.

A8 $\{\epsilon_k(x,\omega), \mathcal{F}_k\}$ is a martingale difference sequence for any $x \in \mathbb{R}^l$, and for some $p \in (1,2]$

$$\mathsf{E}(\|\epsilon_{k+1}(x,\omega)\|^p|\mathcal{F}_k) \triangleq \sigma_{k+1}(x) < \infty, \quad a.s. \quad \forall x, \tag{3.42}$$

$$\sup_k \sup_{\|x\|\leq N} \sigma_{k+1}(x) \triangleq \sigma(N) < \infty, \quad \forall N, \tag{3.43}$$

where $\{\mathcal{F}_k\}$ is a family of nondecreasing σ-algebras independent of x.

We first give an example satisfying A8. Let $\{w_k, \mathcal{F}_k\}$ be an m-dimensional ($m \geq 1$) martingale difference sequence with $\sup_k(\|w_{k+1}\|^p|\mathcal{F}_k) < \infty$ for some $p \in (1,2]$ and let $g(x) : (\mathbb{R}^l, \mathcal{B}^l) \to (\mathbb{R}^{l\times m}, \mathcal{B}^{l\times m})$ be a measurable and locally bounded function. Then $\epsilon_{k+1}(x,\omega) = g(x)w_{k+1}(\omega)$ satisfies A8, because

$$\mathsf{E}(\|\epsilon_{k+1}(x,\omega)\|^p|\mathcal{F}_k) = \|g(x)\|^p\mathsf{E}(\|w_{k+1}\|^p|\mathcal{F}_k) < \infty, \quad \forall x,$$

and

$$\sup_k \sup_{\|x\|\leq N} \sigma_{k+1}(x) = \sup_{\|x\|\leq N} \|g(x)\|^p \sup_k \mathsf{E}(\|w_{k+1}\|^2|\mathcal{F}_k) < \infty, \quad \forall N,$$

by assumption.

Theorem 3.9 Let $\{x_k\}$ be given by (3.4)–(3.6) for a given initial value. Assume A1, A2, A4 and A8 hold and $\sum_{k=1}^{\infty} a_k^p < \infty$ for p given in A8. Then $d(x_k, J^*) \to 0$ *a.s.*, where J^* is a connected subset contained in $\bar{J}$.

Proof. Since $\epsilon_{k+1}(x,\omega)$ is measurable and x_k is $\mathcal{F}_k$-measurable, it follows that $(\epsilon_{k+1}(x_k(\omega),\omega),\mathcal{F}_{k+1})$ is adapted. Approximating $\epsilon_{k+1}(x,\omega)$ by simple functions, it is seen that

$$\mathsf{E}(\epsilon_{k+1}(x_k,\omega)|\mathcal{F}_k) = \mathsf{E}(\epsilon_{k+1}(x,\omega)|\mathcal{F}_k)|_{x=x_k} = 0.$$

Hence, $(\epsilon_{k+1}(x_k(\omega),\omega),\mathcal{F}_{k+1})$ is a martingale difference sequence, and

$$\mathsf{E}(\|\epsilon_{k+1}(x_k(\omega),\omega)\|^p I_{[\|x_k(\omega)\|\le N]}|\mathcal{F}_k) \le I_{[\|x_k(\omega)\|\le N]}\sigma_{k+1}(x_k) \le \sigma(N), \ a.s.$$

By the convergence theorem for martingale difference sequences, the series

$$\sum_{k=1}^{\infty} a_k\epsilon_{k+1}(x_k(\omega),\omega)I_{[\|x_k(\omega)\|\le N]}$$

converges a.s., which implies that Ω_N with $\Pr\Omega_N = 1$ exists such that for each $\omega\in\Omega_N$

$$\sum_{k=n}^{m} a_k\epsilon_{k+1}(x_k(\omega),\omega)I_{[\|x_k(\omega)\|\le N]}$$

converges to zero as $n\to\infty$ uniformly in m. This means that A8 holds, and the conclusion of the theorem follows from Theorem 3.1. □

In applications it may happen that $f(\cdot)$ is not directly observed. Instead, the time-varying functions $g_k(\cdot)$ are observed, and the observations y_{k+1} may be made not at x_k, but at $x_k + r_k$, i.e., at x_k with bias r_k,

$$y_{k+1} = g_k(x_k + r_k) + \epsilon_{k+1}(x_k + r_k,\omega). \tag{3.44}$$

Theorem 3.10 Let $\{x_k\}$ be given by (3.4)–(3.6) for a given initial value. Assume that A1, A2, A4 and A8 hold and $\sum_{k=1}^{\infty} a_k^p < \infty$ for p given in A8. Further, assume $(r_k,\mathcal{F}_k)$ is an adapted sequence, $\{r_k\}$ is bounded by a constant and for any sufficiently large integer $N(\ge N_0)$ there exists Ω_N with $\Pr\Omega_N = 1$ such that for any $\omega\in\Omega_N$ and any $T_k\in[0,T)$,

$$\lim_{T\to 0}\limsup_{k\to\infty}\frac{1}{T}\sum_{i=n_k}^{m(n_k,T_k)} a_i(g_i(x_i+r_i) - f(x_i))I_{[\|x_i\|\le N]}\| = 0, \tag{3.45}$$

for any $\{n_k\}$ such that $\{x_{n_k}\}$ converges. Then, $d(x_k,J^*)\to 0$ a.s., where J^* is a connected subset contained in $\bar{J}$.

Proof. By assumption $\|r_k\| < c, \quad \forall k$, where c is a constant. Then

$$\begin{aligned}&\mathsf{E}(\|\epsilon_{k+1}(x_k(\omega)+r_k(\omega),\omega)\|^p I_{[\|x_k(\omega)\|\le N]}|\mathcal{F}_k)\\ &= I_{[\|x_k(\omega)\|\le N]}\sigma_{k+1}(x_k+r_k) \le \sigma(N+c),\end{aligned}$$

and again by the convergence theorem for the martingale difference sequences, the series

$$\sum_{k=1}^{\infty} a_k \epsilon_{k+1}(x_k(\omega) + r_k(\omega), \omega) I_{[\|x_k(\omega)\| \le N]}$$

convergence a.s.. Consequently, there exists Ω' with $\Pr \Omega' = 1$ such that for any $\omega \in \Omega'$ the convergence indicated in (3.45) holds and for any integer $N \ge N_0$

$$\sum_{k=n}^{m} a_k \epsilon_{k+1}(x_k(\omega) + r_k(\omega), \omega) I_{[\|x_k(\omega)\| \le N]}$$

tends to zero as $n \to \infty$ uniformly in m. Therefore, A7 is fulfilled and the conclusion of the theorem follows from Theorem 3.8. □

Remark 3.11 The obvious sufficient condition for (3.45) is

$$\sup_{\|x\| \le N} \|g_k(x + r_k) - f(x)\| \to 0,$$

which in turn is satisfied if $g_k(\cdot) \equiv f(\cdot)$, $f(\cdot)$ is continuous and $r_k \to 0$.

Remark 3.12 Theorems 3.9 and 3.10 with A2 and A4 replaced by A5 and A6, respectively, remain valid if $f(\cdot)$ is replaced by time-varying $f_k(\cdot)$.

3.4 Applications

3.4.1 Application to Optimization

Let $L(\cdot)$ be an unknown $\mathbb{R}^l \to R$ function. Based on the noisy measurements of $L(\cdot)$ the well-known Kiefer-Wolfowitz (KW) algorithm is used to find the local maximizers of $L(\cdot)$. In order to reduce the number of observations at each step the algorithms with randomized differences are introduced. For the conventional KW algorithm $2l$ measurements at each step are needed, while for the algorithms with randomized differences only 2 measurements per step are used.

We now consider the algorithm with randomized differences, and the measurement noise is now allowed to depend on locations of the measurement.

At each iteration, two measurements are taken:

$$y_{k+1}^{+} = L(x_k + c_k\Delta_k) + \xi_{k+1}^{+}(x_k + c_k\Delta_k, \omega), \tag{3.46}$$

$$y_{k+1}^{-} = L(x_k - c_k\Delta_k) + \xi_{k+1}^{-}(x_k - c_k\Delta_k, \omega), \tag{3.47}$$

where $\Delta_k \triangleq [\Delta_k^1, \ldots, \Delta_k^l]^\tau$ is defined as follows:

$(\Delta_k^i, i = 1, \ldots, l,\ k = 1, 2, \ldots)$ is a sequence of mutually independent and identically distributed random variables with $|\Delta_k^i| < a$, $|1/\Delta_k^i| < b$ and $\mathsf{E}(1/\Delta_k^i) = 0$ for all $i \in \{1, \ldots, l\}$, $k = 1, 2, \ldots$, where a and b are positive real numbers.

$(\Delta_k^i, i = 1, \ldots, l, k = 1, 2, \ldots,)$ is independent of both sequences $\{\xi_k^+(x, \omega)\}$ and $\{\xi_k^-(x, \omega)\}$, $\forall x \in I\!\!R^l$.

It is important to note that both ξ_{k+1}^- and ξ_{k+1}^+ depend on Δ_k.

Define

$$y_{k+1} = \frac{(y_{k+1}^+ - y_{k+1}^-)}{2c_k} \Delta_k^{-1}, \tag{3.48}$$

$$\xi_{k+1} = \xi_{k+1}^+(x_k + c_k \Delta_k, \omega) - \xi_{k+1}^-(x_k - c_k \Delta_k, \omega), \tag{3.49}$$

where $\Delta_k^{-1} \triangleq [\frac{1}{\Delta_k^1}, \ldots, \frac{1}{\Delta_k^l}]^\tau$. It is clear that

$$y_{k+1} = \frac{L(x_k + c_k \Delta_k) - L(x_k - c_k \Delta_k)}{2c_k} \Delta_k^{-1} + \frac{\xi_{k+1} \Delta_k^{-1}}{2c_k}. \tag{3.50}$$

Let $\{x_k\}$ be still defined by (3.4)–(3.6) but with y_{k+1} given by (3.48) or (3.50).

We need the following conditions:

H1 The function $\nabla L = f$ is locally Lipschitz continuous. There is a unique maximum of L at x^0 so that $f(x^0) = 0$ and $f(x) \neq 0$, $\forall x \neq x^0$. There is a $c_0 > 0$ such that $\|x^*\| < c_0$ and $\sup_{\|x\| = c_0} L(x) < L(x^*)$ where x^* is the one used in (3.4).

H2 $a_k > 0$, $c_k > 0$, $c_k \to 0$ as $k \to \infty$, $\sum_{k=1}^\infty a_k = \infty$ and there is a $p \in (1, 2]$ such that $\sum_{k=1}^\infty (\frac{a_k}{c_k})^p < \infty$.

H3 Both $\xi_k^+(x, \omega)$ and $\xi_k^-(x, \omega)$ are measurable functions: $(I\!\!R^l \times \Omega, \mathcal{B}^l \times \mathcal{F}) \to (R, \mathcal{B})$, $\forall k$, satisfying the following conditions

$$\xi_k^+(x, \omega) \in \mathcal{F}_k, \quad \xi_k^-(x, \omega) \in \mathcal{F}_k,$$

$$\mathsf{E}(\xi_{k+1}^+(x, \omega) | \mathcal{F}_k) = 0, \quad \mathsf{E}(\xi_{k+1}^-(x, \omega) | \mathcal{F}_k) = 0, \quad \forall x \in I\!\!R^l,$$

$$\mathsf{E}(\|\xi_{k+1}^+(x, \omega)\|^p | \mathcal{F}_k) \triangleq \sigma_{k+1}^+(x) < \infty,$$

$$\mathsf{E}(\|\xi_{k+1}^-(x, \omega)\|^p | \mathcal{F}_k) \triangleq \sigma_{k+1}^-(x) < \infty,$$

$$\sup_k \sup_{\|x\| \le N} [\sigma_{k+1}^+(x) + \sigma_{k+1}^-(x)] \triangleq \sigma(N) < \infty, \quad \forall N, \tag{3.51}$$

where $\{\mathcal{F}_k\}$ is a family of nondecreasing σ-algebras independent of both x and $\{\Delta_k^i, i = 1, \ldots, l, k = 1, 2, \ldots\}$.

Theorem 3.13 Assume H1, H2 and H3 hold. Then $\{x_k\}$ defined by (3.4)–(3.6) with y_{k+1} given by (3.48) converges to x^0 a.s.

Proof. Introduce σ-algebra $\mathcal{F}'_k$ generated by $\{\Delta_i, i \le k\}$ and $\{\mathcal{F}_j, j \le k\}$, i.e., $\mathcal{F}'_k = \sigma\{\Delta_i, 1 \le i \le k, \mathcal{F}_j, 0 \le j \le k\}$.

It is clear that x_k is measurable with respect to $\sigma\{\Delta_j, 1 \le j \le k-1, \mathcal{F}_j, 0 \le j \le k\}$, and hence x_k is $\mathcal{F}'_k$-measurable. Both $\xi^+_{k+1}(x_k + c_k\Delta_k, \omega)$ and $\xi^-_{k+1}(x_k - c_k\Delta_k, \omega)$ are $\mathcal{F}'_{k+1}$-measurable. Similar to the proof of Theorem 3.9, approximating by simple functions leads to

$$\mathsf{E}(\xi^+_{k+1}(x_k + c_k\Delta_k, \omega) I_{[\|x_k\| \le N]} | \mathcal{F}'_k) = 0,$$

$$\mathsf{E}(\xi^-_{k+1}(x_k - c_k\Delta_k, \omega) I_{[\|x_k\| \le N]} | \mathcal{F}'_k) = 0,$$

and

$$\mathsf{E}(\|\xi^+_{k+1}(x_k + c_k\Delta_k, \omega)\Delta_k^{-1}\|^p I_{[\|x_k\| \le N]} | \mathcal{F}'_k) \le \gamma\sigma^+_{k+1}(x_k + c_k\Delta_k),$$

$$\mathsf{E}(\|\xi^-_{k+1}(x_k - c_k\Delta_k, \omega)\Delta_k^{-1}\|^p I_{[\|x_k\| \le N]} | \mathcal{F}'_k) \le \gamma\sigma^-_{k+1}(x_k - c_k\Delta_k),$$

where $\gamma \triangleq (b\sqrt{l})^p \ge \|\Delta_k^{-1}\|^p$.

Therefore, by (3.49) $(\xi_k \Delta_{k-1}^{-1} I_{[\|x_{k-1}\| \le N]}, \mathcal{F}'_k)$ is a martingale difference sequence for any integer N and

$$\begin{aligned} &\mathsf{E}(\|\xi_{k+1}\Delta_k^{-1}\|^p I_{[\|x_k\| \le N]} | \mathcal{F}'_k) \\ \le\ & \gamma 2^{p-1} I_{[\|x_k\| \le N]} (\sigma^+_{k+1}(x_k + c_k\Delta_k) + \sigma^-_{k+1}(x_k - c_k\Delta_k)). \end{aligned}$$

Noticing $\{\|\Delta_k\|\}$ is bounded and $c_k \to 0$, by (3.51) and the convergence theorem for martingale difference sequences we have for any integer $N > 0$

$$\sum_{i=1}^{\infty} \frac{a_i}{c_i} \xi_{i+1} \Delta_i^{-1} I_{[\|x_i\| \le N]} < \infty \quad \text{a.s.} \tag{3.52}$$

By an elementary manipulation, y_{k+1} given by (3.50) can be written as

$$y_{k+1} = f(x_k) + \epsilon_{k+1}, \tag{3.53}$$

where

$$\epsilon_{k+1} = \frac{1}{2}(w_k(1) + w_k(2) + u_k(1) + u_k(2) + \frac{1}{c_k}\xi_{k+1}\Delta_k^{-1}). \tag{3.54}$$

The ith component of $w_k(1)$ and $u_k(1)$ are respectively given by

$$w^i_{k,i} = f_i(x_k + c_k\Delta_k + \delta_i(i)) - f_i(x_k), \tag{3.55}$$

$$u^i_{k,i} = f_i(x_k - c_k\Delta_k + \delta_i(i)) - f_i(x_k), \tag{3.56}$$

where $f_i(\cdot)$ denotes the ith component of $f(\cdot)$, and $\delta_k(i)$ denotes an $\mathbb{R}^l$-valued random vector such that

$$\delta_k(i) = [\underbrace{0,\ldots,0}_{i-1},\delta_k^i,\underbrace{0,\ldots,0}_{l-i}]^\tau, \quad |\delta_k^i| \le ac_k, (|\Delta_k^i| < a). \tag{3.57}$$

The ith components of $w_k(2)$ and $u_k(2)$ are

$$\sum_{j\neq i}^{l} h_{kj}\frac{1}{\Delta_j^i} \qquad \text{and} \qquad \sum_{j\neq i}^{l} g_{kj}\frac{1}{\Delta_j^i}, \tag{3.58}$$

respectively, where h_{kj} and g_{kj} are independent of Δ_k^i $\forall k \ge 1, j = 1,\ldots,i-1, i+1,\ldots,l$, and $\{|h_{kj}|I_{[\|x_k\|\le N]}\}$ and $\{|g_{kj}|I_{[\|x_k\|\le N]}\}$ are bounded by a constant possibly depending on N for any $j = 1,\ldots,i-1,i+1,\ldots,l$.

Since $\sum_{i=1}^{\infty} a_i^p < \infty$ for $p \in (1,2]$, by the convergence theorem for martingale difference sequences it follows that

$$\sum_{k=1}^{\infty} a_k(w_k(2) + u_k(2))I_{[\|x_k\|\le N]} < \infty \qquad \text{a.s.} \tag{3.59}$$

for any integer N.

By the local Lipschitz continuity from (3.55),(3.56) it is seen that

$$(w_k(1) + u_k(1))I_{[\|x_k\|\le N]} \to 0 \quad \text{as} \quad k \to \infty. \tag{3.60}$$

From (3.52), (3.59) and (3.60) it follows that for ϵ_{k+1} given by (3.54) Condition A7 is fulfilled.

Then paying attention to (3.53) by Theorem 3.8 we conclude that $x_k \to x^0$ a.s., since in the present case A2 automatically holds by taking $v(\cdot) = -L(\cdot)$. □

Remark 3.14 Similar to Remark 3.12, extension from the time-invariant $f(\cdot)$ to the time-varying $f_k(\cdot)$ can also be made for Theorem 3.13.

3.4.2 Application to Signal Processing

We now apply the obtained results to adaptive filtering.

Let $\{X_n\} \in \mathbb{R}^r$ and $\{s_n\} \in R$ be the measured input and reference signal, respectively. Assume $[s_n, X_n^T]^T$ is stationary and ergodic. The vector H is adaptively adjusted so that $H^T X_n$ best matches s_n in a certain sense. If the cost function $L(H)$ to be minimized is $\mathsf{E}|s_n - H^T X_n|$, then the following sign-algorithm

$$H_{k+1} = H_k + a_k X_k \text{sign}(s_k - H_k^T X_k), \quad a_k = \frac{1}{k} \tag{3.61}$$

is used to estimate the minimizer of $L(H)$. If the cost function is $L(H) = \mathsf{E}|s_n - H^T X_n|^2$, then the following algorithm

$$H_{k+1} = H_k + a_k X_k (s_k - H_k^T X_k) \tag{3.62}$$

has extensively been studied.

According to (3.4)–(3.6), instead of (3.61), (3.62) we consider the corresponding versions with expanding truncations.

Take $\{M_k\}$, $M_k > M_{k-1}$, $\forall k$, $M_k \to \infty$ as $k \to \infty$. The algorithms (3.61), (3.62) are modified to (3.63), (3.64), respectively

$$H_{k+1} = (H_k + a_k X_k \text{sign}(s_k - H_k^T X_k)) I_{\{\|H_k + a_k X_k \text{sign}(s_k - H_k^T X_k)\| \le M_{\sigma_k}\}}, \tag{3.63}$$

$$\sigma_k = \sum_{i=1}^{k-1} I_{\{\|H_i + a_i X_i \text{sign}(s_i - H_i^T X_i)\| > M_{\sigma_i}\}}, \quad a_k = \frac{1}{k},$$

and

$$H_{k+1} = (H_k + a_k X_k (s_k - H_k^T X_k)) I_{\{\|H_k + a_k X_k (s_k - H_k^T X_k)\| \le M_{\sigma_k}\}}, \tag{3.64}$$

$$\sigma_k = \sum_{i=1}^{k-1} I_{\{\|H_i + a_i X_i (s_i - H_i^T X_i)\| > M_{\sigma_i}\}}, \quad a_k = \frac{1}{k}.$$

Notice that x^* in (3.4) is now set to be zero in (3.63) and (3.64).

Theorem 3.15 Assume $[s_k, X_k^T]^T$ is stationary and ergodic such that

$$\mathsf{E} \begin{bmatrix} s_1 \\ X_1 \end{bmatrix} [s_1 X_1^T] > 0.$$

Then as $k \to \infty$, $d(H_k, J) \to 0$ a.s. for H_k defined by both (3.63) and (3.64), where J denotes the set of minimizers of $\mathsf{E}|s_1 - H^T X_1|$ for (3.63) and $\mathsf{E}|s_1 - H^T X_1|^2$ for (3.64).

It is clear that to minimize $L(H) = \mathsf{E}(s_1 - H^T X_1)^2$ is equivalent to finding the roots of its gradient

$$f(H) = \mathsf{E}(X_1(s_1 - H^T X_1)).$$

Therefore, the corresponding RM algorithm should be

$$H_{k+1} = H_k - a_k(\mathsf{E}(X_1(s_1 - H_k^T X_1))) \quad (= H_k - a_k f(H_k)). \tag{3.65}$$

Comparing (3.65) with (3.62) we rewrite (3.62) as

$$H_{k+1} = H_k - a_k f(H_k) + a_k \epsilon_{k+1}, \tag{3.66}$$

where

$$\epsilon_{k+1} = X_k(s_k - H_k^T X_k) + \mathsf{E}(X_1(s_1 - H_k^T X_1)). \tag{3.67}$$

Thus, (3.66) is an RM algorithm with observation noise (3.67), which is state-dependent. The similar situation takes place when minimizing $\mathsf{E}|s_n - H^T X_n|$.

The proof of Theorem 3.15 can be carried out by applying Theorem 3.1 incorporating with stationarity of $(s_k, X_k^T)^T$. For details we refer to [17].

3.5 Notes

The SA algorithm was first introduced by Robbins and Monro in [1], where the convergence of the estimate x_k to the sought-for root in the mean square sense was established. After this pioneer work, research on SA has mostly been concentrating on the path-wise convergence analysis, although other types of convergence, e.g., weak convergence [2, 3], have also drawn attention from some researchers. Until mid-seventies of twentieth century the probabilistic methods were the main approach to convergence analysis for SA algorithms. By this method the key assumptions include the linear growth rate of $f(\cdot)$ and probabilistic properties such as independence or martingale property for the observation noise, in addition to the usual condition concerning existence of a Lyapunov function, which normally has to be assumed in all analysis methods for SA algorithms [4]. This type of works was very well summarized in [5].

Starting from mid-seventies the ordinary differential equation (ODE) method was developed for convergence analysis of SA algorithms [6, 7]. The name ODE for the method comes from the following observation that the tail part of the interpolating function of $\{x_k\}$ with step size used in the algorithm as its interpolation length satisfies the ODE $\dot{x} = f(x)$. However, to derive this, one has to *a priori* assume that $\{x_k\}$ is bounded. Because the ODE method requires much weaker conditions on the observation noise in comparison with the probabilistic approach, it is widely applied [3, 8] in different areas.

The boundedness assumption on $\{x_k\}$ in essence is a growth rate restriction on $f(\cdot)$. To remove this kind of assumptions, the expanding truncation technique was proposed in [9], and was further developed in [10, 11]. The convergence analysis is carried out at a fixed trajectory along convergent subsequences, so it is called TS (trajectory-subsequence) method. Since the noise may depend on state (estimate) [12], to verify noise condition along convergent subsequences of $\{x_k\}$ is much easier in comparison with verifying it along the whole sequence $\{x_k\}$. This is justified by many applications described in [11]. Two of them are given in this chapter.

Applying SA algorithms to solve the optimization problem was first discussed in [13]. In order to reduce the number of observations per iteration, which are used to form finite differences serving as estimates of partial derivatives, the random direction techniques were separately introduced in [14] and [15]. As shown in [16] conditions used for optimization can greatly be weakened by applying TS method for SA algorithms with expanding truncations. Similar improvement can be achieved for problems arising from signal processing [17, 18]. For more application examples and for more detailed asymptotic properties of SA algorithms with expanding truncations we refer to [11].

3.6 References

[1] Robbins, H. and Monro, S., A stochastic approximation method, *Ann. Math, Statist.*, **22** (1951), 400–407.

[2] Kushner, H.J., *Approximation and Weak Convergence Methods for Random Processes, with applications to Stochastic Systems Theory*, MIT Press, Cambridge, MA, 1984.

[3] Kushner, H.J. and Yin, G., *Stochastic Approximation Algorithms and Applications*, Springer-Verlag, Berlin, 1997.

[4] Fang, H.T. and Chen, H.F., Stability and Instability of limit points for stochastic approximation algorithms, *IEEE Trans. Autom. Control*, **45** (2000), 413–420.

[5] Nevelson, M.B. and Hasminskii, R.A., *Stochastic Approximation and Recursive Estimation*, AMS Transl. Math. Monographs, **49**, 1976.

[6] Ljung, L., Analysis of recursive stochastic algorithms, *IEEE Trans. Autom. Control*, **22** (1977), 551–575.

[7] Kushner, H.J. and Clark, D.S., *Stochastic Approximation Methods for Constrained and Unconstrained Systems*, Springer-Verlag, Berlin, 1978.

[8] Benveniste, A., Métivier, M. and Priouret, P., *Adaptive Algorithms and Stochastic Approximations*, Springer-Verlag, Berlin, 1990.

[9] Chen, H.F. and Zhu, Y.M., Stochastic approximation procedures with randomly varying truncations, *Scientia Sinica* (Series A), **29** (1986), 914-926.

[10] Chen, H.F., Stochastic approximation and its new applications, *Proc. Hong Kong Int. Workshop on New Directions of Control and Manufacturing*, 2–12, 1994.

[11] Chen, H.F., *Stochastic Approximation and Its Applications*, Kluwer, to appear in 2002.

[12] CHEN, H.F., Stochastic approximation with state-dependent noise, *Science in China* (Series E), **43** (2000), 531–541.

[13] KIEFER, J. AND WOLFOWITZ, J., Stochastic approximation of a regression function, *Ann. Math. Stat.*, **23** (1952), 552–558.

[14] KORONASKI, J., Random-seeking methods for the stochastic unconstrained optimization, *Int. J. Control*, **21** (1975), 517–527.

[15] SPALL, J.C., Multivariate stochastic approximation using a simultaneous perturbation gradient approximation, *IEEE Trans. Autom. Control*, **37** (1992), 332–341.

[16] CHEN, H.F., DUNCAN, T.E. AND PASIK-DUNCAN, B., A Kiefer-Wolfowitz algorithm with randomized differences, *IEEE Trans. Autom. Control*, **44** (1999), 442–453.

[17] CHEN, H.F. AND YIN, G., Asymptotic properties of sign algorithms for adaptive filtering, submitted for publication.

[18] EWEDA, E., Convergence of the sign algorithm for adaptive filtering with correlated data, *IEEE Trans. Inform. Theory*, **IT-37** (1991), 1450–1457.

4

Performance Potential Based Optimization and MDPs

Xi-Ren Cao

This chapter presents some recent results in the area of optimization and Markov decision processes (MDPs). The work starts with performance sensitivity analysis of Markov processes. It is a continuation of the research in the past two decades on the optimization of discrete event dynamic systems, especially the theory of perturbation analysis (PA). We approach the MDP problem from a sensitivity point of view. This new perspective leads to some new insights and offers a clear and concise explanation for the basic concepts and results in MDPs.

Two types of sensitivities are studied: sensitivities with respect to continuous and discrete parameters. In the continuous case, the policies are randomized, with a parameter δ indicating the probability of applying one policy and $1 - \delta$ the probability of the other, the derivative of the performance with respect to δ is found; this is the derivative along the direction from one policy to the other in the policy space. The continuous sensitivity formula was first developed by PA. In the discrete case, the difference of the performance of the Markov processes under two different policies is found. Both sensitivities can be obtained by using a quantity called the *performance potential* defined by a Poisson equation. With the discrete sensitivity formula, the standard policy iteration algorithm in MDP can be very easily derived. Compared with the continuous sensitivity formula, it is clear that at each step the policy iteration algorithm simply chooses the "steepest" direction (with the largest absolute value of the directional derivative) to go for its next policy.

The performance potentials can be estimated on a sample path of a Markov process; thus, performance derivatives can be obtained and policy iteration can be implemented with a single sample path. Therefore, there are two ways to implement potential-based on-line performance optimization: by perturbation analysis using performance derivatives, or by policy iteration. Stochastic approximation methods have to be used in these two cases when the sample paths used to estimate the potentials are relatively short and stochastic errors exist. (For details, see Section 4.4 and [12], [17].) These on-line optimization schemes are of practical importance to real-world engineering problems such as the optimization of manufacturing and communication systems.

Inspired by the potential-based sensitivity view of MDPs and the single sample path-based approach, a number of new research topics emerge. One of them is the time aggregation of MDPs. When the number of controllable states is small, this approach may save computation or the number of transitions of the sample path used in optimization. Other topics include gradient-based policy iteration.

We first study MDPs with long-run average costs. Section 4.1 starts with presenting sensitivities with respect to continuous parameters and then introduces the concept of performance potentials. Maclaurin series for the performance of a particular policy is derived by using the fundamental matrix. Section 4.2 starts with discrete sensitivity, i.e., the difference of performance of any two policies. From this, the MDP policy iteration algorithm can be easily derived. Section 4.3 extends the results to MDPs with discounted performance criteria. All the results for long-run average MDPs can be viewed as special cases when the discount factor α goes to one. Thus, potential theory provides a unified approach to both types of MDPs. Section 4.4 offers a physical explanation of potentials and derives sample path-based estimates. With these algorithms, performance optimization and policy iteration can be implemented online. Section 4.5 develops the time aggregation approach. Section 4.6 presents connections to perturbation analysis. We show that policy iteration can be directly derived from perturbation analysis. Section 4.7 contains two examples illustrating the application of the results presented in this chapter. Section 4.8 provides some supplementary material to help the readers to further understand the relation between PA and MDP and the significance of the results.

4.1 Sensitivity Analysis and Performance Potentials

In this section, we derive the first and high-order derivatives of the performance with respect to a continuous parameter, and introduce the concept of performance potentials [6]. The results will be used in the next section to develop the policy iteration solution to MDPs.

Denote a finite Markov chain as $\mathbf{X} = \{X_0, X_1, \cdots, X_k, \cdots\}$. Let $P = [p(i,j)]$ be the $M \times M$ transition matrix of the Markov chain, with M denoting the number of states in the state space $\mathcal{S} = \{1, 2, \cdots, M\}$. Let $e = (1, 1, \cdots, 1)^T$, where the superscript "T" denotes transpose, be an M-dimensional column vector in which every component equals 1. Then,

$$Pe = e.$$

Let $f : \mathcal{S} \to \mathcal{R}$, where $\mathcal{R} = (-\infty, \infty)$ is the space of real numbers. We also write $f = (f(1), f(2), \cdots, f(M))^T$ as a column vector; i.e., we use f for both a function and a vector; therefore, $f(i)$ denotes either the value of the function f at $i \in \mathcal{S}$ or the ith component of the vector f.

Assume that the Markov chain is ergodic (i.e., P is irreducible and aperiodic). Let $\pi = (\pi(1), \pi(2), \cdots, \pi(M))$ be the row vector representing the steady-state probability of $\mathbf{X}$. We have

$$\pi e = 1, \qquad \pi = \pi P.$$

We consider the average-cost problem; the performance measure is defined as the expected value of f with respect to π:

$$\eta = \mathsf{E}_\pi(f) = \sum_{i=1}^{M} \pi(i) f(i) = \pi f,$$

where E_π denotes the expectation with respect to the steady-state measure π.

Let $A = I - P$. It is well known that if P is irreducible, then $(I - P + e\pi)$ is invertible. Its inverse is called the *fundamental matrix* in [18]. Define

$$A^{\#} = (I - P + e\pi)^{-1} - e\pi.$$

$A^{\#}$ is called the *group inverse* of A [19]. It is easy to verify that A satisfies

$$AA^{\#} = A^{\#}A = I - e\pi,$$

and

$$A^{\#}e = 0, \quad \pi A^{\#} = 0.$$

Suppose that P changes to $P(\delta) = P + \delta Q$, with $Q = [q(i,j)]$ satisfying $Qe = 0$. We assume that δ and Q are such that $P(\delta)$ is also an irreducible Markov matrix. Let $\pi(\delta)$ and $\eta(\delta)$ be the steady-state probability and the performance of the Markov chain with $P(\delta)$, respectively. With these notation, we have $P = P(0)$, $A = A(0)$, $\pi = \pi(0)$, and $\eta = \eta(0)$.

For any irreducible finite Markov chain with transition matrix $P(\delta)$, we have

$$\pi(\delta)[I - P(\delta)] = 0.$$

Taking derivative on both sides of this equation with respect to δ, we have

$$\frac{d\pi(\delta)}{d\delta}[I - P(\delta)] = \pi(\delta)Q.$$

Continuing taking derivatives on both sides of the resulting equations, we obtain, for any $n \geq 1$,

$$\frac{d^n \pi(\delta)}{d\delta^n}[I - P(\delta)] = n \frac{d^{n-1}\pi(\delta)}{d\delta^{n-1}} Q.$$

Setting $\delta = 0$, multiplying both sides of the above equation on the right with $A^{\#}$, and noting that $AA^{\#} = I - e\pi$ and $\pi e = 1$, we get

$$\frac{d^n \pi}{d\delta^n}|_{\delta=0} = n \frac{d^{n-1}\pi}{d\delta^{n-1}}|_{\delta=0} QA^{\#}.$$

The subscript $\delta = 0$ will be omitted in the subsequent discussions. Thus,

$$\frac{d^n \pi}{d\delta^n} = n! \pi (QA^{\#})^n.$$

Finally, for any performance function f, we have

$$\begin{aligned} \frac{d^n \eta}{d\delta^n} &= n! \pi (QA^{\#})^n f \\ &= n! \pi [Q(I - P + e\pi)^{-1}]^n f. \end{aligned} \tag{4.1}$$

In addition, $\eta(\delta)$ is an analytical function of δ. (More precisely, it is a rational function of δ whose denominator and numerator are both polynomials of δ with finite degrees. This can be verified by solving $\pi(\delta)[I - P(\delta)] = 0$ and $\pi(\delta)e = 1$.) Thus, $\eta(\delta)$ has a Maclaurin expansion at $\delta = 0$, which is

$$\begin{aligned} \eta(\delta) - \eta &= \sum_{n=1}^{\infty} \frac{1}{n!} \frac{d^n \eta}{d\delta^n} \delta^n \\ &= \pi \{\sum_{n=1}^{\infty} (\delta Q A^{\#})^n\} f, \end{aligned} \tag{4.2}$$

or equivalently,

$$\eta(\delta) = \pi \sum_{n=0}^{\infty} \{(QA^{\#})^n f \delta^n\}. \tag{4.3}$$

Denote the spectral radius of a matrix W as $\rho(W)$ (i.e., the largest absolute value of the eigenvalues of W). Define

$$r = \frac{1}{\rho(QA^{\#})} = \frac{1}{\rho(Q(I - P + e\pi)^{-1})}.$$

Then for $\delta < r$, the summation in (4.3) converges.

Setting $f(i) = 1$ and $f(j) = 0$, $j \neq i$, we get $\eta = \pi f = \pi(i)$. Thus, (4.3) leads to

$$\pi(\delta) = \pi \sum_{n=0}^{\infty} \{(QA^{\#})^n \delta^n\}. \tag{4.4}$$

From (4.4) and (4.2), we obtain

$$\eta(\delta) - \eta = \pi(\delta) Q A^{\#} f \delta. \tag{4.5}$$

In addition, we have a general form:

$$\eta(\delta) = \pi \sum_{k=0}^{n} (\delta Q A^{\#})^k f + \pi(\delta)(\delta Q A^{\#})^{n+1} f, \quad \text{for any } n = 0, 1, 2, \cdots. \tag{4.6}$$

The last term in (4.6), $\pi(\delta)(\delta Q A^{\#})^{n+1} f$, is the error in taking the first $n+1$ terms in the Maclaurin series as an estimate of $\eta(\delta)$.

Setting $n = 1$ in (4.1) leads to the first order derivative of η with respect to δ:

$$\frac{d\eta}{d\delta} = \pi(Q A^{\#}) f.$$

Since $Qe = 0$, we have

$$Q A^{\#} = Q(I - P + e\pi)^{-1}.$$

Define the performance potential as

$$g = (I - P + e\pi)^{-1} f,$$

which satisfies the Poisson equation:

$$(I - P + e\pi) g = f. \tag{4.7}$$

Each component $g(i)$ is called the performance potential in state i. Then we have

$$Q A^{\#} f = Q g.$$

Thus,

$$\frac{d\eta}{d\delta} = \pi Q g. \tag{4.8}$$

In a more general case, f also changes to $f(\delta) = f + \delta h$ for Markov chain with $P(\delta) = P + \delta Q$. We simply have $\frac{df(\delta)}{d\delta} = h$ and

$$\frac{d\eta}{d\delta} = \pi(Q g + h). \tag{4.9}$$

This is the derivative with respect to a continuous parameter, which will be compared with the discrete sensitivity (the difference) derived in the next section. Multiplying both sides of (4.7) by π, we get

$$\pi g = \pi f = \eta. \tag{4.10}$$

4.2 Markov Decision Processes

In Markov decision problems, we need to compare the performance of two policies, corresponding to two different transition matrices [5]. Therefore, we set $\delta = 1$ and denote $P' = P(1) = P + Q$, $\pi' = \pi(1)$, and $\eta' = \eta(1)$. Then (4.5) becomes

$$\eta' - \eta = \pi' Q g, \tag{4.11}$$

and more general, we have

$$\eta' - \eta = \pi'(Qg + h). \tag{4.12}$$

We shall refer to these as discrete sensitivity formulas.

Compared with (4.9), the performance difference can be obtained by simply replacing π in the performance derivative by π'. This nature will be used to explain the direction of choosing the policy in the next iteration in policy iteration later. (4.12) can also be obtained by simply multiplying both sides of (4.7) by π', which yields

$$\pi' Q g = \pi'(P' - P)g = \pi'(I - P)g = \pi' f - \pi g = \pi' f - \eta.$$

For two M-dimensional vectors a and b, we define $a = b$ if $a(i) = b(i)$ for all $i = 1, 2 \cdots, M$; $a \leq b$ if $a(i) < b(i)$ or $a(j) = b(j)$ for all $i = 1, 2 \cdots, M$; $a < b$ if $a(i) < b(i)$ for all $i = 1, 2 \cdots, M$; and $a \preceq b$ if $a(i) < b(i)$ for at least one i, and $a(j) = b(j)$ for other components. The relation $\leq$ includes $=$, $\preceq$, and $<$. Hence, similarly define $>$, $\succeq$, and $\geq$.

Next, we note that $\pi'(i) > 0$ for all $i = 1, 2, \cdots, M$. Thus, from (4.12), we know that if $Qg + h = (P' - P)g + (f' - f) \succeq 0$ then $\eta' - \eta > 0$. From (4.12) and the fact $\pi' > 0$, the proof of the following lemma is straightforward.

Lemma 4.1 If $Pg + f \preceq P'g + f'$, then $\eta < \eta'$.

It is interesting to note that in the lemma, we use only the potential associated with one Markov chain, i.e., g. Thus, to compare the performance measures under two policies, only the potentials with one policy is needed.

In an MDP, at any transition instant $n \geq 0$ of a Markov chain $\mathbf{X} = \{X_n, n \geq 0\}$, an action is chosen from an action space $\mathcal{A}$ and is applied to the Markov chain. We assume that the number of actions is finite, and we only consider stationary policies. The actions that are available in state $i \in \mathcal{S}$ form a nonempty subset $A(i) \subseteq \mathcal{A}$. A stationary policy is a mapping $\mathcal{L} : \mathcal{S} \to \mathcal{A}$; i.e., for any state i, $\mathcal{L}$ specifies an action $\mathcal{L}(i) \in A(i)$. Let $\mathcal{E}$ be the policy space. If action α is taken in state i, then the state transition probabilities in state i are denoted as $p^\alpha(i,j)$, $j = 1, 2, \cdots, M$. With a policy $\mathcal{L}$, the Markov process evolves according to the transition matrix $P^\mathcal{L} = [p^{\mathcal{L}(i)}(i,j)]_{i=1}^M|_{j=1}^M$. We use the superscript $\mathcal{L}$ to denote the quantities associated with policy $\mathcal{L}$.

For most stable systems in real-world applications, the Markov chains under any policy are positive recurrent. For example, in any stable communication systems, from any state it is always possible to reach the null state where there are no packets in the system. Therefore, in this chapter we assume that the Markov chain is recurrent under any policy. In addition, for simplicity we assume that it is aperiodic; although all the results in this chapter hold for periodic chains if we replace the steady-state value by the corresponding time average value [20]. Therefore, the Markov chains considered are ergodic [11]. This corresponds to the problems classified in [20] as the recurrent case.

The steady-state probabilities corresponding to policy $\mathcal{L}$ is denoted as a vector $\pi^{\mathcal{L}} = (\pi^{\mathcal{L}}(1), \cdots, \pi^{\mathcal{L}}(M))$. Suppose that at each stage with state i and control action $\alpha \in A(i)$, a cost $f(i, \alpha) = f(i, \mathcal{L}(i))$ is incurred. The long-run expected value of the average cost per stage corresponding to policy $\mathcal{L}$ is then

$$\eta^{\mathcal{L}} = \lim_{N \to \infty} \frac{1}{N} \mathsf{E}\{\sum_{n=0}^{N-1} f[X_n, \mathcal{L}(X_n)]\},$$

where "E" denotes the expectation on the underlying probability space of the Markov chain. For ergodic chains, the above limit exists and does not depend on the initial state. Our objective is to minimize this average cost per stage over the policy space $\mathcal{S}$, i.e., to obtain $\min_{\mathcal{L} \in \mathcal{E}} \eta^{\mathcal{L}}$.

Define $f^{\mathcal{L}} = (f[1, \mathcal{L}(1)], \cdots, f[M, \mathcal{L}(M)])^T$. (4.7) and (4.10) becomes

$$(I - P^{\mathcal{L}} + e\pi^{\mathcal{L}})g^{\mathcal{L}} = f^{\mathcal{L}}, \tag{4.13}$$

and

$$\pi^{\mathcal{L}} g^{\mathcal{L}} = \pi^{\mathcal{L}} f^{\mathcal{L}}.$$

The following optimality theorem follows almost immediately from Lemma 4.1, which is derived directly from (4.12).

Theorem 4.2 A policy $\mathcal{L}$ is optimal if and only if

$$P^{\mathcal{L}} g^{\mathcal{L}} + f^{\mathcal{L}} \leq P^{\mathcal{L}'} g^{\mathcal{L}} + f^{\mathcal{L}'}, \tag{4.14}$$

for all $\mathcal{L}' \in \mathcal{E}$.

The optimality condition (4.14) is, of course, equivalent to the other conditions in the literature. To see this, we rewrite (4.13) in the following form:

$$\eta^{\mathcal{L}} e + g^{\mathcal{L}} = f^{\mathcal{L}} + P^{\mathcal{L}} g^{\mathcal{L}}. \tag{4.15}$$

Then Theorem 4.2 becomes: A policy $\mathcal{L}$ is optimal if and only if

$$\eta^{\mathcal{L}} e + g^{\mathcal{L}} = \min_{\mathcal{L}' \in \mathcal{E}} \{P^{\mathcal{L}'} g^{\mathcal{L}} + f^{\mathcal{L}'}\}. \tag{4.16}$$

The minimum is taken for every component of the vector. (4.16) is the optimality equation, or the Bellman equation. From (4.16), $g^{\mathcal{L}}$ is equivalent to the "differential" or "relative cost vector" in [2], or the "bias" in [20]. In our approach, g is directly related to the long-run average performance, and many results, such as the existence and the uniqueness of the solution to (4.15), the optimality condition (4.16), and the convergence of the optimal algorithms, become almost obvious. In addition, as shown in Section 4.4, $g^{\mathcal{L}}$ can be easily estimated based on a single sample path. This is an important feature which allows optimization algorithms to be implemented online for real-world systems.

Policy iteration algorithms to determine the optimal policy can be easily developed by combining Lemma 4.1 and Theorem 4.2. Roughly speaking, at the kth step with policy $\mathcal{L}_k$, we set the policy for the next step (the $(k+1)$th step) as $\mathcal{L}_{k+1} = arg\{\min[P^{\mathcal{L}}g^{\mathcal{L}_k} + f^{\mathcal{L}}]\}$, with $g^{\mathcal{L}_k}$ being the solution to the Poisson equation for $P^{\mathcal{L}_k}$. Lemma 4.1 implies that performance improves at each iteration. Theorem 4.2 shows that the minimum is reached when no performance improvement can be achieved. We shall not state the details here because they are standard.

Finally, since $P^{\mathcal{L}_k}$ and $f^{\mathcal{L}_k}$ are fixed, we note that $(P^{\mathcal{L}} - P^{\mathcal{L}_k})g^{\mathcal{L}_k} + (f^{\mathcal{L}} - f^{\mathcal{L}_k})$ takes the minimal value (component wise) at $P^{\mathcal{L}_{k+1}}$. Thus, because $\pi^{\mathcal{L}_k} > 0$, from (4.9) the performance derivative also reaches minimum at $P^{\mathcal{L}_{k+1}}$ (i.e., the largest absolute value, since the derivative is negative). Therefore, we conclude that policy iteration in fact chooses the steepest direction to go in the policy space.

4.3 Problems with Discounted Performance Criteria

In this section, we extend the above results to MDPs with discounted performance criteria [8]. We shall see, with performance potentials, both the average- and the discounted-cost problems can be solved in the same way, and the average-cost problem can be treated as a special case when the discount factor α goes to 1.

Let $f(i)$, $i \in \mathcal{S}$, be a performance function and α, $0 < \alpha \leq 1$, be a discount factor. For $0 < \alpha < 1$, the performance cost is defined as a column vector $\eta_\alpha = (\eta_\alpha(1), \eta_\alpha(2), \cdots, \eta_\alpha(M))^T$ with

$$\eta_\alpha(i) = (1-\alpha)\mathsf{E}\{\sum_{n=0}^{\infty} \alpha^n f(X_n) | X_0 = i\}. \tag{4.17}$$

The factor $(1-\alpha)$ in (4.17) is used to obtain the continuity of η_α at $\alpha = 1$. In fact, we define

$$\eta_1 = \lim_{\alpha \to 1^-} \eta_\alpha. \tag{4.18}$$

It is proved in (4.19) and (4.23) that the above limit exists.

Lemma 4.3 $\eta_1 = \eta e$ with η being the average-cost performance:

$$\eta = \pi f = \lim_{N\to\infty} \{\mathsf{E}[\frac{1}{N}\sum_{n=0}^{N-1} f(X_n)]\}.$$

Proof. In a matrix form, we have

$$\eta_\alpha = (1-\alpha)\sum_{n=0}^{\infty} \alpha^n P^n f = (1-\alpha)(I-\alpha P)^{-1} f, \qquad 0<\alpha<1. \quad (4.19)$$

The second equation in (4.19) holds because for $0<\alpha<1$ all the eigenvalues of αP are within the unit circle [1]. Next, it is easy to verify

$$(I-\alpha P)^{-1} = (I-\alpha P+\alpha e\pi)^{-1} + \frac{\alpha}{1-\alpha} e\pi, \quad \alpha<1. \qquad (4.20)$$

Thus,

$$\lim_{\alpha\to 1-} (1-\alpha)(I-\alpha P)^{-1} = e\pi. \qquad (4.21)$$

The lemma then follows directly from (4.19) and (4.21). □

Similar to (4.7), the *α-Poisson equation* is defined as

$$(I-\alpha P+\alpha e\pi) g_\alpha = f. \qquad (4.22)$$

g_α is called the *α-potential*, which is the same as what is defined in the classical potential theory for $0<\alpha<1$ (see (4.34) in Section 4.4 and [11]). Moreover, (4.20) and (4.19) leads to

$$\eta_\alpha = (1-\alpha) g_\alpha + \alpha\eta e. \qquad (4.23)$$

It becomes obvious that (4.18) does converge.

When $\alpha = 1$, it is the standard Poisson equation. From (4.22), we have

$$\begin{aligned} g_\alpha &= (I-\alpha P+\alpha e\pi)^{-1} f \\ &= \{\sum_{n=0}^{\infty} \alpha^n (P-e\pi)^n\} f \\ &= \{I + [\sum_{n=1}^{\infty} \alpha^n (P^n - e\pi)]\} f, \qquad 0<\alpha\le 1. \end{aligned}$$

In particular,

$$g_1 = (I-P+e\pi)^{-1} f = \{I + \sum_{n=1}^{\infty} (P^n - e\pi)\} f.$$

This is the same as the performance potentials defined for the average cost MDPs. We have

$$\lim_{\alpha \to 1-} g_\alpha = g_1,$$

$$\pi g_\alpha = \pi f.$$

Now suppose that P changes to P', then π and η will change to π' and η', respectively. Let $Q = P' - P$. From (4.19), we get

$$\begin{aligned} \eta'_\alpha - \eta_\alpha &= \alpha(P'\eta'_\alpha - P\eta_\alpha) \\ &= \alpha(P' - P)\eta_\alpha + \alpha P'({\eta'}_\alpha - \eta_\alpha). \end{aligned}$$

This leads to

$${\eta'}_\alpha - \eta_\alpha = \alpha(I - \alpha P')^{-1} Q \eta_\alpha.$$

Substituting (4.23) into the right-hand side of the above equation and noting that $Qe = 0$, we obtain

$$\eta'_\alpha - \eta_\alpha = \alpha(1-\alpha)(I - \alpha P')^{-1} Q g_\alpha, \qquad 0 < \alpha < 1, \tag{4.24}$$

Letting $\alpha \to 1-$, we obtain (4.11) as a special case.

Similar to Lemma 4.1, if we assume that f also changes to f', then we have

Lemma 4.4 If $\alpha P' g_\alpha + f' \preceq \alpha P g_\alpha + f$, then ${\eta'}_\alpha < \eta_\alpha$, $0 < \alpha \leq 1$.

Proof. For $0 < \alpha < 1$, we have

$$(I - \alpha P')^{-1} = I + \alpha P' + \alpha^2 P'^2 + \cdots.$$

Since the Markov chain is positive recurrent, every item in $(I - \alpha P')^{-1}$ is positive. The lemma for $0 < \alpha < 1$ then follows directly from (4.24); for $\alpha = 1$, it is the same as Lemma 4.1. □

Now, policy iteration algorithms for discounted cost can be derived easily. At the kth step, it simply chooses the policy that minimizes $\alpha P^{\mathcal{L}} g_\alpha^{\mathcal{L}_k} + f^{\mathcal{L}_k}$ as the policy for the next step.

Next, we study the performance sensitivity. let $P(\delta) = P + Q\delta$. Thus, $P(1) = P'$, $P(0) = P$. Then from (4.24), we have

$$\frac{d\eta_\alpha}{d\delta} = \alpha(1-\alpha)(I - \alpha P)^{-1} Q g_\alpha, \qquad 0 < \alpha < 1. \tag{4.25}$$

From (4.24) and (4.25), it is clear that policy iteration for discounted problems also chooses the next policy along the steepest direction. We have

$$\frac{d\eta_1}{d\delta} = e\frac{d\eta}{d\delta} = \lim_{\alpha \to 1} \frac{d\eta_\alpha}{d\delta}.$$

4.4 Single Sample Path Based Implementations

In this section, we present more properties of performance potentials, which play an crucial role in sensitivity analysis and MDPs [5], [7], [9]. Expanding $(I - P + e\pi)^{-1}$, we get

$$(I - P + e\pi)^{-1} = \sum_{k=0}^{\infty}(P - e\pi)^k = I + \sum_{k=1}^{\infty}(P^k - e\pi).$$

Note that the (i,j)th entry of P^k is

$$P^k(i,j) \equiv \mathsf{P}(X_k = j | X_0 = i).$$

Thus, we have

$$g(i) = \lim_{N\to\infty} \{\mathsf{E}[\sum_{n=0}^{N-1} f(X_n)|X_0 = i] - N\eta\}. \tag{4.26}$$

Taking N as a large integer, we have the approximation

$$g(i) \approx \mathsf{E}[\sum_{n=0}^{N-1} f(X_n)|X_0 = i] - N\eta. \tag{4.27}$$

From both sensitivity equations, (4.9) (4.24) for continuous and (4.12) (4.25) for discrete parameters, the potential g can be replaced by $g + ce$, with c being any real number. (Note that $Qe = 0$). Thus, only the relative values of g are important. This is the same as the potential energy in physics. Thus, we can ignore the constant term $N\eta$ in (4.27) and get

$$g(i) \approx \mathsf{E}[\sum_{n=0}^{N-1} f(X_n)|X_0 = i].$$

Clearly, $g(i)$ measures the average long-run sum of the cost of the system starting from state i, which in a sense represents the "potential" contribution of state i to the performance η.

Next, we define (called *realization factors*)

$$d(i,j) = g(j) - g(i), \qquad i,j = 1,2,\cdots,M. \tag{4.28}$$

From (4.26), we have

$$d(j,i) = \lim_{N\to\infty} \{\mathsf{E}[\sum_{n=0}^{N} f(X_n)|X_0 = i] - \mathsf{E}[\sum_{n=0}^{N} f(X_n)|X_0 = j]\}. \tag{4.29}$$

To simplify (4.29), consider two independent chains, $\mathbf{X}$ and $\mathbf{X}'$, having the same state space $\mathcal{S}$ and the same transition matrix P, but different initial states $X_0 = i$, $X'_0 = j$. Define $N_{ij} = min\{n : n \geq 0, X_n = X'_n\}$; i.e., at $n = N_{ij}$ the two chains merge together for the first time. Then [9]

$$d(i,j) = \mathsf{E}\{\sum_{n=0}^{N_{ij}}[f(X'_n) - f(X_n)|X_0 = i, X'_0 = j\}. \tag{4.30}$$

Both (4.26) and (4.29) relate $g(i)$ to the sample paths of a Markov chain, but both of them require infinitely many transitions. (4.30) essentially requires two sample paths. There is another way to estimate $g(i)$ which is based on a finite sample path. On a Markov chain $\mathbf{X} = \{X_n, n \geq 0\}$ with $X_0 = i$, we define $L_i(j) = \min\{n : n \geq 0, X_n = j\}$; i.e., at $n = L_i(j)$, the Markov chain reaches state j for the first time. We have $\mathsf{E}[L_i(j)|X_0 = i] < \infty$ [11], and from [9]

$$d(j,i) = \mathsf{E}\{\sum_{n=0}^{L_i(j)-1}[f(X_n) - \eta]|X_0 = i\}. \tag{4.31}$$

(4.31) relates $d(i,j)$ to a finite portion of the sample paths of $\mathbf{X}$.

The preceding results provide a number of ways to estimate the performance potentials. First, from (4.26), we can choose a fixed N and use $g_N(i) = \mathsf{E}[\sum_{n=0}^{N-1} f(X_n)|X_0 = i] - N\eta$ as an estimate of $g(i)$. Define $\epsilon_i(x) = 1$, if $x = i$, and $\epsilon_i(x) = 0$, otherwise. Given a long sample path $(X_0, X_1, \cdots, X_K)$, we can prove

$$g_N(i) = \lim_{K\to\infty}\left\{\frac{\sum_{k=0}^{K-N+1}\{\epsilon_i(X_k)[\sum_{j=0}^{N-1} f(X_{k+j})]\}}{\sum_{k=0}^{K-N+1}\epsilon_i(X_k)} - \frac{N}{K}\sum_{k=0}^{K-1} f(X_k)\right\}, \quad \text{w.p.1.} \tag{4.32}$$

Next, we can apply (4.31) to get an estimate. Define $u_0 = 0$, and $u_{k+1} = \min\{n : n > u_k, X_n = i\}$, $k \geq 0$, where i is a fixed state; u_k, $k \geq 0$ are regenerative points. For any $j \neq i$, define $v_k(j) = \min\{n : u_{k+1} > n > u_k, X_n = j\}$ and $\chi_k(j) = 1$, if $\{u_{k+1} > n > u_k, X_n = j\} \neq \emptyset$; and $\chi_k(j) = 0$, otherwise. From (4.31), we have

$$\begin{aligned} d(i,j) &= \lim_{K\to\infty}\frac{1}{\sum_{k=0}^{K-1}\chi_k(j)}\left\{\left[\sum_{k=0}^{K-1}\{\chi_k(j)\sum_{n=v_k(j)}^{u_{k+1}-1} f(X_n)\right]\right. \\ &\quad \left. - \left[\sum_{k=0}^{K-1}\chi_k(j)[u_{k+1} - v_k(j)]\right]\eta\right\}, \quad \text{w.p.1,} \end{aligned} \tag{4.33}$$

where η can be simply estimated by

$$\eta = \lim_{N\to\infty}\frac{1}{N}\sum_{n=0}^{N-1} f(X_n), \qquad \text{w.p.1.}$$

Given a sample path of a system run under the current policy, we can estimate the potentials using (4.32) or (4.33). Then the performance derivatives in a continuous parameter space can be obtained from (4.9). The derivatives can be used together with stochastic approximation techniques to generate the optimal performance. Algorithms and proofs of convergence and other researches in this direction is reported in [17]. We can also use the potentials estimated on a sample path to implement policy iteration. When the new policy is found, we run the system under the new policy for another sample path and update the policy again until some criterion is met. We shall refer to this procedure as the *sample path-based MDP algorithm.* Stochastic approximation techniques can also be used in this procedure to provide fast algorithms. Research in this direction is reported in [12].

The above discussion applies to discounted MDPs as well. In particular, ignoring the constant term, the α potential can be written as

$$g_\alpha(i) = \mathsf{E}\{\sum_{n=0}^{\infty} \alpha^n f(X_n)|X_0 = i\}, \qquad 0 < \alpha < 1. \tag{4.34}$$

which can be estimated on a sample path.

4.5 Time Aggregation

In this section, we study the problem of reducing computational complexity for systems with certain structures [10]. Consider two complementary subsets $\mathcal{S}_1$ and $\mathcal{S}_2 = \mathcal{S} - \mathcal{S}_1$. Without loss of generality, we let $\mathcal{S}_1 = \{1, \cdots |\mathcal{S}_1|\}$ and $\mathcal{S}_2 = \{|\mathcal{S}_1| + 1, \cdots, |\mathcal{S}|\}$. Consider a sample path $\omega = (X_0, X_1, X_2, \cdots) \in \Omega$ with $X_0 \in \mathcal{S}_1$ and let P be the probability measure on Ω.

Let $t_0 = 0$ and $t_i = \min_t\{t > t_{i-1},\ X_t \in \mathcal{S}_1\}$, $i = 1, 2, \cdots$. Then, $\{X_{t_i}, i = 0, 1, 2, \cdots\}$ forms an embedded Markov chain that is also ergodic (Theorem 1 in [13] and [21], p.15). Let $\boldsymbol{Y} \triangleq \{Y_i = X_{t_i}, i \geq 0\}$, $Y_i \in \mathcal{S}_1$. Let $\tilde{P}$ and $\tilde{\pi}$ denote the transition matrix and steady-state probability (row) vector of the embedded chain, which satisfy

$$\tilde{\pi} = \tilde{\pi}\tilde{P}.$$

The sample path of the Markov chain is divided into segments, called $\mathcal{S}_1$-segments, by the embedded chain. Denote $\xi_i = (X_{t_i}, X_{t_i+1}, \cdots, X_{t_{i+1}-1})$; ξ_i is then an $\mathcal{S}_1$-segment. We can write $\omega = (\xi_0, \xi_1, \xi_2, \cdots)$, where ξ_i, $i = 1, 2, \cdots$, are random sequences defined on Ω.

Let Ξ be the (countable) set of all feasible $\mathcal{S}_1$-segments. $(\xi_0, \xi_1, \cdots)$ defines a Markov chain with state space Ξ, which is called a *segmented Markov chain.* We use a generic notation $\xi = [\xi(1), \xi(2), \cdots, \xi(n(\xi))]$ to denote an

$\mathcal{S}_1$-segment, where $n(\xi)$ is the length of ξ. The infinite segmented chain is irreducible, aperiodic, and positive recurrent, thus ergodic. Let E denote the expectation with respect to P on Ω. Define the quantities

$$h_f(\xi) = \sum_{j=1}^{n(\xi)} f(\xi(j)),$$

and

$$H_f(i) = \mathsf{E}[h_f(\xi)|\xi(1) = i], \qquad i = 1, 2, .., |\mathcal{S}_1|. \tag{4.35}$$

Let $H_f = [H_f(1), H_f(2), \cdots, H_f(|\mathcal{S}_1|)]^T$. We will use the notation h_1 and H_1 above for the case $f(i) = 1, \forall i \in \mathcal{S}$. Thus, $h_1(\xi) = n(\xi)$ and $H_1(i) = \mathsf{E}[n(\xi)|\xi(1) = i]$. From ergodicity, it is clear that $|H_f(i)| < \infty$ for any finite function f. Applying the strong law of large numbers to the segmented chain, we have (cf. Eq. (19) in [13])

$$\begin{aligned} \eta &= \lim_{T\to\infty} \eta_T = \lim_{M\to\infty} \frac{\frac{1}{M}\sum_{m=0}^{M-1} h_f(\xi_m)}{\frac{1}{M}\sum_{m=0}^{M-1} h_1(\xi_m)} \\ &= \frac{\sum_\xi \tilde{\pi}(\xi) h_f(\xi)}{\sum_\xi \tilde{\pi}(\xi) n(\xi)} = \frac{\tilde{\pi} H_f}{\tilde{\pi} H_1} = \frac{1}{\bar{n}} \tilde{\pi} H_f, \quad \text{w.p.1}, \end{aligned} \tag{4.36}$$

where $\bar{n} = \tilde{\pi} H_1$ is the mean length of the $\mathcal{S}_1$-segments.

From (4.36), if we define a new performance vector for the embedded Markov chain,

$$\tilde{f} = \frac{1}{\bar{n}} H_f, \tag{4.37}$$

then the embedded chain has the same long-run average performance $\eta = \tilde{\pi}\tilde{f}$ as the original one. The values of H_f and $\bar{n}$ depend on the $\mathcal{S}_1$-segments, and can be calculated as follows. First, we partition P and f according to $\mathcal{S}_1$ and $\mathcal{S}_2$:

$$P = \begin{bmatrix} P_{\mathcal{S}_1\mathcal{S}_1} & P_{\mathcal{S}_1\mathcal{S}_2} \\ P_{\mathcal{S}_2\mathcal{S}_1} & P_{\mathcal{S}_2\mathcal{S}_2} \end{bmatrix}, \qquad f = \begin{bmatrix} f_{\mathcal{S}_1} \\ f_{\mathcal{S}_2} \end{bmatrix}.$$

Then, we can prove ([13], [10])

$$\begin{aligned} \tilde{P} &= \begin{bmatrix} P_{\mathcal{S}_1\mathcal{S}_1} & P_{\mathcal{S}_1\mathcal{S}_2} \end{bmatrix} \begin{bmatrix} I \\ (I - P_{\mathcal{S}_2\mathcal{S}_2})^{-1} P_{\mathcal{S}_2\mathcal{S}_1} \end{bmatrix} \\ &= P_{\mathcal{S}_1\mathcal{S}_1} + P_{\mathcal{S}_1\mathcal{S}_2}(I - P_{\mathcal{S}_2\mathcal{S}_2})^{-1} P_{\mathcal{S}_2\mathcal{S}_1}, \end{aligned} \tag{4.38}$$

and

$$H_f = f_{S_1} + P_{S_1 S_2}(I - P_{S_2 S_2})^{-1} f_{S_2}. \tag{4.39}$$

In a standard policy iteration for the embedded Markov chain, at the kth iteration we first solve the Poisson equation

$$(I - \tilde{P}^{\mathcal{L}_k} + e\tilde{\pi}^{\mathcal{L}_k})\tilde{g}^{\mathcal{L}_k} = \tilde{f}^{\mathcal{L}_k},$$

for the potential vector $\tilde{g}^{\mathcal{L}_k}$, where $\mathcal{L}_k$ is the policy at the kth iteration. The policy improvement step is carried out by setting $\mathcal{L}_{k+1} = \arg_{\mathcal{L}\in\mathcal{E}}\{\min[\tilde{P}^{\mathcal{L}}\tilde{g}^{\mathcal{L}_k} + \tilde{f}^{\mathcal{L}}]\}$ as the policy for the $(k+1)$th iteration.

However, there is one major obstacle here: According to (4.37), the performance function $\tilde{f}^{\mathcal{L}}$ depends on the average length $\bar{n}^{\mathcal{L}}$. In other words, $\tilde{f}^{\mathcal{L}}(i)$ depends on actions taken in states other than i. Therefore, standard policy iteration cannot be implemented on the embedded Markov chain using $\tilde{f}^{\mathcal{L}}(i)$. The solution based on this aggregated performance requires calculating $\bar{n}^{\mathcal{L}}$ for every policy, which is not practically feasible. To overcome this obstacle, in the following, we introduce a new performance function; the policy iteration algorithm based on this performance function reaches the same optimal policy without encountering the aforementioned problem.

Recall that the embedded Markov chain is $\mathbf{Y} = \{Y_m = X_{t_m}, m \geq 0\}$, $Y_m \in \mathcal{S}_1$. An action $\alpha \in \mathcal{A}(i)$ determines transition probabilities $\tilde{p}^{\alpha}(i,j), j = 1, 2, \cdots, |\mathcal{S}_1|$. Define a performance function

$$r_\delta(i,\alpha) = H_f(i,\alpha) - \delta H_{\mathbf{1}}(i,\alpha) \tag{4.40}$$

at state i of $\mathbf{Y}$, where δ is a real parameter, and α indicates the dependency on action. Since the performance function $r_\delta(i,\alpha)$ "aggregates" the average performance of a segment on $\mathbf{X}$ onto $\mathbf{Y}$, this approach is called *time aggregation* (this approach was first used for performance sensitivity in [22]).

A policy $\mathcal{L} \in \mathcal{E}$ determines a transition probability matrix denoted as $\tilde{P}^{\mathcal{L}} = \left[\tilde{p}^{\mathcal{L}(i)}(i,j)\right]_{i,j=1}^{|\mathcal{S}_1|}$, and a performance function vector

$$r_\delta^{\mathcal{L}} = H_f^{\mathcal{L}} - \delta H_{\mathbf{1}}^{\mathcal{L}}$$

for the Markov chain Y. The steady-state probability (row) vector is $\tilde{\pi}^{\mathcal{L}}$, and along with (4.36), the steady-state performance is

$$\lambda_\delta^{\mathcal{L}} = \tilde{\pi}^{\mathcal{L}} r_\delta^{\mathcal{L}} = \tilde{\pi}^{\mathcal{L}} H_{\mathbf{1}}^{\mathcal{L}}(\eta^{\mathcal{L}} - \delta).$$

The next theorem (see [10] for a proof) shows that using the performance function $r_\delta^{\mathcal{L}}$ with $\delta = \eta^{\mathcal{L}}$, we can implement standard policy iteration on $\mathbf{Y}$ to obtain the optimal policy for the original problem.

Theorem 4.5 (i) If $\mathcal{L}'$ is a policy better than $\mathcal{L}$ for the embedded Markov chain with $\delta = \eta^{\mathcal{L}}$ in the performance function in (4.40), then $\mathcal{L}'$ is also better than $\mathcal{L}$ for the original Markov chain.

(ii) Policy $\mathcal{L}^*$ is optimal for the original Markov chain if and only if $\mathcal{L}^*$ is optimal for the embedded chain with $\delta = \eta^{\mathcal{L}^*}$ in the performance function in (4.40).

Time aggregation saves considerable computation if $|\mathcal{S}_1|$ is small. The major computation involved in the algorithm consists of three parts: (i) Calculating $(I - P_{\mathcal{S}_2,\mathcal{S}_2})^{-1}$ (i.e., the inverse of a $|\mathcal{S}_2| \times |\mathcal{S}_2|$ matrix) once; (ii) Solving $\tilde{\pi}\tilde{P}^{\mathcal{L}_k} = \tilde{\pi}$ for the embedded chain (for $|\mathcal{S}_1|$ unknowns) at each iteration; and (iii) Solving the Poisson equation to get the potential (again for $|\mathcal{S}_1|$ unknowns) at each iteration. The total computation is roughly of order $(|\mathcal{S}_2|^3 + 2L_1|\mathcal{S}_1|^3)$, where L_1 denotes the number of iterations required for convergence. On the other hand, the complexity for routine policy iteration is of order $(2L_2(|\mathcal{S}|)^3)$, where L_2 is the number of iterations required in finding an optimal policy. If $|\mathcal{S}_1|$ is small, then Algorithm 1 would require a computation of order $|\mathcal{S}_2|^3$ (inverting the matrix $(I - P_{\mathcal{S}_2,\mathcal{S}_2})$ only once), while a routine policy iteration would require $2L_2|\mathcal{S}|^3$ (solving the Poisson equation at every iteration). Single sample path-based algorithms are developed in [10]. The readers are referred to it for further discussions.

4.6 Connections to Perturbation Analysis

It is interesting to note the relation between perturbation analysis and MDPs. In fact, the concept of potentials was first developed by using the idea of realization in perturbation analysis.

Define the realization matrix using (4.28):

$$D = [D(i,j)] = eg^T - ge^T, \tag{4.41}$$

and

$$F = ef^T - fe^T.$$

We can prove that D satisfies the following Lyapunov equation [9]

$$D - PDP^T = F, \tag{4.42}$$

and

$$\frac{d\eta}{d\delta} = \pi(QD^T\pi^T + h), \tag{4.43}$$

where

$$g^T = \pi D.$$

For discounted costs with $0 < \alpha < 1$, the above becomes the *α-Lyapunov equation*

$$D_\alpha - \alpha P D_\alpha P^T = F,$$

where

$$D_\alpha = e g_\alpha^T - g_\alpha e^T,$$

and

$$\begin{aligned}\frac{d\eta_\alpha}{d\delta} &= \Pi_\alpha(\alpha Q D^T \pi^T + h), \\ &= \Pi_\alpha(\alpha Q g_\alpha + h), \qquad (4.44)\end{aligned}$$

where Π_α is a matrix defined as follows:

$$\Pi_\alpha = (1-\alpha)(I - \alpha P)^{-1}.$$

We have,

$$\Pi(i,k) = (1-\alpha)E\{\sum_{n=0}^{\infty} \alpha^n \epsilon_k(X_n) | X_0 = i\},$$

which can also be estimated on a sample path. From (4.21), it holds that

$$\lim_{\alpha \to 1-} \Pi_\alpha = e\pi.$$

Therefore, as $\alpha \to 1-$, (4.44) becomes (4.43).

With the concept of realization, (4.43) has an intuitive explanation. To see this, let us consider a simplified case where $q(l,s) = -1$, $q(l,t) = 1$, $q(i,j) = 0$ for $i \neq l$ or $j \neq t, s$, and $f = f'$. For this case, (4.43) becomes

$$\begin{aligned}\frac{d\eta}{d\delta} &= \pi(l) \sum_{j=1}^{M} \sum_{k=1}^{M} \pi(j) q(l,k) d(j,k) \\ &= \pi(l) \sum_{j=1}^{M} \pi(j)(q(l,s)d(j,s) + q(l,t)d(j,t)) \\ &= \pi(l) \sum_{j=1}^{M} \pi(j)(d(j,t) - d(j,s)) \\ &= \pi(l) d(s,t). \qquad (4.45)\end{aligned}$$

Thus, $\frac{d\eta}{d\delta}$ equals the expected value of $d(s,t)$, and (4.45) can be derived using the following intuitive argument. Observe a sample path $\{X_n; 0 \leq n < L\}$. Among the L transitions on the sample path, the Markov chain X visits state l, on the average, $L\pi(l)$ times. Because the probability that a transition from l to s changes to a transition from l to t is δ, on the sample path there are, on the average, $L\pi(l)\delta$ times when state s changes to state t. Let $F_L = \sum_{n=0}^{L-1} f(X_n)$. Each time the state changes from s to t, F_L changes, on the average and approximately, by the amount $d(s,t)$ (see (4.29)). (In the terminology of PA, we say that the perturbation (a jump from state s to state t) is realized by a factor $d(s,t)$.)

In addition, because δ can be chosen arbitrarily small and N_{ij} in (4.30) is always finite, the probability of two changes from s to t occurring within N_{ij} state transitions is of order δ^2 and is hence negligible. This implies that the effect of each change can be treated separately. By the principle of performance realization theory in perturbation analysis of queueing networks (see [4]), the total change in F_L due to the changes in P equals the product of the effect of one jump from state s to t and the probability of such a jump; that is

$$\Delta\eta = \frac{1}{L}\Delta F_L = \frac{1}{L}\{L\pi(l)\delta \times d(s,t)\} = \pi(l)\delta d(s,t).$$

Finally,

$$\frac{d\eta}{d\delta} = \lim_{\delta\to 0}\frac{\eta' - \eta}{\delta} = \pi(l)d(s,t),$$

which is the same as (4.45). Thus, the performance sensitivities of Markov processes can be explained by using the same concept of perturbation realization as for the PA of queueing systems.

4.7 Application Examples

The proposed sample path-based approach applies well to systems such as communication and manufacturing systems. A number of cases have been studied, and here we briefly introduce two of them.

Example 4.6 (CDMA Admission Control) Consider a cell in a CDMA-based wireless system. Suppose that the base station is located in the center of each cell, which processes a call admission control algorithm to determine whether a new call in the cell is accepted or not. The system provides K classes of services with different quality of service (QoS) requirements. Let $\mathbf{X} = (x_1, ..., x_K)$ be the load of the considered cell at time t, where x_i is the number of class k on-going calls in the cell. The state $(x_1, ..., x_K)$ changes when a call is terminated or a new coming call is accepted in the cell. Assume that the arrival process of class i calls follows a Poisson process with rate λ_i and have exponentially distributed service times with mean $1/\mu_i$, then $\mathbf{X}$ is a Markov process. Let the Markov chain embedded in the call arrival or termination time be $\mathbf{Y}$. When the number of on-going calls becomes large, the admission policy will block new calls, so that the QoS of current calls can be guaranteed. In CDMA, a new call will interfere with the existing calls and hence reduces their QoS; on the other hand, blocking an incoming call also induces a cost. With properly defined performance function and some extensions of the MDP model, it can be shown that the sample path-based algorithm yields the optimal admission policy. We did a simulation for $K = 2$ (representing handover calls and normal ones) and obtained a near-optimal policy (due to estimation errors) which performs better than the complete sharing policy studied in the literature (see [15]). In this approach, the system parameters λ_i and μ_i need not to be estimated.

Example 4.7 (A Manufacturing System) Consider a manufacturing system with two machines and N parts circulating between the two machines, as shown in Figure 4.1. Machine 1 (M_1) can perform three operations (1, 2, and 3); their service times are exponentially distributed with rates λ_1, λ_2, and λ_3, respectively. Some parts need to go through all three operations in the sequences of 1,2,3; others only require operations 2 and 3; the rest only need operation 3. The probabilities that a part belongs to these three types are p_1, p_2, and p_3, respectively. Machine 1 works on one part at a time. Machine 2 (M_2) has only one operation; its service time is exponential with rate λ_4.

The system can be modeled as a Markov process with states denoted as (n, i), where n is the number of parts at machine 1 and $i = 1, 2, 3$ denotes the operation that machine 1 is performing. We further assume that the manufacturing procedure requires that $100 * p^0$ percent of parts $(0 < p^0 < 1)$ will go back to machine 1 after leaving it, and $100 * p^1$ percent will enter machine 2 after finishing the services at machine 1. The remaining $100 * (1 - p^0 - p^1)$ percent can go to either machine 1 or 2, depending on the actions denoted as α. When $\alpha = 0$ (or 1) is taken, the remaining parts go to machine 2 (or machine 1). Thus, the transition probabilities associated with these two actions (see Figure 4.1) are $p^0(1) = p^0(2) = p^0(3) = 1 - p^0$, and $p^1(1) = p^1(2) = p^1(3) = p^1$. The cost (control-action unrelated) in state (n, i) is $f((n, i)) = -k_1 f_1((n, i)) - k_2 f_2((n, i))$, where $f_1((n, i)) = 1$ if $n = 0; 0$ otherwise, $f_2((n, i)) = 1$ if $n = N; 0$ otherwise, and k_1, k_2 are assigned weighting factors. (The problem is to minimize the weighted machine idle time, which is equivalent to maximizing the weighted sum of the two machines' throughputs).

In this example, the state space is $\mathcal{S} = \{(n, i)\} \cup \{0\}$, $n = 1, \cdots, N$, $i = 1, 2, 3$ (in general, $i = 1, 2, \cdots, k$, for any integer k), and the controllable states are in $\mathcal{S}_1 = \{(n, 3)\}$, $n = 1, \cdots, N$. Therefore, we can apply both the standard MDP or the time aggregation MDPs to the problem. In simulation, we choose $\lambda_1 = \lambda_2 = \lambda_3 = \lambda_4 = 1$; $N = 3$; $p_1 = 0.2, p_2 = 0.2, p_3 = 0.6$; $p^0(1) = p^0(2) = p^0(3) = 1 - p^0 = 0.8, p^1(1) = p^1(2) = p^1(3) = p^1 = 0.2$; and $k_1 = 0.9, k_2 = 0.1$.

Both the standard and the time-aggregated sample path-based policy iterations yield the same policy sequence as shown in Table 4.1. (the number at the nth place in the second column is the control action for state $[n, 3]$, $n = 1, 2, 3$.) However, with the time aggregation approach each iter-

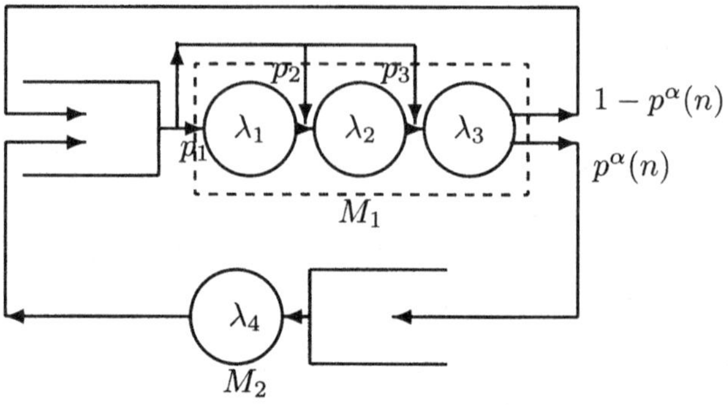

FIGURE 4.1. The Queueing Network for the Manufacturing Example

ation only requires 5000 transitions to reach the right decision; while with the standard approach each iteration requires 7000 transitions. For more details, see [10].

Iteration#	Policy	η
0	000	-0.89
1	111	-0.91
2	110	-0.93

η: -1 × weighted combined machine throughput

TABLE 4.1. Policy iteration for both the traditional and the new policy iteration algorithms

4.8 Notes

We have presented a sensitivity approach to the optimization of Markov processes. The sensitivity analysis in continuous parameter space is the perturbation analysis which provides performance gradients; the sensitivity analysis in discrete parameter (policy) spaces leads to the fundamental theory of MDP.

Performance potentials play a crucial role in both continuous and discrete sensitivities. Just as the potential energy in physics, only the relative values (i.e., the differences) of the performance potentials at different states are meaningful, i.e., the potential is determined only up to an additive constant. The difference between the potentials at two states i and j, $g(i)-g(j)$, measures the effort of a jump from state j to state i in the long-run average or discounted performance. This offers an intuitive explanation to performance sensitivity and MDPs (in view of discrete sensitivity).

A brief introduction to PA of queueing networks and a few historical remarks may further help the readers to understand the relation between perturbation analysis and MDPs, and that of the realization factors and performance potentials. Perturbation analysis is a single sample path-based sensitivity analysis for discrete event systems [16], and realization factors are the building blocks for steady-state performance sensitivities [4].

Consider an M-server closed Jackson (Gordon-Newell) network with exponentially distributed service times with means $\bar{s}_i$, $i = 1, 2, ...M$, where M is the number of servers in the network. The system state is $\mathbf{n} = (n_1, ..., n_M)$, with n_l being the number of customers at server l. Let $\mathbf{N}(t)$ denote a sample path ($\mathbf{N}(t)$ is the system state at time t). The system performance is defined as

$$\eta^{(f)} = \lim_{L\to\infty} \frac{1}{L} \int_0^{T_L} f(\mathbf{N}(t))dt = \lim_{L\to\infty} \frac{F_L}{L}, \quad F_L = \int_0^{T_L} f(\mathbf{N}(t))dt, \quad (4.46)$$

where f is a performance function, and T_L is the Lth transition time of the system.

Now, we study the average effect of a small perturbation on F_L. Suppose that at time $t = 0$ the system is in state $\mathbf{n}$ and server i obtains a small perturbation Δ, which means that the service completion time is delayed by a small amount Δ. This perturbation will be "propogated" to other servers [16]). We denote the "perturbed" sample path as $\mathbf{N}'(t)$. The realization factor of a perturbation Δ of server i in state $\mathbf{n}$ is defined as

$$c^{(f)}(\mathbf{n}, i) = \lim_{L\to\infty} \mathsf{E}\left[\frac{1}{\Delta}\left(\int_0^{T'_L} f(\mathbf{N}'(t))dt - \int_0^{T_L} f(\mathbf{N}(t))dt\right)\right]. \quad (4.47)$$

where the prime " $'$ " represents variables on the perturbed path. It is clear that the realization factor $c^{(f)}(\mathbf{n}, i)$ measures the average effect of a perturbation at $(\mathbf{n}, i)$ on F_L in (4.46) as $L \to \infty$. Moreover, it is proved that $c^{(f)}(\mathbf{n}, i)$ can be uniquely determined by the following set of linear equations [4].

1. If $n_i = 0$, then $c^{(f)}(\mathbf{n}, i) = 0$;
2. $c^{(f)}(\mathbf{n}, \Gamma) = f(\mathbf{n})$, with $\Gamma = \{1, 2, \cdots, M\}$;
3. If $V_1 \cap V_2 = \emptyset$ and $V_1 \cup V_2 = V_3$, V_1, V_2, $V_3 \subset \Gamma$, then
$$c^{(f)}(\mathbf{n}, V_1) + c^{(f)}(\mathbf{n}, V_2) = c^{(f)}(\mathbf{n}, V_3);$$
4. $\sum_{i=1}^M c^{(f)}(\mathbf{n}, i) = f(\mathbf{n})$; and,
5. Let $\mathbf{n}_{i,j} = (n_1, \cdots, n_i - 1, \cdots, n_j + 1, \cdots, n_M)$ be a neighboring state of $\mathbf{n}$, then
$$\{\sum_{i=1}^M \epsilon(n_i)\mu_i\} c^{(f)}(\mathbf{n}, k) = \sum_{i=1}^M \sum_{j=1}^M \epsilon(n_i)\mu_i q_{i,j} c^{(f)}(\mathbf{n}_{i,j}, k)$$

$$+\sum_{j=1}^{M}\mu_k q_{k,j}\{(1-\epsilon(n_j))c^{(f)}(\mathbf{n}_{k,j},j)+f(\mathbf{n})-f(\mathbf{n}_{k,j})\}$$

$$n_k>0,\quad k\in\Gamma,$$

where $q_{i,j}$ is the routing probability from server i to j in the closed network.

Next, suppose that the mean service time of a server, say server i, changes from $\bar{s}_i$ to $\bar{s}_i+\Delta\bar{s}_i$. Every customer's service time, denoted as s, will change to $s\frac{\bar{s}_i+\Delta\bar{s}_i}{\bar{s}_i}=s+s\frac{\Delta\bar{s}_i}{\bar{s}_i}$. Let $p(\mathbf{n})$ be the steady-state probability of state $\mathbf{n}$. Consider a time period $[0,T_L]$ with $L>>1$. The length of the total time that the system is in state $\mathbf{n}$ in $[0,T_L]$ is $T_Lp(\mathbf{n})$. The total perturbation induced in this period due to the change $\Delta\bar{s}_i$ in the mean service time is $T_Lp(\mathbf{n})\frac{\Delta\bar{s}_i}{\bar{s}_i}$. Since each perturbation on the average has an effect of $c^{(f)}(\mathbf{n},i)$ on F_L, the overall effect on F_L of the total perturbation induced when the system state is $\mathbf{n}$ is $(T_Lp(\mathbf{n})\frac{\Delta\bar{s}_i}{\bar{s}_i})c^{(f)}(\mathbf{n},i)$. Finally, the total effect of the mean service time change, $\Delta\bar{s}_i$, on F_L is

$$\Delta F_L=\sum_{all\ \mathbf{n}}T_Lp(\mathbf{n})\frac{\Delta\bar{s}_i}{\bar{s}_i}c^{(f)}(\mathbf{n},i).$$

From this, we have

$$\frac{\bar{s}_i}{T_L/L}\frac{\Delta F_L/L}{\Delta\bar{s}_i}=\sum_{all\ \mathbf{n}}p(\mathbf{n})c^{(f)}(\mathbf{n},i).$$

Letting $L\to\infty$ and $\Delta\bar{s}_i\to 0$, we obtain the steady-state performance sensitivity as follows.

$$\frac{\bar{s}_i}{\eta^{(I)}}\frac{\partial\eta^{(f)}}{\partial\bar{s}_i}=\sum_{all\ \mathbf{n}}p(\mathbf{n})c^{(f)}(\mathbf{n},i),\tag{4.48}$$

where

$$\eta^{(I)}=\lim_{L\to\infty}\frac{T_L}{L}=\lim_{L\to\infty}\frac{1}{L}\int_0^{T_L}I(\mathbf{n})dt,$$

in which $I(\mathbf{n})=1$ for all $\mathbf{n}$. Thus, the *normalized* performance sensitivity (the left-hand side of (4.48)) equals the steady-state expectation of the realization factor. A rigorous proof involves exchanging the order of $\lim_{L\to\infty}$ and $\Delta\bar{s}_i\to 0$, see [4].

The above idea was first extended to Markov processes by defining $d(i,j)$ as in (4.30) which represents the meaning of a realization of a discrete perturbation (i.e., the effort of a jump from state i to j on the performance). Then the sensitivity equation (4.43) was first developed by the intuition presented in Section 4.6 (which is similar to the intuition presented above for queueing networks) and was proved rigorously later. Furthermore, performance potential g was defined as (4.41) because of the skew-symmetry of D. The term "potential" came from the fact that g is determined only upto an additive constant. The Lyapunov equation can be derived from the definition of D directly. Substituting (4.41) into the Lyapunov equation leads to the Poisson equation for g, and then the sensitivity equation (4.8) can be derived easily. Finally, the discrete sensitivity (4.11) (the base for MDPs) follows naturally from the Poisson equation. The above procedure shows that MDP can be developed from a new perspective, i.e., from a sensitivity point of view, which may lead to some new insights as well as new research topics.

The performance potential in state i can be approximated by the mean sum of the performance at the first N states on a sample path starting from state i, and hence it can be estimated online. A number of other single sample path-based estimation algorithms for potentials have been derived. Therefore, performance derivatives can be obtained and policy iteration can be implemented based on a single sample path. The single sample path-based approach is practically important because it can be applied online to real engineering systems.

The advantages of the potential-based on-line policy iteration approach have been discussed in a few previous papers (e.g., [7]). First, we shall emphasize that with this approach, most system parameters do not have to be estimated and matrix inversion can be avoided. Second, this approach offers some flexibility and may save computation. For example, if we know some states are more important (e.g., more frequently visited) than others, we can update the actions only at those important states and get a close-to-optimal policy. Third, to implement policy iteration the system does not need to try all actions in each state.

A number of new research topics are motivated by this sensitivity and potential-based approach. One of them is the time aggregation discussed in section 4.5, which may save computation or the number of transitions in real time implementation for systems where only a subset of the states are controllable. Other topics include the MDPs with hidden state components, distributed MDPs, and the extensions to other processes such as semi-Markov processes and general semi-Markov processes. $TD(\lambda)$, neuro-dynamic programming, and Q-learning [3] are similar approaches in estimating the potential and its variant Q-factor. Similar algorithms may be developed for estimating the performance derivatives in (4.8) and (4.9).

Finally, we point out that (4.34) is the same as the α potential defined in the standard potential theory [11]. When $\alpha = 1$, the above expression goes to infinity. Thus, (4.26) can be viewed as the extension of the α-potential to the case of $\alpha = 1$.

Acknowledgment

The author is supported in part by a grant from Hong Kong RGC.

4.9 References

[1] Berman, A. and Plemmons, R. J., *Nonnegative Matrices in the Mathematical Sciences, SIAM*, Philadelphia, 1994.

[2] Bertsekas, D. P., *Dynamic Programming and Optimal Control,* Vols. I, II, Athena Scientific, Belmont, Massachusetts, 1995.

[3] Bertsekas, D. P., and Tsitsiklis, J. N., *Neuro-Dynamic Programming,* Athena Scientific, Belmont, Massachusetts, 1996.

[4] Cao, X. R., *Realization Probabilities: The Dynamics of Queueing Systems,* Springer-Verlag, New York, 1994.

[5] Cao, X. R., The relation among potentials, perturbation analysis, Markov decision processes, and other topics, *Journal of Discrete Event Dynamic Systems,* **8** (1998), 71-87.

[6] Cao, X. R., The Maclaurin series for performance functions of Markov chains, *Advances in Applied Probability,* **30** (1998) ,676-692.

[7] Cao, X. R., Single sample path-based optimization of Markov chains, *Journal of Optimization Theory and Application,* **100** (1999), 527-548.

[8] Cao, X. R., A unified approach to Markov decision problems and performance sensitivity analysis, *Automatica,* **36** (2000), 771-774.

[9] Cao, X. R. and Chen, H. F., Potentials, perturbation realization, and sensitivity analysis of Markov processes, *IEEE Transactions on Automatic Control,* **42** (1997), 1382-1393.

[10] Cao, X. R., Ren, Z. Y., Bhatnagar, S., Fu, M., and Marcus, S., A time aggregation approach to Markov decision processes, *Automatica,* to appear, (2001).

[11] Çinlar, E., *Introduction to Stochastic Processes,* Prentice Hall, Englewood cliffs, NJ, 1975.

[12] Fang, H. T. and Cao, X. R., Potential-based on-line policy iteration algorithms for Markov decision processes, *IEEE Trans. on Automatic Control,* submitted.

[13] Forestier, J. P. and Varaiya, P., Multilayer control of large Markov chains, *IEEE Transactions on Automatic Control,* **23** (1978), 298-305.

[14] Glynn P. W. and Meyn, S. P., A Lyapunov bound for solutions of Poisson's equation, *Ann. Probab.,* **24** (1996), 916-931.

[15] HE, M., *Performance Optimization in Wireless Communication Networks - Online Algorithms*, MPhil. Thesis, Dept. of EEE, Hong Kong University of Science and Technology, 2001.

[16] HO, Y. C. AND CAO, X. R., *Perturbation Analysis of Discrete-Event Dynamic Systems*, Kluwer Academic Publisher, Boston, 1991.

[17] MARBACH, P. AND TSITSIKLIS, J. N., Simulation-based optimization of Markov reward processes, *IEEE Transactions on Automatic Control*, **46** (2001), 191-209.

[18] KEMENY, J. G. AND SNELL, J. L., *Finite Markov Chains*, Van Nostrand, New York, 1960.

[19] MEYN S. P. AND TWEEDIE, R. L., *Markov Chains and Stochastic Stability*, Springer-Verlag, London, 1993.

[20] PUTERMAN, M. L., *Markov Decision Processes: Discrete Stochastic Dynamic Programming*, Wiley, New York, 1994.

[21] REVUZ, D., *Markov Chains*, North-Holland, New York, 1984.

[22] ZHANG B. AND HO, Y. C., Performance gradient estimation for very large finite Markov chains, *IEEE Transactions on Automatic Control*, **36** (1991), 1218-1227.

5

An Interior-Point Approach to Multi-Stage Stochastic Programming

Shuzhong Zhang

Let us start our discussion with a famous textbook example: the newsboy problem. The story goes like this. Every morning a newsboy has to decide how many newspapers to buy from a newspaper publisher. Let us assume that the publisher sells the newspaper to the boy at the price of \$2 each paper, and the boy then sells the newspaper along the street at the price of \$5 per copy. In the end of the day, the newsboy may return any unsold copies to the publisher at \$1 for each copy. The profit of the newsboy in this business depends, obviously, on the success of the sales and his initial decision on the order quantity. Unfortunately, the problem is that, as it is always the case, one cannot really predict the future with certainty. To make our analysis simple, let us further assume that there are only two possible scenarios: (1) the newsboy can sell 100 copies a day, or, (2) in the case of a boring day, he can only sell 50 copies. Furthermore, let us assume that the chance for a day with some exciting news is lower. Hence, (1) occurs with probability 0.25, and (2) occurs with probability 0.75. In fact, for a person with mathematical background, it is more convenient to introduce a random variable, ω, to denote the demand for this purpose,

$$\omega = \begin{cases} 100, & \text{with probability } 0.25 \\ 50, & \text{with probability } 0.75. \end{cases}$$

After some struggling, one can figure out that we need to solve the following optimization problem

$$
\begin{array}{llll}
(NB) & \text{minimize} & 2x & + \; \mathsf{E}\,[-5y_\omega - z_\omega] \\
& \text{subject to} & x \geq 0, & \\
& & & y_\omega \leq \omega, \\
& & & y_\omega + z_\omega = x, \\
& & & y_\omega \geq 0,\ z_\omega \geq 0.
\end{array}
$$

Let us explain what is going on here. In the above model, x stands for the quantity of newspapers to order in the morning, and y_ω is the amount of sold newspapers, which is bounded by the demand ω. Finally, z_ω is the amount of newspapers to be returned to the publisher at the end of the day. A closer look suggests that the objective is actually to maximize the expected profit, which is the revenue minus the costs.

The model looks like a pretty ordinary optimization problem, except that there is a random variable ω in the constraint. The model is known as *two-stage stochastic linear programming*, as it involves decisions at two stages: (1) the decision at the beginning of the day, namely x; (2) the decision to be made during the day, namely y_ω and z_ω, given that x is already decided. Moreover, all the relationships happen to be linear. Clearly, one may consider extended models where more than two stages of decisions have to be made sequentially, depending on the newly arrived information on the uncertain factors. That more general case is naturally termed *multi-stage stochastic linear programming.*

Since ω has only two possibly outcomes in our case, we may write (NB) equivalently as a usual deterministic linear programming problem as follows

$$
\begin{array}{llll}
(DNB) & \text{minimize} & 2x & + \; 0.25 \times [-5y_1 - z_1] + 0.75 \times [-5y_2 - z_2] \\
& \text{subject to} & x \geq 0, & \\
& & & y_1 \leq 100, \\
& & & y_1 + z_1 = x, \\
& & & y_1 \geq 0,\ z_1 \geq 0, \\
& & & y_2 \leq 50, \\
& & & y_2 + z_2 = x, \\
& & & y_2 \geq 0,\ z_2 \geq 0.
\end{array}
$$

If one uses a linear programming solver, then the solution can be found to be

$$[x^*, y_1^*, z_1^*, y_2^*, z_2^*] = [64.2569, 64.2569, 0, 50, 14.2569].$$

The interpretation is: one should buy 64.2569 copies of the newspaper in the morning from the publisher, and then sell them all if it turns out to be a good day, and sell 50 copies and return 14.2569 copies to the publisher if it is a bad day.

The second stage problem is known as *recourse*. In the above example, it is to decide on y_ω and z_ω, provided that x was already done and the state ω is just observed. Its analog in control theory is called the feedback control: observing the true state of the world and then decide the best action to take. In fact, in the above newsboy's situation, the recourse problem is extremely simple. It can even be explicitly solved and written out as: $y_\omega = \max\{\omega, x\}$ and $z_\omega = \max\{x - \omega, 0\}$. In most other applications, of course, this is not possible.

So, the good news for the newsboy is that the decision problem can be solved by some ordinary LP solver. But why stop? There are so many real decision problems where the future uncertainty will have a great impact on the quality of the decision we make now. So the idea of the above approach should be extended. This is exactly what happened. But there is an intrinsic difficulty with the extension, known as the dimensionality curse. That is to say, as we introduce reasonably fine (but finite) approximations of the underlying random variables (or stochastic process), and try to go beyond two stages, then the total amount of scenarios will explode very quickly, in fact, exponentially. As a result, even though we end up with a deterministic optimization problem to solve, the size of the problem, for most real applications, will be far too large to be plugged into an existing optimization solver. We know that the situation will not improve if we just wait for more powerful computers to appear, as the problem is structural. In stochastic programming, people therefore look for better solution methods, exploiting the very nature of multiple-stage stochastic programming.

In this chapter we shall introduce a particular method of this type, using the so called homogeneous self-dual embedding technique and the central path following method. The results presented in this chapter are based on the author's earlier papers, [5] and [6], and are made simpler for the expository purpose. In [6], a convex objective function is allowed.

5.1 Two-Stage Stochastic Linear Programming

Before we talk about the solution methods, let us first consider the models. We start with two-stage stochastic linear programming. Two stage stochastic linear programming is cast as

$$
\begin{array}{llll}
(2SLP) & \text{minimize} & c_0^T x_0 & + \ \mathsf{E}\left[c_\omega^T x_\omega\right] \\
& \text{subject to} & W_0 x_0 = h_0, & \\
& & x_0 \geq 0, & \\
& & & B_\omega x_0 + W_\omega x_\omega = h_\omega, \\
& & & x_\omega \geq 0,
\end{array}
$$

where x_0 is the first stage decision variable, and ω is the uncertain factor (a random variable), and x_ω is the corresponding recourse action.

Clearly, the newsboy problem is just a special case of this model.

Now let us further assume that ω takes discrete values, say, there are in total N different scenarios. Then, we can explicitly write down the deterministic equivalent of $(2SLP)$ just like in the newsboy's case

$$
\begin{array}{llll}
(2DSLP) & \text{minimize} & c_0^T x_0 & + \ \sum_{n=1}^N \pi_n \left[c_n^T x_n\right] \\
& \text{subject to} & W_0 x_0 = h_0, & \\
& & x_0 \geq 0, & \\
& & & B_n x_0 + W_n x_n = h_n, \\
& & & x_n \geq 0, \ n = 1, \ldots, N,
\end{array}
$$

where π_n is the probability of the event $\omega = n$, $n = 1, ..., N$.

If one examines closely the constraint matrix in $(2DSLP)$, then one can observe an L-shaped block structure. That is to say, most part of the matrix contains only zero elements, and the possible non-zero blocks form a fallen L letter viewed from a far distance. Based on this observation, a famous simplex-type method was introduced by Van Slyke and Wets [23]. For a detailed account of the method, one is referred to, e.g., the recent textbooks on stochastic programming, [9] and [17]. In this chapter however, we will concentrate on a quite different approach, based on the interior point method. No matter what, the key idea here is always to make use of the structure so that a decomposition of some sort can be applied. In fact there are, broadly speaking, two types of decomposition approaches: (1) the scenario-based decomposition, and (2) the recourse-based decomposition. The L-shaped method belongs to the second category. That is to say, we employ an existing method, say the simplex method, to solve $(2DSLP)$, and in the process of implementing the method we make use of the special structure of $(2DSLP)$, and decompose the computation of the data as much as possible. Another good example of this type is Birge and Qi's decomposition algorithm, [10], based on Karmarkar's original method for linear programming. Our method to be introduced is in this category as well.

A typical scenario-based method is based on the so-called Lagrangian multiplier approach, or, the augmented Lagrangian multiplier approach; see [21]. The key observation is that, $(2DSLP)$ can be rewritten as

$$
\begin{array}{lll}
(2DSLP)' & \text{minimize} & \sum_{n=1}^{N} \pi_n \left[c_0^T x_{0n} + c_n^T x_n\right] \\
& \text{subject to} & W_0 x_{0n} = h_0, \\
& & x_{0n} \geq 0, \\
& & B_n x_{0n} + W_n x_n = h_n, \\
& & x_n \geq 0, \; n = 1, ..., N, \\
& & x_{0n} - \sum_{n=1}^{N} \pi_n x_{0n} = 0, \; n = 1, ..., N.
\end{array}
$$

The last constraint in $(2DSLP)'$ is called *non-anticipativity*, i.e., one is not allowed, however desirable it may be, to use the second stage information in the first stage decision. This reformulation suggests that, if the optimal Lagrangian multipliers for the last set of constraints would be known, say, to be equal to y_n, $n = 1, ..., N$, then the problem could be solved by simply putting them in the objective, i.e.,

$$
\begin{array}{ll}
\text{minimize} & \sum_{n=1}^{N} \left\{ \pi_n \left[c_0^T x_{0n} + c_n^T x_n\right] + y_n \left[x_{0n} - \sum_{n=1}^{N} \pi_n x_{0n}\right] \right\} \\
\text{subject to} & W_0 x_{0n} = h_0, \\
& x_{0n} \geq 0, \\
& B_n x_{0n} + W_n x_n = h_n, \\
& x_n \geq 0, \; n = 1, ..., N.
\end{array}
$$

The above problem is completely separable in terms of scenarios and hence can be solved by quickly solving N small size linear programs. This approach also works when one introduces an additional pure quadratic penalty term of the constraints $x_{0n} - \sum_{n=1}^{N} \pi_n x_{0n} = 0$, $n = 1, ..., N$, in the objective, yielding the so-called augmented Lagrangian approach. The remaining task is to come up with a good scheme to update the Lagrangian multipliers y_n in order to accelerate convergence, for which existing theory for the Lagrangian method can be used. We note that the scenario-based approach extends easily to the multiple stage framework, even when the objective is nonlinear. One drawback, however, is the slow convergence. In the bid to a faster convergence speed, we turn to the state-of-the-art interior point method, where the number of iterations is known to be insensitive to the problem dimension. Certainly this will be an extremely important property in the real application.

Before we introduce the new method, let us first consider an application of the two-stage stochastic programming model to motivate the solution method.

5.2 A Case Study

We consider the following two-period problem. An investor can invest in a bank account, a stock index, and European (exchange listed) options on this index with different maturities. We denote the stock index by S. Current time is denoted by t_0, and the expiration dates of the options by t_1 and t_2 with $t_0 < t_1 < t_2$. At t_0 the investor forms a portfolio consisting of some amount of money invested in the bank account, some in the stock index, and a set of options on the stock index. At time t_1 he/she may revise his/her portfolio, depending on the value of the index at t_1, i.e. he/she can change some of the existing positions in the options and/or buy new options starting from t_1 and maturing at t_2. The investor's goal is to guarantee that the value of the portfolio is always above a given level depending on the index at t_2, and that the expected value of the portfolio is maximized at the horizon of the investment.

Assume that the level of the stock index is S_0 at time t_0, S_1 at time t_1, and S_2 at time t_2. Moreover, there are n European puts and calls struck at K_l^j with $l = 1, 2, ..., n$, respectively, where $j = 1, 2$ denotes the expiration of the options t_j. Let $Q^p_{t_i t_j}(S) \in \Re^n$ denote the n-dimensional vector whose l-th component represents the price of buying a put option at time t_i maturing at t_j with strike price K_l, while the stock index at t_i is S. Similarly, denote $Q^c_{t_i t_j}(S) \in \Re^n$ to be the n-dimensional vector which l-th component represent the price of buying a call option at time t_i maturing at t_j with strike price K_l while the stock index at t_i is S. The risk-free interest rate from t_0 to t_1 is denoted by r_1, the risk-free interest rate from t_0 to t_2 is denoted by r_2, and the forward rate from t_1 to t_2 is denoted by f_2. Now, let $x^p_{t_i t_j} \in \Re^n$ denote the amount of put options purchased at time t_i maturing at t_j, and $x^c_{t_i t_j} \in \Re^n$ be the amount of call options purchased at time t_i maturing at t_j. Let x_0^s be the amount invested in the stock index, and x_0^f be the amount invested at t_0 in the money-market account. Similarly, let x_1^s be the amount invested in the stock index and x_1^f be the amount invested in the money-market account at t_1. The decision variables $x^p_{t_0 t_j}$ and $x^c_{t_0 t_j}$ with $j = 1, 2$, and x_0^s and x_0^f denote the first-stage variables. The decision variables $x^p_{t_1 t_2}$ and $x^c_{t_1 t_2}$, and x_1^s and x_1^f denote the second-stage variables. Suppose that the initial budget for the investment is B.

Clearly, the following initial budget equation should hold:

$$B = x_0^s S_0 + x_0^f + \sum_{j=1}^{2} \langle x^p_{t_0 t_j}, Q^p_{t_0 t_j}(S_0) \rangle + \sum_{j=1}^{2} \langle x^c_{t_0 t_j}, Q^c_{t_0 t_j}(S_0) \rangle, \qquad (5.1)$$

where for notational simplicity in this context we used $\langle x, y\rangle$ as the inner product between x and y, i.e., $\langle x, y\rangle = x^T y$.

At t_1 the value of the portfolio is given by:

$$\begin{aligned} V(t_1, S_1; x^s, x^f, x^p, x^c) &= x_0^s S_1 + x_0^f \exp(r_1(t_1 - t_0)) \\ &+\langle(\mathcal{K}_1 - S_1 e)^+, x^p_{t_0 t_1}\rangle + \langle(S_1 e - \mathcal{K}_1)^+, x^c_{t_0 t_1}\rangle \\ &+\langle Q^p_{t_1 t_2}(S_1), x^p_{t_0 t_2}\rangle + \langle Q^c_{t_1 t_2}(S_1), x^c_{t_0 t_2}\rangle, \end{aligned} \tag{5.2}$$

where $\mathcal{K}_1 = (K_1^1, ..., K_n^1)^T$, and for given $y \in \Re^n$, y^+ denotes the vector

$$(\max\{y_1, 0\}, ..., \max\{y_n, 0\})^T.$$

The second-stage recourse problem is as follows. First, there is an intermediate budget constraint:

$$\begin{aligned} & V(t_1, S_1; x^s, x^f, x^p, x^c) \\ = \;& x_1^s S_1 + x_1^f + \langle Q^p_{t_1 t_2}(S_1), x^p_{t_1 t_2}\rangle + \langle Q^c_{t_1 t_2}(S_1), x^c_{t_1 t_2}\rangle. \end{aligned} \tag{5.3}$$

Second, the value of the portfolio at the horizon is given by:

$$\begin{aligned} & V(t_2, S_2; x^s, x^f, x^p, x^c) \\ = \;& x_1^s S_2 + x_1^f \exp(f_2(t_2 - t_1)) \\ &+\langle(\mathcal{K}_2 - S_2 e)^+, x^p_{t_1 t_2}\rangle + \langle(S_2 e - \mathcal{K}_2)^+, x^c_{t_1 t_2}\rangle. \end{aligned} \tag{5.4}$$

We require the value of the portfolio at the horizon never to be less than $c_0 S_2 + c_1$ with $c_0 \geq 0$ and $c_1 > 0$. Using the piecewise linearity of $V(t_2, S_2; x^p, x^c)$, this yields:

$$V(t_2, K_i^2; x^s, x^f, x^p, x^c) \geq c_0 K_i^2 + c_1 \text{ for } i = 1, ..., n, \tag{5.5}$$

$$V(t_2, 0; x^s, x^f, x^p, x^c) \geq c_1,$$

and

$$V'_{S_2}(t_2, S_2; x^s, x^f, x^p, x^c) \mid_{S_2 = K_n^2 +} \geq c_0.$$

These constraints are all linear in terms of x^s, x^f, x^p and x^c.

Finally, we require the probability that the portfolio value will be above a given threshold value $c_2 > 0$ to be at least λ $(0 < \lambda < 1)$. This, again by piecewise linearity, can be modelled by selecting a given I $(1 \leq I \leq n)$, and adding the following constraints:

$$V(t_2, K_i; x^s, x^f, x^p, x^c) \geq c_2, \quad \text{for } i = I, I+1, ..., n. \tag{5.6}$$

Similar constraints can be added to the model at t_1.

The expected value of the portfolio at t_2 is given by:

$$\begin{aligned} & \mathsf{E}\left[V(t_2, S_2; x^s, x^f, x^p, x^c)\right] \\ = \;& x_1^s \mathsf{E}[S_2] + x_1^f \exp(r(t_2 - t_1)) \\ &+\langle \mathsf{E}\left[(\mathcal{K}_2 - S_2 e)^+\right], x^p_{t_1 t_2}\rangle + \langle \mathsf{E}\left[(S_2 e - \mathcal{K}_2)^+\right], x^c_{t_1 t_2}\rangle. \end{aligned} \tag{5.7}$$

The optioned portfolio selection problem is now well defined as a two-stage stochastic linear program:

$$\begin{aligned} \max \quad & w_1 \mathsf{E}\left[V(t_1, S_1; x^s, x^f, x^p, x^c)\right] + w_2 \mathsf{E}\left[V(t_2, S_2; x^s, x^f, x^p, x^c)\right], \\ \text{s.t.} \quad & (5.1), (5.3), (5.5) \text{ and } (5.6), \end{aligned}$$

where w_1 and w_2 ($w_2 > w_1$) are weights for the first and second stage expected values.

Problems of this nature exist widely in the investment world. The above model is discussed in Berkelaar, Dert, Oldenkamp and Zhang [5]; two of the authors work for the ABN-AMRO Bank and manage financial products similar to the above described model.

5.3 Multiple Stage Stochastic Programming

There is of course no need to restrict ourselves to the two stage models only. It is very natural to extend them to *multiple stage stochastic programming*. For simplicity, we consider here the problem with a linear objective.

Let us assume that there are K decision stages. At stage 0, which is the current time, no information about the future is known for certain. However, we can model the future development using a *scenario tree*; see Figure 5.1. Each node in this tree is layered according to the stage, where it may occur. At stage t, we let the set of nodes be $\mathcal{F}_t$. Technically, $\mathcal{F}_t$ is also known as filtration.

The multi-stage stochastic programming model can thus be described as

$$\begin{array}{llll} (MSLP) & & & \\ \min & c_0^T x_0 & + \sum_{t=1}^{K} \mathsf{E}\left[c_t(\omega)^T x_t(\omega)\right] & \\ \text{s.t.} & W_0 x_0 & & = h_0, \\ & x_0 \geq 0, & & \\ & B_t(\omega) x_{t-1}(\omega) & +W_t(\omega) x_t(\omega) & = h_t(\omega), \\ & x_t(\omega) \geq 0, & \omega \in \mathcal{F}_t,\ t = 1, \dots, K. & \end{array}$$

More specifically, we used finite scenarios in the tree, and so each node has a predecessor, except for node 0, which is the current stage, and each node has possibly several successors, except for the last layer at final stage K. Variables defined on ω are denoted by subscripts n and t, where $n \in \mathcal{F}_t$.

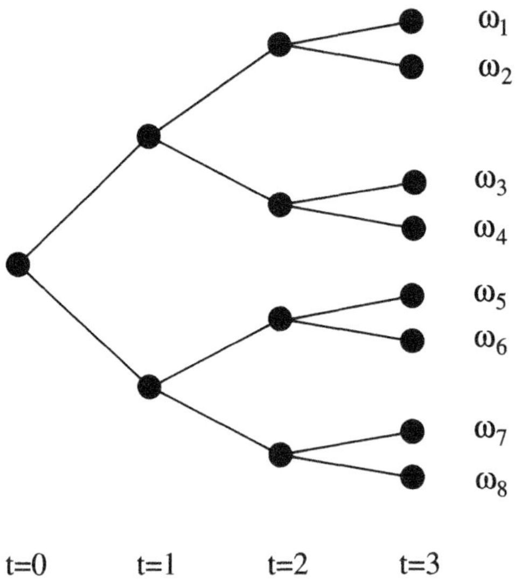

FIGURE 5.1. Scenario Tree

The deterministic equivalent problem can now be cast as

$$
\begin{array}{llll}
(MDSLP) & & & \\
\min & c_0^T x_0 & & + \sum_{t=1}^{K} \sum_{n \in \mathcal{F}_t} \pi_{nt} c_{nt}^T x_{nt} \\
\text{s.t.} & W_0 x_0 & & = h_0, \\
& & B_{nt} x_{a(n),t-1} & + W_{nt} x_{nt} = h_{nt}, \\
& x_0 \geq 0, & x_{nt} \geq 0, & n \in \mathcal{F}_t,\ t = 1, \ldots, K,
\end{array}
$$

where subscript (n, t) stands for scenario n at stage t, π_{nt} is its probability, $(a(n), t-1)$ is the ancestor of (n, t), and $(C(n), t+1)$ will be denoted the set of children of (n, t). As a convention we let $n = 0$ denote the current state. Therefore, the decision variables are x_0 (the immediate action) and x_{nt} (the recourse action to be taken if scenario n unfolds at stage t).

In our notation, each node on the scenario tree (at different stages) is associated with a unique n. In this sense, the association t in the notation (n, t) is redundant. Nevertheless, we shall keep it in places where confusion is possible.

In general, the above formulation $(MDSLP)$ can be a large size linear program. The number of variables is of the order $O(C^K)$ where C represents the number of children following each scenario at an intermediate stage. For practical purposes we may assume that each of the matrices B_{nt}, and W_{nt} are reasonably sized. For technical reasons we also assume that all the matrices W_{nt} have full row ranks.

This problem, $(MDSLP)$, has the following dual problem

$$\begin{aligned}
\max \quad & h_0^T y_0 + \sum_{t=1}^{K} \sum_{n \in \mathcal{F}_t} h_{nt}^T y_{nt} \\
\text{s.t.} \quad & W_0^T y_0 + \sum_{m \in C(0)} B_{m1}^T y_{m1} + s_0 = c_0, \\
& W_{nt}^T y_{nt} + \sum_{m \in C(n)} B_{m,t+1}^T y_{m,t+1} + s_{nt} = \pi_{nt} c_{nt},\ t = 1, ..., K-1,\ n \in \mathcal{F}_t, \\
& W_{nK}^T y_{nK} + s_{nK} = \pi_{nK} c_{nK},\ n \in \mathcal{F}_K, \\
& x_0 \geq 0,\ x_{nt} \geq 0,\ s_0 \geq 0,\ s_{nt} \geq 0,\ n \in \mathcal{F}_t,\ t = 1, ..., K.
\end{aligned}$$

5.4 An Interior Point Method

Before we offer our decomposition solution method for solving $(MDSLP)$, let us first introduce a particular type of interior point method, known as the *homogeneous self-dual and path-following* algorithm. To start with, consider a standard linear program

$$\begin{aligned}
(P) \quad \text{minimize} \quad & c^T x \\
\text{subject to} \quad & Ax = b, \\
& x \geq 0,
\end{aligned}$$

where $A \in \Re^{m \times n}$, $b \in \Re^m$ and $c \in \Re^n$. Moreover, without loss of generality, let us assume that the rank of A is m. The dual of (P) is

$$\begin{aligned}
(D) \quad \text{maximize} \quad & b^T y \\
\text{subject to} \quad & A^T y + s = c, \\
& s \geq 0.
\end{aligned}$$

The famous analytic central path is defined to be

$$\begin{aligned}
\{(x(\mu), y(\mu), s(\mu)) \mid\ & Ax(\mu) = b,\ x(\mu) > 0, \\
& A^T y(\mu) + s(\mu) = c,\ s(\mu) > 0,\ x_i(\mu) s_i(\mu) = \mu,\ i = 1, 2, ..., n\}.
\end{aligned}$$

It is well known that if (P) and (D) both have interior solutions, i.e.

$$\{x \mid Ax = b,\ x > 0\} \neq \emptyset \text{ and } \{(y, s) \mid A^T y + s = c,\ s > 0\} \neq \emptyset, \tag{5.8}$$

then the analytic central path exists, and vice versa.

The limit, $\lim_{\mu\downarrow 0}(x(\mu), y(\mu), s(\mu)) =: (x(0), y(0), s(0))$, exists, and it is the analytic center of the optimal set. Hence it is optimal. For detailed proofs of those nowadays well known facts, one is referred to the book of Roos, Terlaky and Vial [22].

Most interior point methods, and certainly by definition the so-called path-following methods, are based on tracing the analytic central path till close to $(x(0), y(0), s(0))$.

All is well, except for the fact that one may not know whether the interior point condition (5.8) is satisfied, and if yes, how to get hold of one initial interior point solution. This difficulty was resolved beautifully by Ye, Todd and Mizuno in [27], and later was simplified by Xu, Hung and Ye [24]. A detailed description of an efficient numerical implementation of the method was documented in [1]. The technique is called *homogeneous self-dual embedding.* In fact, the iterates produced by the algorithm in [27] will be identical to that produced by the one in [24], provided that they have identical parameters and initial points. Hence, in a sense, these two algorithms are simply identical for linear programming, and below we shall choose to present the method based on [24].

The idea of dealing with homogeneous self-dual systems can be traced back to Goldman and Tucker [14]. In [14] the following system is considered:

$$\begin{array}{rrrcl} & Ax & -b\tau & = & 0, \\ -A^T y & & +c\tau & \geq & 0, \\ b^T y & -c^T x & & \geq & 0, \\ & & x \geq 0, \quad \tau \geq 0. & & \end{array}$$

Clearly, this system is homogeneous and has a skew-symmetric constraint matrix leading to the notion of self-duality. For convenience, additional variables are introduced to replace the inequality constraints, yielding

$$(H)\left\{\begin{array}{rrrrrcl} & Ax & & -b\tau & & = & 0 \\ -A^T y & & -s & +c\tau & & = & 0 \\ b^T y & -c^T x & & & -\kappa & = & 0 \\ & x \geq 0, & s \geq 0, & \tau \geq 0, & \kappa \geq 0. & & \end{array}\right.$$

If system (H) has a solution $(y^*, x^*, s^*, \tau^*, \kappa^*)$ such that $\tau^* > 0$ and $\kappa^* = 0$, then an optimal solution to (P) is simply x^*/τ^* and an optimal solution to (D) is $(y^*/\tau^*, s^*/\tau^*)$.

However, (H) also contains trivial solutions such as

$$(y, x, s, \tau, \kappa) = (0, 0, 0, 0, 0),$$

from which no information concerning solutions for (P) and (D) can be deduced. To avoid trivial solutions, we note the following fundamental result concerning (H) due to Goldman and Tucker [14].

Theorem 5.1 There exists a solution $(y^*, x^*, s^*, \tau^*, \kappa^*)$ for (H) such that

$$x^* + s^* > 0 \text{ and } \tau^* + \kappa^* > 0.$$

It is elementary to check that any solution (y, x, s, τ, κ) to (H) necessarily satisfies

$$x^T s + \tau\kappa = 0.$$

That is why the Goldman-Tucker type solution is called a strictly complementary solution, since it implies that either x_i^* or s_i^* is zero (and not both) for all i, and either τ^* or κ^* is zero (and not both). Based on a strictly complementary solution for (H), solutions for the original linear programming problems (P) and (D) can easily be found, as the next lemma demonstrates.

Lemma 5.1 Let $(y^*, x^*, s^*, \tau^*, \kappa^*)$ be as in Theorem 5.1. If $\tau^* > 0$, then x^*/τ^* is an optimal solution to (P) and $(y^*/\tau^*, s^*/\tau^*)$ is an optimal solution to (D). If $\tau^* = 0$, then $\kappa^* > 0$, i.e. $b^T y^* - c^T x^* > 0$. In this case, if $b^T y^* > 0$, then (P) is infeasible, and if $c^T x^* < 0$, then (D) is infeasible.

One may argue, however, that (H) is nothing but a reformulation of the problem, as finding the desirable $(y^*, x^*, s^*, \tau^*, \kappa^*)$ as stipulated in Lemma 5.1 is equally difficult as solving (P) and (D).

This view proves to be a little narrow minded. Let us see why.

Take any vectors $0 < x^0 \in \Re^n$, $y^0 \in \Re^m$, $0 < s^0 \in \Re^n$, $\tau^0 > 0$ and $\kappa^0 > 0$, such that $x_i^0 s_i^0 = 1$, $i = 1, ..., n$, and $\tau^0\kappa^0 = 1$. In practice, one may simply let $x^0 = e$, $y^0 = 0$ and $s^0 = e$, where e is the all-one vector, and $\tau^0 = \kappa^0 = 1$. Denote

$$\left\{ \begin{array}{rcl} r_p^0 & = & Ax^0 - b\tau^0 \\ r_d^0 & = & -A^T y^0 + c\tau^0 - s^0 \\ r_g^0 & = & b^T y^0 - c^T x^0 - \kappa^0. \end{array} \right. \tag{5.9}$$

Now we introduce a parameterized system of nonlinear equations

$$(H_\mu) \left\{ \begin{array}{rcl} Ax - b\tau & = & \mu r_p^0 \\ -A^T y - s + c\tau & = & \mu r_d^0 \\ b^T y - c^T x - \kappa & = & \mu r_g^0 \\ x_i s_i & = & \mu, \; i = 1, ..., n, \\ \tau\kappa & = & \mu. \end{array} \right.$$

Obviously, (H_1) has a solution, e.g.,

$$(y, x, s, \tau, \kappa) = (y^0, x^0, s^0, \tau^0, \kappa^0).$$

Now we argue that (H_μ) always has a unique solution

$$(y(\mu), x(\mu), s(\mu), \tau(\mu), \kappa(\mu)),$$

for any fixed $\mu > 0$.

To see this we consider a linear programming problem

$$\begin{array}{ll} \text{minimize} & -\mu(r_p^0)^T y - \mu(r_d^0)^T x - \mu r_g^0 \tau \\ \text{subject to} & Ax - b\tau = \mu r_p^0, \\ & -A^T y + c\tau \geq \mu r_d^0, \\ & b^T y - c^T x \geq \mu r_g^0, \\ & x \geq 0,\ \tau \geq 0. \end{array}$$

This problem has an interior feasible solution, e.g.

$$(y, x, \tau) = \mu(y^0, x^0, \tau^0).$$

Moreover, it is self-dual, i.e., its dual problem is the same as itself. Hence, the dual also has an interior feasible solution, and therefore the primal-dual analytic central path exists. The point on the central path with parameter μ, after introducing the slack variables s and κ for the second and third constraints, is the unique solution for (H_μ). This leads naturally to the idea of following the central path by reducing μ, starting from $\mu = 1$. As is standard nowadays in primal-dual interior point methods, we solve the system (H_μ) by applying Newton's method. In particular, we use the predictor-corrector approach proposed by Mizuno, Todd and Ye in [19]. To facilitate the method, we introduce a concept called the neighborhood of the central path with size $0 < \beta < 1$

$$\mathcal{N}(\beta) = \left\{ \begin{bmatrix} x \\ s \\ \tau \\ \kappa \end{bmatrix} \middle| \ \delta(x, s, \tau, \kappa) := \left\| e - \frac{n+1}{x^T s + \tau\kappa} \begin{bmatrix} Xs \\ \tau\kappa \end{bmatrix} \right\| \leq \beta \right\}, \tag{5.10}$$

where X stands for the diagonal matrix whose elements are taken from the vector x. This notation is standard in interior point methods and will be used throughout the chapter

We now wish to update (x, s, τ, κ) in such a way that it is in a good position, i.e., the iterate stays inside a certain neighborhood of the central path, and at the same time, the duality gap, namely the value of $x^T s + \tau\kappa$, is reduced. These two wishes, however, are often contradicting to each other. A compromise is proposed in the predictor-corrector approach, in which we simply separate these two tasks by using two neighborhoods, one is bigger than the other, say one is $\mathcal{N}(1/4)$ and the other one is $\mathcal{N}(1/2)$. Once we are inside the smaller neighborhood, then we consider ourselves well positioned, and can afford to sacrifice the centrality to a certain degree while concentrating on reducing the duality gap at the step. This step is termed *predictor*. As soon as we are on the boundary of the big neighborhood, this is the sign of alarming and we try to move back into the smaller neighborhood without damaging the duality gap. This is the so called *corrector* step.

Specifically, the predictor step is to take a Newton direction, $(d_x, d_s, d_\tau, d_\kappa)$, based on linearizing (H_μ) while targeting at $\mu = 0$. The corresponding system of linear equations is

$$(Predictor) \left\{ \begin{array}{rcl} Ad_x - bd_\tau & = & -r_p \\ -A^T d_y - d_s + cd_\tau & = & -r_d \\ b^T d_y - c^T d_x - d_\kappa & = & -r_g \\ s_i d_{x_i} + x_i d_{s_i} & = & -x_i s_i, \;\; i = 1, ..., n, \\ \kappa d_\tau + \tau d_\kappa & = & -\tau\kappa, \end{array} \right.$$

where the new residuals are defined similarly as in (5.9)

$$\left\{ \begin{array}{rcl} r_p & = & Ax - b\tau \\ r_d & = & -A^T y + c\tau - s \\ r_g & = & b^T y - c^T x - \kappa. \end{array} \right.$$

The corrector step aims at reaching the solution of (H_μ) with $\mu = (x^T s + \tau\kappa)/(n+1)$. Hence, the corresponding system of Newton equations is

$$(Corrector) \left\{ \begin{array}{rcl} Ad_x - bd_\tau & = & 0 \\ -A^T d_y - d_s + cd_\tau & = & 0 \\ b^T d_y - c^T d_x - d_\kappa & = & 0 \\ s_i d_{x_i} + x_i d_{s_i} & = & \mu - x_i s_i, \;\; i = 1, ..., n, \\ \kappa d_\tau + \tau d_\kappa & = & \mu - \tau\kappa. \end{array} \right.$$

The key to note here is that, if one follows either the predictor step, or the corrector step, then it always holds that

$$\begin{cases} r_p = \mu r_{p_0} \\ r_d = \mu r_{d_0} \\ r_g = \mu r_{g_0}. \end{cases}$$

In other words, the direction of residuals will not change, only the norm will be reduced, proportional to μ. In general, we need to solve the following linear equations

$$(Newton) \begin{cases} Ad_x - bd_\tau & = & -\rho r_p \\ -A^T d_y - d_s + cd_\tau & = & -\rho r_d \\ b^T d_y - c^T d_x - d_\kappa & = & -\rho r_g \\ s_i d_{x_i} + x_i d_{s_i} & = & (1-\rho)\mu - x_i s_i, \; i = 1, ..., n, \\ \kappa d_\tau + \tau d_\kappa & = & (1-\rho)\mu - \tau\kappa, \end{cases} \tag{5.11}$$

where $\rho = 1$ yields the predictor direction, and $\rho = 0$ yields the corrector direction.

Let us solve the linear system (5.11) and update the iterates based on the Newton directions with a steplength $t > 0$, i.e. we let

$$\begin{cases} x(t) & := & x + td_x > 0 \\ y(t) & := & y + td_y \\ s(t) & := & s + td_s > 0 \\ \tau(t) & := & \tau + td_\tau > 0 \\ \kappa(t) & := & \kappa + td_\kappa > 0. \end{cases}$$

Next we study what happens to the new iterate $(x(t), y(t), s(t), \tau(t), \kappa(t))$. First let us study the new duality gap

$$\begin{aligned} x(t)^T s(t) + \tau(t)\kappa(t) &= (x + td_x)^T(s + td_s) + (\tau + td_\tau)(\kappa + td_\kappa) \\ &= x^T s + \tau\kappa + t(x^T d_s + s^T d_x + \tau d_\kappa + \kappa d_\tau) \\ &\quad + t^2(d_x^T d_s + d_\tau d_\kappa) \\ &= (1 - t\rho)(n+1)\mu + t^2(d_x^T d_s + d_\tau d_\kappa), \end{aligned} \tag{5.12}$$

where we used the fourth and the fifth equations in (5.11). It remains to estimate the second term on the right hand side of (5.12). For this purpose we note from the second equation in (5.11) that

$$d_s = -A^T d_y + cd_\tau + \rho r_d,$$

and therefore

$$\begin{aligned} d_x^T d_s &= -(Ad_x)^T d_y + c^T d_x d_\tau + \rho r_d^T d_x \\ &= -(bd_\tau - \rho r_d)^T d_y + c^T d_x d_\tau + \rho r_d^T d_x \\ &= \rho(r_p^T d_y + r_d^T d_x) + (c^T x - b^T y) d_\tau. \end{aligned} \tag{5.13}$$

Hence,

$$\begin{aligned} & d_x^T d_s + d_\tau d_\kappa \\ &= \rho(r_p^T d_y + r_d^T d_x) + (c^T x - b^T y + d_\kappa) d_\tau \\ &= \rho(r_p^T d_y + r_d^T d_x + r_g d_\tau) \\ &= \rho\left[(Ax - b\tau)^T d_y + (-A^T y + c\tau - s)^T d_x + (b^T y - c^T x - \kappa) d_\tau\right] \\ &= \rho(x^T A^T d_y - \tau b^T d_y - y^T A d_x + \tau c^T d_x \\ &\quad - s^T d_x + b^T y d_\tau - c^T x d_\tau - \kappa d_\tau) \\ &= \rho[x^T(-d_s + c d_\tau + \rho r_d) - \tau b^T d_y - y^T(b d_\tau - \rho r_p) \\ &\quad + \tau c^T d_x - s^T d_x + b^T y d_\tau - c^T x d_\tau - \kappa d_\tau] \\ &= \rho\left[-x^T d_s - s^T d_x + \rho x^T r_d + \rho y^T r_p + \tau(c^T d_x - b^T d_y) - \kappa d_\tau\right] \\ &= \rho\left[\rho(x^T r_d + y^T r_p + \tau r_g) - (x^T d_s + s^T d_x + \kappa d_\tau + \tau d_\kappa)\right]. \end{aligned} \tag{5.14}$$

On one hand, adding up the terms in the fourth and the fifth equations of (5.11) we obtain

$$\begin{aligned} & x^T d_s + s^T d_x + \kappa d_\tau + \tau d_\kappa \\ &= (1-\rho)(n+1)\mu - (n+1)\mu = -\rho(n+1)\mu. \end{aligned} \tag{5.15}$$

On the other hand,

$$\begin{aligned} & x^T r_d + y^T r_p + \tau r_g \\ &= x^T(-A^T y + \tau c - s) + y^T(Ax - \tau b) + \tau(b^T y - c^T x - \kappa) \\ &= -x^T s - \tau\kappa \\ &= -(n+1)\mu. \end{aligned} \tag{5.16}$$

Substituting (5.15) and (5.16) into (5.14) yields

$$d_x^T d_s + d_\tau d_\kappa = 0. \tag{5.17}$$

Now we obtain from (5.12) that

$$x(t)^T s(t) + \tau(t)\kappa(t) = (1 - t\rho)(n+1)\mu. \tag{5.18}$$

Having established how the duality gap, viz. the value of $x(t)^T s(t) + \tau(t)\kappa(t)$, will be improved, next we wish to see how the centering measure,

$$\delta(x(t), s(t), \tau(t), \kappa(t)) = \left\| e - \frac{n+1}{x(t)^T s(t) + \tau(t)\kappa(t)} \begin{bmatrix} X(t)s(t) \\ \tau(t)\kappa(t) \end{bmatrix} \right\|$$

(see (5.10)), is affected by such a move.

Using (5.18), and the fourth and fifth equations in (5.11) it follows that

$$(1 - t\rho)\delta(x(t), s(t), \tau(t), \kappa(t)) = \left\| e - \frac{1}{\mu}\begin{bmatrix} Xs \\ \tau\kappa \end{bmatrix} - t\left((1-\rho)e - \frac{1}{\mu}\begin{bmatrix} Xs \\ \tau\kappa \end{bmatrix}\right) - \frac{t^2}{\mu}\begin{bmatrix} d_x \circ d_s \\ d_\tau \circ d_\kappa \end{bmatrix} \right\|, \tag{5.19}$$

where $\circ$ stands for the componentwise product of two vectors.

In order to further study the change in the centrality measure, it is crucial to estimate the term $\left\| \begin{bmatrix} d_x \circ d_s \\ d_\tau \circ d_\kappa \end{bmatrix} \right\|$. The last two equations in (5.11) can be rewritten as

$$\begin{aligned} (XS)^{0.5}\left[X^{-0.5}S^{0.5}d_x + X^{0.5}S^{-0.5}d_s\right] &= (1-\rho)\mu e - Xs, \\ (\tau\kappa)^{0.5}\left[\tau^{-0.5}\kappa^{0.5}d_\tau + \tau^{0.5}\kappa^{-0.5}d_\kappa\right] &= (1-\rho)\mu - \tau\kappa. \end{aligned}$$

Let

$$\begin{bmatrix} \bar{d}_x \\ \bar{d}_\tau \end{bmatrix} := \begin{bmatrix} X^{-0.5}S^{0.5}d_x \\ \tau^{-0.5}\kappa^{0.5}d_\tau \end{bmatrix} \text{ and } \begin{bmatrix} \bar{d}_s \\ \bar{d}_\kappa \end{bmatrix} := \begin{bmatrix} X^{0.5}S^{-0.5}d_s \\ \tau^{0.5}\kappa^{-0.5}d_\kappa \end{bmatrix}.$$

We then have

$$\begin{bmatrix} \bar{d}_x + \bar{d}_s \\ \bar{d}_\tau + \bar{d}_\kappa \end{bmatrix} = \begin{bmatrix} (XS)^{-0.5} & 0 \\ 0 & (\tau\kappa)^{-0.5} \end{bmatrix} \begin{bmatrix} (1-\rho)\mu e - Xs \\ (1-\rho)\mu - \tau\kappa \end{bmatrix}. \tag{5.20}$$

Moreover,

$$(\bar{d}_x)_i(\bar{d}_s)_i = (d_x)_i(d_s)_i \text{ for } i = 1, ..., n, \text{ and } \bar{d}_\tau\bar{d}_\kappa = d_\tau d_\kappa.$$

Hence

$$\bar{d}_x^T\bar{d}_s + \bar{d}_\tau\bar{d}_\kappa = d_x^T d_s + d_\tau d_\kappa = 0.$$

Therefore

$$\begin{aligned} \left\| \begin{bmatrix} d_x \circ d_s \\ d_\tau \circ d_\kappa \end{bmatrix} \right\| &\le \left\| \begin{bmatrix} \bar{d}_x \circ \bar{d}_s \\ \bar{d}_\tau \circ \bar{d}_\kappa \end{bmatrix} \right\| \\ &\le \left\| \begin{bmatrix} \bar{d}_x \circ \bar{d}_s \\ \bar{d}_\tau \circ \bar{d}_\kappa \end{bmatrix} \right\|_1 \end{aligned}$$

$$
\begin{aligned}
&\leq \left\| \begin{bmatrix} \bar{d}_x \\ \bar{d}_\tau \end{bmatrix} \right\| \cdot \left\| \begin{bmatrix} \bar{d}_s \\ \bar{d}_\kappa \end{bmatrix} \right\| \\
&\leq \left(\left\| \begin{bmatrix} \bar{d}_x \\ \bar{d}_\tau \end{bmatrix} \right\|^2 + \left\| \begin{bmatrix} \bar{d}_s \\ \bar{d}_\kappa \end{bmatrix} \right\|^2 \right) /2 \\
&= \left\| \begin{bmatrix} \bar{d}_x + \bar{d}_s \\ \bar{d}_\tau + \bar{d}_\kappa \end{bmatrix} \right\|^2 /2, \qquad (5.21)
\end{aligned}
$$

where in the third inequality follows from the Cauchy-Schwarz inequality, and the last equality follows from the orthogonality.

Since $(x, s, \tau, \kappa) \in \mathcal{N}(\beta)$, we have $x_i s_i \geq (1-\beta)\mu$, $i = 1, ..., n$, and $\tau\kappa \geq (1-\beta)\mu$. This yields, using (5.21), that

$$
\left\| \begin{bmatrix} d_x \circ d_s \\ d_\tau \circ d_\kappa \end{bmatrix} \right\| \leq \frac{1}{(1-\beta)\mu} \left\| \begin{bmatrix} (1-\rho)\mu e - Xs \\ (1-\rho)\mu - \tau\kappa \end{bmatrix} \right\|^2 /2. \qquad (5.22)
$$

Using (5.22), we obtain from (5.19) that

$$
\begin{aligned}
&(1-t\rho)\delta(x(t), s(t), \tau(t), \kappa(t)) \\
&\leq (1-t)\delta(x, s, \tau, \kappa) + t\rho\sqrt{n+1} + \frac{t^2}{2(1-\beta)\mu^2} \left\| \begin{bmatrix} (1-\rho)\mu e - Xs \\ (1-\rho)\mu - \tau\kappa \end{bmatrix} \right\|^2 . \qquad (5.23)
\end{aligned}
$$

Now we are in the position to present the pure form of predictor-corrector homogeneous self-dual interior point algorithm.

Algorithm PCHSD

Step 0 Let $k = 0$.

Step 1 If $k = 0$ or k is even, go to Step 2; otherwise go to Step 3.

Step 2 Let $\rho = 0$. Solve the linear system (5.11) to get the predictor search directions. Let

$$
\begin{cases}
x^{k+1} := x^k + t^* d_x > 0 \\
y^{k+1} := y^k + t^* d_y \\
s^{k+1} := s^k + t^* d_s > 0 \\
\tau^{k+1} := \tau^k + t^* d_\tau > 0 \\
\kappa^{k+1} := \kappa^k + t^* d_\kappa > 0,
\end{cases}
$$

where $t^* > 0$ is chosen to be the maximum step length such that

$$(x^k + t^* d_x, s^k + t^* d_s, \tau^k + t^* d_\tau, \kappa^k + t^* d_\kappa) \in \mathcal{N}(1/2).$$

Let $k := k+1$ and return to Step 1.

Step 3 Let $\rho = 1$. Solve the linear system (5.11) to get the corrector search directions. Let

$$\begin{cases} x^{k+1} & := & x^k + d_x > 0 \\ y^{k+1} & := & y^k + d_y \\ s^{k+1} & := & s^k + d_s > 0 \\ \tau^{k+1} & := & \tau^k + d_\tau > 0 \\ \kappa^{k+1} & := & \kappa^k + d_\kappa > 0. \end{cases}$$

Let $k := k+1$ and return to Step 1.

Now we show that each corrector step brings the iterate from $\mathcal{N}(1/2)$ to $\mathcal{N}(1/4)$.

Lemma 5.2 If $(x^k, s^k, \tau^k, \kappa^k) \in \mathcal{N}(1/2)$ and $\rho = 0$, then

$$(x^{k+1}, s^{k+1}, \tau^{k+1}, \kappa^{k+1}) \in \mathcal{N}(1/4).$$

Proof. In the corrector step, we use unitary Newton step, i.e. $t = 1$. Moreover, $\beta = 1/2$, and so by (5.23) we get

$$\delta((x^{k+1}, s^{k+1}, \tau^{k+1}, \kappa^{k+1})) \leq \frac{1}{2(1-\beta)} \delta((x^k, s^k, \tau^k, \kappa^k))^2 \leq 1/4.$$

□

Next we shall prove that the predictor step guarantees a steplength at least of order $1/\sqrt{n+1}$.

Lemma 5.3 *If* $(x^k, s^k, \tau^k, \kappa^k) \in \mathcal{N}(1/4)$ *and* $\rho = 1$, *then* $t^* \geq \frac{1}{8\sqrt{n+1}}$.

Proof. We let $t = \frac{1}{8\sqrt{n+1}}$. Since $(x^k, s^k, \tau^k, \kappa^k) \in \mathcal{N}(1/4)$, it follows that $x_i^k s_i^k \leq (1+1/4)\mu^k$, $i = 1, ..., n$, and $\tau^k \kappa^k \leq (1+1/4)\mu^k$. By (5.23) we get

$$\begin{aligned} & (1-t\rho)\delta(x(t), s(t), \tau(t), \kappa(t)) \\ \leq \; & (1-t)\delta(x^k, s^k, \tau^k, \kappa^k) \\ & +t/\sqrt{n+1} + \frac{t^2}{2(1-1/4)}(1+1/4)(n+1) \\ \leq \; & (1-t)\delta(x^k, s^k, \tau^k, \kappa^k) + 0.15. \end{aligned}$$

Because $t \leq 1/8$, we have $1 - t \geq 7/8$, and so

$$\delta(x(t), s(t), \tau(t), \kappa(t)) \leq 1/4 + \frac{0.15}{7/8} \leq 1/2.$$

This shows that the maximum steplength must be at least $\frac{1}{8\sqrt{n+1}}$. □

Theorem 5.2 Algorithm PCHSD requires to take at most $O(\sqrt{n}L)$ number of steps to reach a solution with duality gap no more than 2^{-L}, where $L \geq \log(n+1)$.

Proof. The duality gap is reduced at the predictor step, while the corrector step will not change the duality gap. Algorithm PCHSD takes a predictor step and a corrector step alternatively. According to (5.18), the rate of duality gap reduction at a predictor step is simply $1 - t \geq 1 - \frac{1}{8\sqrt{n+1}}$. Note that the initial duality gap is $n + 1$. The theorem thus follows. □

A careful analysis shows that the algorithm actually converges quadratically; see [26]. We remark that, despite of its simple form, Algorithm PCHSD works extremely well in practice. In particular, one can observe that it requires only a very few number (almost insensitive to the dimension n) of iterations to reach a highly accurate solution. Since the deterministic equivalence of multiple stage stochastic linear programming can be cast as a large size, however well structured, linear program. Therefore it is natural to consider applying Algorithm PCHSD to solve it. The key to success, of course, lies in the speed of solving the direction finding subproblem (5.11). This becomes the topic of next section.

5.5 Finding Search Directions

As we have mentioned in the previous section, it is crucial to be able to efficiently compute the search directions, based on the Newton equation (5.11), when the homogeneous self-dual embedding technique is applied to the deterministic-equivalent of multistage stochastic programming problem. To be precise, the linear system of concern is as follows, which we will denote as $(\bar{S})$ for later reference,

$$
\left\{
\begin{array}{llcl}
W_0 d_{x_0} & -h_0 d_\tau & = & -\rho r_{p_0} \\
B_{nt} d_{x_{a(n),t-1}} + W_{nt} d_{x_{nt}} & -h_{nt} d_\tau & = & -\rho r_{p_{nt}}, \\
 & \multicolumn{3}{r}{t = 1, ..., K,\ n \in \mathcal{F}_t} \\
-W_{nt}^T d_{y_{nt}} - \sum_{m \in C(n)} B_{m,t+1}^T d_{y_{m,t+1}} - d_{s_{nt}} & +\pi_{nt} c_{nt} d_\tau & = & -\rho r_{d_{nt}}, \\
 & \multicolumn{3}{r}{t = 0, ..., K-1,\ n \in \mathcal{F}_t} \\
-W_{nK}^T d_{y_{nK}} - d_{s_{nK}} & +\pi_{nK} c_{nK} d_\tau & = & -\rho r_{d_{nK}}, \\
 & & & n \in \mathcal{F}_K \\
S_0 d_{x_0} + X_0 d_{s_0} & & = & (1-\rho)\mu e \\
 & & & -X_0 s_0 \\
S_{nt} d_{x_{nt}} + X_{nt} d_{s_{nt}} & & = & (1-\rho)\mu e \\
 & & & -X_{nt} s_{nt}, \\
 & \multicolumn{3}{r}{t = 1, ..., K,\ n \in \mathcal{F}_t} \\
 & \kappa d_\tau + \tau d_\kappa & = & (1-\rho)\mu \\
 & & & -\tau\kappa \\
\sum_{t=0}^{K} \sum_{n \in \mathcal{F}_t} \left[h_{nt}^T d_{y_{nt}} - \pi_{nt} c_{nt}^T d_{x_{nt}} \right] & -d_\kappa & = & -\rho r_g,
\end{array}
\right.
$$

where

$$\begin{aligned}
r_{p_0} &= W_0 x_0 - \tau h_0, \\
r_{p_{nt}} &= B_{nt} x_{a(n),t-1} + W_{nt} x_{nt} - \tau h_{nt},\ t = 1, ..., K,\ n \in \mathcal{F}_t, \\
r_{d_{nt}} &= -W_{nt}^T y_{nt} - \sum_{m \in C(n)} B_{m,t+1}^T y_{m,t+1} - s_{nt} + \pi_{nt} c_{nt}, \\
& \quad t = 0, ..., K-1,\ n \in \mathcal{F}_t, \\
r_{d_{nK}} &= -W_{nK}^T d_{y_{nK}} - s_{nK} + \pi_{nK} c_{nK},\ n \in \mathcal{F}_K, \\
r_g &= \sum_{t=0}^{K} \sum_{n \in \mathcal{F}_t} \left[h_{nt}^T y_{nt} - \pi_{nt} c_{nt}^T x_{nt} \right] - \kappa.
\end{aligned}$$

Consider a given $m \in \mathcal{F}_K$. From the sixth equation in $(\bar{S})$ we obtain

$$d_{s_{mK}} = -X_{mK}^{-1} S_{mK} d_{x_{mK}} + X_{mK}^{-1}((1-\rho)\mu e - X_{mK}^{-1} s_{mK}). \tag{5.24}$$

Substituting (5.24) into the fourth equation in $(\bar{S})$ yields

$$\begin{aligned}
& -W_{mK}^T d_{y_{mK}} + X_{mK}^{-1} S_{mK} d_{x_{mK}} - X_{mK}^{-1}((1-\rho)\mu e - X_{mK}^{-1} s_{mK}) \\
& + \pi_{mK} c_{mK} d_\tau = (1-\rho) r_{d_{mK}},
\end{aligned}$$

and therefore

$$\begin{aligned}
d_{x_{mK}} &= (X_{mK}^{-1} S_{mK})^{-1} W_{mK}^T d_{y_{mK}} - \pi_{mK} (X_{mK}^{-1} S_{mK})^{-1} c_{mK} d_\tau \\
& + (X_{mK}^{-1} S_{mK})^{-1} \left[-\rho r_{d_{mK}} + X_{mK}^{-1}((1-\rho)\mu e - X_{mK}^{-1} s_{mK}) \right].
\end{aligned}$$

Let

$$M_{mK} := X_{mK}^{-1} S_{mK}. \tag{5.25}$$

We have

$$\begin{aligned}
& d_{x_{mK}} = M_{mK}^{-1} W_{mK}^T d_{y_{mK}} - \pi_{mK} M_{mK}^{-1} c_{mK} d_\tau \\
& + M_{mK}^{-1} \left[-\rho r_{d_{mK}} + X_{mK}^{-1}((1-\rho)\mu e - X_{mK}^{-1} s_{mK}) \right].
\end{aligned} \tag{5.26}$$

Now consider an $n \in \mathcal{F}_{K-1}$ such that $m \in C(n)$, i.e., $n = a(m)$. Using (5.26) and the second equation in $(\bar{S})$ we get

$$\begin{aligned}
& B_{mK} d_{x_{n,K-1}} + W_{mK} M_{mK}^{-1} W_{mK}^T d_{y_{mK}} - W_{mK} M_{mK}^{-1} W_{mK}^T q_{mK} d_\tau \\
= \ & W_{mK} M_{mK}^{-1} W_{mK}^T v_{mK},
\end{aligned}$$

and consequently

$$d_{y_{mK}} = -(W_{mK} M_{mK}^{-1} W_{mK}^T)^{-1} B_{mK} d_{x_{n,K-1}} + q_{mK} d_\tau + v_{mK}, \tag{5.27}$$

where

$$q_{mK} := (W_{mK} M_{mK}^{-1} W_{mK}^T)^{-1} \left[h_{mK} + \pi_{mK} W_{mK} M_{mK}^{-1} c_{mK} \right], \tag{5.28}$$

and

$$v_{mK} := (W_{mK} M_{mK}^{-1} W_{mK}^T)^{-1} \{-\rho r_{p_{mK}} - W_{mK} M_{mK}^{-1} [-\rho r_{d_{mK}} + X_{mK}^{-1} ((1-\rho)\mu e - X_{mK} s_{mK})]\}. \tag{5.29}$$

Now we substitute (5.27) back into the third equation in $(\bar{S})$. Further using the sixth equation in $(\bar{S})$ yields

$$-W_{n,K-1}^T d_{y_{n,K-1}} + \left[X_{n,K-1}^{-1} S_{n,K-1} + \sum_{m \in C(n)} B_{mK}^T (W_{mK} M_{mK}^{-1} W_{mK}^T)^{-1} B_{mK} \right] d_{x_{n,K-1}} + \left[\pi_{n,K-1} c_{n,K-1} - \sum_{m \in C(n)} B_{mK}^T q_{mK} \right] d_\tau = -\rho r_{d_{n,K-1}} + \sum_{m \in C(n)} B_{mK}^T v_{mK} + X_{n,K-1}^{-1} ((1-\rho)\mu e - X_{n,K-1} s_{n,K-1}).$$

Letting

$$M_{n,K-1} := X_{n,K-1}^{-1} S_{n,K-1} + \sum_{m \in C(n)} B_{mK}^T (W_{mK} M_{mK}^{-1} W_{mK}^T)^{-1} B_{mK}, \tag{5.30}$$

we may rewrite the above expression as

$$d_{x_{n,K-1}} = M_{n,K-1}^{-1} W_{n,K-1}^T d_{y_{n,K-1}} - M_{n,K-1}^{-1} \left[\pi_{n,K-1} c_{n,K-1} - \sum_{m \in C(n)} B_{mK}^T q_{mK} \right] d_\tau + M_{n,K-1}^{-1} \cdot \left[-\rho r_{d_{n,K-1}} + \sum_{m \in C(n)} B_{mK}^T v_{mK} + X_{n,K-1}^{-1} ((1-\rho)\mu e - X_{n,K-1} s_{n,K-1}) \right]. \tag{5.31}$$

We trace one more stage back, and let k be at stage $K-2$ such that $n \in C(k)$, i.e., $k = a(n)$. Similar as before, we substitute (5.31) into the second equation in $(\bar{S})$, which is

$$B_{n,K-1} d_{x_{k,K-2}} + W_{n,K-1} d_{x_{n,K-1}} - h_{n,K-1} d_\tau = -\rho r_{p_{n,K-1}},$$

obtaining

$$d_{y_{n,K-1}} = -(W_{n,K-1} M_{n,K-1}^{-1} W_{n,K-1}^T)^{-1} B_{n,K-1} d_{x_{k,K-2}} + q_{n,K-1} d_\tau + v_{n,K-1}, \tag{5.32}$$

where

$$\begin{aligned} q_{n,K-1} &:= (W_{n,K-1}M_{n,K-1}^{-1}W_{n,K-1}^T)^{-1} \cdot \{h_{n,K-1} + \\ & W_{n,K-1}M_{n,K-1}^{-1}[\pi_{n,K-1}c_{n,K-1} - \sum_{m \in C(n)} B_{mK}^T q_{mK}]\}, \end{aligned} \tag{5.33}$$

and

$$\begin{aligned} v_{n,K-1} &:= -(W_{n,K-1}M_{n,K-1}^{-1}W_{n,K-1}^T)^{-1} \cdot \{\rho r_{p_{n,K-1}} \\ & +W_{n,K-1}M_{n,K-1}^{-1}[-\rho r_{d_{n,K-1}} \\ & +X_{n,K-1}^{-1}((1-\rho)\mu e - X_{n,K-1}s_{n,K-1}) + \sum_{m \in C(n)} B_{mK}^T v_{mK}]\}. \end{aligned} \tag{5.34}$$

Once again, we substitute (5.32) back into the third equation in $(\bar{S})$ and use the sixth equation at the same time. This yields an equation from which we can solve out $d_{x_{k,K-2}}$ as follows

$$\begin{aligned} d_{x_{k,K-2}} &= M_{k,K-2}^{-1}W_{k,K-2}^T d_{y_{k,K-2}} - \\ & M_{k,K-2}^{-1}[\pi_{k,K-2}c_{k,K-2} - \sum_{n \in C(k)} B_{k,K-1}^T q_{k,K-1}]d_\tau + \\ & M_{k,K-2}^{-1}[-\rho r_{d_{k,K-2}} + \sum_{n \in C(k)} B_{n,K-1}^T v_{n,K-1} + \\ & X_{k,K-2}^{-1}((1-\rho)\mu e - X_{k,K-2}s_{k,K-2})]. \end{aligned} \tag{5.35}$$

Repeating this procedure, we get the following recursive formula

$$d_{y_{mt}} = -(W_{mt}M_{mt}^{-1}W_{mt}^T)^{-1}B_{mt}d_{x_{n,t-1}} + q_{mt}d_\tau + v_{mt}, \tag{5.36}$$

where $n = a(m)$, and

$$\begin{aligned} d_{x_{mt}} &= M_{mt}^{-1}W_{mt}^T d_{y_{mt}} - M_{mt}^{-1}[\pi_{mt}c_{mt} - \sum_{k \in C(m)} B_{k,t+1}^T q_{k,t+1}]d_\tau + M_{mt}^{-1} \cdot \\ & \cdot[-\rho r_{d_{mt}} + \sum_{k \in C(m)} B_{k,t+1}^T v_{k,t+1} + X_{mt}^{-1}((1-\rho)\mu e - X_{mt}s_{mt})], \end{aligned} \tag{5.37}$$

for $t = K, K-1, ..., 1$ and $m \in \mathcal{F}_t$. As a notational convention, we assume that $C(m) = \emptyset$ for $m \in \mathcal{F}_K$. The matrices M_{mt} and vectors q_{mt} and v_{mt}

are generated according to the following recursive formula

$$(R_1)\begin{cases} M_{n,t-1} &= X_{n,t-1}^{-1}S_{n,t-1} + \sum_{m\in C(n)} B_{mt}^T(W_{mt}M_{mt}^{-1}W_{mt}^T)^{-1}B_{mt} \\ q_{n,t-1} &= (W_{n,t-1}M_{n,t-1}^{-1}W_{n,t-1}^T)^{-1}\cdot \\ &\quad \{h_{n,t-1} + W_{n,t-1}M_{n,t-1}^{-1}[\pi_{n,t-1}c_{n,t-1} - \sum_{m\in C(n)} B_{mt}^T q_{mt}]\} \\ v_{n,t-1} &= -(W_{n,t-1}M_{n,t-1}^{-1}W_{n,t-1}^T)^{-1}\cdot \\ &\quad \{\rho r_{p_{n,t-1}} + W_{n,t-1}M_{n,t-1}^{-1}[-\rho r_{d_{n,t-1}} + \\ &\quad X_{n,t-1}^{-1}((1-\rho)\mu e - X_{n,t-1}s_{n,t-1}) + \sum_{m\in C(n)} B_{mt}^T v_{mt}]\}. \end{cases}$$

with $m \in C(n)$ and $m \in \mathcal{F}_t$; see (5.30), (5.33) and (5.34). The terminal values are given by the following formula

$$(T)\begin{cases} M_{mK} &= X_{mK}^{-1}S_{mK} \\ q_{mK} &= (W_{mK}M_{mK}^{-1}W_{mK}^T)^{-1}\left[h_{mK} + \pi_{mK}W_{mK}M_{mK}^{-1}c_{mK}\right] \\ v_{mK} &= (W_{mK}M_{mK}^{-1}W_{mK}^T)^{-1}\cdot\{-\rho r_{p_{mK}} - W_{mK}M_{mK}^{-1}[-\rho r_{d_{mK}} + \\ &\quad X_{mK}^{-1}((1-\rho)\mu e - X_{mK}s_{mK})]\}, \end{cases}$$

where $m \in \mathcal{F}_K$; see (5.25), (5.28) and (5.29).

To further simplify the notation let us rewrite (5.37) as

$$d_{x_{mt}} = M_{mt}^{-1}W_{mt}^T d_{y_{mt}} - p_{mt}d_\tau + u_{mt} \tag{5.38}$$

with

$$p_{mt} = M_{mt}^{-1}[\pi_{mt}c_{mt} - \sum_{k\in C(m)} B_{k,t+1}^T q_{k,t+1}], \tag{5.39}$$

$$\begin{aligned} u_{mt} &= M_{mt}^{-1}[-\rho r_{d_{mt}} + \sum_{k\in C(m)} B_{k,t+1}^T v_{k,t+1} + \\ &\quad X_{mt}^{-1}((1-\rho)\mu e - X_{mt}s_{mt})]. \end{aligned} \tag{5.40}$$

In particular we have

$$d_{x_0} = M_0^{-1}W_0^T d_{y_0} - p_0 d_\tau + u_0. \tag{5.41}$$

Substituting this into the first equation in $(\bar{S})$ we have

$$W_0(M_0^{-1}W_0^T d_{y_{mt}} - p_0 d_\tau + u_0) - h_0 d_\tau = -\rho r_{p_0}.$$

Equivalently,

$$d_{y_0} = (W_0 M_0^{-1} W_0^T)^{-1}(W_0 p_0 + h_0)d_\tau + (W_0 M_0^{-1} W_0^T)^{-1}(-\rho r_{p_0} - W_0 u_0).$$

We denote

$$\begin{aligned} \alpha_0 &= (W_0 M_0^{-1} W_0^T)^{-1}(W_0 p_0 + h_0), && (5.42)\\ \beta_0 &= (W_0 M_0^{-1} W_0^T)^{-1}(-\rho r_{p_0} - W_0 u_0), && (5.43) \end{aligned}$$

and so

$$d_{y_0} = \alpha_0 d_\tau + \beta_0. \quad (5.44)$$

Substituting (5.44) into (5.41) we get

$$\begin{aligned} d_{x_0} &= M_0^{-1} W_0^T d_{y_0} - p_0 d_\tau + u_0 \\ &= M_0^{-1} W_0^T (\alpha_0 d_\tau + \beta_0) - p_0 d_\tau + u_0 \\ &=: \psi_0 d_\tau + \phi_0 && (5.45) \end{aligned}$$

with

$$\psi_0 = M_0^{-1} W_0^T \alpha_0 - p_0, \quad (5.46)$$

and

$$\phi_0 = M_0^{-1} W_0^T \beta_0 + u_0. \quad (5.47)$$

Now we wish to find a general expression for $d_{x_{mt}}$ and $d_{y_{mt}}$ in terms of d_τ. Let us write them as

$$\begin{aligned} d_{y_{mt}} &= \alpha_{mt} d_\tau + \beta_{mt}, && (5.48)\\ d_{x_{mt}} &= \psi_{mt} d_\tau + \phi_{mt}, && (5.49) \end{aligned}$$

where $m \in \mathcal{F}_t$ and $t = 0, 1, ..., K$.

A forward recursive formula for α_{mt}, β_{mt}, ψ_{mt} and ϕ_{mt} can be found, based on (5.38) and (5.36), as follows

$$(R_2) \begin{cases} \alpha_{mt} = -(W_{mt} M_{mt}^{-1} W_{mt}^T)^{-1} B_{mt} \psi_{n,t-1} + q_{mt} \\ \beta_{mt} = -(W_{mt} M_{mt}^{-1} W_{mt}^T)^{-1} B_{mt} \phi_{n,t-1} + v_{mt} \\ \psi_{mt} = M_{mt}^{-1} W_{mt}^T \alpha_{mt} - p_{mt} \\ \phi_{mt} = M_{mt}^{-1} W_{mt}^T \beta_{mt} + u_{mt}, \end{cases}$$

where $m \in \mathcal{F}_t$, $n \in \mathcal{F}_{t-1}$ and $n = a(m)$. Putting together (5.42), (5.43), (5.46) and (5.47), we have the following initial values for the recursion

$$(I) \begin{cases} \alpha_0 &= (W_0 M_0^{-1} W_0^T)^{-1}(W_0 p_0 + h_0) \\ \beta_0 &= (W_0 M_0^{-1} W_0^T)^{-1}(-\rho r_{p_0} - W_0 u_0) \\ \psi_0 &= M_0^{-1} W_0^T \alpha_0 - p_0 \\ \phi_0 &= M_0^{-1} W_0^T \beta_0 + u_0. \end{cases}$$

Now, we eliminate d_κ using the seventh and the eighth equations in $(\bar{S})$. Then we substitute $d_{x_{nt}}$ and $d_{y_{nt}}$, according to (5.48) and (5.49), into the resulting equation. This yields

$$\sum_{t=0}^{K} \sum_{n \in \mathcal{F}_t} [h_{nt}^T(\alpha_{nt} d_\tau + \beta_{nt}) - \pi_{nt} c_{nt}^T(\psi_{nt} d_\tau + \phi_{nt})]$$
$$-(-\tau^{-1}\kappa d_\tau + \tau^{-1}((1-\rho)\mu - \tau\kappa)) = -\rho r_g,$$

and so

$$d_\tau = \frac{-\rho r_g + \tau^{-1}((1-\rho)\mu - \tau\kappa) - \sum_{t=0}^{K} \sum_{n \in \mathcal{F}_t} \left[h_{nt}^T \beta_{nt} - \pi_{nt} c_{nt}^T \phi_{nt}\right]}{\sum_{t=0}^{K} \sum_{n \in \mathcal{F}_t} \left[h_{nt}^T \alpha_{nt} - \pi_{nt} c_{nt}^T \psi_{nt}\right] + \tau^{-1}\kappa}. \tag{5.50}$$

All other search directions can be solved using (5.50). In particular, from the seventh equation in $(\bar{S})$ we have

$$d_\kappa = -\tau^{-1}\kappa d_\tau + \tau^{-1}((1-\rho)\mu - \tau\kappa), \tag{5.51}$$

and from (5.48) and (5.49) we get the values for $d_{y_{mt}}$ and $d_{x_{mt}}$. Furthermore, the value for $d_{s_{mt}}$ can be obtained from the sixth equation in $(\bar{S})$, i.e.

$$d_{s_{mt}} = -X_{mt}^{-1} S_{mt} d_{x_{mt}} + X_{mt}^{-1}\left((1-\rho)\mu e - X_{mt} s_{mt}\right). \tag{5.52}$$

To summarize, the search directions implied by the Newton equation $(\bar{S})$ can be solved in the following way. First, we generate the working matrices and vectors M_{mt}, q_{mt}, v_{mt}, p_{mt} and u_{mt} according to the recursive scheme (R_1) with the terminal values determined by (T). This works in a backward manner. Then, we generate another set of working vectors α_{mt}, β_{mt}, ψ_{mt} and ϕ_{mt} using the recursive scheme (R_2) with the initial values given by (I). Finally, the search directions are computed by the formulae (5.50), (5.51), (5.48), (5.49) and (5.52). Clearly, the time and space complexity of such a decomposition scheme is linear with respect to the total number of scenarios.

The crux of this computational scheme is to generate M_{mt}, q_{mt}, v_{mt}, p_{mt} and u_{mt}. This, however, can be accelerated if we have a parallel computational environment, because the computation of the quantities $(M_{n,t-1}, q_{n,t-1}, v_{n,t-1}, p_{n,t-1}, u_{n,t-1})$ from $(M_{mt}, q_{mt}, v_{mt}, p_{mt}, u_{mt})$ can be done independently. The same can be said about the computation of α_{mt}, β_{mt}, ψ_{mt} and ϕ_{mt}. In the computation, one actually only needs to store M_{mt}, q_{mt}, v_{mt}. All the other quantities can easily be derived from this information.

Summarizing, we have the following conclusion.

Theorem 5.3 Let the total number of scenarios in $(MDSLP)$ be $N := \sum_{t=1}^{K} |\mathcal{F}_t|$. Suppose that that matrices W_{nt} and B_{nt} in $(MDSLP)$ are of size s. Then, solving the Newton equation $(\bar{S})$ requires no more than $O(s^3 N)$ elementary operations according to the decomposition method described above.

5.6 Model Diagnosis

Any optimization model is only an approximation of the reality. This means in particular that if a model returns with an undesirable solution, or declares that no solution exists at all, then it might be: 1) the model is not appropriate; or 2) the model is good, but some 'soft constraints' must be relaxed in order to produce a sensible solution. In either case, it is of crucial importance to understand why the model does not have an optimal solution. For stochastic programming, testing the model is especially relevant, because very often the model is only a rough approximation of the real situation in the presence of uncertainty.

A major advantage of using a self-dual embedded model is that if the primal or the dual problem is infeasible, then, instead of getting insignificant output, such as "Infeasible Model", we always get a Farkas type certificate. As we will see in this section, this information is useful in order to analyze the cause of the infeasibility. To see this, consider a linear programming problem given as

$$\begin{aligned} \max \quad & b^T y \\ \text{s.t.} \quad & A^T y \le c. \end{aligned}$$

Suppose that the problem is infeasible. Then by the Farkas lemma and a well known result of Goldman and Tucker [14], there exists $x^* \ge 0$ with maximal cardinality of positive components, such that $Ax^* = 0$ and $c^T x^* < 0$. Let the positive support of x^* be I, i.e., $I = \{i \mid x_i^* > 0\}$. Now we claim that if we remove the set of constraints indexed by I, then the remaining problem must be feasible. The proof is as follows. Let the complement set of I be J. Further let us assume, for the sake of obtaining a contradiction, that the system $A_J^T y \leq c_J$ is still infeasible. Then, by applying the Farkas lemma once more, we conclude that there is $u \geq 0$ such that $A_J u = 0$ and $c_J^T u < 0$. As a by-product, we also know $u \neq 0$. Let

$$\bar{x}_i = \begin{cases} u_i, & \text{if } i \in J \\ 0, & \text{if } i \notin J. \end{cases}$$

This implies that $x^* + \bar{x} \geq 0$, $A(x^* + \bar{x}) = 0$, and $c^T(x^* + \bar{x}) < 0$. Moreover, $x^* + \bar{x}$ has a positive support set which is larger in cardinality than that of x^*. This yields a contradiction. As a result, the claimed fact is proven.

One implication of the above result is the following. Suppose that we have a Farkas type infeasibility certificate with maximal support. Then, its positive support part corresponds to the set of constraints that are causing the infeasibility. In other words, this is the set of scenarios that requires scrutiny. Numerically, if all the data are properly scaled, then we may interpret a higher-valued dual multiplier (Farkas type certificate) to correspond to the scenario that is likely to have more responsibilities for causing the infeasibility. Certainly, due to the primal-dual symmetricity, exactly the same thing can be said about a linear program in the primal form

$$\begin{aligned} \min \quad & c^T x \\ \text{s.t.} \quad & Ax = b, \\ & x \geq 0. \end{aligned}$$

In this case, if the problem is infeasible then the corresponding Farkas type certificate is $-A^T y^* \geq 0$ such that $b^T y^* > 0$.

Let us consider one example to illustrate this point. Suppose that there is a risky asset, say a common stock, whose value may go up, or down, or remain unchanged in the next period. To be specific, assume that with 40% chance it will go up 10%, with 30% chance it will go down 4%, and with 30% chance it will be unchanged. Let us concentrate on a two-period situation. A riskless investment will always yield a 2% return rate in the same period, no matter what happens. Suppose we have a notional 1$ to invest. Now we wish to find an investment plan, allocating our wealth and possibly updating our portfolio at the end of period 1, in such a way that our wealth will never go below value g, and at the same time, the expected return rate is maximized at the investment horizon, i.e., at the end of period 2.

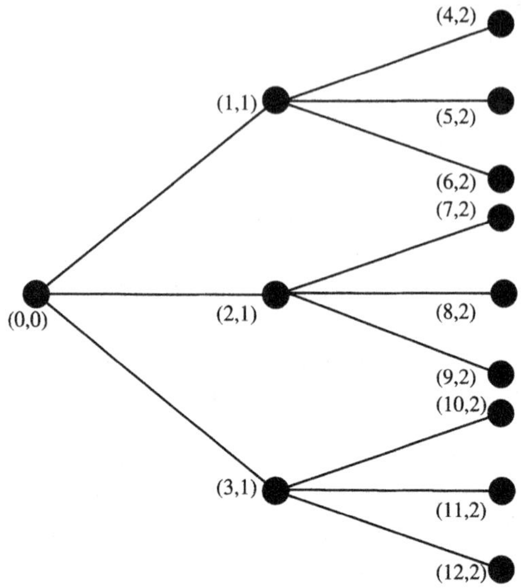

FIGURE 5.2. Scenario tree: Example

Clearly, in this case the scenario tree involves 13 nodes (see Figure 5.2). We denote the current stage to be (0,0), and denote the set of scenarios at the first stage (up till down) be (1,1), (2,1) and (3,1), each followed by three children. In particular, (1,1) is followed by (4,2), (5,2) and (6,2), (2,1) is followed by (7,2), (8,2) and (9,2), and (3,1) is followed by (10,2), (11,2) and (12,2); see Figure 5.2.

Let the uncertain return rate of the stock (over one period) be ω. In our model, $\omega_u = 0.1$ (up), $\omega_e = 0$ (equal), and $\omega_d = -0.04$ (down). The riskless rate over one period is $r = 0.02$. Let x^s_{nt} be the amount invested in the stock, and x^b_{nt} be the amount invested in the riskless asset, at stage t if scenario n unfolds. Then, the constraints in the model are:

$$x^s_0 + x^b_0 = 1,$$

$$(1+\omega)x^s_{a(n),0} + (1+r)x^b_{a(n),0} = x^s_{n1} + x^b_{n1}, \text{ for } n = 1,2,3,$$

$$x^s_{n2} + x^b_{n2} \geq g, \text{ for } n = 4,5,...,12.$$

We further assume that no short selling is allowed. In our notation, this is equivalent to the following constraint matrices

$$W_0 = [1,1],\ h_0 = 1,$$

$$B_{n1} = -[1+\omega, 1+r],\ W_{n1} = [1,1],\ h_{n1} = 0,$$

$$B_{n2} = \begin{bmatrix} -(1+\omega), & -(1+r) \\ 0, & 0 \end{bmatrix},\ W_{n2} = \begin{bmatrix} 1, & 1, & 0 \\ 1, & 1, & -1 \end{bmatrix},\ h_{n2} = \begin{bmatrix} 0 \\ g \end{bmatrix}.$$

Note here that we have introduced 9 slack variables for the guaranteed return constraints: $x^s_{n2} + x^b_{n2} \ge g$, $n = 4, 5, ..., 12$.

Now we run our algorithm on this model. When the investor invests his wealth in the riskless asset over the two-periods in our model, his final wealth will be 1.0404 $(1 * 1.02 * 1.02)$. A higher guaranteed return is not possible due to the uncertainty of the risky asset. Hence, for $g \le 1.0404$, the model is solvable. For instance, if $g = 1$, then $x^s_0 = 0.6601$ and $x^b_0 = 0.3399$, with an expected return rate at the horizon being 5.03%. In comparison, the riskless return rate at the horizon is 4.04%. As soon as $g > 1.0404$, the model becomes demanding a guaranteed return rate higher than the riskless rate. This is certainly impossible in the world where no arbitrage opportunity exists, and the model should return as being infeasible. Indeed, this turns out to be the case. Set for instance $g = 1.05$. The model is infeasible. The point here is, our algorithm also provides a dual certificate

$$
\begin{aligned}
&-A^T y^* = \\
&[0.3548, 0.0068, 0.1983, 0.0138, 0.2513, 0.0120, 0.2972, 0.0077, \\
&0.0248, 0.0248, 1.7230, 0.0243, 0.0243, 2.8122, 0.0242, 0.0242, 4.4356, \\
&0.0240, 0.0240, 2.1821, 0.0238, 0.0238, 3.7830, 0.0239, 0.0239, 5.6368, \\
&0.0239, 0.0239, 2.5476, 0.0237, 0.0237, 4.5298, 0.0237, 0.0237, 6.7116].
\end{aligned}
$$

If we examine the values in the certificate carefully, then we will find that there are 9 numbers that are significantly higher than 1, and they correspond to the constraints:

$$x^s_{n2} + x^b_{n2} \ge g, \text{ for } n = 4, 5, ..., 12.$$

As we know, these constraints are indeed the ones that are causing the infeasibility.

Similar analysis applies to the situation when the problem has an unbounded optimum value. In portfolio applications, this phenomenon is known as the existence of an arbitrage opportunity. In particular, the primal Farkas type certificate plays the role of displaying one such arbitrage opportunity. In this case, the meaning of the solution produced by our algorithm is quite obvious.

5.7 Notes

There is a large body of literature on multistage stochastic programming. Below we shall point out a few representative references related to our discussions in this chapter for the benefit of the interested reader.

Along the direction of (parallel) implementations of the Benders type decomposition, one may consult [7, 13] and [15].

For the augmented Lagrangian approach, one may further read [4, 20] and [21].

As for the decomposition type approach based on the interior point methods, one is referred to [8, 11, 18, 28] for serial implementations, and [12, 16, 25] for parallel implementations. Finally, for the cutting plane type approach combined with the interior point methodology, see [2] and [3].

Acknowledgments

This project has been supported by the Foundation 'Vereniging Trustfonds Erasmus Universiteit Rotterdam' in The Netherlands, and by RGC Earmarked Grant CUHK 4233/01E.

5.8 References

[1] E.D. Andersen, and K.D. Andersen, *The MOSEK interior point optimizer for linear programming: an implementation of the homogeneous algorithm*, High Performance Optimization Techniques, pp. 197-232, eds. J.B.G. Frenk, K. Roos, T. Terlaky and S. Zhang, Kluwer Academic Publishers, 1999.

[2] K.A. Ariyawansa, and P.L. Jiang, Polynomial cutting plane algorithms for two-stage stochastic linear programs based on ellipsoids, volumetric centers and analytic centers, *Working Paper*, Department of Pure and Applied Mathematics, Washington State University, Pullman, USA, 1996.

[3] O. Bahn, O. du Merle, J.-L. Goffin, and J.P. Vial, A cutting plane method from analytic centers for stochastic programming, *Mathematical Programming*, **69** (1995), 45-73.

[4] A.J. Berger, J.M. Mulvey and A. Ruszczynski, An extension of the DQA algorithm to convex stochastic programs, *SIAM Journal on Optimization*, **4** (1994), 735-753.

[5] A. Berkelaar, C. Dert, C. Oldenkamp, and S. Zhang, A Primal-Dual Decomposition-Based Interior Point Approach to Two-Stage Stochastic Linear Programming, Report 9918/A, Econometric Institute, Erasmus University Rotterdam, 1999. (To appear in *Operations Research*).

[6] A. Berkelaar, R. Kouwenberg, and S. Zhang, A Primal-Dual Decomposition Algorithm for Multistage Stochastic Convex Programming, Technical Report SEEM2000-07, Department of Systems Engineering & Engineering Management, The Chinese University of Hong Kong, 2000. (Submitted for publication).

[7] J.R. Birge, C.J. Donohue, D.F. Holmes, and O.G. Svintsitski, A parallel implementation of the nested decomposition algorithm for multistage stochastic linear programs, *Mathematical Programming*, **75** (1996), 327-352.

[8] J.R. Birge, and D.F. Holmes, Efficient solution of two-stage stochastic linear programs using interior point methods, *Computational Optimization and Applications*, **1** (1992), 245-276.

[9] J.R. Birge, and F. Louveaux, *Introduction to Stochastic Programming*, Springer, New York, 1997.

[10] J.R. Birge, and L. Qi, Computing block-angular Karmarkar projections with applications to stochastic programming, *Management Science*, **34**(1988), 1472-1479.

[11] I.C. Choi, and D. Goldfarb, Exploiting special structure in a primal-dual path-following algorithm, *Mathematical Programming*, **58** (1993), 33-52.

[12] J. Czyzyk, R. Fourer and S. Mehrotra, Using a massively parallel processor to solve large sparse linear programs by an interior-point method, *SIAM Journal on Scientific Computing*, **19** (1988), 553-565.

[13] E. Fragnière, J. Gondzio, and J.-P. Vial, Building and Solving Large-scale Stochastic Programs on an Affordable Distributed Computing System, forthcoming in *Annals of Operations Research*, 1998.

[14] A.J. Goldman and A.W. Tucker, Polyhedral convex cones, in H.W. Kuhn and A.W. Tucker eds., *Linear Inequalities and Related Systems*, Princeton University Press, New Jersey, 19-40, 1956.

[15] J. Gondzio, and R. Kouwenberg, High Performance Computing for Asset Liability Management, forthcoming in *Operations Research*, 2000.

[16] E.R. Jessup, D. Yang, and S.A. Zenios, Parallel factorization of structured matrices arising in stochastic programming, *SIAM Journal on Optimization*, **4** (1994), 833-846.

[17] P. Kall, and S.W. Wallace, *Stochastic Programming*, John Wiley and Sons, Chichester, 1994.

[18] I.J. Lustig, J.M. Mulvey and T.J. Carpenter, The formulation of stochastic programs for interior point methods, *Operations Research*, **39** (1991), 757-770.

[19] S. Mizuno, M.J. Todd, and Y. Ye, On adaptive-step primal-dual interior-point algorithms for linear programming, *Mathematics of Operations Research*, **18** (1993), 964-981.

[20] J.M. Mulvey, and A. Ruszczynski, A new scenario decomposition method for large-scale stochastic optimization, *Operations Research* , **43** (1995), 477-490.

[21] R.T. Rockafellar, and R.J.-B. Wets, Scenarios and policy aggregation in optimization under uncertainty, *Mathematics of Operations Research*, **16** (1991), 119-147.

[22] C. Roos, T. Terlaky, and J.-Ph. Vial, *Theory and Algorithms for Linear Optimization*, John Wiley & Sons, chapter 19, 1997.

[23] R. Van Slyke, and R.J.-B. Wets, L-shaped linear programs with applications to optimal control and stochastic linear programs, *SIAM Journal on Applied Mathematics*, **17** (1969), 638-663.

[24] X. Xu, P.F. Hung, and Y. Ye, A simplified homogeneous self-dual linear programming algorithm and its implementation, *Annals of Operations Research*, **62** (1996), 151-171.

[25] D. Yang, and S.A. Zenios, A Scalable parallel interior point algorithm for stochastic linear programming and robust optimization, *Computational Optimization and Applications*, **7** (1997), 143-158.

[26] Y. Ye, and K. Anstreicher, On quadratic and $O(\sqrt{n}L)$ convergence of a predictor-corrector algorithm for LCP, *Mathematical Programming*, **62** (1993), 537-551.

[27] Y. Ye, M.J. Todd, and S. Mizuno, An $\mathcal{O}(\sqrt{n}L)$-iteration homogeneous and self-dual linear programming algorithm, *Mathematics of Operations Research*, **19** (1994), 53-67.

[28] G. Zhao, A log-barrier method with Benders decomposition for solving two-stage stochastic linear programs, *Mathematical Programming*, **90** (2001), 507-536.

6

A Brownian Model of Stochastic Processing Networks

Hong Chen

Stochastic processing network and its special case, queueing network, has long been widely used in assisting design, performance analysis and optimal control of manufacturing systems, and computer and telecommunication networks. However, exact analysis of such networks is inherently difficult and has been limited to some very simple instances. Hence, in the past 30 years, considerable efforts have been focused on studying the simpler approximation models such as fluid models and Brownian models.

This chapter is motivated by and closely related to a recent work of Harrison [10], in which the Brownian model is developed via the heavy traffic regime. We take the view point of Chen and Mandelbaum [2, 3] that the strong approximation provides a framework for the hierarchical approach to the stochastic processing network: namely, it links the microscopic detailed stochastic processing network model with its macroscopic highly aggregated fluid network and mesoscopic moderately aggregated Brownian network. The distinctive advantage of the strong approximation over the heavy traffic approximation is that the strong approximation requires neither the heavy traffic condition nor the scaling in time and space, thus making the presentation of the strong approximation approach much simpler. In addition, with some mild additional conditions, the strong approximation, when it exists, implies the heavy traffic approximation. This point of view is supported by Glynn [13], where the pedagogical advantage of the strong approximation approach over the heavy traffic approach is illustrated through a GI/G/1 queue.

The stochastic processing network under consideration is the same as the one considered in [10]. It is a very general model. For example, it can model the scheduling control and the routing control of a multiclass queueing network, and it can also model an assemble-to-order production-inventory system.

In this chapter, following the strong approximation heuristic, we develop a new Brownian network model of the stochastic processing network. By removing the random element from this Brownian model, we also obtain a fluid model. (Note that the fluid model can also be developed independently under the framework of the functional law-of-iterated logarithm.) The simplification of this fluid model under a heavy traffic condition will lead to the static planning problem in [10]. A simple example is used to illustrate the potential advantage of the Brownian model developed here over the one in [10]. However, a thorough comparison remains an interesting future research.

The chapter is organized as follows. In the next section, we present some preliminary results on the strong approximation. Of particular importance is the strong random-time-change theorem. The stochastic processing network is described in section 6.2. In section 6.3, we present a number of examples and show how to identify various modeling parameters of the network. In section 6.4, the Brownian model of the stochastic processing network is presented, and its link to the Brownian model in [10] is discussed. As a simple corollary, we obtain a fluid model of the stochastic processing network. The extension to the stochastic model with input control is also discussed. The heuristic derivation of the Brownian model is presented in section 6.5. In the Appendix, we discuss the relationship between the heavy traffic limit regime and the strong approximation approach. This puts our Brownian model in perspective visa vis the one developed by Harrison [10].

6.1 Preliminaries

Let $\{\xi_n, n \geq 1\}$ be a sequence of i.i.d. random variables; the summation process $X = \{X(t), t \geq 0\}$ is defined by

$$X(t) = 0, \quad t \in [0,1); \qquad X(t) = \sum_{n=1}^{\lfloor t \rfloor} \xi_n, \quad t \geq 1.$$

Let m and σ be the mean and the standard deviation of ξ_1.

If ξ_1 has a finite second moment, then the functional law-of-iterated logarithm (FLIL) holds for X, namely,

$$\sup_{0 \leq t \leq T} |X(t) - \bar{X}(t)| \overset{a.s.}{=} O\left(\sqrt{T \log\log T}\right) \quad \text{as } T \to \infty,$$

where $\bar{X}(t) = mt$. In general, for any two processes X and $\bar{X}$ satisfying the above equality, we shall write

$$X(t) \overset{lil}{\approx} \bar{X}(t), \qquad \text{or equivalently,} \qquad X \overset{lil}{\approx} \bar{X}.$$

If, further, ξ_1 has a finite moment of order $r > 2$, then the strong approximation holds for X (see, for example, Csörgő and Horváth [8]); specifically, there exists a Wiener process $W = \{W(t), t \geq 0\}$ such that

$$\sup_{0 \leq t \leq T} |X(t) - \tilde{X}(t)| \stackrel{a.s.}{=} o(T^{1/r}) \text{ as } T \to \infty, \tag{6.1}$$

where

$$\tilde{X}(t) = mt + \sigma W(t), \quad t \geq 0.$$

(More precisely, there exists a probability space on which a version of X is defined such that the above holds; for simplicity, we shall assume through the chapter that all the concerned processes have already been defined on the appropriate probability space.) In general, for any two processes X and $\tilde{X}$ satisfying (6.1), we shall write

$$X(t) \stackrel{r}{\approx} \tilde{X}(t), \qquad \text{or equivalently,} \qquad X \stackrel{r}{\approx} \tilde{X}.$$

Let $W = \{W(t), t \geq 0\}$ be a standard Brownian motion and let $\tau = \{\tau(t), t \geq 0\}$ be a random time-change process defined on the same probability space. Assume that τ has an FLIL approximation; namely, $\tau(t) \stackrel{\ell i\ell}{\approx} \alpha t$ for some nonnegative constant α. Then for any $r \in (2, 4)$,

$$\sup_{0 \leq t \leq T} |X(\tau(t)) - X(\alpha t)| \stackrel{a.s.}{=} o(T^{1/r}), \qquad \text{as } T \uparrow \infty;$$

see Lemma 6.21 of Chen and Yao [7]. This result resembles the random-time-change theorem in the weak convergence, and we shall refer to it a strong random-time-change theorem.

Next we generalize this strong random-time-change result. A stochastic process $X = \{X(t), t \geq 0\}$ is said to be *strongly continuous* with degree r, or r-strongly continuous, for some $r \in (2, 4)$, if

$$\sup_{\substack{0 \leq u,v \leq T \\ |u-v| \leq h(T)}} |X(u) - X(v)| \stackrel{a.s.}{=} o(T^{1/r}), \text{ as } T \to \infty, \tag{6.2}$$

where $h(T) \equiv \sqrt{T \log\log T}$, and it is simply said to be strongly continuous if it is r-strongly continuous for all $r \in (2, 4)$. Note, in particular, that the standard Brownian motion is strongly continuous. We say a stochastic process X has *a strong approximation* if for some $r \in (2, 4)$, there exists a probability space on which a version of X (for simplicity we still write it as X) and an r-strongly continuous stochastic process $\hat{X}$ are defined such that

$$\sup_{0 \leq t \leq T} |X(t) - mt - \hat{X}(t)| \stackrel{a.s.}{=} o(T^{1/r}),$$

where m is a (deterministic) constant. When the above equality holds, we also say that X has a strong approximation $\tilde{X} = \{\tilde{X}(t), t \geq 0\}$ with $\tilde{X}(t) = mt + \hat{X}(t)$.

Theorem 6.1 (i) A Wiener process is r-strongly continuous;
(ii) (Strong Random-Time-Change Theorem) Suppose that

$$X(t) \overset{r}{\approx} at + \hat{X}(t) \quad \text{and} \quad \tau(t) \overset{\ell i \ell}{\approx} bt.$$

If $\hat{X}$ is r-strongly continuous, then

$$X(\tau(t)) \overset{r}{\approx} a\tau(t) + \hat{X}(bt).$$

The role of this theorem in deriving the strong approximation is as important as the role played by the random-time-change theorem in deriving the diffusion approximation for the stochastic processing network. (cf. Lemma 9.6 of Chen and Yao [7].)

6.2 Stochastic Processing Network Model

The stochastic processing network is the same as the one considered in Harrison [10]. Specifically, the network consists of J stations or resources; these stations support N different processing activities that produce and consume K types of materials (or process K different classes of jobs). Processing stations are related to activities by a $J \times N$ matrix $A = (A_{jn})$, where A_{jn} denotes the average amount of station j capacity consumed per unit of activity n. Material types (or job classes) are related to activities by a $K \times N$ matrix $R = (R_{kn})$, where R_{kn} denotes the average amount of material k consumed per unit of activity n, with a negative value indicating a unit of activity n produces R_{kn} units of material k. (Note that R_{kn} can also be interpreted as the average number of class k job departures generated per unit of activity n, with a negative value indicating job arrivals.) We denote by $c = (c_j)$ the J-dimensional capacity vector with c_j indicating the capacity of station j. Following [11], we call matrix A the *capacity consumption matrix*, matrix R the *input-output matrix*, and vector c the *capacity vector.*

We take as given $(N+1)$ K-dimensional flow processes, $F_n = \{F_n(t), t \geq 0\}$ with $F_n(t) = (F_{kn}(t))_{k=1}^K$, $n = 0, 1, \ldots, N$. The quantity $F_{k0}(t)$ indicates the cumulative amount of exogenous input of material k, and for $n = 1, \ldots, N$, $F_{kn}(t)$ indicates the cumulative amount of material k consumed by the first t units of activity n, with a negative value indicating a net production. We assume that these flow processes have the following strong approximations:

$$F_0(t) \overset{r}{\approx} \lambda t + \hat{F}_0(t), \tag{6.3}$$

$$F_n(t) \overset{r}{\approx} R_n t + \hat{F}_n(t), \qquad n = 1, \ldots, N, \tag{6.4}$$

where $\lambda = (\lambda_k)_{k=1}^K$, $R_n = (R_{kn})_{k=1}^K$ is the n-th column of matrix R, and $\{\hat{F}_n, n = 0, \ldots, N\}$ are $(N+1)$ K-dimensional driftless Brownian motions.

We denote by $Q = \{Q(t), t \geq 0\}$ the inventory process, where $Q_k(t)$, the k-th component of $Q(t)$, indicates the inventory of material k (or the number of class k jobs). The allocation of the resources to activities is described by an N-dimensional (resource) allocation process $T = \{T(t), t \geq 0\}$, where $T_n(t)$, the n-th component of $T(t)$, is the cumulative amount of activity n undertaken during the time interval $[0, t]$. Then, the material flow must satisfy the following flow-balance relation:

$$Q(t) = Q(0) + F_0(t) - \sum_{n=1}^{N} F_n(T_n(t)), \tag{6.5}$$

and the allocation process must satisfy the following resource capacity constraint:

$$0 \leq A[T(t) - T(s)] \leq c[t - s]. \tag{6.6}$$

We do *not* assume that the inventory process is nonnegative; a negative value can model, for example, a backlog of materials. Hence, in general, we assume that Q lies in a general state space, $\mathcal{S}$, a subset of $\Re^K$.

The allocation process is in general a decision process, which is chosen to maximize the expected profit that can be generated by the stochastic processing network. Let v_n denote the net revenue generated per unit of activity n, and let $h(\cdot)$ be an inventory cost function. The network optimization problem is to choose the allocation process T, so as to maximize the expected profit,

$$\mathsf{E} \int_0^t e^{-rs} \Big[v' dT(s) - h(Q(s)) ds \Big], \tag{6.7}$$

where $r \geq 0$ is the discount rate and the time horizon t can be either finite or infinity. For the special case $r = 0$ and $t < \infty$, the above objective function has the form,

$$\mathsf{E} \left[v'T(t) - \int_0^t h(Q(s)) ds \right], \tag{6.8}$$

In many applications, the allocation process may be constrained. For example, in the queueing network context, the allocation process corresponds to the service discipline, and one may be interested in studying the network under a particular service discipline (such as a particular priority service discipline or a first-in-first-out service discipline). In this case, the allocation problem is uniquely specified, and the problem of interest is the performance evaluation, i.e., finding the value of the expected profit (6.7) or its approximation under the specified allocation.

6.3 Examples of Stochastic Processing Networks

6.3.1 Scheduling Control of Multiclass Queueing Network

Consider a queueing network with J single server stations processing K classes of jobs. Each station may process (or serve) more than one class of jobs, but each job class is processed at one station. The service rate for a class k job is μ_k, or equivalently its mean service time is $1/\mu_k$. A proportion p_{ki} of class k jobs, after their service completion, becomes class i jobs (for further service), and the remaining proportion, $1 - \sum_{i=1}^{K} p_{ki}$, leaves the network. The optimal allocation problem is to determine the order at which jobs of different classes are served at each station.

This problem is a special case of the stochastic processing network problem introduced in the previous section. In this case, the state space $\mathcal{S} = \{x \in \Re^K : x \geq 0\}$ is the nonnegative orthant, the capacity vector $c = e$ (a vector of ones), and the number of activities N is the same as the number of job classes (or distinct materials) K. The (j, k)-th component of the $J \times K$ matrix A equals 1 if class k job is processed at station j and 0 otherwise. The matrix $R = (I - P')\text{diag}(\mu)$.

The flow processes F_n, $n = 0, 1, \ldots, K$, in this case can be constructed from the primitive processes: the exogenous arrival process, service process and routing process. Specifically, let $A_k(t)$ indicate the number of class k jobs that have arrived exogenously up to time t; let $S_k(t)$ denote the number of class k jobs that can be processed for the first t units of processing time; and let $\xi_i^k(\ell)$ indicate the event that the ℓ-th class k job that completes the service becomes a class i job. Without loss of generality, assume that $\xi_k^k(\ell) = 0$ almost surely for all $k = 1, ..., K$ and $\ell = 1, 2, \ldots$, i.e., no immediate feedback; otherwise, we can simply redefine S_k. Then we have

$$
\begin{aligned}
F_{0k}(t) &= A_k(t), \\
F_{kk}(t) &= S_k(t), \qquad k = 1, ..., K, \\
F_{ki}(t) &= -\sum_{\ell=1}^{S_i(t)} \xi_k^i(\ell), \qquad i \neq k \text{ and } i, k = 1, ..., K.
\end{aligned}
$$

Suppose that the primitive arrival process, the service process and the routing sequence satisfy the following strong approximation,

$$
\begin{aligned}
A(t) &\overset{r}{\approx} \lambda t + \hat{A}(t), \\
S(t) &\overset{r}{\approx} \mu t + \hat{S}(t), \\
\sum_{\ell=1}^{\lfloor t \rfloor} \xi^k(\ell) &= P_k t + \hat{R}^k(t),
\end{aligned}
$$

where P_k is the transpose of the k-th row of matrix $P = (p_{ki})$ and $\xi^k(\ell) = (\xi_i^k(\ell))_{i=1}^K$. (A sufficient condition for the above approximation is that the routing sequence $\{\xi^k(\ell), \ell \geq 1\}$ is an i.i.d. sequence with finite r-th moment, $k = 1, \ldots, K$, and both A and S are renewal processes whose interarrival times have finite r-th moments.) Then the strong approximations (6.3) and (6.4) hold with

$$\begin{aligned} \hat{F}_0(t) &= \hat{A}(t), \\ \hat{F}_{kk}(t) &= \hat{S}_k(t), \qquad k = 1, \ldots, K, \\ \hat{F}_{ki}(t) &= -\hat{R}_k^i(\mu_i t), \qquad i \neq k \text{ and } i, k = 1, \ldots, K. \end{aligned}$$

Suppose that each class k job that completes service and becomes a class i generates a net reward r_{ki}, and each class k job that completes service and leaves the network generates a reward r_{k0}. A negative reward means a cost, which may happen when a class k job becoming a class i job represents a defect requiring rework/repair. Note that the activity k generates the class k departure at the rate μ_k. Hence, the net reward generated by activity k is given by

$$v_k = \mu_k \sum_{i=0}^{K} p_{ki} r_{ki}, \qquad k = 1, \ldots, K,$$

where $p_{k0} = 1 - \sum_{i=1}^{K} p_{ki}$, $k = 1, \ldots, K$.

6.3.2 A Simple Queueing Network with both Scheduling and Routing

Consider the queueing network as depicted by figure 6.1. The network has $J = 3$ single server stations. There are three classes of jobs entering the network exogenously, which are indexed by $k = 1, 2, 3$. Jobs of classes 1 and 2 are processed at stations 1 and 2 with mean service times $1/\mu_1$ and $1/\mu_2$, respectively. Jobs of class 2 leaves the network after service completion; however, jobs of class 1 after service completion requires further service at station 3 with mean service time $1/\mu_5$, before exiting the network. Jobs of class 3 can be processed at either station 1 with mean service time $1/\mu_3$ or station 2 with mean service time $1/\mu_4$. While a job of class 3 after service completion at station 1 will become a job of class 2, a job of class 3 after service completion at station 2 requires further service at station 3 with mean service time μ_6 before exiting the network.

This problem is a special case of the stochastic network described in the previous section, with $J = 3$, $K = 5$ and $N = 6$. In this case, the state space $\mathcal{S} = \{x \in \Re^5 : x \geq 0\}$, the capacity vector $c = e$, and the capacity consumption matrix A and the input-output matrix R are as given below.

$$A = \begin{pmatrix} 1 & 0 & 1 & 0 & 0 & 0 \\ 0 & 1 & 0 & 1 & 0 & 0 \\ 0 & 0 & 0 & 0 & 1 & 1 \end{pmatrix},$$

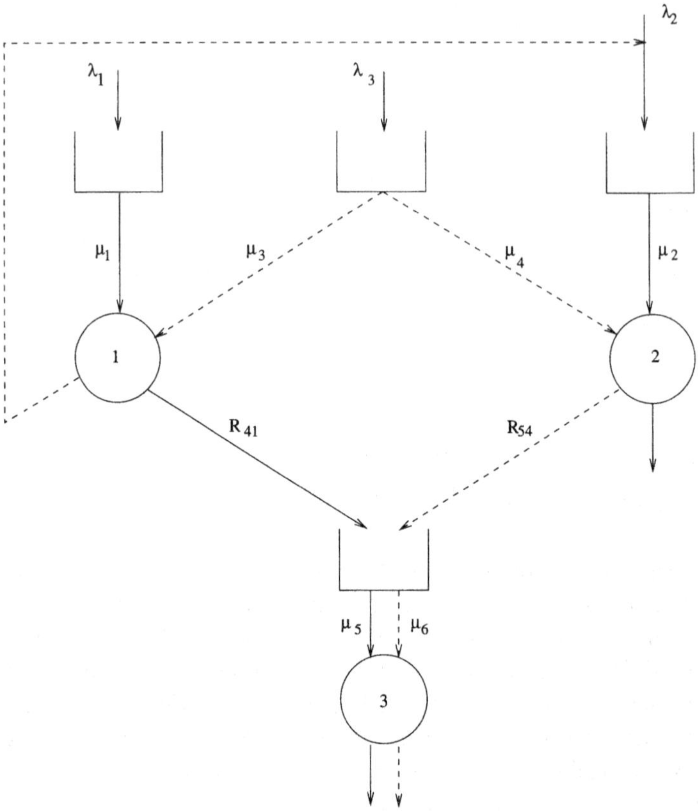

FIGURE 6.1. A simple queueing network subject to both scheduling and routing control.

$$R = \begin{pmatrix} \mu_1 & 0 & 0 & 0 & 0 & 0 \\ 0 & \mu_2 & -\mu_3 & 0 & 0 & 0 \\ 0 & 0 & \mu_3 & \mu_4 & 0 & 0 \\ -\mu_1 & 0 & 0 & 0 & \mu_5 & 0 \\ 0 & 0 & 0 & -\mu_4 & 0 & \mu_6 \end{pmatrix}.$$

In this network, the exogenous input flow $F_{0k}(t) = 0$ for $k = 4, 5$, and for $k = 1, 2, 3$, $F_{0k}(t)$ indicates the number of class k jobs that have arrived up to time t. Let S_k be the renewal process that models the service activity k,

$k = 1, \ldots, 6$. Then the other flow processes are given by

$$(F_1, \ldots, F_6) = \begin{pmatrix} S_1 & 0 & 0 & 0 & 0 & \\ 0 & S_2 & -S_3 & 0 & 0 & 0 \\ 0 & 0 & S_3 & S_4 & 0 & 0 \\ -S_1 & 0 & 0 & 0 & S_5 & 0 \\ 0 & 0 & 0 & -S_4 & 0 & S_6 \end{pmatrix}.$$

The strong approximation assumptions (6.3) and (6.4) hold, if the exogenous arrival processes and the service processes are renewal processes with the interarrival times having finite r-th moments.

6.3.3 An Assemble-To-Order System

Consider a multi-component, multi-product, production-inventory system, in which the final products are assembled to order but the components are made to stock. There are K different types of components that are to be assembled into I types of products. Each type of components is produced and stored at its own production facility (or station). Hence, in this case, we have $J = K = N$. Figure 6.2 depicts an assemble system with $K = 4$ components and $I = 2$ final products. Note that a type k

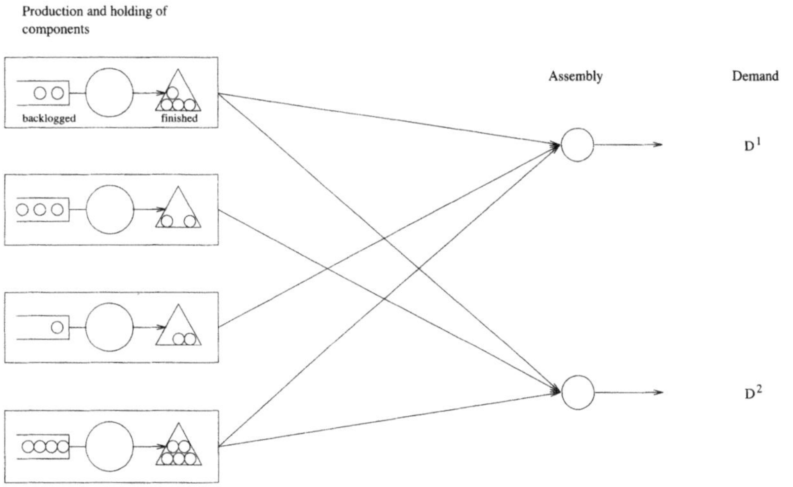

FIGURE 6.2. An assemble-to-order system with four components and two final products.

component corresponds to material k or a class k job in the general model described in section 6.2. The matrix A is a $J \times J$ identity matrix, and matrix $R = \text{diag}(\mu)$ is a diagonal matrix whose k-th diagonal component is μ_k (the k-th component of μ) representing the production rate for component k. Associated with each product i is a bill-of-material $b^i = (b^i_k)_{k=1}^K$, where b^i_k is the number of units of component k in one unit of product i.

In this case, the inventory process Q is used to model the inventory outstanding; specifically, when $Q_k(t)$ is positive, $Q_k(t)$ represents back-ordered quantity of component k (which can also be considered as the number of component k waiting to be produced at its production facility), and when $Q_k(t)$ is negative, $|Q_k(t)|$ represents on-hand inventory of component of k. The state space $\mathcal{S}$ can be taken as $\Re^K$ if the inventory buffer sizes for all components are infinite and there are no upper bounds on the number of component orders that can be backlogged. Otherwise, $\mathcal{S}$ can be taken as an appropriate subset of $\Re^K$ to reflect the inventory buffer size and the backlog limit. With the above state definition, $F_{0k}(t)$ models the cumulative demand for component k during $[0, t]$. Let $D^i(t)$ denote the number of orders of product i placed up to time t, and let $\xi^i(\ell)$ denote the order quantity (the number of units) for the ℓ-th order of product i. Then $F_0(t) = (F_{0k}(t))_{k=1}^K$ is given by

$$F_0(t) := \sum_{i=1}^{I} \sum_{\ell=1}^{D^i(t)} \xi^i(\ell) b^i.$$

Let $S_k(t)$ denote the number of component k that can be produced for the first t units of production time. Then

$$F_{kk}(t) = S_k(t) \quad \text{and} \quad F_{ki}(t) = 0, \qquad i \neq k \text{ and } i, k = 1, \ldots, J.$$

The waiting time for product i can be obtained as follows. Consider a unit demand of component k at time t. If $Q_k(t-) < 0$, then the waiting time for this demand is zero. If $Q_k(t-) \geq 0$, then the number of back-orders at time t is $Q_k(t)$ and the current demand becomes the last unit to be filled. Hence, the waiting time for this component equals the total service times of all $Q_k(t)$ components, which is approximately $Q_k(t)/\mu_k$ following Little's Law. (Note that this becomes exact in the Brownian model.) Let $\pi^i = \{k : b^i_k > 0\}$ be the set of components required by product i. Then the waiting time for a unit demand of product i arriving at time t is

$$W^i(t) := \max_{k \in \pi^i} \frac{[Q_k(t)]^+}{\mu_k}, \qquad i = 1, \ldots, I.$$

In this case, the inventory holding cost function $h(Q(t))$ would take the form, $h(Q(t)) = g([-Q(t)]^+, W(t))$. The net reward associated with the allocation process T can be taken as zero, i.e., $v = 0$.

Finally, we note that the above formulation may also be extended to the case where each production facility may produce more than one type of components. In this case, in addition to the inventory replenishment decision, there is also a scheduling problem (in terms of the order at which the components are produced) at the production facilities.

6.4 Brownian Model for Stochastic Processing Network

In this section, we formulate a Brownian model for the stochastic processing network described in section 6.2. Then we make a brief comparison of our formulation with that given by Harrison [10], and we conclude with a discussion of the extension to incorporate the input control.

Recall that the network is described by a $J \times N$ capacity consumption matrix A, a $K \times N$ input-output matrix R, a J-dimensional capacity vector c, and $(N+1)$ K-dimensional flow processes F_n, $n = 0, 1, \ldots, N$, where the flow processes are assumed to have the strong approximations (6.3) and (6.4). Then the Brownian model is formulated by the following relations:

$$Z(t) = Z(0) + X(t) + RY(t) \in \mathcal{S}, \tag{6.9}$$

$$Y_n(t) \downarrow \text{ if } \beta_n = 0, \qquad n = 1, \ldots, N, \tag{6.10}$$

$$U(t) \equiv \nu t + AY(t) \uparrow, \tag{6.11}$$

where

$$X(t) \equiv \hat{F}_0(t) - \sum_{n=1}^{N} \hat{F}_n(\beta_n t),$$

and the vector parameter β satisfies

$$R\beta = \lambda, \tag{6.12}$$

$$\nu \equiv c - A\beta \geq 0. \tag{6.13}$$

The downward arrow in (6.10) means that Y_n is non-increasing, and the upward arrow in (6.11) means that U is non-decreasing. Relation (6.9) corresponds to the flow-balance relation (6.5), and both relations (6.10) and (6.11) correspond to the resource capacity constraint (6.6) (respectively to the first and the second inequalities in (6.6)). The vector parameter β in the Brownian model formulation and the conditions in (6.12) and (6.13) will be interpreted shortly.

The K-dimensional process Z is the inventory process in the Brownian model, and it can be considered being a Brownian approximation for the inventory process Q. The process X is a driftless Brownian motion. The parameter β_n, the n-th component of the N-dimensional vector β, is the long-run rate at which activity n has been performed over time. Hence, the equality condition $R\beta = \lambda$ in (6.12) is to require that all exogenous input flows be processed, and the inequality $\nu \geq 0$ in (6.13) is the first-order capacity constraint. The quantity $Y_n(t)$, the n-th component of the N-dimensional vector $Y(t)$, models the deviation between the idea amount of activity n (based on the long-run rate) and the actual amount of activity n that have been performed up to time t. Hence, $Y(t)$ can be considered being a Brownian approximation for the difference $[\beta t - T(t)]$. Note that both β and Y are controllable. The J-dimensional process U is the unused capacity process, and its j-th component, U_j, can be considered being a Brownian approximation for the unused capacity of resource j, $j = 1, \ldots, J$.

Note that given any pair (β, Y), relation (6.9) determines the inventory process Z. If a pair (β, Y) is such that relations (6.9)-(6.13) hold and that Y is adapted to the filtration generated by the Brownian motion X, then we call the pair an admissible control. The Brownian optimal control problem is to find an admissible control so that the following is optimized:

$$\max_{\beta \geq 0, Y} \mathsf{E}\left(\left[1 - e^{-rt}\right]\frac{v'\beta}{r} - \int_0^t e^{-rs}\left[v'dY(s) + h(Z(s))ds\right]\right) \quad (6.14)$$

$$\text{subject to} \qquad (6.9) - (6.13). \quad (6.15)$$

This objective function corresponds to the objective function given by (6.7). This optimization problem can be decomposed into an equivalent hierarchical optimization problem: at the upper level is the long-run capacity allocation decision (usually made by the top management) with the anticipation of the lower level performance response, and at the lower level is the short-run production control (usually made by the lower level management or the operators) in response to the stochastic variability of the demand and the production. Specifically, fix any $\beta \geq 0$ satisfying $R\beta = \lambda$ and $A\beta \leq c$ (i.e., the equality in (6.12) and the inequality in (6.13), respectively), and let $\mathcal{U}_\beta$ denote the set of Y such that (β, Y) is an admissible control. The optimal control problem at the lower level is

$$\max_{Y \in \mathcal{U}_\beta} \mathsf{E}\left(\left[1 - e^{-rt}\right]\frac{v'\beta}{r} - \int_0^t e^{-rs}\left[v'dY(s) + h(Z(s))ds\right]\right) \quad (6.16)$$

$$\text{subject to} \qquad (6.9) - (6.11).$$

Denote the value of this optimal control problem by $V(\beta)$; then the optimal control problem at the upper level is

$$\max_{\beta \geq 0} \quad V(\beta) \quad (6.17)$$

$$\text{subject to} \quad R\beta = \lambda,$$

$$A\beta \leq c.$$

In the special case $r = 0$ and t is finite, the objective function (6.14) can be simplified to

$$v'\beta t - \mathsf{E}\left[v'Y(t) + \int_0^t h(Z(s))ds\right]. \quad (6.18)$$

For some applications, additional constraints may need to be imposed on the control process Y; for example, in a queueing network, one may wish only to schedule the order of service at some stations, and impose some specific service policies at the other stations.

The Brownian model we formulated above is based on the framework of the strong approximation. Following Chen and Mandelbaum [2, 3], by removing the random quantity (i.e., the Brownian motion X) in (6.9)-(6.11), we expect to obtain a macroscopic model, known as the fluid model. Specifically, we have

$$Z(t) = Z(0) + RY(t) \in \mathcal{S}, \tag{6.19}$$
$$\beta t - Y(t) \uparrow, \tag{6.20}$$
$$U(t) \equiv \nu t + AY(t) \uparrow, \tag{6.21}$$

where $R\beta = \lambda$ and $\nu \geq 0$. Note that the condition in (6.20) is slightly different from the condition in (6.10). This should become clear from the derivation in section 6.5. Roughly speaking, we would require the monotone condition (6.20) in the Brownian model as well since $[\beta t - Y(t)]$ approximates $T(t)$ (which is nondecreasing). However, in the Brownian model, $Y_n(t)$ is expected to be of order $o(t)$; when $\beta_n > 0$, its contribution to $[\beta_n t - Y_n(t)]$ is negligible, and hence, no related monotone condition is imposed in this case. We note that this fluid model formulation is the same as the one in §4 of Harrison [11]. A special case of this fluid model was proposed by Chen and Yao [6] to study the scheduling control of a multiclass queueing network.

6.4.1 Comparison to Harrison's Brownian Model

In formulating the Brownian model, Harrison [10] proposed first to solve a static planning problem:

$$\max_{\beta \geq 0} \quad v'\beta$$
$$\text{subject to} \quad R\beta = \lambda^* \text{ and } A\beta \leq c^*,$$

where λ^* and c^* are the exogenous input rate and the capacity vector of some "limiting network". The condition $R\beta = \lambda^*$ is to require that all exogenous input flows be processed, and the condition $A\beta \leq c^*$ is the capacity condition. If we ignore the superscript star (*) for the moment, the above static planning problem can be obtained by simplifying the fluid model (6.19)-(6.21) with the objective function (6.18) as follows: assume that the initial fluid level $Z(0) = 0$, and require that the fluid network be weakly stable, namely, $Z(t) \equiv 0$ and $Y(t) \equiv 0$ for all $t \geq 0$.

To derive the Brownian model through the heavy traffic limit regime, Harrison [10] made the "balanced loading assumption" (or the heavy traffic condition). This is to assume that the static planning problem has a unique optimal solution β^* that satisfies $A\beta^* = c^*$. This β^* is referred to as the system manager's nominal processing plan. When the original stochastic processing network satisfies this balanced loading assumption, the following Brownian control problem was proposed:

$$\begin{aligned} \min_{Y} \quad & \mathsf{E}\left[v'Y(t) + \int_0^t h(Z(s))ds\right] \\ \text{subject to} \quad & (6.9) - (6.11), \end{aligned}$$

where the parameter ν in (6.11) is set at zero.

When it is not perfectly balanced, the original network is viewed as one in a sequence of networks such that this sequence of networks approaches a limiting network which is perfectly balanced. This limiting network is also referred to as a reference model in [10]. Then the Brownian control problem for this balanced reference network is used to obtain the Brownian control problem of the original network. However, the choice of the sequence and hence the reference model is not unique. Different choices of the reference models can lead to completely different optimal static processing plan and completely different Brownian control problem. In particular, the choice of the optimal static plan does not depend on the net revenue vector v. In addition, the proposed Brownian control problem also varies with the choice of a scaling constant. At this point, selecting the right reference model and the right scaling parameter remain an art.

On the other hand, formulating the Brownian model proposed in this chapter involves neither a reference network nor a scaling constant. The right nominal production plan β results from the upper level optimization problem (6.17), with the full consideration of the lower level optimal control problem (6.16).

It is our belief that the proposed Brownian model would have an advantage in a non-balanced network. However, further studies are needed.

6.4.2 Extensions

First, the resource capacity can be modeled by a nonlinear stochastic process, which allows modeling of machines subject to breakdowns. This extension only involves the changes in the capacity vector c and the covariance matrix of the Brownian motion X in (6.9). The Reader may follow the treatment of the stochastic capacity in Chen, et al [5] for a possible derivation.

The resource capacity c can be a decision variable. This decision can be viewed as the staffing decision, for example, to determine how many servers of each type are needed at a call center. This decision, together with the choice of β, is the upper level optimization problem with appropriate cost function added to the objective function.

There are two possible ways to introduce an input control. One is simply to introduce a control ξ to the exogenous input flow process, by modifying the approximation (6.3) to

$$F_0(t) \overset{r}{\approx} \lambda t + \hat{F}_0(t) + \xi(t).$$

Then the corresponding Brownian model is obtained by adding a control term $\xi(t)$ to the right-hand-side of the flow-balance relation (6.9). This approach was used in Wein [15]. The second approach is to interpret $F_{0k}(t)$ as the potential, or maximum possible, exogenous input of material k up to time t, and introduce an input control process T_k^0, with $F_{0k}(T_k^0(t))$ indicating the actual exogenous input of material k up to time t. In this case, the flow-balance relation (6.9) will be changed to

$$Z(t) = Z(0) + X(t) - \Lambda\xi(t) + RY(t) \in \mathcal{S},$$

where Λ is a $K \times K$ diagonal matrix whose k-th diagonal component is λ_k, the condition $R\beta = \lambda$ is to be modified to $R\beta = \Lambda\alpha$, and the definition of X is to be modified to

$$X(t) = \hat{F}_0(\alpha t) - \sum_{n=1}^{N} \hat{F}_n(\beta_n t).$$

In this formulation, α is a control vector whose k-th component is the long-run proportion of time that the exogenous input of material k is turned on; hence, α is constrained by $0 \le \alpha \le e$ with e being a vector of ones. The control ξ models the deviation between the nominal and the actual turned-on time for the exogenous input, namely, $\xi(t)$ is the Brownian approximation for the difference $[\alpha t - T^0(t)]$. We note that in both input control extensions, the objective functions need to be modified to reflect the associated profit (or cost) contributed by the input control.

6.5 Brownian Approximation via Strong Approximation

In this section, we provide a heuristic derivation of the proposed Brownian model in the previous section as an approximation to the stochastic processing network in section 6.2.

The starting point of the approximation is that the primitive data, the flow processes, of the stochastic processing network, satisfy the strong approximations (6.3) and (6.4). Next, for the purpose of the approximation, we assume that the allocation process T under consideration has a functional law-of-iterated-logarithm limit:

$$T(t) \overset{\ell i\ell}{\approx} \beta t,$$

for some $\beta = (\beta_i)_{i=1}^I$. Note that this assumption imposes some restriction to the set of the allocation process that can be modeled by the (approximating) Brownian model. Therefore, we should be aware of the possibility (though we believe it is rather rare) that the optimal control obtained through the Brownian model may not capture the optimal control in the original stochastic processing network.

Let $Y(t) = \beta t - T(t)$. By the strong random-time-change theorem and in view of (6.5), we have

$$\begin{aligned} Q(t) &\overset{r}{\approx} Q(0) + \lambda t + \hat{F}_0(t) - \sum_{i=1}^{I}[R_i T_i(t) + \hat{F}_i(T_i(t))] \\ &\overset{r}{\approx} Q(0) + (\lambda - R\beta)t + \hat{F}_0(t) - \sum_{i=1}^{I}\hat{F}_i(\beta_i t) + \sum_{i=1}^{I} R_i Y_i(t). \end{aligned}$$

This establishes

$$Q(t) \overset{r}{\approx} Z(t),$$

with Z being as given by (6.9). Note that $T(t) = \beta t - Y(t)$; hence, the constraint that $ct - AT(t)$ is non-decreasing translates into:

$$U(t) \equiv \nu t + AY(t) \uparrow,$$

which is exactly the same as (6.11). Now consider the constraint $T(t) = \beta t - Y(t)$ is nondecreasing in $t \geq 0$. If $\beta_n > 0$, $Y_n(t)$ is of order $o(t)$; hence, whether Y_n is monotone is inconsequential. If $\beta_n = 0$, then $T_n(t) = -Y_n(t)$; we would require Y_n to be non-increasing; this leads to (6.10).

Finally, given the approximation of Q by Z and given $T(t) = \beta t - Y(t)$, the approximation of the objective function (6.7) by (6.14) is clear. However, we note that we cannot claim that the approximation of the objective function has a strong approximation bound, since the integration in the objective function may amplify the approximation error.

6.6 Notes

There has been more than 30 years' research on establishing the Brownian model as a limit of a stochastic network (or a queueing network). Most of the earlier work has been focused on establishing the rigorous limit theorems, restricted to simpler queueing networks (either single station or single class). Harrison [9] is the first to propose a Brownian network model for a general multiclass queueing network. In addition to the performance analysis, this more general Brownian model enables it to be used as an approximation model to study the optimal control of the original queueing network (e.g., Harrison and Wein [12] and Wein [15]). The more recent papers on introducing more general Brownian models include Chen, et al [5] and Harrison [10]. Refer to the Notes sections of Chen and Yao [7] for the literature review on establishing the Brownian model (or the reflected Brownian motion) as a limit of the queueing network, and to Harrison [10, 11] for the literature review on using the Brownian model for the optimization of the original stochastic processing network. Refer to Song and Zipkin [14] for a literature review on the research of assemble-to-order systems.

6.7 Appendix: Strong Approximation vs. Heavy Traffic Approximation

In Chen and Mandelbaum [3], it was argued that whenever the strong approximation exists, it is a better alternative than the heavy traffic limit theorem since the former requires no scaling and it implies the latter by providing a rate of convergence. These points were made through i.i.d. summations, renewal processes, and a generalized Jackson network. (We note that the strong approximation requires more assumptions on the primitive data. For example, the strong approximation for an i.i.d. summation requires the finiteness of the r-th moment of the random variable for $r > 2$, while the functional central limit theorem (which the heavy traffic limit theorem is based upon) requires the finiteness of the second moment. However, we feel this is not critical from most practical application.) This point of view is echoed and elaborated by Glynn [13], although stated somewhat differently. In [13], it is also shown how the strong approximation establishes that the reflected Brownian motion approximation for an G/G/1 queue is valid over a broader set of temporal and spatial scales than generally recognized through the heavy traffic approximation. Referr to Glynn [13] for additional references and a thorough discussion on the pedagogical advantages of the strong approximation approach.

In this appendix, we recapitulate and elaborate on the argument in [3] through a GI/G/1 queue. In particular, we show that the strong approximation theorem (if hold) implies the heavy traffic limit theorem. At the end of the section, we also discuss their drawbacks when interpreted as an approximation for a GI/G/1 queue under light traffic.

Let $Q(0)$, A and S be the initial queue length, the arrival process and the service process, respectively. Suppose that A and S satisfy the strong approximation: as $T \to \infty$,

$$\sup_{0 \le t \le T} |A(t) - \lambda t - \hat{A}(t)| = o(T^{1/r}), \tag{6.22}$$

$$\sup_{0 \le t \le T} |S(t) - \mu t - \hat{S}(t)| = o(T^{1/r}), \tag{6.23}$$

for some $2 < r < 4$, where $\lambda > 0$, $\mu > 0$, and $\hat{A}$ and $\hat{S}$ are driftless Brownian motions with standard deviation $\lambda^{1/2}c_a$ and $\mu^{1/2}c_s$; c_a and c_s are the coefficients of variations for the interarrival times and the service times, respectively.

It is shown in [3] that the queue length process Q satisfies the following strong approximation:

$$\sup_{0\le t\le T} |Q(t) - \tilde{Q}(t)| = o(T^{1/r}), \tag{6.24}$$

where

$$\begin{aligned} \tilde{Q}(t) &= \tilde{X}(t) + \tilde{Y}(t), \\ \tilde{X}(t) &= Q(0) + (\lambda - \mu)t + \hat{A}(t) - \hat{S}((\rho \wedge 1)t), \\ \tilde{Y}(t) &= \Psi(\tilde{X})(t) \equiv \sup_{0\le u\le t} [-\tilde{X}(u)]^+, \end{aligned} \tag{6.25}$$

$\rho = \lambda/\mu$ and Ψ is known as the one-dimensional reflection mapping.

To see that the strong approximation (6.24) implies the heavy traffic limit, we replace t by nt and T by nT in (6.24) and then divide both sides by $\sqrt{n}$. Next, we consider T being fixed; then we have as $n \to \infty$,

$$\sup_{0\le t\le T} |\frac{1}{\sqrt{n}}Q(nt) - \frac{1}{\sqrt{n}}\tilde{Q}(nt)| = o(n^{-(r-2)/(2r)}), \tag{6.26}$$

In the case $\rho \le 1$, we have the following weak convergence (referring to Billingsley [1] for a reference on the weak convergence in a functional space).

$$\frac{1}{\sqrt{n}}\tilde{Q}(n\cdot) \Rightarrow \hat{Q}(\cdot) \qquad \text{as } n \to \infty, \tag{6.27}$$

where $\hat{Q} \equiv 0$ when $\rho < 1$ and $\hat{Q}$ is a reflected Brownian motion as defined below when $\rho = 1$:

$$\begin{aligned} \hat{Q}(t) &= \hat{X}(t) + \hat{Y}(t), \\ \hat{X}(t) &= \hat{A}(t) - \hat{S}(t), \\ \hat{Y} &= \Psi(\hat{X}). \end{aligned} \tag{6.28}$$

(For the case $\rho > 1$, $\tilde{Q}(n\cdot)/\sqrt{n}$ needs to be centered, and we refer the details to [3].) Note that the above convergence follows from the self-similarity (or the scaling) property of Brownian motion and the continuity of the one-dimensional reflection mapping. (In particular, it requires no condition on the original data.)

By the Convergence-Together Theorem, (6.26) and (6.27) implies the weak convergence of $Q(n\cdot)/\sqrt{n}$ to $\hat{Q}$ as $n \to \infty$ (with $\hat{Q}$ being defined by (6.28)), which is the diffusion approximation, or the heavy traffic limit theorem when $\rho = 1$. From (6.26), we have also shown that the rate of convergence is of order $o(n^{-(r-2)/(2r)})$.

More generally, consider a sequence of queues indexed by n. Let $Q^n(0)$, A^n and S^n be the corresponding primitive data. (One possible construction is that $Q^n(0) = \lfloor \sqrt{n} Q(0) \rfloor$, $A^n(t) = A((\lambda_n/\lambda)t)$ and $S^n(t) = S((\mu_n/\mu)t)$ for some $\lambda_n > 0$ and $\mu_n > 0$.) Similar to (6.22) and (6.23), we assume for each n that as $T \to \infty$,

$$\sup_{0\le t\le T} |A^n(t) - \lambda_n t - \hat{A}^n(t)| = o(T^{1/r}), \tag{6.29}$$

$$\sup_{0\le t\le T} |S^n(t) - \mu_n t - \hat{S}^n(t)| = o(T^{1/r}), \tag{6.30}$$

for some $2 < r < 4$, where $\lambda_n > 0$, $\mu_n > 0$, and $\hat{A}^n$ and $\hat{S}^n$ are driftless Brownian motions with standard deviation σ_a^n and σ_s^n. (With the above special construction, $\sigma_a^n = (\lambda_n/\lambda)^{1/2}\sigma_a$ and $\sigma_s^n = (\mu_n/\mu)^{1/2}\sigma_s$.)

Then for the nth network, we have [cf. (6.24)]

$$\sup_{0\le t\le T} |Q^n(t) - \tilde{Q}^n(t)| = o(T^{1/r}), \tag{6.31}$$

where

$$\begin{aligned} \tilde{Q}^n(t) &= \tilde{X}^n(t) + \tilde{Y}^n(t), \\ \tilde{X}^n(t) &= Q^n(0) + (\lambda_n - \mu_n)t + \hat{A}^n(t) - \hat{S}^n((\rho_n \wedge 1)t), \\ \tilde{Y}^n &= = \Psi(\tilde{X}^n), \end{aligned}$$

and $\rho_n = \lambda_n/\mu_n$. This immediately implies that [cf. (6.26)]

$$\sup_{0\le t\le T} \left|\frac{1}{\sqrt{n}} Q^n(nt) - \frac{1}{\sqrt{n}} \tilde{Q}^n(nt)\right| = o(n^{-(r-2)/(2r)}), \tag{6.32}$$

as $n \to \infty$.

Now suppose this sequence of queues satisfies $\lambda_n \to \lambda$, $\mu_n \to \mu$, and the heavy traffic condition

$$\sqrt{n}(\lambda_n - \mu_n) \to \theta, \qquad \text{as } n \to \infty.$$

Also assume that $Q^n(0)/\sqrt{n} \to Q(0)$, $\sigma_a^n \to \sigma_a$ and $\sigma_s^n \to \sigma_s$ as $n \to \infty$. (These latter conditions are automatically satisfied under the previous conditions on A and S if we use the specific construction of $Q^n(0)$, A^n and S^n mentioned earlier. Otherwise, they need also be assumed for the heavy traffic limit theorems.) Under these conditions, we have

$$\frac{1}{\sqrt{n}} \tilde{Q}^n(n\cdot) \Rightarrow \hat{Q}(\cdot) \qquad \text{as } n \to \infty, \tag{6.33}$$

where

$$\begin{aligned}\hat{Q}(t) &= \hat{X}(t)+\hat{Y}(t), \\ \hat{X}(t) &= \hat{Q}(0)+\theta t+\hat{A}(t)-\hat{S}(t), \\ \hat{Y} &= \Psi(\hat{X}).\end{aligned} \tag{6.34}$$

(Note that the convergence in distribution is considered here, which allows us to use $\hat{A}$ and $\hat{S}$ for the ease of notation.)

Again, by the Convergence-Together Theorem, (6.32) and (6.33) implies the weak convergence of $Q(n\cdot)/\sqrt{n}$ to $\hat{Q}$ as $n \to \infty$ (with $\hat{Q}$ being defined by (6.34)), which is the more general form of the heavy traffic limit theorem for GI/G/1 queue.

In summary, we have established that whenever it exists, *the strong approximation implies the heavy traffic limit theorem* for GI/G/1 queue under the heavy traffic condition. (In fact, this has already been established in Chen and Mandelbaum [3] for a generalized Jackson network.)

Finally, we discuss some problems associated with the strong approximation and the heavy traffic approximation, when they are used to approximate the original queues.

It was shown as Theorem 2.5 in Chen and Shen [4] that when $\rho < 1$ (i.e., $\lambda < \mu$), the reflected Brownian motion $\tilde{Q}$ defined by (6.25) is of order $\log T$, i.e.,

$$\sup_{0\le t\le T} |\tilde{Q}(t)| = O(\log T).$$

This suggests that the strong approximation (6.24) hold for any $\tilde{Q}$ that is of order no more than $o(T^{1/r})$, for example, we can take $\tilde{Q} \equiv 0$. This would argue against the use of the strong approximation to derive heuristic approximations for the queueing process. However, with $\tilde{Q} \equiv 0$, we would not be able to derive the heavy traffic limit theorem from the strong approximation. In other words, the selected $\tilde{Q}$ in (6.25) is robust with scaling and taking limit to heavy traffic, and more importantly it is consistent with the previous approximations heuristically derived from the heavy traffic limit theorem. (See also Remark 2 after Theorem 3.5 in Chen and Shen (2000).) To fully overcome this difficulty, it would be necessary to improve the strong approximation bound in (6.24) to $o(\log T)$ when $\rho < 1$. This seems to be very difficult, and at this point, we even doubt whether this would be true in general.

On the other hand, the heavy traffic approximation has the similar drawback. In interpreting the weak convergence $Q^n(n\cdot)/\sqrt{n} \to \hat{Q}(\cdot)$ as $n \to \infty$, one uses the approximation: for a fixed large n,

$$Q^n(t) \approx \sqrt{n}\hat{Q}(t/n).$$

Consider the sequence $x_n(t) = n^{1/4}$ and $x(t) = 0$ for all $t \ge 0$. Clearly

$x_n(n\cdot)/\sqrt{n} \to x(\cdot)$ as $n \to \infty$. But

$$n^{1/4} \equiv x_n(t) \approx \sqrt{n}x(t/n) \equiv 0$$

for large n is clearly a very poor approximation.

Given that the strong approximation implies the heavy traffic limit theorem and given that the strong approximation does not require scaling, it is our belief that the strong approximation (when it exists) is a better alternative to derive an approximation heuristically.

Acknowledgments

I would like to thank Peter Glynn for the helpful comments on the strong approximation via the heavy traffic approximation, and thank Avi Mandelbaum for the helpful discussion. In particular, it is pointed out to me by Avi Mandelbaum the application of the model to the staffing and the control of call centers. This research is partially supported by grants from NSERC (Canada) and NSFC (China). Part of the research is done while the author is visiting the Department of Systems Engineering and Engineering Management at the Chinese University of Hong Kong.

6.8 References

[1] Billingsley, P., *Convergence of Probability Measures*, Wiley-Interscience, New York, 2nd edition, 1999.

[2] Chen, H. and Mandelbaum, A., Hierarchical modelling of stochastic networks, part I: fluid models. In: *Stochastic Modelling and Analysis of Manufacturing Systems,* D.D. Yao (ed.), Springer Series in Operations Research, 1994, 47-106.

[3] Chen, H. and Mandelbaum, A., Hierarchical modelling of stochastic networks, part II: strong approximations. In: *Stochastic Modelling and Analysis of Manufacturing Systems,* D.D. Yao (ed.), Springer Series in Operations Research, 1994, 107-131.

[4] Chen, H. and Shen, X., Strong approximations for multiclass feedforward queueing networks. *Annals of Applied Probability,* **10** (2000), 822-876.

[5] Chen, H., Shen, X. and Yao, D.D., Brownian approximations of open queueing networks. *Operations Research,* 2002.

[6] Chen, H. and Yao, D.D., Dynamic scheduling of a multi-class fluid network. *Operations Research,* **41** (1993), 1104-1115.

[7] CHEN, H. AND YAO, D.D., *Fundamentals of Queueing Networks: Performance, Asymptotics and Optimization*, Springer-Verlag, New York, 2001.

[8] CSÖRGŐ, M. AND HORVÁTH, L., *Weighted Approximations in Probability and Statistics*, Wiley, New York, 1993.

[9] HARRISON, J.M., Brownian models of queueing networks with heterogeneous customer populations. In: *Stochastic Differential Systems, Stochastic Control Theory and Applications, IMA Volumes in Mathematics and its Applications*, volume 10, W. Flemming and P.L. Lions (eds.), 147-186, Springer-Verlag, New York, 1988.

[10] HARRISON, J.M., A broad view of Brownian network. Preprint, 2000.

[11] HARRISON, J.M., Stochastic networks and activity analysis. In: *Analytic Methods in Applied Probability, In Memory of Fridrih Karpelevich,* Y. Suhov (ed.), American Mathematical Society, Providence, RI, to appear.

[12] HARRISON, J.M. AND WEIN, L., Scheduling networks of queues: heavy traffic analysis of a two-station closed network. *Operations Research*, **38** (1990), 1052-1064.

[13] GLYNN, P., Strong approximations in queueing theory. In: *Asymptotic Methods in Probability and Statistics*, B. Szyszkowitz (ed.), 135-150, North-Holland, 1998.

[14] SONG, J.-S. AND ZIPKIN, P., Supply chain operations: assemble-to-order and configure-to-order systems. In: *Handbooks in Operations Research and Management Science*, **Vol. XXX:** *Supply Chain Management*, T. De Kok and S. Graves (eds.), North-Holland, forthcoming.

[15] WEIN, L., Scheduling networks of queues: heavy traffic analysis of a two-station network with controllable inputs. *Operations Research*, **38** (1990), 1065-1078.

7

Stability of General Processing Networks

Jim Dai and Otis B. Jennings

Multiclass queueing networks are effective tools for capturing the dynamics of complex manufacturing systems. For instance, one can model multiple product lines as well as processes with highly reentrant flows. In addition to their industrial relevance, multiclass queueing networks present theoretical challenges absent from their single class precursors. Not surprisingly, the research community has devoted considerable effort recently to the study of such network models, both as an academic exercise as well as for practical purposes.

Still, there remain some aspects of manufacturing that escape the modelling scope of multiclass queueing networks, which we henceforth refer to as *standard networks*. Consider the following three examples. An explicit assumption of the standard network is that each work area is populated by a single server, whereas manufacturing environments often have parallel machines with similar processing capabilities. Furthermore, in many systems, servers are able to process multiple jobs simultaneously, i.e., in batches, without a negative effect on processing speed. In the standard network, servers process *one* job at a time. Lastly, consider a machine that has a job class-specific configuration, such as a baking temperature, die, mask, toolset, or physical location of personnel. Before the server can switch its processing efforts from one class to another, the associated configuration must first be set. The resulting loss in potential processing effort is referred to as a setup (delay). As far as the standard network is concerned, all setup times are assumed to be zero. As such, there is no disincentive in the standard network for excessive switching.

In response to such application shortcomings, some of the recent research efforts in queueing theory focus on extending the standard network model. In this chapter, we continue this trend by inserting all three features into a single model. That is, in our family of stochastic processing networks, we include batch processing operations, setup times and multi-server workstations. Each workstation in the model may contain any combination of these three elements. The formal model is presented in Section 7.2.

Beyond the presentation of our generalized queueing network model, the primary concern of the chapter is the stability of such networks. Our notion of stability, formalized in Section 7.2.4, is analogous to stability of standard networks. In short, a network is stable if the long-run input rate of the system is equalled by the long-run output rate. As suggested by studies of the stability of standard networks, stability for our generalized network is highly affected by the manner in which scheduling throughout the network is conducted.

Production policies govern the scheduling of servers in stochastic processing networks. Dispatch policies perform the analogous task for standard networks. The difference is that, to be effective, production policies must take the additional features of our stochastic processing network model into account. With this in mind, we define a family of "sensible" production policies that are adaptations of dispatch policies. In the following section, we justify the restriction to sensible production policies through instructive simulation examples.

In Section 7.3 we provide a framework for proving the stability of a stochastic processing network operating under a sensible production policy. Central to this framework is the artificial fluid model of a stochastic processing network. The artificial fluid model presented here is an extension of the identically termed model presented in Dai and Jennings [15], which, in turn, is a generalization of fluid models of standard networks. Having achieved widespread acceptance in the literature, fluid models of standard networks will be referred to as standard fluid models. Unlike their standard fluid model counterparts, artificial fluid models do not arise directly from a limiting procedure of some discrete network process; hence the "artificial" qualifier. Nevertheless, stability of the artificial fluid model implies stability of the stochastic processing network, a connection paralleling the relation between standard fluid models and standard networks.

Drawing on the similarities between artificial fluid models and standard fluid models, one can often exploit methods used to prove the stability of the latter to perform the same task for the former. In particular, Lyapunov functions that help demonstrate the stability of standard fluid models may also work for artificial fluid models. Unfortunately, for our most general model, this is as strong a statement as we are prepared to make. A more precise statement can be made for fluid model analogs of networks with no setup times. Such networks, referred to as batch processing networks, and their corresponding scheduling rules, referred to as full batch policies, were studied by Dai and Li [17]. Lyapunov functions and some nuances of employing them in demonstrating the stability of artificial fluid models is the focus of the discussion in Section 7.4.

As an exercise in using Lyapunov functions and the artificial fluid model framework for proving stability of stochastic processing networks, we investigate two production policies in Section 7.5. The policies, adapted from the early-steps-first and generalized round robin dispatch policies, stabilize a stochastic processing network as long as the traffic intensity at each station is less than one.

7.1 Motivating Simulations

Our performance measure of concern in this chapter is the throughput rate, or the long run rate at which jobs leave the system. As demonstrated through the well-known standard network examples of Lu and Kumar [31] and Rybko and Stolyar [33], the issue of throughput rates is not a simple matter. In this section we provide additional examples to argue that matters are further complicated in the presence of batch operations and setup times.

But first, we revisit the major takeaways from the Lu-Kumar and Rybko-Stolyar examples. In the networks under consideration therein, the nominal server utilizations are less than 100%. Yet, because of ill-conceived dispatch policies, the number of jobs in the systems grows linearly with time. Such unbounded growth in jobs hints at a discrepancy between the throughput rate and the arrival rate. Dispatch policies that produce such pathological behavior, when it is possible for the throughput rate to equal the arrival rate, are considered *inefficient.* As the examples imply, inefficient policies are not necessarily the product of needless idling of servers when jobs are present and work can be done. Notice we have yet to discuss the effect of setups and batch operations!

Setups and batch processing operations have the obvious efficiency ramifications. Nevertheless, we suggest an extra dose of caution when addressing the effects of such features. Given a server that is subject to setups, excessive switching of the processing effort between different job classes results in a significant loss in potential server effort, measured in units of time. Successful scheduling of such servers will result in sufficiently long stretches of time during which the server dedicates its full effort to a single job class. Exhaustive service is a rather extreme example of such scheduling rules-of-thumb. Under exhaustive service, the server processes jobs of the same class until there are no longer jobs of that class present.

Although this method of exhaustive service before switching is highly effective in single station systems, blind adherence to the policy can be detrimental in the network setting. Consider the following simulation study, devised by Jennings [26]. As illustrated in Figure 7.1, there are three single-server processing stations, labelled 1, 2 and 3. Each job must go through six steps, labelled 1 through 6, as they pass through the network. Jobs being processed or waiting to be processed in step k are called class k jobs. Class k jobs that are awaiting processing are said to reside in buffer k. Steps 1 and 4 are conducted at station 1, steps 2 and 5 at station 2, and steps 3

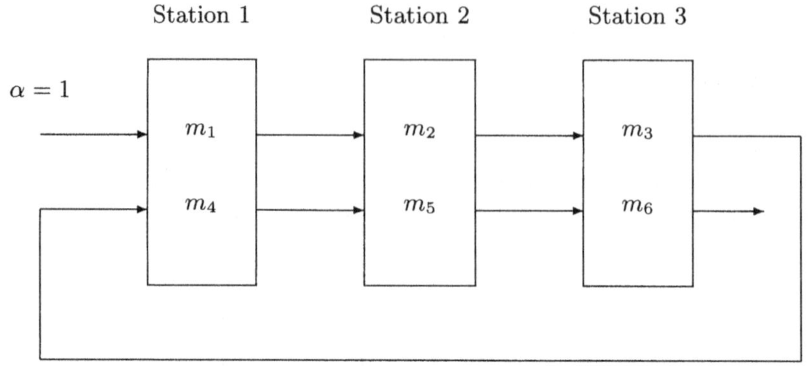

FIGURE 7.1. A three-station, six-class standard network with setups

and 6 at station 3. In the simulation, jobs arrive according to a Poisson process with rate $\alpha = 1$ job per hour. Processing times and setup times are all exponentially distributed. Processing steps 2, 4 and 6 each require 45 minutes on average. Processing steps 1, 3 and 5 require six minutes of effort on average. Setup times for a server switching between any pair of classes are 30 minutes on average. Nominally speaking, each server should be busy processing jobs a fraction

$$\frac{1 \text{ job}}{60 \text{ minutes}} \times \left(\frac{45 + 6 \text{ minutes}}{\text{job}} \right) = .85$$

of the total simulation time. Notice, this quantity is independent of setup times.

The servers employ an exhaustive service policy. Suppose a server has just processed all of the jobs of one buffer. If there are no jobs in the other buffer at the station, the server sits idle and waits for an arrival. Otherwise, the server performs a setup and then exhausts the other buffer at the station. The scheduling policy just described exhibits the so-called *non-idling* property; that is, no server idles when there are jobs at its corresponding station. The graph in Figure 7.2 depicts the total number of jobs in the system, or work-in-progress (WIP) as a function of time. One can see that the WIP exhibits a roughly linear growth trend, again a strong sign of instability.

To better understand what is going wrong in this system, we track how potential server effort is being used over the course of the simulation. The results, summarized in Table 7.1, reveal that no server is busy 85% of the time, as expected from the pre-simulation calculations. The implication is that the servers are not working enough to process the arriving jobs. What is perhaps more disturbing is the reason, or lack thereof, for why the servers are not working. It is *not* the case that servers are spending too much time performing setups. To the contrary, setup time is insignificant in terms of distribution of server effort. Neither are servers idling when jobs

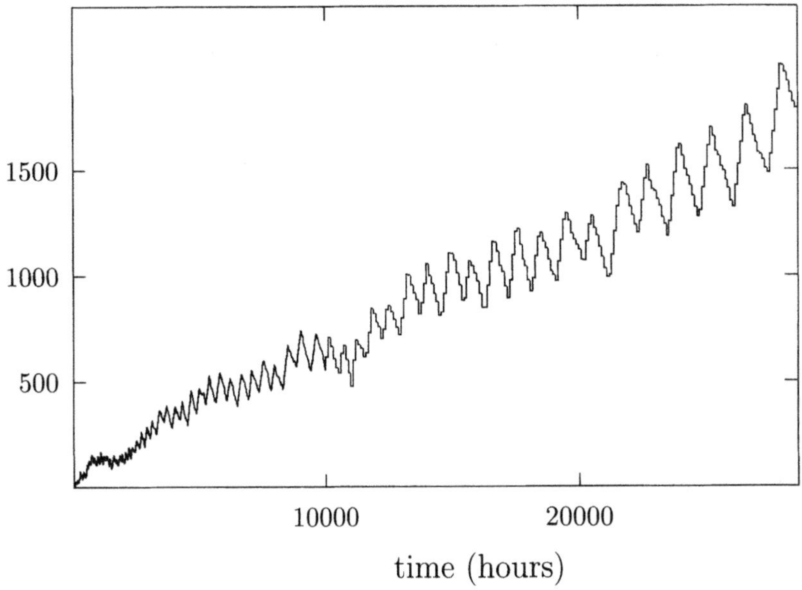

FIGURE 7.2. WIP as a function of time.

are present, as dictated by the non-idling condition.

For more evidence of what is wrong with the system, see Figure 7.3, where we plot the job counts for steps 2, 4 and 6 as functions of time. It is rarely the case that buffers 2, 4 and 6 are all simultaneously strictly positive. The implication is that, throughout the majority of the simulation, at most two of these buffers can be processed simultaneously. This fact would not be alarming if the steps were processed at two servers. But the steps reside at three distinct stations.

The behavior just described is an example of a so-called *pseudo-station*, also referred to by Bertsimas [2] as a *K-virtual station*. For more on pseudo-stations, see Hasenbein [25]. The behavior is related to an example in Dai,

Utilization Profile			
	Server 1	Server 2	Server 3
Idle	.181	.189	.202
in-Setup	.003	.002	.002
in-Service	.816	.809	.796

TABLE 7.1. Summary of server usage.

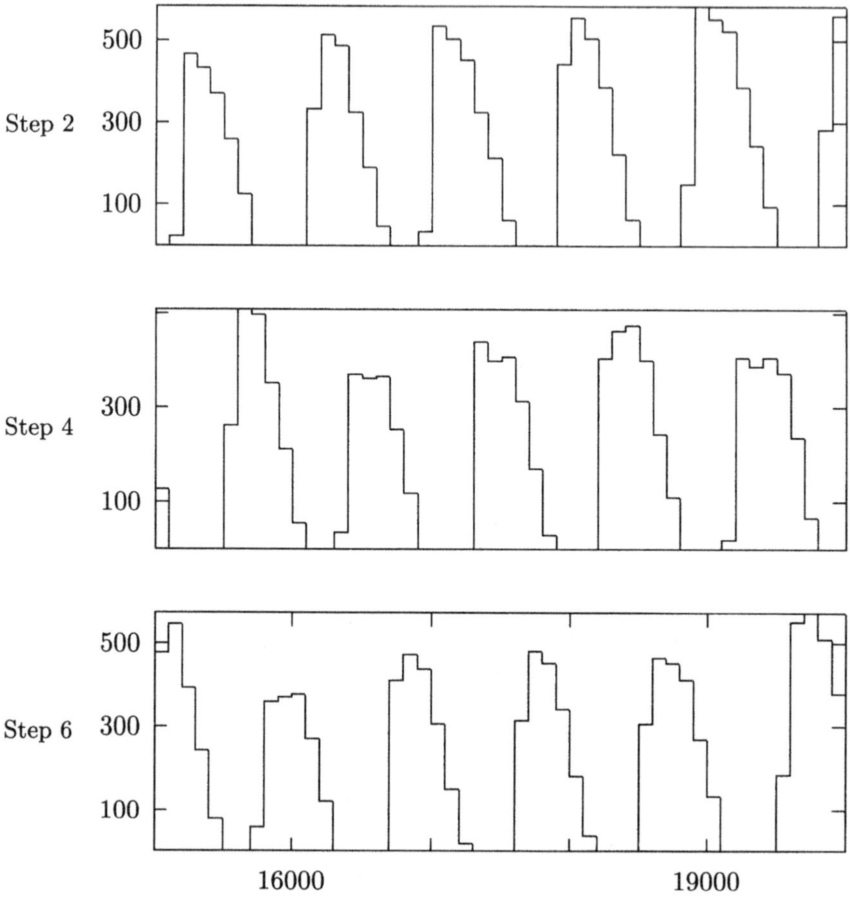

FIGURE 7.3. Job count at steps 2, 4 and 6.

Hasenbein and Vande Vate [14], the origin of the network in Figure 7.1. The fundamental differences in [14] are that jobs are replaced by units of fluid and that there are no setups. Not surprisingly, further simulation studies reveal that the same pathological behavior in our discrete network can occur when the setup times are equal to zero.

In this first simulation example, exhaustive service led to instability. In many other cases, namely those in Lu and Kumar [31], Rybko and Stolyar [33], and Dai, Hasenbein and Vande Vate [14], instability may be caused by so-called static buffer priority (SBP) policies. Under an SBP policy, jobs of some buffers have strict processing priority over jobs in other buffers. Byproducts of SBP policies are long production runs during which a single class of jobs is processed. These long production runs may starve downstream buffers which, once they finally receive a very large number of jobs during a relatively short interval, ultimately starve the buffers responsible

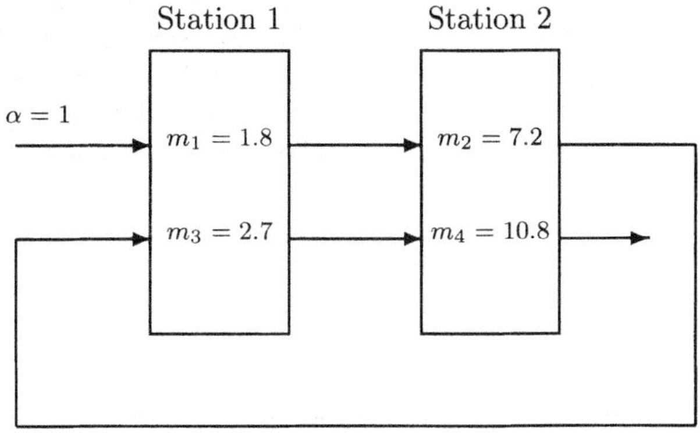

FIGURE 7.4. A two-station, four-class batch processing network

for starving them. This cycle of "starve and be starved" continues as oscillations of WIP grow larger and larger. In terms of the dynamics leading to instability, exhaustive service effectively mimics SBP policies.

One lesson to be gleaned from this example is that, in the presence of setups, the duration of dedicated server effort should be *long, but not too long.* For single-station systems with setups, or polling stations, this notion is embodied in the ℓ-limited service scheduling policy; see, for example, Takagi [34]. For such policies, the number of consecutively processed class k jobs is forced to equal ℓ_k, unless the corresponding buffer is exhausted first. Policies of this limited variety will resurface when we define our family of "sensible" policies.

A second lesson is that dispatch policies that prove successful in scheduling standard networks may inform our decision-making process when scheduling the more general processing networks. This is true whether stations are populated by setup servers, batch servers or combinations thereof. Indeed, this is the approach employed by both Dai and Jennings [15] for queueing networks with setups and Dai and Li [17] for batch processing networks.

With servers that process multiple jobs simultaneously, forming large batches is desirable. Not doing so can also be considered an inefficient use of potential server effort. Empowered by the second lesson above and equipped with a dispatch policy proven to work well for a standard network, one might reach erroneous conclusions. Specifically, one might declare that as long as a smart dispatch policy is used to select the class to be processed next and the subsequently constructed batch is as large as possible, throughput will be maximized. However, as demonstrated by Dai and Li [17], this reasoning is flawed. Consider their simulation.

The network has two single-server stations serving four job classes, as illustrated in Figure 7.4. As with the previous simulation, this network is an example of a reentrant line, where all jobs follow the same deterministic route through the network. Each job follows four processing steps, corresponding with the four classes and alternating between stations 1 and 2. The batches for steps at station 1 contain at most five jobs and the maximum batch size at station 2 is 20. Jobs arrive from the outside according to a Poisson process with rate $\alpha = 1$ job per minute. The processing times for class k batches are independent, exponentially distributed with mean m_k, $k = 1, 2, 3, 4$. The mean service times are set to be $m_1 = 1.8$, $m_2 = 7.2$, $m_3 = 2.7$, and $m_4 = 10.8$ minutes, as shown in the figure.

As in the previous simulation, we compute the traffic intensity as the product of the arrival rate and the mean service time. In the presence of batches, the mean processing time can be amortized over all of the jobs in a maximum sized batch. Hence, the traffic intensities are given by

$$\rho_1 = \alpha(m_1 + m_3)/5 = 0.9 \quad \text{and} \quad \rho_2 = \alpha(m_2 + m_4)/20 = 0.9.$$

A more formal definition of traffic intensities will be given in (7.5) in Section 2. Clearly, the usual traffic condition (7.6) is satisfied for the parameter set. Intuitively, the batch processing network should have enough capacity to handle all incoming jobs, achieving a throughput of 1 job per minute.

The last-buffer-first-serve (LBFS) dispatch policy maximizes throughput in a standard reentrant line (i.e., without batch operations); see, for example Dai and Weiss [19] or Kumar and Kumar [29]. Under LBFS, steps 3 and 4 have priority over steps 1 and 2, respectively. It is reasonable to believe that, with a few modifications, LBFS can perform the dispatching duties for reentrant lines with batch operations. The simulation carries out this idea. Namely, once the LBFS policy selects a buffer, we process the largest batch possible. Under this modified LBFS policy, referred to as the LBFS batch policy, each server always selects jobs from the highest priority nonempty buffer to form a batch, even though the selected buffer may have only one job in it. The following table shows the average times in system.

Number of jobs leaving the system	50	500	5000	50000
Average time in system	54.2	208.4	1057.3	6831.6

Figure 7.5 plots the total number of jobs in the system as time increases. Clearly, the system is unstable, thus it cannot handle the traffic loads in long run. On the other hand, the same simulation shows that, after completing 50000 jobs, server 1 is busy 96% of the time with average batch size 4.19 jobs and server 2 is busy 99.97% of the time with average batch size 16.41 jobs. In contrast with our first simulation, the servers this time are apparently heavily utilized, yet the system is unstable. Under the LBFS batch policy, server 2 keeps serving class 4 batches that may consist of only five jobs, sent recently from class 3 by server 1. Moreover, this takes place even when class 2 has a large number of jobs waiting. This example shows that, in the presence of batch operations, a naive adaptation of a dispatch policy may prove inefficient, although the policy performs well in a standard network.

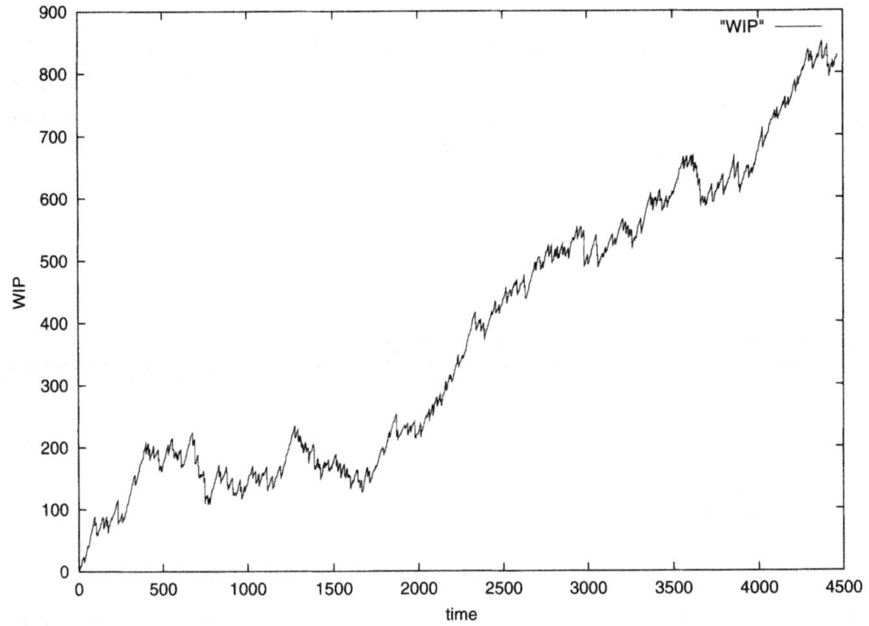

FIGURE 7.5. The total number of jobs in system

This brings us to our third and final lesson. For a batch server, it is not enough to form the largest batch *after* the class has already been chosen. Similarly, for a server subject to setups and operating under an ℓ-limited policy, it is not enough to attempt to process ℓ_k jobs once class k is already chosen. Instead, when the class is being selected, we should restrict attention to those buffers from which a sufficiently large batch can be formed, or a sufficiently long production run can be performed, if such a buffer exists.

7.2 Open Processing Networks

In this section we present the family of networks under study throughout the chapter. The model is an extension of open multiclass queueing networks, as presented by Harrison [23]. For the purpose of this chapter, we will refer to that model as the standard network. Our model, which adds setup times, batching operations, and multiple-server workstations to the standard network, will be referred to as the *open stochastic processing network*, or processing network for short. Readers should be warned that Harrison [24] uses that term with a somewhat different meaning. We adopt the name nonetheless.

In the remainder of the section, we derive the corresponding standard network of a given processing network. After drawing connections between scheduling the resources of both types of networks, we conclude with a formal definition of rate stability.

7.2.1 Network Description

Consider a network of stations, labelled $j = 1, \ldots, J$. The stations are populated by classes, labelled $k, k' = 1, \ldots, K$, where class k is associated with a unique station $\sigma(k)$. Whenever k and j appear together, it is implied that $j = \sigma(k)$. The collection $\mathcal{C}(j)$ of all classes associated with station j is referred to as the station's constituents.

The basic processing unit is a *batch*, which consists of one or more *jobs*, the basic unit of flow. Jobs enter the network exogenously, and change classes as they move through the network. While awaiting processing, class k jobs are said to reside in buffer k. We use the terms "buffer" and "class" interchangeably. For the most part, buffer is used to connote physical location, as in a storage place for jobs awaiting processing. The class label has a more metaphysical interpretation. For instance, one removes a job from a buffer for processing, but the job retains its class designation until processing is complete.

Just prior to being processed, jobs are removed from the buffer and grouped to form a batch. Batches consist exclusively of jobs from a single class. (In some industrial settings, jobs from multiple classes can be included in the same batch, provided they share the same "recipe".) The number of jobs in a class k batch is at most B_k; the minimum is one. When a class k batch is formed, it consists of the oldest jobs in the buffer, measured in terms of their most recent arrival time to the buffer. In this sense, forming batches from jobs is first-come-first-batched. Immediately after the formation of a batch, processing commences and no additional jobs may be added to the batch. The batch grouping remains intact until processing is complete. Post-processing, the batch grouping is dissolved and the component jobs are individually routed through the network, perhaps to different buffer locations. (When a batch is dissolved, we assume an order is assigned to the departing jobs.) All jobs eventually leave the network.

Each station has a pool of servers that share the processing duties at the station. We denote the number of servers residing at station j by p_j, a quantity which generally varies from station to station. The servers of a given station are identical in the sense that each has the capability of processing any batch formed at the station. Moreover, the length of the processing time is independent of which server performs the work. A server may process at most one batch at a time and batches may be processed by at most one server. That is, servers cannot combine efforts in the processing of any batch. Once a server starts processing a batch, it cannot be interrupted. In other words, there is no preemption of service. (There may be occasions where it makes sense to allow preemption of service. We exclude it for modelling ease.) We will occasionally say that a server is processing a class or buffer, meaning, processing the batches from that class.

Suppose a server last processed class k and is about to process a batch from a different class $k' \neq k$. Before the actual batch formation can begin, the server must perform a *setup*. That is, a delay is incurred whenever a server switches its processing efforts between classes. Setups are denoted by the pair (k, k'), where k signifies the class just processed and k' is the subsequent class. Generally, the duration of a type (k, k') setup, or *setup time*, depends on both k and k', as well as their order. In this sense, setups are *sequence-dependent.* With the inclusion of setups in our model, servers are always in one of three states: *idle*, *in-service* or *in-setup*. In addition to not being interrupted while processing batches, servers are never preempted while performing a setup.

For each class k, we have the cumulative processes $E_k = \{E_k(t), t \geq 0\}$, $V_k = \{V_k(n) : n = 1, 2, ...\}$, and $\Phi^k = \{\Phi^k(n) : n = 1, 2, ...\}$. For each time $t \geq 0$, $E_k(t)$ counts the number of external arrivals to class k in $[0, t]$. For each positive integer n, $V_k(n)$ records the total service time requirement for the first n class k batches (regardless of batch size or processing server). For each positive integer n, $\Phi^k(n)$ is a K-dimensional vector with each component being a nonnegative integer. For each class k', $\Phi^k_{k'}(n)$ records the number of the first n *processed* class k jobs that are routed to buffer k' upon completion of service. When $\Phi^k(n-1) = \Phi^k(n)$, the nth processed class k job immediately leaves the system. By convention, we assume

$$E_k(0) = 0, \quad V_k(0) = 0, \quad \text{and} \quad \Phi^k(0) = 0.$$

For each time $t \geq 0$, we extend the definitions of $V_k(t)$ and $\Phi^k(t)$ as follows:

$$V_k(t) = V_k(\lfloor t \rfloor) \quad \text{and} \quad \Phi^k(t) = \Phi^k(\lfloor t \rfloor),$$

where $\lfloor t \rfloor$ denotes the largest integer less than or equal to t.

In addition, we define the cumulative process $F_{kk'} = \{F_{kk'}(n) : n = 1, 2, \ldots\}$, associated with type (k, k') setups. When k and k' appear together in the same subscript of a setup quantity, it is implied that the classes are distinct, $k \neq k'$, and that they reside at the same station, $\sigma(k) = \sigma(k')$. For each positive integer n, $F_{kk'}(n)$ records the total time required for the first n setups from class k to k'. Again we assume $F_{kk'}(0) = 0$. Furthermore, for each $t \geq 0$, we extend the definition so that $F_{kk'}(t) = F_{kk'}(\lfloor t \rfloor)$.

We call (E, V, Φ, F) the set of primitive processes, where $E = \{E(t), t \geq 0\}$, $V = \{V(t), t \geq 0\}$, $\Phi = \{\Phi(t), t \geq 0\}$, and $F = \{F(t), t \geq 0\}$, with $E(t) = (E_1(t), E_2(t), ..., E_K(t))'$, $V(t) = (V_1(t), V_2(t), ..., V_K(t))'$, $\Phi(t) = (\Phi^1(t), \Phi^2(t), \ldots, \Phi^K(t))$, and $F(t) = \{F_{kk'}(t), \, k, k' \in \mathcal{C}(j), \, j = 1, \ldots, J\}$. When appearing with a vector quantity, a prime symbol denotes a column vector.

We assume that the strong law of large numbers holds for the primitive processes; namely, with probability one,

$$\lim_{t \to \infty} \frac{E_k(t)}{t} = \alpha_k, \quad \lim_{t \to \infty} \frac{V_k(t)}{t} = m_k,$$

$$\lim_{t\to\infty}\frac{\Phi^k_{k'}(t)}{t}=P_{kk'},\quad\text{and}\quad\lim_{t\to\infty}\frac{F_{kk'}(t)}{t}=s_{kk'}. \tag{7.1}$$

The parameter set (α, m, P, s) with $\alpha = (\alpha_1, \ldots, \alpha_K)'$, $m = (m_1, \ldots, m_K)'$, $P = (P_{kk'})$, and $s = \{s_{kk'}, k, k' \in \mathcal{C}(j), j = 1, \ldots, J\}$ has the following interpretation: For each k, α_k is the external job arrival rate to buffer k and m_k is the mean service time for class k batches. (Recall that the processing time of a batch is independent of its batch size.) For classes k and k', $P_{kk'}$ is the long-run fraction of class k jobs that immediately become class k' jobs after being processed. It is also called the routing probability from class k to class k'. The $K \times K$ matrix P is called the routing matrix. We assume that the network is open, i.e., the matrix

$$Q = I + P' + (P')^2 + \ldots$$

is finite, which is equivalent to the fact that $(I - P')$ is invertible such that $Q = (I - P')^{-1}$. A reentrant line is a special type of processing network in which all jobs follow the same deterministic route of K steps, and jobs may visit some stations multiple times. The transition matrix of a reentrant line consists of $K - 1$ ones and the remaining entries are all zeros. For each pair (k, k'), $s_{kk'}$ is the mean setup time when a server switches from class k to class k'. Let $s_k = \max_{k' \in \mathcal{C}(j)} s_{k'k}$ be the maximum possible mean setup time *to* class k.

For future purposes, we introduce the counting processes $\Psi = \{\Psi(t) : t \geq 0\}$ associated with the primitive service process V and $\Upsilon = \{\Upsilon(t) : t \geq 0\}$ associated with the primitive setup process F. For each time $t \geq 0$, $\Psi(t) = (\Psi_1(t), \ldots, \Psi_K(t))'$ with

$$\Psi_k(t) = \max\{n : V_k(n) \leq t\}, \quad k = 1, \ldots, K$$

and $\Upsilon(t) = \{\Upsilon_{kk'}(t),\ k, k' \in \mathcal{C}(j),\ j = 1, \ldots, J\}$ with

$$\Upsilon_{kk'}(t) = \max\{n : F_{kk'}(n) \leq t\}, \quad k, k' \in \mathcal{C}(j), \quad j = 1, \ldots, J.$$

The process $\Upsilon_{kk'}$ is only defined when $s_{kk'} > 0$. It follows from the strong law of large numbers (7.1) that

$$\lim_{t\to\infty}\frac{\Psi_k(t)}{t} = \mu_k, \quad k = 1, \ldots, K, \tag{7.2}$$

where $\mu_k = 1/m_k$, and that, for each setup pair (k, k') such that $s_{kk'} > 0$,

$$\lim_{t\to\infty}\frac{\Upsilon_{kk'}(t)}{t} = 1/s_{kk'}. \tag{7.3}$$

Let $\lambda = (\lambda_1, \ldots, \lambda_K)'$ be the vector of nominal total arrival rates. It is defined by the following system of equations

$$\lambda_k = \alpha_k + \sum_{k'=1}^{K} \lambda_{k'} P_{k'k}, \quad \text{for each} \quad k = 1, \ldots, K. \tag{7.4}$$

In vector form, $\lambda = \alpha + P'\lambda$. Since P is transient, the unique solution to (7.4) is given by $\lambda = Q\alpha$. We define the traffic intensity ρ_j for station j as

$$\rho_j = (1/p_j) \sum_{k \in \mathcal{C}(j)} \lambda_k (m_k/B_k), \quad j = 1, \ldots, J, \tag{7.5}$$

with $\rho = (\rho_1, \rho_2, \ldots, \rho_J)'$ being the corresponding vector. In the absence of batch operations ($B_k = 1$), this definition matches previous definitions of traffic intensity in the presence of multiple, identical servers; see, for example, Gross and Harris [22]. One can think of ρ_j as the average of the nominal utilizations of the p_j servers at station j if every batch is of the maximum size. Because class k batch sizes can be smaller than B_k, the fraction of actual effort dedicated to processing batches may very well exceed ρ_j. When, for each station $j = 1, ..., J$,

$$\rho_j < 1, \tag{7.6}$$

we say that the usual traffic condition is satisfied for the processing network.

7.2.2 The Standard Network and Dispatch Policies

We now define the corresponding *standard network* of a processing network. Our notational convention is to use a tilde when referring to a quantity associated with the standard network. The corresponding standard network is identical to the processing network except that (a) the maximum batch size is one, or equivalently, jobs are the basic processing unit; (b) there is only one server at each station; (c) the setup times are zero such that the server exists in either the idle state or the in-service state; and (d) the primitive service process is given by $\tilde{V}_k = \{\tilde{V}_k(n) : n = 1, \ldots\}$, where $\tilde{V}_k(n) = V_k(n)/(B_k p_j)$. As a result, the counting process $\tilde{\Psi}$ associated with the primitive service process $\tilde{V}$ is described by $\tilde{\Psi}_k(t) = \max\{n : \tilde{V}_k(n) \leq t\} = \Psi_k(B_k p_j t)$, for each $k = 1, \ldots, K$. Accordingly, the strong law of large numbers (7.1) yields

$$\lim_{n\to\infty} \frac{\tilde{V}_k(n)}{n} = \frac{m_k}{B_k p_j} \equiv \tilde{m}_k \quad \text{and} \quad \lim_{t\to\infty} \frac{\tilde{\Psi}_k(t)}{t} = B_k p_j \mu_k \equiv \tilde{\mu}_k, \tag{7.7}$$

for each $k = 1, \ldots, K$. Moreover, the traffic intensity for station j in the standard network is the same as in the processing network, i.e., $\tilde{\rho}_j = \rho_j$. It goes without saying that the usual traffic condition for the standard network holds if and only if the traffic condition for the processing network (7.6) holds. In short, in the standard network, stations process one job at a time, there is no loss in potential server effort from switching, and when class k jobs are in service, the station $j = \sigma(k)$ processing rate increases by a factor of $B_k p_j$ over a corresponding server's speed in the processing network.

For a stochastic processing network driven by the primitive processes (E, V, Φ, F) with maximum batch sizes $(B_1, \ldots, B_K)'$, the corresponding standard network is driven by the primitive processes $(E, \tilde{V}, \Phi)$ with maximum batch sizes that are equal to one.

Whenever multiple jobs reside at a station, there is discretion in the processing order of those jobs. For standard networks, the *dispatch policy* $\tilde{\pi}$ is the sole mechanism by which servers are assigned to classes. That is, when a server becomes available for processing, the dispatch policy selects the class from which the next job will be processed. Given the class assignment, the oldest job, based on arrival to the associated buffer, is processed. In this sense, jobs within a single buffer are processed in a first-come-first-served (FCFS) fashion. A dispatch policy is said to be *non-idling* if a server is never in the idle state when jobs are present at the station.

Not surprisingly, scheduling of servers in a standard network is less complex than in a processing network. To stress this point, an alternative term is used to distinguish between the two scheduling tasks. A *production policy* is to the processing network what a dispatch policy is to a standard network. One of the main themes of this chapter is that dispatch policies that work well for standard networks are useful in crafting effective production policies for processing networks.

7.2.3 Production Policies and Sensible Policies

The decision process governing the formation and processing of batches in a processing network is embodied in the *production policy*. Because of the complex nature of each station, we envision most "useful" production policies having the following three-tiered approach. When a server requires an assignment to a class for processing, one first filters the set of constituent classes into a subset of *eligible* classes. Secondly, the server is *dispatched* to one of the eligible classes. The final decision involves setting the termination time of the assignment as well as determining how batches are formed in the interim.

Throughout this chapter, we assume production policies have the form $\pi = (\theta, \tilde{\pi}, \ell)$. (Exceptions are discussed in Section 7.6.) The K-dimensional vectors $\theta = (\theta_1, \theta_2, \ldots, \theta_K)$ and $\ell = (\ell_1, \ell_2, \ldots, \ell_K)$ of positive integers enforce the filtering function. When a server at station j requires an assignment, the constants determine which constituent classes are eligible. Class k is eligible if the number of class k jobs is equal to or greater than

$$\theta_k + \ell_k B_k \times \text{ the \# of servers already assigned to class } k.$$

(Exceptions that relax the eligibility requirement can be found in Section 7.6.) It is possible that the collection of eligible classes at station j is empty. In cases where no class passes the first eligibility test, the criterion is relaxed. Under the relaxed test, any class with a nonempty buffer is eligible. From the standpoint of the first eligibility test, if the number of class

k jobs is less than θ_k and no server is assigned to it, that class is effectively empty. In this sense, the components of θ can be thought of as *thresholds*.

The second component of the production policy π is the dispatch policy $\tilde{\pi}$. This terminology is borrowed from literature on standard networks. The idea here is the same. Given the current state of the system, dispatch policies perform the actual assignment of servers to classes. With the standard network, the assignment lasts at most through the processing of a single job. For processing networks, however, the assignment lasts (potentially) for the processing of several batches. We assume that the largest possible batches are formed and processed until a new assignment is sought.

The stretch of time during which the dispatched assignment holds is referred to as the *production run*. The length of the production run or, equivalently, the number of batches processed before seeking a new assignment, is the third and final decision to be made. The vector ℓ determines the length of the production run. If the assignment is for a class k that passed the first eligibility test, then we process ℓ_k batches before seeking a new assignment (if possible). If the class only passed the relaxed eligibility test, we form and process the largest possible batch and terminate the production run immediately, allowing the opportunity to make another production run decision based on updated system information.

The non-idling property for processing networks is not as straightforward as for the standard network. Part of the complication stems from having multiple servers in one of three states. As the name suggests, non-idling now refers to both the in-service and in-setup states. A server in the processing network is said to be *busy* if it is in-service or in-setup. A production policy is said to be *non-idling* if each server is allowed to idle only when there are either no jobs at the corresponding station or all of the jobs present are currently being processed in batches. (This is not to say that non-idling policies are ideal. Indeed, in some instances it may beneficial to delay processing the available jobs until a larger batch can be formed.) A consequence of the non-idling condition is that the number of busy station j servers at time t is at least

$$\min\left(p_j, \sum_{k\in C(j)} \lceil Z_k(t)/B_k \rceil\right), \tag{7.8}$$

where $\lceil x \rceil$ is the smallest integer greater than or equal to x and $Z_k(t)$ denotes the number of class k jobs at time t. More concisely, a server can idle only when the constituent buffers are empty. A precise mathematical description would require knowledge of the number of jobs in each buffer k (excluding those in service), whereas we only track the aggregate class k job count process Z_k; for more, see Section 7.3.1.

The non-idling condition does not rule out small batches or short production runs, whose occurrences are limited by the eligibility requirements. A natural quantity of interest is the maximum number of servers at a given station that can simultaneously be assigned to eligible classes, a reflection of how efficiently the resident servers must work. Consider the following index:

$$I_j(t) = \min\left(p_j, \sum_{k\in C(j)} \left(1_{\{Z_k(t)\geq\theta_k\}} + \lfloor (Z_k(t)-\theta_k)^+/(\ell_k B_k)\rfloor\right)\right), \quad (7.9)$$

for each $j = 1, \ldots, J$, where $1_{\{.\}}$ is the indicator function, $(x)^+ = \max(0, x)$ and $\lfloor x \rfloor$, as stated earlier, is the integer part of x. Suppose a server from station j is free at time t and requires dispatching. If $I_j(t) = p_j$ then this server will be assigned to a class that passes the more stringent eligibility test. To see this, first notice that if the class k jobcount exceeds its threshold ($Z_k(t) \geq \theta_k$) then at least one server can be assigned to this class. To each class k, for which $Z_k(t) \geq \theta_k$ holds and to which no server has been dispatched, there is the potential for assigning a server. Furthermore, for each additional $\ell_k B_k$ class k jobs, an additional server can be assigned. Therefore, $I_j(t)$ bounds from below the number of servers that *can* be assigned to eligible classes at time t. We offer a tighter bound for the number of servers processing eligible classes in Section 7.3.1.

Assuming that a server is assigned to a class k that passed the first round eligibility test, we are assured that at least $\min(\theta_k, \ell_k B_k)$ class k jobs are available at the beginning of the production run to be processed in one of at least $\min(\lceil\theta_k/B_k\rceil, \ell_k)$ batches. If we assume the thresholds are set such that they exceed the maximum number of jobs processed in a production run, i.e.,

$$\theta_k \geq \ell_k B_k, \quad (7.10)$$

then the above quantities $\min(\theta_k, \ell_k B_k)$ and $\min(\lceil\theta_k/B_k\rceil, \ell_k)$ reduce to $\ell_k B_k$ and ℓ_k, respectively. One can amortize the setup time incurred from switching to class k over all of the batches processed during the subsequent production run. The result is a *setup-adjusted mean processing time* and *rate*, which, when (7.10) holds, are respectively equal to

$$\breve{m}_k = m_k + s_k/\ell_k \quad \text{and} \quad \breve{\mu}_k = 1/\breve{m}_k, \quad k = 1, \ldots, K. \quad (7.11)$$

With the adjusted mean processing times, one can compute a *setup-adjusted traffic intensity*

$$\breve{\rho}_j = (1/p_j) \sum_{k\in C_j} \lambda_k(\breve{m}_k/B_k), \quad j = 1, \ldots, J. \quad (7.12)$$

Throughout this chapter, the convention $\breve{x}$ signifies an inflation of the quantity x or simply an adjustment to the quantity x related to the influence of setups. The inflation can be due to the amortization of setup time, the presence of multiple servers, or the performance of batching operations.

We conclude with a further restriction of the type of production policies under consideration. A production policy is said to be *sensible* if it is non-idling, equation (7.10) holds, and $\breve{\rho}$ obeys the usual traffic condition; i.e.,

$$\breve{\rho}_j < 1, \quad \text{for each } j = 1, \ldots, J. \tag{7.13}$$

The reasoning behind the restriction to sensible policies is straightforward. By ensuring that the servers avoid spending an inordinate amount of time performing setups, sensible policies eliminate setups (and non-maximal batches) as sources of instability. Furthermore, implicit in the sensible policy condition is the ability to make tradeoffs of production run lengths. For instance, it is simple to compensate for setting a relatively short production run length for one class with a long production run for another class at the same station. One might use such tradeoffs as a mechanism for accounting for high priority "hot lots," for which production run length (and batch size) considerations are secondary at best. The last reason for restricting to sensible policies is more subtle. Having the non-idling condition and (7.13) ensures that no job can be ignored indefinitely. Moreover, there exists a finite constant such that the number of jobs at a station falls below the constant infinitely often. This last feature helps to demonstrate positive Harris recurrence, a stronger notion of stability investigated by Dai [12] for standard networks and Jennings [27] for queueing networks with setups. The latter also employed a similar notion of sensible policies. Lastly, we point out that, unlike the analogous condition for standard networks, the traffic condition part of the sensible policy definition (7.13) is not necessary for stability; for more details, see Section 7.6.

7.2.4 Rate Stability

We now define rate stability for open stochastic processing networks. Let $D_k(t)$ denote the number of jobs in the processing network that have departed class k during the interval $[0, t]$. In the following definition, the term *state* is used. The precise definition of a state depends on the particular production policy used. The system state typically includes, but is not limited to, the number of jobs in each class, the status and assignment of each server, the remaining processing times and the sizes of the batches being processed, the remaining interarrival times for jobs arriving from outside, the lengths so far of the current production runs, and the remaining setup time for each server. We do not attempt a precise definition of state here. Roughly speaking, a state is a snapshot of the network at any given time. It should contain enough information such that once the current state of the network is given, the future evolution of the network is completely determined in distribution. Readers are referred to Dai [12] and Bramson [5] for examples and additional discussions of states in standard networks under various dispatch policies.

Definition 7.1 A processing network is *rate stable* if, for each fixed initial state, with probability one,

$$\lim_{t\to\infty} \frac{D_k(t)}{t} = \lambda_k, \quad \text{for each } k = 1, ..., K. \tag{7.14}$$

The processing network is rate stable if the throughput rate or departure rate from each class is equal to the nominal total arrival rate to that class. Rate stability has been advanced by Stidham and his co-authors (see El-Taha and Stidham [21] and references therein). This notion of stability was first introduced for the standard network setting in Chen [7]. As in a standard network, the usual traffic condition is necessary for rate stability of a processing network; see Dai [13]. As stated earlier, even though we specify that the setup-adjusted traffic intensities for sensible policies must obey the usual traffic condition, this is not necessary for stability of the processing network. There are other definitions of stability, such as positive Harris recurrence; see Dai [12]. The results in this chapter can be extended to those settings as well.

As mentioned earlier, the main message of this chapter is that, in some cases, a dispatch policy of a standard network can be transformed into an efficient production policy. A precise statement to this effect cannot be articulated in general. However, Dai and Li [17] successfully provide a precise result for a subclass of dispatch policies adapted for networks with zero setup times, that is, for so-called *batch processing networks*. Their result, demonstrated for networks of single-server stations, also holds for networks of multi-server stations. We will resume this discussion on batch processing networks in Section 7.4.1.

7.3 Network and Fluid Model Equations

In this section, we define fluid models of both processing networks and standard networks. Fluid models are continuous, deterministic analogs of discrete networks and are defined through a set of equations. To describe the fluid models, we start with equations governing the dynamics of the discrete networks. Unless explicitly stated otherwise, we assume that the processing network is operated under a sensible production policy π and the standard network is operated under a non-idling dispatch policy $\tilde{\pi}$.

7.3.1 Network Dynamics

The dynamics of the processing network can be captured by the process $\mathbb{X} = (A, D, S, T, U, Y, Z)$. The components $A = \{A(t), t \geq 0\}$, $D = \{D(t), t \geq 0\}$, $T = \{T(t), t \geq 0\}$, $S = \{S(t), t \geq 0\}$, and $Z = \{Z(t), t \geq 0\}$ are K-dimensional. For each class k, $A_k(t)$ denotes the number of jobs that have arrived to class k (from external and internal sources) in $[0,t]$, $D_k(t)$ denotes the number of jobs that have departed from class k in $[0,t]$, $S_k(t)$ denotes the amount of time that servers at station $j = \sigma(k)$ have *collectively* spent setting up for class k during the interval $[0,t]$, $T_k(t)$ denotes the amount of time that the servers at station $j = \sigma(k)$ have *collectively* spent processing class k batches during interval $[0,t]$, and $Z_k(t)$ denotes the total number of class k jobs that are buffered or being served at station j at time t. (One should keep in mind that the servers at a given station work as a team. Even though servers individually dedicate portions of their potential effort to the processing of batches and the performance of setups, the quantities S and T reflects the total effort among all of the servers.) The processes A, D, S, T, and Z are called the arrival, departure, setup allocation, service allocation, and jobcount processes, respectively. The components $U = \{U(t), t \geq 0\}$ and $Y = \{Y(t), t \geq 0\}$ are J-dimensional. For each station j, $U_j(t)$ denotes the total number of jobs at station j that are buffered or being served at time t, and $Y_j(t)$ denotes the collective total amount of time that servers at station j have been idle in the time interval $[0,t]$. (Again, idle time is summed over all of the resident servers at station j.) The process Y is called the cumulative idle time process. The process $\mathbb{X} = (A, D, S, T, U, Y, Z)$ satisfies the following set of equations:

$$A(t) = E(t) + \sum_k \Phi^k(D_k(t)), \quad t \geq 0, \tag{7.15}$$

$$Z(t) = Z(0) + A(t) - D(t), \quad t \geq 0, \tag{7.16}$$

$$Z(t) \geq 0, \quad t \geq 0, \tag{7.17}$$

$$U(t) = CZ(t), \quad t \geq 0, \tag{7.18}$$

$$C(S(t) + T(t)) + Y(t) = pt, \quad t \geq 0, \tag{7.19}$$

$$Y_j(t) \text{ can increase only if } \sum_{k \in \mathcal{C}(j)} \lceil Z_k(t)/B_k \rceil < p_j, \tag{7.20}$$

additional equations associated with the particular production policy π. (7.21)

Here C is the constituency matrix defined as

$$C_{jk} = \begin{cases} 1 & \text{if } k \in \mathcal{C}(j), \\ 0 & \text{otherwise,} \end{cases}$$

and $p = (p_1, p_2, \ldots, p_J)'$ denotes the J-dimensional vector indicating the number of servers at each station.

We provide a brief interpretation of equations (7.15)–(7.21); where convenient, the interpretation is component-wise. Equation (7.15) implies the cumulative arrivals to buffer ℓ consists of those jobs arriving to ℓ from the outside ($E_\ell(t)$), and those jobs routed to class ℓ after being processed in some other class. As for (7.16), the class ℓ jobcount process at time t, $Z_\ell(t)$, is equal to the number of class ℓ jobs present initially, plus all jobs that have arrived to buffer ℓ thus far, net those class ℓ jobs that have been processed. Expression (7.17), referred to as the nonnegativity constraint, is self-explanatory. For each j, the station-wide jobcount quantities are computed in (7.18). Equation (7.19) tracks, for each station j, how the total server time has been distributed, up until time t, i.e., between performing setups, processing batches, and idling. Note that the non-idling condition (7.20) contains the expression from (7.8). For additional interpretation, a version of (7.20) appeared in Chen and Shanthikumar [9]. Finally, the production policy π, used to govern the scheduling of servers, will have a major effect on system dynamics, hence the provision in (7.21).

Condition (7.20) gives an adequate, if not completely accurate, account of what we actually mean by non-idling. However, it may be more interesting to consider the number of servers assigned to eligible classes at time t. A first pass at this quantity was presented in (7.9), where, for each $j = 1, \ldots, J$,

$$I_j(t) = \min\left(p_j, \sum_{k \in C(j)} \left(1_{\{Z_k(t) \geq \theta_k\}} + \lfloor (Z_k(t) - \theta_k)^+ / (\ell_k B_k) \rfloor \right) \right).$$

We can tighten the definition of non-idling through the introduction of the K-dimensional process Z^*, referred to as the "uncommitted" jobcount process. When a server is dispatched to buffer k it reserves $\ell_k B_k$ jobs. Thus the uncommitted number of class k jobs Z_k^* decreases by $\ell_k B_k$ upon receiving a dispatched server. The uncommitted jobcount processes increases with arrivals, just as Z_k increases. Some extra care must be taken when a server is dispatched to a class with an insufficient number of uncommitted jobs. Assuming this special case is handled properly, the new non-idling condition reads

$$Y_j(t) \text{ does not increase if } Z_k^*(t) \geq \theta_k \quad \text{for some} \quad k \in C(j).$$

As mentioned earlier, the selection of jobs in the formation of batches follows a first-come-first-batched order. That is, jobs that have been in buffer k the longest have priority when batches are formed. When there are multiple servers at a station, batches from the same class may be processed in parallel. If the processing time of batches from the same class differs, it is possible for the departures of jobs from a class to violate the first-in-first-out protocol that the formation of batches obeys. To account for this feature, we have the following additional equations. For $0 \leq t_1 < t_2$ and $k = 1, \ldots, K$,

$$\Psi_k(T_k(t_2)) - \breve{\Psi}_k(T_k(t_1)) \leq \frac{1}{B_k}(D_k(t_2) - D_k(t_1)) + p_j - 1 \qquad (7.22)$$

when $I_j(s) = p_j$ for $s \in [t_1, t_2]$, where $j = \sigma(k)$, and

$$\breve{\Psi}_k(T_k(t_2)) - \Psi_k(T_k(t_1)) \geq \frac{1}{B_k}(D_k(t_2) - D_k(t_1)) - p_j + 1. \tag{7.23}$$

The function $\breve{\Psi}_k(t)$, defined fully in the appendix, is an inflated version of $\Psi_k(t)$. The inflated process $\breve{\Psi}_k(t)$ reflects the fact that the completion sequence of processed class k batches may diverge from the batch-formation sequence. However, as expressed formally in Lemma 7.19 of the appendix, under fluid scaling, the difference between $\breve{\Psi}_k$ and Ψ_k is negligible. Equations (7.22) and (7.23) follow from (7.75) and (7.76), both found in the appendix. For the lower bound on departures provided by (7.22), the condition $I_j(s) = p_j$ for each $s \in [t_1, t_2]$ ensures that servers at station j are making the largest batches possible. The upper bound on departures embodied in (7.23) captures the fact that class k batch sizes are no greater than B_k.

When enough jobs are present at the station, a sensible production policy restricts the frequency of class k setups to one for every ℓ_k batches. To facilitate capturing the restriction mathematically, we expand the definition of $S_k(t)$ to $S_{k'k}(t)$, the cumulative server time spent performing setups from k' to k. Clearly $S_k(t) = \sum_{k'} S_{k'k}(t)$. Suppose (7.10) holds. Then for every $t_1 < t_2$ such that $Z_k(s) \geq \theta_k + (p_j - 1)\ell_k B_k$ for every $s \in [t_1, t_2]$ we have

$$\begin{aligned} &D_k(t_2) - D_k(t_1) \\ &\quad \geq \ell_k B_k \left(\textstyle\sum_{k' \in \mathcal{C}(j)} [\Upsilon_{k'k}(S_{k'k}(t_2)) - \Upsilon_{k'k}(S_{k'k}(t_1))] - 2p_j \right), \end{aligned} \tag{7.24}$$

for each $k \in \mathcal{C}(j)$ and $j = 1, \ldots, J$. The left hand side of (7.24) is the number of processed jobs within the interval. Because class k is eligible throughout the interval, each occurrence where both a setup for class k and the subsequent production run lie entirely within the interval, the production run consists of exactly ℓ_k batches, each with the maximum B_k number of jobs. Hence the $\ell_k B_k$ term on the right hand side. The $2p_j$ term accounts for the fact that, for each server in the pool, a setup could be in process for class k both at time t_1 and at time t_2. The setups at the beginning of the time horizon could have been initiated when no class passed the first (stricter) eligibility test. Hence, the setup is not necessarily followed by the processing of ℓ_k batches. The setups at the end of the time interval may not have a chance to be followed by the ℓ_k batches because of running out of time. We call equations (7.15)-(7.24) the *processing network equations*. Note that S, T and Y are continuous, and that A, D, and Z are right continuous with left limits. All variables are nonnegative in each component, with A, D, S, T, and Y being non-decreasing. By assumption,

$$A(0) = D(0) = S(0) = T(0) = Y(0) = 0.$$

For each processing network driven by (E, V, Φ, F), the corresponding standard network driven by $(E, \tilde{V}, \Phi)$ has similar processes. In contrast with the processing network process $\mathbb{X} = (A, D, S, T, U, Y, Z)$, the standard network process is denoted $\tilde{\mathbb{X}} = (\tilde{A}, \tilde{D}, \tilde{T}, \tilde{U}, \tilde{Y}, \tilde{Z})$. (Again, the tilde is specific to standard networks.) Note that $\tilde{S}$ is missing from $\tilde{\mathbb{X}}$ because there are no setups in the standard network. The equations governing the standard network process are almost the same as the ones for processing network. The exceptions are that equation (7.19) lacks the $S(t)$ term and the vector p is replaced by the J-dimensional vector $e = (1, \ldots, 1)'$; equation (7.20) is replaced by

$$Y_j(t) \text{ can only increase if } U_j(t) = 0; \tag{7.25}$$

equations (7.22) and (7.23) are reduced to

$$\tilde{\Psi}_k(\tilde{T}_k(t)) = \tilde{D}_k(t), \text{ for all } t \geq 0,\ k = 1, \ldots, K,$$

which is well-known for standard networks operating under a head-of-line dispatch policy; equation (7.24) is removed; and equation (7.21) is replaced by

$$\text{additional equations associated with the particular dispatch policy } \tilde{\pi}. \tag{7.26}$$

7.3.2 *Fluid Models*

Let $\hat{\mathbb{X}} = (\hat{A}, \hat{D}, \hat{T}, \hat{U}, \hat{Y}, \hat{Z})$ be the formal deterministic analog of the standard network process $\tilde{\mathbb{X}} = (\tilde{A}, \tilde{D}, \tilde{T}, \tilde{U}, \tilde{Y}, \tilde{Z})$. Consider the following collection of equations:

$$\hat{A}(t) = \alpha t + P'\hat{D}(t), \quad t \geq 0, \tag{7.27}$$
$$\hat{Z}(t) = \hat{Z}(0) + \hat{A}(t) - \hat{D}(t), \quad t \geq 0, \tag{7.28}$$
$$\hat{Z}(t) \geq 0, \quad t \geq 0, \tag{7.29}$$
$$\hat{U}(t) = C\hat{Z}(t), \quad t \geq 0, \tag{7.30}$$
$$C\hat{T}(t) + \hat{Y}(t) = et, \quad t \geq 0, \tag{7.31}$$
$$\hat{Y}_j(t) \text{ can increase only if } \hat{U}_j(t) = 0, \quad j = 1, ..., J, \tag{7.32}$$
$$\hat{D}_k(t) = \tilde{\mu}_k \hat{T}_k(t), \quad k = 1, \ldots, K, \tag{7.33}$$
$$\text{additional equations associated with the particular, dispatch policy } \tilde{\pi}, \tag{7.34}$$

where $\tilde{\mu}_k$ is the class k service rate for the corresponding standard network, defined in (7.7), and, as before, e is the J-dimensional column vector of 1's. Equations (7.27)-(7.34), which define the standard fluid model, are referred to as the standard fluid model equations. As with the equations describing the processing network and the corresponding standard network, we assume that the the components of the processes $\hat{T}$ and $\hat{Y}$ are zero at time zero and are nondecreasing thereafter. Any process $\hat{\mathbb{X}} = (\hat{A}, \hat{D}, \hat{T}, \hat{U}, \hat{Y}, \hat{Z})$ satisfying (7.27)-(7.34) is called a standard fluid model solution. The component processes $\hat{A}$, $\hat{D}$, $\hat{T}$, and $\hat{Z}$ are called the fluid arrival, departure, service allocation, and buffer level processes, respectively. The quantity $\hat{U}_j(t)$ denotes the total amount of fluid at station j at time t. The process $\hat{Y}_j$ is referred to as the server idle time for station j. Standard fluid models and their solutions are fairly well-known; see, for instance, Dai [13]. Such models arise from taking fluid limits of standard networks, a topic we explore in the following section.

One can make qualitative comparisons between the standard fluid model and the equations governing the standard network. The major difference is that jobs are (discrete) units of flow in the standard network whereas flow in the fluid model is continuous; hence, the term "fluid." Along these same lines, the non-idling condition in the fluid model states that the station j cumulative idling process $\hat{Y}_j$ cannot increase in the presence of any positive amount of fluid at the station, as opposed to any jobs in the corresponding standard network equation.

Next, we present a generalization of the standard fluid model. Consider the process $\breve{\mathbb{X}} = (\breve{A}, \breve{D}, \breve{T}, \breve{U}, \breve{Y}, \breve{Z})$ and the following set of equations:

$$\breve{A}(t) = \alpha t + P'\breve{D}(t), \quad t \geq 0, \tag{7.35}$$
$$\breve{Z}(t) = \breve{Z}(0) + \breve{A}(t) - \breve{D}(t), \quad t \geq 0, \tag{7.36}$$
$$\breve{Z}(t) \geq 0, \quad t \geq 0, \tag{7.37}$$
$$\breve{U}(t) = C\breve{Z}(t), \quad t \geq 0, \tag{7.38}$$
$$C\breve{T}(t) + \breve{Y}(t) = pt, \quad t \geq 0, \tag{7.39}$$
$$\breve{Y}_j(t) \text{ can increase only if } \breve{U}_j(t) = 0, \quad j = 1, ..., J, \tag{7.40}$$
$$\breve{D}_k(t_2) - \breve{D}_k(t_1) \leq B_k\mu_k(\breve{T}_k(t_2) - \breve{T}_k(t_1)), \quad t_1 < t_2, \forall k, \tag{7.41}$$
$$\breve{D}_k(t_2) - \breve{D}_k(t_1) \geq B_k\breve{\mu}_k(\breve{T}_k(t_2) - \breve{T}_k(t_1)) \tag{7.42}$$
$$\text{if } \breve{U}_j(s) > 0 \ \forall s \in [t_1, t_2], \quad 0 \leq t_1 < t_2,$$

additional equations associated with the particular production policy π, (7.43)

where $\breve{\mu}_k$ is the setup-adjusted quantity, defined in (7.11), and, as before, $p = (p_1, \ldots, p_J)'$. Equations (7.35)-(7.43) are called artificial fluid model equations, and they define the *artificial fluid model* of the stochastic processing network. Any process $\breve{\mathbb{X}} = (\breve{A}, \breve{D}, \breve{T}, \breve{U}, \breve{Y}, \breve{Z})$ satisfying (7.35)-(7.43) is called an artificial fluid model solution. As with the standard fluid model, the components of $\breve{T}$ and $\breve{Y}$ are initially zero at time zero and are nondecreasing for all $t > 0$.

Although the connection between the standard fluid model and the standard (discrete) network is straightforward, we cannot claim a direct derivation of the artificial fluid model from a limiting procedure on the processing network. The dubious origin of this particular fluid model partially explains the moniker "artificial". In fact there are additional peculiarities yet to be resolved. In particular, notice the setup allocation process S of the processing network is conspicuously missing from the artificial fluid model counterpart. Yet another questionable feature of the model is the component process $\breve{T}$, which, in some way, is analogous to the service allocation process T as well as the setup allocation process S. With intentional ambiguity, we refer to $\breve{T}$ as the artificial *server* allocation process. Justification of the artificial fluid model is delayed until the following section. The remaining processes, $\breve{A}$, $\breve{D}$, $\breve{Z}$, $\breve{U}$, and $\breve{Y}$, retain their interpretations and terms from the standard fluid model.

Even without a formal justification of the model, it is nevertheless possible and, more importantly, instructive to interpret the equations of the artificial fluid model. Especially interesting are those equations in the artificial fluid model that differ from their analogous standard fluid model counterparts. For instance, compare equations (7.31) and (7.39). The difference is that by time t each station in the standard fluid network has t units of potential server effort, whereas station j in the artificial fluid network has $p_j t$ potential units of *artificial* server effort. Secondly, equation (7.33) in the standard fluid model is replaced by (7.41) and (7.42) in the artificial fluid model. In the standard fluid model, departures are directly proportional to server effort. However, in the artificial fluid model, the returns on artificial server effort are not as straightforward. Equation (7.41) gives the maximum rate at which effort is converted to departing units of fluid. There is an analogous interpretation from the (discrete) processing network. To see this, consider the average rate at which jobs depart a class when maximum-sized batches are being processed and no setup delays are incurred. Equation (7.42) provides a lower bound on the departure rate of fluid as a function of the rate of artificial server allocation when there is positive fluid at the station. This particular interpretation is related to the enforcement of the sensible policy rules. For one, maximum-sized batches are being formed. Secondly, at least ℓ_k batches are being processed per class k setup. Hence, μ_k is replaced by its setup-adjusted quantity $\breve{\mu}_k = 1/\breve{m}_k \le \mu_k$, as defined in (7.11). Finally, additional standard fluid model equations associated with the dispatch policy $\tilde{\pi}$, (7.34), may differ from artificial fluid model equations associated with the production policy π, (7.43).

Definition 7.2 An artificial fluid model is said to be *weakly stable* if for each artificial fluid model solution $\breve{\mathbb{X}}$ with $\breve{Z}(0) = 0$, $\breve{Z}(t) = 0$ for $t \ge 0$.

Weak stability of a standard fluid model can be defined similarly; see, for example, Chen [7].

7.3.3 Connection between Processing Networks and Fluid Models

The criterion for including an equation in the standard fluid model is that the equation is satisfied by a *fluid limit*. As suggested in the previous section, the inclusion of an equation in the artificial fluid model has an additional step. In this section we provide the details of both the fluid limits and the additional steps.

A fluid limit of a standard network is obtained through a law-of-large-numbers limiting procedure on the standard network process. Identically, a fluid limit of a processing network is obtained through a law-of-large-numbers limiting procedure on the processing network process. Note that the processing network process $\mathbb{X}$ (resp. standard network process $\tilde{\mathbb{X}}$) is random, depending on the sample path ω in an underlying probability space. To denote such dependence explicitly, we sometimes use $\mathbb{X}(\omega)$ to denote the network process with sample path ω. For an integer d, $\mathbb{D}^d[0,\infty)$ denotes the set of functions $x : [0,\infty) \to \mathbb{R}^d$ that are right continuous on $[0,\infty)$ and have left limits on $(0,\infty)$. An element x in $\mathbb{D}^d[0,\infty)$ is sometimes denoted by $x(\cdot)$ to emphasize that x is a function of time. For each ω, $\mathbb{X}(\omega)$ is an element in $\mathbb{D}^{5K+2J}[0,\infty)$.

For each $r > 0$, define

$$\bar{\mathbb{X}}^r(t,\omega) = r^{-1}\mathbb{X}(rt,\omega), \quad t \geq 0. \tag{7.44}$$

Again, note that for each $r > 0$, $\bar{\mathbb{X}}^r(\cdot,\omega)$ is an element in $\mathbb{D}^{5K+2J}[0,\infty)$. The scaling in (7.44) is called the fluid or law-of-large-numbers scaling.

Definition 7.3 A function $\bar{\mathbb{X}} \in \mathbb{D}^{5K+2J}[0,\infty)$ is said to be a *fluid limit* of the processing network if there exists a sequence $r_n \to \infty$ and a sample path ω satisfying (7.1) such that

$$\lim_{n\to\infty} \bar{\mathbb{X}}^{r_n}(\cdot,\omega) \to \bar{\mathbb{X}}(\cdot),$$

where, throughout this chapter, the convergence is interpreted as the uniform convergence on compact sets (u.o.c.).

Uniform convergence on compact sets, as it pertains to fluid limits of networks, is discussed, for instance, in Chen and Mandelbaum [8].

The existence of fluid limits is well-known. A standard argument like the one in Dai [12] shows that for any $r \to \infty$ and any sample path ω, there is a subsequence r_n such that $\bar{S}^{r_n}(\cdot,\omega)$ and $\bar{T}^{r_n}(\cdot,\omega)$ converge as $n \to \infty$. Fix an ω that satisfies (7.1). The convergence of $\bar{T}^{r_n}$, together with equation (7.23), condition (7.1) and Lemma 7.19, implies that $\bar{D}^{r_n}$ converges. This latter convergence, together with equation (7.15) and condition (7.1), implies that $\bar{A}^{r_n}$ converges. The convergence of other components of $\bar{\mathbb{X}}^{r_n}$ then readily follows. Thus, $\bar{\mathbb{X}}^{r_n}$ converges to a fluid limit as $n \to \infty$.

We now convert the fluid limit $\bar{\mathbb{X}}$ into an artificial "fluid limit" $\breve{\mathbb{X}}$. We use the term "limit" facetiously. In fact, the process $\breve{\mathbb{X}}$ is not a limit at all. Our true intentions are embodied in the following proposition. By the expression $\breve{T} = \bar{S} + \bar{T}$ we mean $\breve{T}(t) = \bar{S}(t) + \bar{T}(t)$ for each $t \geq 0$.

Proposition 7.4 Given a fluid limit $\bar{\mathbb{X}} = (\bar{A}, \bar{D}, \bar{S}, \bar{T}, \bar{U}, \bar{Y}, \bar{Z})$ of a processing network operating under a sensible production policy π, the process $\breve{\mathbb{X}} = (\breve{A}, \breve{D}, \breve{T}, \breve{U}, \breve{Y}, \breve{Z}) = (\bar{A}, \bar{D}, \bar{T} + \bar{S}, \bar{U}, \bar{Y}, \bar{Z})$ is an artificial fluid model solution.

Proof. Fix the fluid limit $\bar{\mathbb{X}}$ and construct $\breve{\mathbb{X}}$ by collapsing the allocation processes $\bar{S}$ and $\bar{T}$ to $\breve{T}$, i.e., $\breve{T} = \bar{S} + \bar{T}$. Equation (7.41) follows from (7.23), Lemma 7.19 and the fact that $\breve{T}_k(t_2) - \breve{T}_k(t_1) \geq \bar{T}_k(t_2) - \bar{T}_k(t_1)$ for every $0 \leq t_1 \leq t_2$. As for equation (7.42), from the strong law of large numbers (7.3), for each $k = 1, \ldots, K$ and $0 \leq t_1 < t_2$,

$$\lim_{r_n \to \infty} \sum_{k' \in \mathcal{C}(j)} \left(\bar{\Upsilon}^{r_n}_{k'k}(\bar{S}^{r_n}_{k'k}(t_2)) - \bar{\Upsilon}^{r_n}_{k'k}(\bar{S}^{r_n}_{k'k}(t_1)) \right) \geq \frac{1}{s_k} \left(\bar{S}_k(t_2) - \bar{S}_k(t_1) \right), \tag{7.45}$$

and, from (7.22), (7.23) and Lemma 7.19,

$$\bar{D}_k(t_2) - \bar{D}_k(t_1) = B_k \mu_k \left(\bar{T}_k(t_2) - \bar{T}_k(t_1) \right), \quad \text{for each } k \in \mathcal{C}(j), \tag{7.46}$$

whenever $\bar{U}_j(s) > 0$ for each $s \in [t_1, t_2]$. By (7.24), (7.45) and (7.46) we have,

$$\bar{S}_k(t_2) - \bar{S}_k(t_1) \leq \frac{s_k}{\ell_k m_k} \left(\bar{T}_k(t_2) - \bar{T}_k(t_1) \right),$$

and, hence, by (7.11)

$$\begin{aligned} \breve{T}_k(t_2) - \breve{T}_k(t_1) &= \bar{T}_k(t_2) + \bar{S}_k(t_2) - \left(\bar{T}_k(t_1) + \bar{S}_k(t_1) \right) \\ &\leq \frac{\breve{m}_k}{m_k} \left(\bar{T}_k(t_2) - \bar{T}_k(t_1) \right), \end{aligned} \tag{7.47}$$

whenever $\bar{U}_j(s) = \breve{U}_j(s) > 0$ for each $s \in [t_1, t_2]$. Equation (7.42) follows from (7.46) and (7.47). Other fluid model equations can be verified as in Dai [12]. □

By now it should be abundantly clear that artificial fluid models are the combination of a law-of-large-numbers limiting procedure, the limitations on setup allocation via sensible policy constraints, and the collapsing of setup and service allocation processes into a single artificial server allocation process.

Theorem 7.5 Let a sensible production policy π be fixed. If the artificial fluid model is weakly stable, then the corresponding processing network is rate stable.

Proof. The theorem was first explicitly stated in Chen [7] for standard networks. The only difference here is recognizing that a fluid limit of the processing network is a solution to the artificial fluid model once the allocation processes are collapsed. The remainder of the proof is identical to one for the standard network. See, for example, Dai [13]. □

7.4 Connection between Artificial and Standard Fluid Models

Our goal is to craft a sensible production policy π such that the processing network operating under the policy π is rate stable, if at all possible. As stated earlier, this pursuit is possible only if the usual traffic condition (7.6) holds. It turns out that when (7.6) holds, coming up with a stabilizing π is always possible.

The standard fluid model has become the conventional tool for demonstrating stability of the standard network operating under a given dispatch policy. The connection between stability of the standard network and its associated fluid model was made concrete through the standard network analog to Theorem 7.5, provided by Chen [7]. A sizeable body of literature is devoted to investigating standard fluid models, primarily in demonstrating some form of stability. One particularly useful technique for demonstrating stability is via Lyapunov functions. Not surprisingly, a large portion of the literature focuses on finding the right Lyapunov function to demonstrate stability, given the dispatch policy in question.

Given the similarities between the standard fluid model and the artificial fluid model, we would like to piggyback on these efforts. In particular, it would be ideal if the Lyapunov function that demonstrates the stability of a standard fluid model also works for the artificial fluid model analog. In this section we explore the connections between the fluid models. We start with the special case of batch processing networks, where setups times are all zero, and conclude with the general case.

7.4.1 Batch Processing Networks and Normal Policies

A batch processing network is the special case of a processing network for which the setup times are all zero. This definition is slightly more general than the one in Dai and Li [17]. In their paper, each station is populated by a single server. Let $\pi = (\theta, \tilde{\pi}, \ell)$ be a sensible production policy used to schedule a batch processing network. Since setup times are zero, we assume that $\ell_k = 1$ for each class k. Borrowing from Dai and Li [17], the resulting sensible production policy is referred to as a *full batch policy.* A dispatch policy $\tilde{\pi}$ is said to *induce* a full batch policy π if $\pi = (\theta, \tilde{\pi}, 1)$. We now consider when dispatch policies that are known to stabilize standard networks induce full batch policies that stabilize batch processing networks.

Let $\bar{\mathbb{X}} = (\bar{A}, \bar{D}, \bar{S}, \bar{T}, \bar{U}, \bar{Y}, \bar{Z})$ be a fluid limit of a batch processing network. We include the component $\bar{S}$, even though it is always zero, to emphasize that this is still the fluid limit of a processing network. We can trivially construct a process $\breve{\mathbb{X}} = (\breve{A}, \breve{D}, \breve{T}, \breve{U}, \breve{Y}, \breve{Z})$ by simply copying all of the nontrivial components of $\bar{\mathbb{X}}$; that is $(\breve{A}, \breve{D}, \breve{T}, \breve{U}, \breve{Y}, \breve{Z}) = (\bar{A}, \bar{D}, \bar{T}, \bar{U}, \bar{Y}, \bar{Z})$. We refer to processes $\breve{\mathbb{X}}$ constructed in this manner as batch fluid limits. The batch fluid model then is defined by (7.35)-(7.41), along with

$$\breve{D}_k(t_2) - \breve{D}_k(t_1) = B_k \mu_k \left(\breve{T}_k(t_2) - \breve{T}_k(t_1) \right),$$
$$\text{if } \breve{U}_j(s) > 0 \quad \forall s \in [t_1, t_2], \quad 0 \le t_1 < t_2, \tag{7.48}$$

and

$$\text{additional equations associated with the particular full batch policy } \pi, \tag{7.49}$$

which replace (7.42) and (7.43), respectively.

In a sense, the batch fluid model bridges the gap between artificial and standard fluid models. For example, as in the artificial fluid model, there are two equations governing the conversion of allocated server effort to departing jobs. In fact, to further match (7.42), equation (7.48) could be expressed as an inequality. Equality holds because the upper bounds and lower bounds agree on intervals during which there is positive fluid at the associated station. Since there are no setup times, the class k inflated mean service times $\breve{m}_k$ (see equation (7.11)) equals the regular class k mean service time m_k. Modulo the server pool constant p_j implicit in the computation of the class k service time in the corresponding standard network $\tilde{m}_k$ (see equation (7.7)), equation (7.48) matches its standard fluid model analog (7.33) on intervals during which $\breve{U}_j(\cdot)$ is positive.

Let $\breve{\mathbb{X}} = (\breve{A}, \breve{D}, \breve{T}, \breve{U}, \breve{Y}, \breve{Z})$ be a batch fluid model solution, i.e., a solution to (7.35)-(7.41), (7.48) and (7.49). We would like to convert it into a standard fluid model solution $\hat{\mathbb{X}} = (\hat{A}, \hat{D}, \hat{T}, \hat{U}, \hat{Y}, \hat{Z})$. We define $\hat{\mathbb{X}}$ as follows: for each $t \ge 0$,

$$\hat{A}(t) = \breve{A}(t), \tag{7.50}$$
$$\hat{D}(t) = \breve{D}(t), \tag{7.51}$$
$$\hat{T}_k(t) = \tilde{m}_k \hat{D}_k(t), \; k = 1, ..., K, \tag{7.52}$$
$$\hat{Y}_j(t) = t - \sum_{k \in C(j)} \hat{T}_k(t), \quad j = 1, \dots, J, \tag{7.53}$$
$$\hat{U}(t) = \breve{U}(t), \tag{7.54}$$
$$\hat{Z}(t) = \breve{Z}(t). \tag{7.55}$$

One can verify that the constructed process $\hat{\mathbb{X}}$ satisfies (7.27)-(7.33) such that it is *almost* a standard fluid model solution. As forwarded by Dai and

Li [17], we define a subclass of dispatch policies $\tilde{\pi}$ such that the constructed process is indeed a standard fluid model solution.

Definition 7.6 A dispatch policy $\tilde{\pi}$ is called *normal* if for any batch fluid model solution $\breve{\mathbb{X}}$ under a full batch policy $\pi = (\theta, \tilde{\pi}, 1)$ induced by $\tilde{\pi}$, $\hat{\mathbb{X}}$ constructed by (7.50)-(7.55) also satisfies (7.34).

For the single-server stations case, Dai and Li [17] show that static buffer priority (SBP) policies, the first-in–first-out (FIFO) policy, and the generalized round robin (GRR) policy are normal. Because we modify dispatch policies in a different manner than in Dai and Li [17], generalized round robin is not normal for our circumstances. It should not be surprising that normal policies for single-server networks are often normal for the multi-server systems.

Proposition 7.7 If a dispatch policy $\tilde{\pi}$ operating in a standard network is normal, then the batch fluid model under the induced full batch policy π is weakly stable if the standard fluid model under policy $\tilde{\pi}$ is weakly stable.

Proof. Consider the fluid model of the batch processing network operating under the full batch policy π induced by the normal policy $\tilde{\pi}$. Through a slight modification of the proof by Dai and Li [17], it can be shown that any solution to the batch fluid model $\breve{\mathbb{X}}$ can be converted to $\hat{\mathbb{X}}$ through equations (7.50)-(7.55), where $\hat{\mathbb{X}}$ is a solution to the standard fluid model associated with the dispatch policy $\tilde{\pi}$. □

With this preparation, we now provide an extension of the main result of Dai and Li [17] to networks with multiple servers. Application of the following theorem can be used in conjunction with Corollary 7.15 of Section 7.5.1.

Theorem 7.8 For a given batch processing network, assume that a dispatch policy $\tilde{\pi}$ is normal for the corresponding standard network. The batch processing network operating under the induced batch policy π is rate stable if the standard fluid model operating under $\tilde{\pi}$ is weakly stable.

Proof. Assume that $\tilde{\pi}$ is a normal dispatch policy in the corresponding standard network. Assume further that the corresponding standard fluid model is weakly stable. By Proposition 7.7, the batch fluid model operating under the induced batch policy π is weakly stable. The proof follows from Theorem 7.5. □

We conclude with an extension of a result by Dai and Li [17] for so-called *static buffer priority (SBP) policies.* Under an SBP policy, each station has a strict priority ranking of the constituent buffers. When a dispatching decision is made, the associated server is dispatched to the highest priority, eligible buffer. When the network under consideration is a standard one, eligible is synonymous with nonempty. The aforementioned extension can easily be shown:

Lemma 7.9 Any SBP policy is normal.

This lemma is critical to Corollary 7.15 at the conclusion of Section 7.5.1.

7.4.2 Stability under Sensible Production Policies

The presence of setups in the processing network renders the arguments of the previous section difficult to apply. That is, extending the definition of normal policies would not be fruitful under our definition of production policy (see the extensions section for discussion of when the extension might be useful). The problem is that, for the artificial fluid model, conversion of allocated server effort to the departure process (embodied in equations (7.41) and (7.42)) is not constant, as it is for a batch fluid model when the associated station has positive fluid (embodied in equation (7.48)). Accordingly, we require a more hands-on approach.

Typically, Lyapunov functions are used to demonstrate stability of standard fluid models. The following is one method of using a Lyapunov function. For more examples of its usage, see Dai [13]. Given a standard fluid model solution $\hat{\mathbb{X}} \in \mathbb{D}^{4K+2J}[0,\infty)$:

- Find a functional L which maps $\hat{\mathbb{X}}$ to a nonnegative function f, where $f(t) = L(\hat{\mathbb{X}})(t)$ is absolutely continuous in t, and
- $f(t) > 0$ if and only if $\hat{Z}(t) \neq 0$, such that
- there exists an $\epsilon > 0$ where $\hat{Z}(t) \neq 0$ implies $\frac{d}{dt} f(t) \leq -\epsilon$.

Given such a functional L with the absolute continuity of f, one has $\hat{Z}(t) = 0$ for all $t \geq f(0)/\epsilon$.

This approach is somewhat overkill. The approach allows one to show that, given some positive fluid at time zero, the fluid model solution drains in a finite time proportional to the Lyapunov function value at time zero. But since our fluid model starts with zero fluid, the Lyaponuv function has an initial value of zero, and the fluid model solution remains as such for all time. There is no overkill if the ϵ in the description is equal to zero.

Clearly, Lyapunov functions can also be used for showing weak stability of artificial fluid models. In fact, in some instances:

The Lyapunov function that shows a given dispatch policy $\tilde{\pi}$ stabilizes a standard network may also show that a corresponding sensible production policy π stabilizes a processing network.

This admittedly nebulous statement is the closest we can come to replicating, for general processing networks, the normal policy paradigm for batch fluid models. However, in the following section, we carry out the details in two examples.

The idea just articulated was previously forwarded in Dai and Jennings [15] for systems with no batch processing operations and single-server stations, also referred to as queueing networks with setups. In [15] they examine so-called Kelly networks with setups. A Kelly network, as described by Bramson [3], is a special class of a standard network, where, for each station, the mean processing times of all constituent classes are identical; that is, $m_k = m_{k'}$ for all $k, k' \in \mathcal{C}(j)$, $j = 1, \dots, J$. A Kelly network with setups, introduced in [15], has the same mean processing time restrictions of Kelly networks, but the mean setup times have the general sequence-dependent structure of processing networks. Adapting the entropy Lyapunov function used by Bramson [3] to prove stability of Kelly networks, two results emerge: the stability of almost-Kelly networks under the FIFO dispatch policy and the stability of Kelly networks with setups operating under a subclass of sensible FIFO production policies $\pi = (\theta, \text{FIFO}, \ell)$.

Paralleling the result in [15] is the analogous result by Dai and Li [17] for so-called batch Kelly networks, whose predictable definition we omit. It turns out that the FIFO dispatch policy for Kelly networks is a normal policy. Hence, by Theorem 7.8, the batch Kelly network, operating under an induce full batch FIFO policy $\pi = (\theta, \text{FIFO}, 1)$, is rate stable. It would not be surprising if the results in [15] and [17] could be combined for would-be Kelly processing networks.

7.5 Examples of Stable Policies

We now demonstrate the power of our framework by proving the stability of two sensible production policies. In the first case we consider a special class of multi-type networks operated under a sensible early-steps-first (ESF) policy. In the second case, the processing network with the most general form discussed in the chapter is operated under a sensible generalized round robin (GRR) production policy.

7.5.1 Early Steps First

Consider a family of standard networks where each job follows some deterministic *route*, or sequence of buffers, through the network. Suppose that for each buffer there is only one manner in which jobs arrive, exogenously or from one (upstream) buffer. Buffers that receive jobs exogenously are referred to as *sources*. Let $\mathcal{E}$ denote the set of all source buffers. Without loss of generality, $\alpha_k > 0$ for each $k \in \mathcal{E}$. Jobs that enter the system through buffer k are said to belong to the same job *type* and follow the same route of buffers through the network. We refer to such networks as *multi-type queue-*

ing networks, or multi-type networks for short. Jobs of different types do not mix; that is, there are no common buffers in the routes of two distinct job types. In a sense, multi-type networks are the superposition of multiple reentrant lines on a single collection of stations.

For a multi-type network, the corresponding transition matrix P and arrival rate vector α exhibit special structure. If some element k of α is nonzero, then the corresponding column of P has all zeros. Otherwise, $\alpha_k = 0$ and exactly one element of the corresponding column of P is one and the other elements are all zero.

To facilitate the description of the ESF dispatch policy we alter the notation slightly. In the new notation, classes are denoted by their (type, step) pair. The types are indexed $q = 1, \ldots, |\mathcal{E}|$ and the steps are indexed $k = 1, \ldots, K_q$, where K_q reports the length of the type q route. All of the class-specific quantities now have this alternative notation. For instance, $m_{(q,k)}$ denotes the class (q,k) mean processing time and $Z_{(q,k)}(t)$ records the number of class (q,k) jobs at time t. In a slight abuse of notation, we replace the arrival rate of type q jobs $\alpha_{(q,1)}$ with the quantity α_q.

We are now equipped to describe the ESF dispatch policy. Suppose the server at station j needs to be dispatched at time t. Among all of the nonempty buffers at the station, the server will be dispatched to a buffer (q,k), where k is the first nonempty step at the station. Any other constituent buffer $(q',k') \in \mathcal{C}(j)$ with $k' < k$, must therefore be empty. Note, it is possible that station j houses multiple classes that are the kth step for their respective routes. The policy does not explicitly say how to choose among such classes; that is, ties are broken arbitrarily.

Theorem 7.10 Under the usual traffic condition (7.6), a multi-type network, operating under the earlier-step-first dispatch policy, is rate stable.

A proof of Theorem 7.10 might use the following linear Lyapunov function:

$$L(t) = \sum_{(q,k)} \beta^{+}_{(q,k)} \hat{Z}_{(q,k)}(t), \tag{7.56}$$

where

$$\beta_{(q,k)} = \begin{cases} m_{(q,k)} \left(1 + \frac{\gamma_{j,k}}{1-\rho_j}\right) & k = 1, \ldots, K_q, \\ 0 & k = K_q + 1, \end{cases} \tag{7.57}$$

$$\gamma_{j,k} = \sum_{(q,k) \in \mathcal{C}(j,k)} \alpha_q \beta^{+}_{(q,k+1)}, \tag{7.58}$$

and

$$\beta^{+}_{(q,k)} = \sum_{k'=k}^{K_q} \beta_{(q,k')}. \tag{7.59}$$

The set $\mathcal{C}(j,k) = \{(q,k') \in \mathcal{C}(j) : k' = k\}$, used in (7.58), contains the constituent classes of station j that are composed of the kth step of some route. The proof of Theorem 7.10 is implied by the proof of Theorem 7.11.

Multi-type processing networks have the same routing and arrival characteristics as their multi-type network analogs. We can adapt the ESF dispatch policy to form a sensible early-steps-first production policy $\pi = (\theta, \tilde{\pi}, \ell)$. The policy evolves as follows: When a server at station j is free for dispatching, we create a list of eligible constituent buffers. The server will be dispatched to a buffer (q,k), where k is the earliest step with an eligible buffer at the station. Any other constituent buffer $(q',k') \in \mathcal{C}(j)$ with $k' < k$, must be ineligible. Moreover, if (q',k') is ineligible, it must be the case that

$$Z_{(q',k')}(t) < \theta_{(q',k')} + (p_j - 1)\ell_{(q',k')}B_{(q',k')}.$$

Assuming (q,k) passed the more stringent eligibility test, a setup is performed (if necessary) for class (q,k) and then $\ell_{(q,k)}$ full batches are processed in a row (if possible) before the server is freed for subsequent dispatching. As in the standard network setting, there may be more than one eligible buffer from step k at station j. Any arbitrarily chosen tie-breaking scheme will yield the same stability results.

Theorem 7.11 A multi-type processing network operating under a sensible early-steps-first production policy $\pi = (\theta, \mathrm{ESF}, \ell)$ is rate stable.

We delay the proof of the following lemma until the appendix.

Lemma 7.12 Under the sensible early-steps-first production policy $\pi = (\theta, \mathrm{ESF}, \ell)$, the artificial fluid model equation (7.43) takes the form

$$\sum_{k=1}^{k'} \sum_{(q,k)\in\mathcal{C}(j)} \breve{Z}_{(q,k)}(t) > 0 \quad \text{implies} \quad \sum_{k=1}^{k'} \sum_{(q,k)\in\mathcal{C}(j)} \dot{\breve{T}}_{(q,k)}(t) = p_j, \tag{7.60}$$

for every step $k' \geq 1$ and each station $j = 1, \ldots, J$.

Proof of Theorem 7.11. Let $\breve{\mathbb{X}}$ be a solution to the artificial fluid model operating under a sensible ESF production policy. We adapt the linear Lyapunov function devised for the ESF dispatch policy to obtain:

$$L(t) = \sum_{(q,k)} \beta^+_{(q,k)} \breve{Z}_{(q,k)}(t), \tag{7.61}$$

where

$$\beta_{q,k} = \begin{cases} \frac{\breve{m}_{(q,k)}}{B_{(q,k)}} \left(1 + \frac{\gamma_{j,k}}{p_j(1-\breve{\rho}_j)}\right) & k = 1, \ldots, K_q, \\ 0 & k = K_q + 1, \end{cases} \tag{7.62}$$

$$\gamma_{j,k} = \sum_{(q,k)\in\mathcal{C}(j,k)} \alpha_q \beta^+_{(q,k+1)}, \tag{7.63}$$

and

$$\beta^+_{(q,k)} = \sum_{k'=k}^{K_q} \beta_{(q,k')}. \tag{7.64}$$

Notice that equations (7.63) and (7.64) are identical to (7.58) and (7.59), respectively.

Assume that $\breve{Z}(t) \neq 0$ and $\breve{\mathbb{X}}$ is differentiable at time t. Let k be the first system-wide, nonempty step. That is, there is some class (q,k) such that $\breve{Z}_{(q,k)}(t) > 0$ and $\breve{Z}_{(q,k')}(t) = 0$ for all classes (q,k') such that $k' < k$. Fix step k and time t for the remainder of the proof. We investigate the derivative of the Lyapunov function in (7.61):

$$\begin{aligned}
\dot{L}(t) &= \sum_{(q',k')} \beta^+_{(q',k')} \dot{\breve{Z}}_{(q',k')}(t) = \sum_{q'=1}^{|\mathcal{E}|} \sum_{k'=1}^{K_{q'}} \beta^+_{(q',k')} \dot{\breve{Z}}_{(q',k')}(t) \\
&= \sum_{q'=1}^{|\mathcal{E}|} \sum_{k'=1}^{K_{q'}} \beta^+_{(q',k')} \left(\dot{\breve{D}}_{(q',k'-1)}(t) - \dot{\breve{D}}_{(q',k')}(t) \right) \\
&= \sum_{q'=1}^{|\mathcal{E}|} \left[\alpha_{q'} \beta^+_{(q',1)} - \sum_{k'=1}^{K_{q'}} \left(\beta^+_{(q',k')} - \beta^+_{(q',k'+1)} \right) \dot{\breve{D}}_{(q',k')}(t) \right] \\
&= \sum_{q'=1}^{|\mathcal{E}|} \left[\alpha_{q'} \beta^+_{(q',1)} - \sum_{k'=1}^{K_{q'}} \beta_{(q',k')} \dot{\breve{D}}_{(q',k')}(t) \right],
\end{aligned}$$

where $\breve{D}_{(q',0)}(t) \equiv \alpha_{q'} t$. By the nonnegativity of $\breve{Z}$, $\breve{Z}_{(q',k')}(t) = 0$ implies $\dot{\breve{Z}}_{(q',k')}(t) = 0$. Since k is the first nonempty step, $\dot{\breve{D}}_{(q',k')}(t) = \alpha_{q'}$ for all classes (q',k') such that $k' < k$ and the derivative of the $L(t)$ equals

$$\begin{aligned}
\dot{L}(t) &= \sum_{q'=1}^{|\mathcal{E}|} \left[\alpha_{q'} \beta^+_{(q',k)} - \sum_{k'=k}^{K_{q'}} \beta_{(q',k')} \dot{\breve{D}}_{(q',k')}(t) \right] \qquad (7.65) \\
&\leq \sum_{j:\mathcal{C}(j,k)\neq\emptyset} \left[\sum_{(q,k)\in\mathcal{C}(j,k)} \alpha_q \beta^+_{(q,k)} - \sum_{(q,k)\in\mathcal{C}(j,k)} \beta_{(q,k)} \dot{\breve{D}}_{(q,k)}(t) \right],
\end{aligned}$$

where, again, $\mathcal{C}(j,k) = \{(q,k') \in \mathcal{C}(j) : k' = k\}$ denotes the constituent classes of station j that are composed of the kth step of some route. The transition embodied in (7.65) has two subtleties. For one, each class (q',k') with step $k' \geq k$ the quantity $\dot{\breve{D}}_{(q',k')}(t)$ appears on the left hand side of the inequality whereas the quantity appears on the right hand side for only those classes with step k. Secondly, each nonzero $\alpha_{q'}\beta^+_{q',k'}$ term on the left hand side of the inequality appears on the right hand side as well. The terms that are equal to zero correspond to job types with strictly fewer than k steps. The nonzero values correspond to routes that have at least k steps. Moreover, the kth step must occur at one of the stations. Hence, on the right hand side we can exclude those stations without a resident class that is the kth step of some route. By (7.42), Lemma 7.12 and the fact that $\dot{\breve{D}}_{(q,k')}(t) = \alpha_q$ if $k' < K$, we have

$$\sum_{(q,k)\in\mathcal{C}(j,k)} \dot{\breve{T}}_{(q,k)}(t) = p_j - \sum_{k'=1}^{k-1} \sum_{(q,k')\in\mathcal{C}(j,k')} \dot{\breve{T}}_{(q,k')}(t) \qquad (7.66)$$
$$\geq p_j - \sum_{k'=1}^{k-1} \sum_{(q,k')\in\mathcal{C}(j,k')} \frac{\alpha_q \breve{m}_{(q,k')}}{B_{(q,k')}}.$$

Hence, by (7.42), (7.64) and (7.65),

$$\dot{L}(t) \leq \sum_{j:\mathcal{C}(j,k)\neq\emptyset} \Bigg[\sum_{(q,k)\in\mathcal{C}(j,k)} \alpha_q\beta_{(q,k)} + \sum_{(q,k)\in\mathcal{C}(j,k)} \alpha_q\beta^+_{(q,k+1)} - \sum_{(q,k)\in\mathcal{C}(j,k)} \beta_{(q,k)} B_{(q,k)} \breve{\mu}_{(q,k)} \dot{\breve{T}}_{(q,k)}(t) \Bigg].$$

It follows from (7.13), (7.62), (7.63) and (7.66) that

$$\begin{aligned}
\dot{L}(t) &\leq \sum_{j:\mathcal{C}(j,k)\neq\emptyset} \Bigg[\sum_{(q,k)\in\mathcal{C}(j,k)} \frac{\alpha_q \breve{m}_{(q,k)}}{B_{(q,k)}} \left(1 + \frac{\gamma_{j,k}}{p_j(1-\breve{\rho}_j)}\right) + \gamma_{j,k} \\
&\qquad - \left(1 + \frac{\gamma_{j,k}}{p_j(1-\breve{\rho}_j)}\right) \left(p_j - \sum_{k'=1}^{k-1} \sum_{(q,k')\in\mathcal{C}(j,k')} \frac{\alpha_q \breve{m}_{(q,k')}}{B_{(q,k')}}\right) \Bigg] \\
&= \sum_{j:\mathcal{C}(j,k)\neq\emptyset} \Bigg[\gamma_{j,k} - \left(1 + \frac{\gamma_{j,k}}{p_j(1-\breve{\rho}_j)}\right) \\
&\qquad \cdot \left(p_j - \sum_{k'=1}^{k} \sum_{(q,k')\in\mathcal{C}(j,k')} \frac{\alpha_q \breve{m}_{(q,k')}}{B_{(q,k')}}\right) \Bigg] \\
&\leq \sum_{j:\mathcal{C}(j,k)\neq\emptyset} -p_j(1-\breve{\rho}_j).
\end{aligned}$$

The derivative of the Lyapunov function is negative as long as there is positive fluid. Hence, the artificial fluid model is stable. □

In addition to implying Theorem 7.10 for standard networks, the proof of Theorem 7.11 for general processing networks implies the intermediate result for processing networks with zero setup times:

Theorem 7.13 A multi-type batch processing network operating under an induced early-steps-first full batch policy $\pi = (\theta, \text{ESF}, 1)$ is rate stable.

Alternatively, suppose we only knew the following result:

Theorem 7.14 The fluid model associated with the ESF dispatch policy is weakly stable.

Given only Theorem 7.14, it is not immediately obvious that Theorem 7.11 holds. However, we could still assert Theorem 7.10 by virtue of the standard network equivalent to Theorem 7.5; see Chen [7]. Moreover, we can also prove Theorem 7.13 through the following argument. First notice that the ESF dispatch policy is an example of a static buffer priority policy, as described in the conclusion of Section 7.4.1. Hence, Lemma 7.9 leads to the following:

Corollary 7.15 ESF is a normal policy.

Theorem 7.13 follows immediate from Corollary 7.15, Theorem 7.14 and Proposition 7.7.

7.5.2 Generalized Round Robin

For standard networks, the generalized round robin (GRR) dispatch policy is parameterized by a set of strictly positive reals $\beta = (\beta_k,\ k = 1, \ldots, K)$. When the constants are integers, the policy works as follows: At station j, the server "visits" the constituent buffers in $\mathcal{C}(j)$ in a fixed cyclic order; hence, the name round robin. In polling station literature, the order in which classes are visited is referred to as the *polling table*, a term we adopt as well. When the server visits buffer k, β_k jobs are processed, if possible. Otherwise, the buffer is exhausted and the server moves on to the next buffer. In this sense, the β_k's can be thought of as nominal allocations. The span of time, from the beginning of the visit to the first buffer in the polling table to the completion of the visit to the last buffer in the table, is referred to as a *cycle*.

When the β_k's are not integers, the spirit of the dispatch policy is the same. However, some requisite bookkeeping is in order. Consider the nth cycle of the server at station j. Let $a_k(n)$ denote the integer-valued nominal allocation for each class $k \in \mathcal{C}(j)$ and $b_k(n)$ denote the nominal residual allocation. The quantities are defined recursively:

$$a_k(n+1) = \lfloor b_k(n) + \beta_k \rfloor, \tag{7.67}$$

$$b_k(n+1) = b_k(n) + \beta_k - a_k(n+1), \tag{7.68}$$

for $n = 0, 1, \ldots$, where $b_k(0) = 0$ and, as before, $\lfloor x \rfloor$ denotes the integer part of x. When the server visits buffer k for the nth time, it processes $a_k(n)$ jobs, if possible, before moving on to the next buffer.

For any vector of positive constants β, the additional standard fluid model equation (7.34) takes the form

$$\dot{\hat{T}}_k(t) \geq \frac{\beta_k \tilde{m}_k}{\sum_{k' \in C(j)} \beta_{k'} \tilde{m}_{k'}}, \quad k = 1, \ldots, K, \tag{7.69}$$

for each t such that $\hat{T}_k(t)$ is differentiable and $\hat{Z}_k(t) > 0$, where, as is our convention, $j = \sigma(k)$. The following theorem is well-known; see, for example, Dai [13].

Theorem 7.16 Under the usual traffic conditions (7.6), a standard network operating under a generalized round robin policy parameterized by β is stable if, for each $k = 1, \ldots, K$,

$$\frac{\beta_k \tilde{m}_k}{\sum_{k' \in C(j)} \beta_{k'} \tilde{m}_{k'}} \geq \lambda_k \tilde{m}_k. \tag{7.70}$$

We now describe how the GRR dispatch policy for standard networks is adapted to form the sensible generalized round robin policy for processing networks, denoted by $\pi = (\theta, \text{GRR}(\beta), \ell)$. Consider the p_j servers at station j. As in the single server case, the servers visit the classes in $C(j)$ in a round robin fashion. In fact, we use the same nominal values computed in (7.67) and (7.68). However, the interpretations of $a_k(n)$ and $b_k(n)$ are slightly different. The nth visit to buffer k can be made by any (and at most) one of the servers, provided the buffer is eligible. (This is not to say that two servers cannot simultaneously process batches from the same class. It does imply that the servers are performing visits of different cycles.) Suppose one of the servers is performing the nth visit to buffer k. If $a_k(n) \geq 1$ and class k is eligible according to the criterion in Section 7.2.3 then a setup for class k is performed. Otherwise the server moves on to the next buffer. The only time a server does not move is when there is an absence of jobs available for processing at the station. Assuming the server has performed a setup for buffer k, $a_k(n)$ determines how many production runs of length ℓ_k the server will perform. During each production run, each batch contains as many jobs as possible. Suppose the last server was assigned to visit class k in cycle n. The next server to complete its assignment will visit the next class according to the polling table. In this sense, each station's servers march through the cycles in unison. (An alternative description is to have each server perform its own cycle, independent of all other servers.)

Theorem 7.17 A processing network operating under a sensible generalized round robin policy $\pi = (\theta, \text{GRR}(\beta), \ell)$ is rate stable if, for each

$k = 1, \ldots, K$, $j = \sigma(k)$,

$$\frac{p_j \beta_k \ell_k B_k}{\sum_{k' \in \mathcal{C}(j)} \beta_{k'} \ell_{k'} \breve{m}_{k'}} \geq \lambda_k. \tag{7.71}$$

It turns out that when $\beta_k > 1$ for each class k, the condition (7.71) can be relaxed. See Section 7.6 for further discussions. The proof of the theorem depends on the additional artificial fluid model constraint (7.43), described in the following lemma.

Lemma 7.18 Consider a processing network operating under a sensible generalized round robin policy $\pi = (\theta, \text{GRR}(\beta), \ell)$. Artificial fluid solutions obey the following:

$$\dot{\breve{D}}_k(t) \geq \frac{p_j \beta_k \ell_k B_k}{\sum_{k' \in \mathcal{C}(j)} \beta_{k'} \ell_{k'} \breve{m}_{k'}}, \quad k = 1, \ldots, K. \tag{7.72}$$

When the β_k's are strictly positive integers, the intuition behind (7.72) is straightforward. If station j had only one server, the typical cycle length on average would be at most $\sum_{k' \in \mathcal{C}(j)} (s_{k'} + \beta_{k'} \ell_{k'} m_{k'}) \leq \sum_{k' \in \mathcal{C}(j)} \beta_{k'} \ell_{k'} \breve{m}_{k'}$. When there are enough class k jobs present, the average time a server spends processing those batches in a given cycle is $\beta_k \ell_k m_k$. Since p_j servers act in unison, the average fraction of time spent processing class k jobs is at least the ratio

$$p_j \beta_k \ell_k m_k \left[\sum_{k' \in \mathcal{C}(j)} \beta_{k'} \ell_{k'} \breve{m}_{k'} \right]^{-1},$$

leading to the right hand side of (7.72). The formal proof is delayed until the appendix.

Proof of Theorem 7.17. Let $\breve{\mathbb{X}}$ be an artificial fluid model solution. By Lemma 7.18, if (7.71) holds, we have $\dot{\breve{D}}_k(t) \geq \lambda_k$ for any t such that $\breve{Z}_k(t) > 0$ and $\breve{\mathbb{X}}$ is differentiable at t. It follows from a slight modification of Theorem 4 of Bramson [5] that $\breve{Z}(t) = 0$ for $t \geq 0$. Thus, the artificial fluid model is weakly stable. Rate stability of the processing network follows from Theorem 7.5. □

7.6 Extensions

The naming convention for production policies, as described in Section 7.2.3, has the form $\pi = (\theta, \tilde{\pi}, \ell)$, where θ performs the filtering function, $\tilde{\pi}$ is the dispatch policy, and ℓ determines the length of the production run. We further defined a family of "sensible" policies, where (7.10) and (7.13) hold, to which we have heretofore restricted ourselves. Under certain circumstances, either of these restrictions can be eliminated, or at least relaxed, while still achieving stability.

Consider the generalized round robin policies introduced in Section 7.5.2. A benefit of round robin policies is that they ensure every class is visited on a regular (i.e. cyclic) basis. This removes the need for the threshold test provided by θ. The result is that some production runs may be shorter than required by ℓ, even when this can be avoided. Still, the major results of Section 7.5.2, namely Theorem 7.17 and Lemma 7.18, hold. The reason this works is that round robin dispatching is static and provides every class the benefit of receiving its designated allocation during each cycle. If a class cannot take advantage of its allotment because of too few jobs, other classes can only benefit. The thresholds are intended to prevent more dynamic policies from favoring classes with too few jobs present.

The pitfalls of exhaustive service were the focus of the first simulation example in the introduction. The idea of exhaustive service, that setups should be avoided as much as possible, is not completely misguided. Indeed excessive setups require too much server effort and can lead to instability. Hence, limited policies were presented as a tradeoff. Rather than effectively eliminating setups as with exhaustive service, limit service reduces the effects of setups to acceptable levels. However, this is not to say that exhaustive service cannot be an effective production run policy. From initial simulation studies, there seems to be a connection between when exhaustive service (with or without setups) leads to instability and when static buffer priority policies cause instability. Under certain conditions all static buffer priority policies are stable. The same should be true for exhaustive service. In fact, we propose the following for investigation: If a system (with or without setups) is unstable under exhaustive service, it is unstable under some static buffer priority policy when setup times are zero. The contrapositive provides a guideline for when exhaustive service should be effective: If there are no static buffer priority policies for which the system without setups is unstable, then the system with setups is stable under exhaustive service.

Full batches, as demanded by our sensible production policy rules, may be an unnecessary extreme. Suppose that station j has a significant amount of excess capacity, i.e., $\breve{\rho}_j$ defined in (7.12) is well below 1. In this case, we can relax full batch size B_k to $\breve{B}_k$, where $\breve{B}_k$ is a target batch size favored by management. We can redefine the inflated traffic intensity (larger than the value from (7.12)) and traffic condition for each station j:

$$\breve{\rho}_j = \frac{1}{p_j} \sum_{k \in C(j)} \lambda_k (\breve{m}_k / \breve{B}_k) < 1.$$

One can also relax the restriction on the threshold (7.10) as well as the eligibility and non-idling conditions. One advantage of this relaxed requirement on batch sizes is that buffers with smaller numbers of jobs do not have to wait so long for the dispatching of servers.

Consider again the generalized round robin policy. In some cases, the choice of the β_k constants can be used to relax the inflated traffic intensity condition. For instance, suppose β_k is equal to 2. Then in each cycle where there is a sufficient number of jobs on hand, the production run for buffer k will have a length of $2\ell_k$. In this sense, the ℓ_k's that satisfy (7.11) may be unnecessarily large. Just as with the less-than-full batches discussed above, we can define a new inflated traffic intensity (7.12), thus relaxing the traffic condition (7.13). In this case, the new values are based on the new inflated mean processing time,

$$\breve{m}_k = m_k + \frac{s_k}{\ell_k \max(\beta_k, 1)}.$$

Finally, consider the maximum setup time means $s_k = \max_{k'} s_{k'k}$. These constants, used to define inflated service times and, ultimately, sensible policies, are sometimes overly conservative. For instance, there could be a pair of classes (k', k) whose mean setup time $s_{k'k}$ is relatively large (compared to other setup times at that station). If this particular transition from k' to k can be avoided most of the time, there is no reason to include it in the calculation of s_k. To see how this works, consider the generalize round robin policy and its fixed polling table, or order in which classes are visited. This fixed order suggests a natural way to define s_k. That is, s_k is equal to the mean setup time from the class just before k in the order. Of course, if the class before k in the order is ineligible, the transition to class k will have occurred from some other buffer k'. It could be the case that for this alternative predecessor class k' the mean setup time $s_{k'k}$ is relatively large and not accommodated in the calculation of s_k. There is no alarm because this class k' does not directly precede k relatively often.

7.7 Appendix

7.7.1 *Departures As a Function of Server Effort*

When there is a single server at some station j, the number of class $k \in C(j)$ batches processed by time t is captured fully by the quantity $\Psi_k(T_k(t))$. However, when there are multiple servers and random processing times, the quantity can either slightly overshoot or dramatically undershoot the total number of processed batches. Batches begin their processing in FIFO order. However, in the presence of multiple servers, the completion of batches is not necessarily FIFO. The reasoning is that the parallel nature of processing afforded by multiple servers allows for batches with short processing times to emerge before those with long processing time, even when the short batches start later. The quantity $\Psi_k^i(t)$ will provide an upper bound on the number of class k batches processed, given t units of work performed collectively by i servers. This section of the appendix is devoted to providing bounds and fluid limits for $\breve{\Psi}_k(t) \equiv \Psi_k^{p_j}(t)$, for each class k; the results are used in Sections 7.3.1 and 7.5.

Before providing the equations governing Ψ^i, the generalization of Ψ, we first provide a motivating example. Consider three systems, each with a single station network: Systems A, B and C have one, two and three servers, respectively. In each of the systems, 120 minutes of work are performed collectively by the servers. Suppose, for each system, the first 12 batch processing times (in minutes) are $\{10, 25, 10, 80, 10, 10, 90, 5, 5, 5, 5, 40\}$. If there is no idling, how many batches are processed in each system?

For each of the systems, the following chart matches each server with the batches it processes. In this particular example, the total number of batches processed increases as the number of servers increases. This is not always the case. It is possible that the number of processed batches decreases as the number of servers increases.

System	Server 1	Server 2	Server 3	Partial	Completed
A	b1,b2,b3			b4	3
B	b1, b3	b2,b5,b6		b4,b7	5
C	b1	b2	b3,b5,b6,b8,b9	b4,b7	7

Case A. At time 45 the single server completes the processing of the third batch and starts to process batch number 4. This batch requires 80 minutes of processing and will be completed at time 125. However, the clock stops at time 120. Thus, the number of processed batches at time 120 in the single server system is three.

Case B. With two servers, after t time units there will be $2t$ units of combined server effort. Therefore, the simulation clock stops at real time 60, when 120 combined minutes of server effort has taken place.

At time 20, having processed batches 1 and 3, server 1 is ready to start processing batch number 4, which has 80 minutes of processing requirements. Hence, the completion of this batch will occur at time $20+80 = 100$, well past the clock stopping time of 60. Meanwhile, after 25 minutes, server 2 is ready to start processing batch 5, having completed batch 2. Batch 5 has a processing requirement of 10 minutes. This batch will complete its processing at time $25+10 = 35$. Therefore, server 2 still has time to process batch 6 which has a processing requirement of 10. Batch 6 is completed by time $25 + 10 + 10 = 45$ and server 2 starts on batch number 7. With its processing requirement of 90 minutes, batch 7 will not be completed until time 135, well past the stopping time of 60 minutes.

Case C. The simulation clock stops at time $120/3 = 40$ minutes. Server 1 completes batch 1 after 10 minutes and starts batch 4, which finishes at time 90. Server 3 finishes batch 3 at time 10 then moves on to batches 5 which it finishes at time 20. Then server 3 starts to process batch 6 which it will finish at time 30. Meanwhile server 2 completes batch 2 after 25 minutes and then starts batch 7, which will finish at time 115. After batch 6, server 3 picks up batches 8 and then 9, finishing the last of these at time 40.

Consider the first two cases. Under both scenarios, a server gets "stuck" on the batch with the 80 minute processing time. In the second case, server 2 moves past batch 4, before it too gets stuck on a long batch. Typically, as the number of servers increases, more and more batches with atypically long processing times can be bypassed. Indeed, in the third case, the 80 and 90-minute long processing times still do not completely impede the processing of additional batches. The question remains: How can we obtain a bound on the number of processed batches given the collective server effort and the sequence of processing times?

Let $\Psi^i(t)$ denote, for a system with i servers, the largest integer value n such that the sum of the processing times of the first n batches is less that t. As stated earlier, when there is one server, the quantity $\Psi(120) = \Psi^1(120) = 3$ captures the number of batches processed, given 120 minutes of effort. With two servers, server 1 gets stuck on batch 4, which happens to be the longest batch experienced thus far, and server 2 gets to look further along in the list of processing times. In the most extreme case, server 1 has just started working on batch 4. In this case, the maximum distance that server 2 can "look into the future" is 80 minutes. (Actually the value is 60, since each server works exactly 60 minutes. But typically, the station-level time is not evenly split.) Hence, we have an upper bound when there are two servers $\Psi^2(120) = \Psi^1(200) - 1 = 5$. We subtract 1 from the quantity to avoid needlessly counting the 80-minute batch.

Next we increase the number of servers to three. In this case, the two largest processing times happen to occupy two of the servers at the horizon's end. Once again, in the worst case, those occupied servers just started processing those long batches. Hence, we need to look 170 minutes into the future to see how many jobs were processed to get $\Psi^3(120) = \Psi^1(290) - 2 = 9$. (Again, this is overkill.) We see here that we have overestimated the number of processed batches. Our objective is for an upper bound, not necessarily a tight one. One should note that we discovered the 90-minute batch, that is skipped over in the 3-server system, only after analyzing the 2-server system. In general, one needs to examine the n-server system in order to provide bounds for the $(n+1)$-server system.

We now provide an iterative method for computing an upper bound on the number of processed batches. Let $\Gamma_k^n(t)$ denote the number of processed class k batches by time t, given there are n servers at station $\sigma(k)$. Recall the formal definition of $\Psi_k = (\Psi_k(t), t \geq 0)$, $\Psi_k(t) = \max\{n \geq 0 : V_k(n) \leq t\}$ and that $\Psi_k^1(T_k(t)) = \Psi_k(T_k(t)) = \Gamma_k^1(t)$. In order to define $\check{\Psi}$ we need to examine the interjump times of V_k. The quantity $\eta_k(n) = V_k(n) - V_k(n-1)$

tracks the amount of server effort required to process the nth class k batch. It follows that $\sum_{i=1}^{n} \eta_k(i) = V_k(n)$. Recall that $\Psi_k^n(t)$ is an upper bound on the number of processed batches, given t units of collective work. With $\Psi_k^1(t) = \Psi_k(t)$ we iteratively define the following quantities:

$$\eta_k^{i,t} = \max\{\eta_k(n) : 1 \leq n \leq \Psi_k^i(t) + 1\} \tag{7.73}$$

and

$$\Psi_k^{i+1}(t) = \max\{n : V_k(n) \leq t + i\eta_k^{i,t}\} - i, \tag{7.74}$$

for each $i = 1, \ldots, p_j - 1$. Set $\breve{\Psi}_k(t) = \Psi_k^{p_j}(t)$. Notice that $\eta_k^{1,t}$ is the largest possible batch processing time that could have been interrupted, given t units of work and 1 server. Now consider adding one server. Since the long batch associated with $\eta_k^{1,t}$ may not have been processed, we do not want its processing time to be deducted from the total server effort t expended by the two servers. Hence, we add $\eta_k^{1,t}$ to t to reflect the virtual server effort for two servers. Once we compute a bound for the number of batches with two servers, we find $\eta_k^{2,t}$, the largest possible batch processing time when there are two servers. To increase to three servers, the virtual server effort would be $\eta_k^{1,t} + \eta_k^{2,t} + t$. However, we simplify matters with the larger quantity $2\eta_k^{2,t} + t$. We iterate until we find $\breve{\Psi}_k(t) = \Psi_k^{p_j}(t)$. Since for each i, $\Gamma_k^i(t) \leq \Psi_k^i(T_k(t))$, it follows that

$$\Gamma_k^{p_j}(t) \leq \breve{\Psi}_k(T_k(t)), \quad t \geq 0. \tag{7.75}$$

So far we have focused exclusively on the fact that extra jobs may get served as a result of the splitting of processing tasks among several servers. It turns out that the splitting of processing effort can lead to fewer jobs being processed. The lower bound on the number of batches processed is quite straightforward. For each additional server, at most one fewer jobs may have been processed by time t. That is, given there are p_j servers performing the work on class k batches:

$$\Gamma_k^{p_j}(t) \geq \Psi_k(T_k(t)) - p_j + 1. \tag{7.76}$$

We are interested in taking fluid limits of the $\breve{\Psi}_k$ processes and comparing them to fluid limits of the Ψ_k processes. We have the following lemma, which is used in the proof of Proposition 7.4.

Lemma 7.19 For any finite integer $i \geq 1$,

$$\lim_{t \to \infty} \frac{\Psi_k^i(t)}{t} = \mu_k, \quad k = 1, \ldots, K; \tag{7.77}$$

in particular, $\breve{\Psi}_k(t)/t \to \mu_k$.

Proof. By the law of large numbers (7.2) we have $\Psi_k(t)/t \to \mu_k$. It follows that $\eta_k^{1,t}/t \to 0$. Suppose $\eta_k^{i,t}/t \to 0$ as $t \to \infty$ for some arbitrarily chosen integer i. Then

$$\frac{\Psi_k^{i+1}(t)}{t} \le \frac{\Psi_k(t + i\eta_k^{i,t})}{t} = \left(\frac{\Psi_k(t + i\eta_k^{i,t})}{t + i\eta_k^{i,t}}\right)\left(\frac{t + i\eta_k^{i,t}}{t}\right) \to \mu_k.$$

It follows that

$$\frac{\eta_k^{i+1,t}}{t} \le \left(\frac{\eta_k^{1,t+i\eta_k^{i,t}}}{t + i\eta_k^{i,t}}\right)\left(\frac{t + i\eta_k^{i,t}}{t}\right) \to 0.$$

The proof follows from induction. □

7.7.2 *Proofs of Lemmas 7.12 and 7.18*

Proof of Lemma 7.12. Let $\bar{\mathbb{X}}$ be a fluid limit of a processing network operating under the sensible early-steps-first production policy $\pi = (\theta, \text{ESF}, \ell)$. Let $\breve{\mathbb{X}}$ be the associated artificial fluid limit, as constructed in Proposition 7.4. Let $\omega \in \Omega$ be a sample path on which (7.1) holds and $\{r_n\}$ be a sequence of positive reals such that $r_n \to \infty$ and $\bar{\mathbb{X}}^{r_n}(\cdot, \omega) \to \bar{\mathbb{X}}(\cdot)$ u.o.c. as $n \to \infty$. Fix station j and a time $t > 0$ such that $\bar{\mathbb{X}}(t)$ and $\breve{\mathbb{X}}(t)$ are differentiable. Suppose that, for some fixed step k_0 we have

$$\sum_{k=1}^{k_0} \sum_{(q,k) \in \mathcal{C}(j)} \breve{Z}_{(q,k)}(t) > 0.$$

Then for some class $(q_1, k_1) \in \mathcal{C}(j)$ with $k_1 \le k_0$ we have $\breve{Z}_{(q_1,k_1)}(t) = \bar{Z}_{(q_1,k_1)}(t) > 0$. By the continuity of $\bar{Z}$, and the uniform convergence $\bar{Z}^{r_n}(\cdot) \to \bar{Z}(\cdot)$ as $n \to \infty$, there exists a $\delta > 0$ and an integer N such that, for each $n \ge N$

$$Z_{(q_1,k_1)}(s) \ge \theta_{(q_1,k_1)} + (p_j - 1)\ell_{(q_1,k_1)} B_{(q_1,k_1)}, \quad \forall s \in [r_n t, r_n(t + \delta)].$$

This condition ensures that, throughout the interval $[r_n t, r_n(t + \delta)]$, the servers at station j are never dispatched to a class $(q, k) \in \mathcal{C}(j)$ such that $k > k_1$. This does not mean that these classes composed of later steps receive no server effort whatsoever. Any of the p_j servers may have been dispatched to one such class just prior to $r_n t$. So for the combined server pool at station j, the total allocation, during $[r_n t, r_n(t + \delta)]$, to a class $(q, k) \in \mathcal{C}(j)$ with $k > k_1$ does not exceed p_j class (q, k) setups and $p_j \ell_{(q,k)}$ class (q, k) batches. Equivalently, for each class (q, k) with $k > k_1$,

$$T_{(q,k)}(r_n(t + \delta)) - T_{(q,k)}(r_n t) \le V_{(q,k)}(M^n_{(q,k)} + p_j \ell_{(q,k)}) - V_{(q,k)}(M^n_{(q,k)}), \tag{7.78}$$

and

$$S_{(q,k)}(r_n(t+\delta)) - S_{(q,k)}(r_n t) \tag{7.79}$$
$$\leq \sum_{(q',k')\in\mathcal{C}(j)} F_{(q',k')(q,k)}(R^n_{(q',k')(q,k)} + p_j) - F_{(q',k')(q,k)}(R^n_{(q',k')(q,k)}),$$

where $M^n_{(q,k)}$ is the number of class (q,k) batches processed by time $r_n t$ and $R^n_{(q',k')(q,k)}$ is the number of setups from class (q',k') to class (q,k) performed by time $r_n t$. By Lemma 7.20, the law of large numbers (7.1), and (7.78) and (7.79)

$$\sum_{k'>k_1} \sum_{(q',k')\in\mathcal{C}(j)} \left[\breve{T}_{(q',k')}(t+\delta) - \breve{T}_{(q',k')}(t)\right]$$
$$= \lim_{n\to\infty} (1/r_n) \sum_{k'>k_1} \sum_{(q',k')\in\mathcal{C}(j)} \left[S_{(q',k')}(r_n(t+\delta)) - S_{(q',k')}(r_n t))\right.$$
$$\left. + T_{(q',k')}(r_n(t+\delta)) - T_{(q',k')}(r_n t)\right] = 0. \tag{7.80}$$

By (7.39), (7.40) and (7.80),

$$\sum_{k'=1}^{k_1} \sum_{(q',k')\in\mathcal{C}(j)} \left[\breve{T}_{(q',k')}(t+\delta) - \breve{T}_{(q',k')}(t)\right] = p_j\delta.$$

We obtain the result by dividing by δ and letting $\delta \downarrow 0$. □

Proof of Lemma 7.18. Let $\bar{\mathbb{X}}$ be a fluid limit of a processing network operating under the sensible generalized round robin production policy $\pi = (\theta, \mathrm{GRR}(\beta), \ell)$. Let $\breve{\mathbb{X}}$ be the associated artificial fluid limit, as constructed in Proposition 7.4. Let $\omega \in \Omega$ be a sample path on which (7.1) holds and $\{r_n\}$ be a sequence of positive reals such that $r_n \to \infty$ and $\bar{\mathbb{X}}^{r_n}(\cdot,\omega) \to \bar{\mathbb{X}}(\cdot)$ as $n \to \infty$. At time t, suppose that for some class k at station j we have $\breve{Z}_k(t) = \bar{Z}_k(t) > 0$. We would like to show that

$$\dot{\breve{D}}_k(t) = \dot{\bar{D}}_k(t) \geq \frac{p_j \beta_k \ell_k B_k}{\sum_{k'\in\mathcal{C}(j)} \beta_{k'}\ell_{k'} m_{k'}}. \tag{7.81}$$

By the continuity of $\bar{Z}$ we know there exists an $\epsilon > 0$ and a $\delta > 0$ such that $\bar{Z}_k(s) > \epsilon$ for each $s \in [t, t+\delta]$. For large enough n, $Z_k(s) \geq \theta_k + \lceil \beta_k \rceil p_j \ell_k B_k$ for each $s \in [r_n t, r_n(t+\delta)]$. Hence, if the ith cycle takes place entirely within the interval $[r_n t, r_n(t+\delta)]$, the server that visits class k during that cycle processes exactly $a_k(i)\ell_k$ batches. It should be clear by (7.67) and (7.68) that, for any class k and positive integers p and q,

$$q\beta_k - 1 < \sum_{i=p+1}^{p+q} a_k(i) < q\beta_k + 1. \tag{7.82}$$

For each n, we refer to cycles that start after $r_n t$ and end before $r_n(t+\delta)$ as *complete*. Let N^n denote the number of complete cycles. The result (7.81) will follow if we can show that, on almost every sample path,

$$\lim_{n\to\infty} \frac{D_k(r_n(t+\delta)) - D_k(r_n t)}{N^n} \to \beta_k \ell_k B_k, \tag{7.83}$$

and that

$$\limsup_{n\to\infty} \frac{r_n \delta}{N^n} \le \frac{1}{p_j} \sum_{k' \in \mathcal{C}(j)} \beta_{k'} \ell_{k'} \breve{m}_{k'}. \tag{7.84}$$

Notice that, in addition to the N^n complete cycles, there may be $2p_j$ *incomplete* cycles. That is, all p_j servers may have been in the middle of a cycle at time $r_n t$ and again at time $r_n(t+\delta)$. By (7.82), we can bound the number of jobs processed in $[r_n t, r_n(t+\delta)]$,

$$(N^n \beta_k - 1)\ell_k B_k < D_k(r_n(t+\delta)) - D_k(r_n t) < ((N^n + 2p_j)\beta_k + 1)\ell_k B_k. \tag{7.85}$$

By Lemma 7.21, $N^n \to \infty$. Dividing both sides of (7.85) by N^n and letting $n \to \infty$ yields (7.83).

Now we demonstrate that (7.84) holds. Clearly none of the p_j servers idle during $[r_n t, r_n(t+\delta)]$ so that

$$p_j r_n \delta = \sum_{k' \in \mathcal{C}(j)} \left[(T_{k'}(r_n(t+\delta)) - T_{k'}(r_n t)) + (S_{k'}(r_n(t+\delta)) - S_{k'}(r_n t)) \right].$$

We investigate how the server effort throughout the interval $[r_n t, r_n(t+\delta)]$ is allocated. Let $M^n_{k'}$ denote the number of class k' batches that have completed service at time $r_n t$. In line with the arguments of Appendix 7.7.1, the $M^n_{k'}$ batches that have completed processing may not be the first $M^n_{k'}$ that started processing in $[0, r_n t)$. In fact the $M^n_{k'}$th batch that completed processing by time $r_n t$ could be as high as the $(M^n_{k'} + p_j - 1)$th batch to have started processing before $r_n t$. Because of the parallel processing capabilities of the station j servers, $p_j - 1$ of the class k' batches between batch number 1 and batch number $M^n_{k'} + p_j - 1$ could be still in-process at time $r_n t$. We can bound the total time dedicated to class k'.

$$\begin{aligned} & T_{k'}(r_n(t+\delta)) - T_{k'}(r_n t) \\ \le\; & V_{k'}(M^n_{k'} + \lceil (N^n + 2p_j)\beta_{k'} + 1 \rceil \ell_{k'}) - V_{k'}(M^n_{k'}) \\ & + (p_j - 1) \max_{i \le M^n_{k'} + p_j} \eta_{k'}(i), \end{aligned} \tag{7.86}$$

where, as before, $\eta_{k'}(i) = V_{k'}(i) - V_{k'}(i-1)$. Similarly,

$$\begin{aligned} & S_{k''k'}(r_n(t+\delta)) - S_{k''k'}(r_n t) \\ \le\; & F_{k''k'}(R^n_{k''k'} + 2p_j + \lambda^n_{k''k'}) - F_{k''k'}(R^n_{k''k'}) \\ & + (p_j - 1) \max_{i \le R^n_{k''k'}} \left(F_{k''k'}(i) - F_{k''k'}(i-1) \right), \end{aligned} \tag{7.87}$$

where $\lambda^n_{k''k'}$ is the number of type (k'', k') setups among the N^n complete cycles and $R^n_{k''k'}$ is the number of completed type (k'', k') setups before time $r_n t$. The reasoning behind (7.86) is as follows. As argued earlier, there are at most $N^n + 2p_j$ cycles, where N^n of them are complete. We can assume the most extreme case, that for each class k' and in each cycle i, exactly $a_{k'}(i)\ell_{k'}$ batches are processed. Using equation (7.82) we bound the total number of processed batches with $((N^n + 2p_j)\beta_{k'} + 1)\ell_{k'}$. For the last term in (7.86), recall that $M^n_{k'} + p_j - 1$ batches may have started processing before time $r_n t$. So far in our discussion of (7.86) we have only accounted for those batches after batch number $M^n_{k'}$. But the (up to) $p_j - 1$ batches interrupted at time $r_n t$ may be any batch between batch numbers 1 through $M^n_{k'} + p_j - 1$. Hence, the last term. As for (7.87), the arguments are similar. It should be clear that $\sum_{k''\in\mathcal{C}(j)} \lambda_{k''k'} \le N^n(\max(\beta_{k'}, 1)) + 1$. Dividing both sides of equations (7.86) and (7.87) by N^n yields

$$\limsup_{n\to\infty} \, (1/N^n) \, [T_{k'}(r_n(t+\delta)) - T_{k'}(r_n t)] \le \beta_{k'}\ell_{k'}m_{k'},$$

and

$$\limsup_{n\to\infty} \, (1/N^n) \, [S_{k'}(r_n(t+\delta)) - S_{k'}(r_n t)] \le \beta_{k'}s_{k'}.$$

Finally, by (7.11),

$$\beta_{k'}\ell_{k'}m_{k'} + \beta_{k'}s_{k'} = \beta_{k'}\ell_{k'}\check{m}_{k'}$$

and, hence, (7.84) follows. □

Lemma 7.20 Consider any processing network. Let $\{r_n\}$ be a sequence of positive reals such that $r_n \to \infty$ as $n \to \infty$. Let M^n_k be the number of class k batches processed by time $r_n t$ and $R^n_{k'k}$ be the number of type $(k'k)$ setups performed by time $r_n t$. We have, on almost every sample path,

$$\limsup_{n\to\infty} \; M^n_k / r_n < \infty,$$

and

$$\limsup_{n\to\infty} \; R^n_{k'k} / r_n < \infty,$$

for each $k, k' = 1, \ldots, K$.

Proof. We prove the first result for class k batches. The result for setups is similar. Suppose $\limsup_{n\to\infty} M^n_k / r_n = \infty$. Without loss of generality, $M^n_k / r_n \to \infty$. By definition,

$$V_k(M^n_k) \le r_n t.$$

Dividing both sides by M^n_k and taking the limit yields $m_k \le 0$, a contradiction. □

Lemma 7.21 Suppose the processing network is operating under a sensible generalized round robin policy. Let $\{r_n\}$ be a sequence of positive reals such that $r_n \to \infty$ as $n \to \infty$. Fix time t, station j and the constant $\delta > 0$. Suppose that, for large enough n, servers at station j never idle during the interval $[r_n t, r_n(t+\delta)]$. Let N^n denote the number of cycles that transpire completely during $[r_n t, r_n(t+\delta)]$. We have

$$\liminf_{n\to\infty} N^n/r_n > 0. \tag{7.88}$$

Proof. The p_j servers at station j provide $p_j r_n \delta$ units of potential effort during the interval $[r_n t, r_n(t+\delta)]$. All of this effort goes to processing batches or performing setups. Since there are K classes there are, at most, $K^2 - K$ types of setups. This yields a maximum of K^2 activities over which the $p_j r_n \delta$ units of server effort is distributed over the interval. There is a subsequence r_{n_q} with $n_q \to \infty$ as $q \to \infty$ such that one of the activities receives at least $r_{n_q} p_j \delta / K^2$ units of server effort during the interval for each q. Assume class k receives this amount of effort and, without loss of generality, that this occurs for each number in the original sequence $\{r_n\}$. That is

$$\begin{aligned} V_k(M_k^n + \lceil (N^n + 2p_j)\beta_k + 1 \rceil \ell_k) &- V_k(M_k^n) \\ &+ (p_j - 1) \max_{i \le M_k^n + p_j} \eta_k(i) \ge r_n p_j \delta / K^2, \end{aligned} \tag{7.89}$$

where M_k^n denotes the number of class k batches completed by time $r_n t$. (The case that a setup activity receives this amount of effort can be argued similarly.) The terms in (7.89) are identical to those in (7.86) of the previous proof. If $N^n/r_n \to 0$ then, by Lemma 7.20, dividing both sides of (7.89) by N^n yields $\delta \le 0$. Hence, (7.88) must hold. □

7.8 Notes

There are recent works in the literature that also extend the standard multiclass queueing network model. For instance, Maglaras and Kumar [32], Kumar and Zhang [30], and Dai and Li [17] all treat the batch processing operations issue. An analogous service was performed by Jennings [27] and Andridotter, Ayhan and Down [1] for queueing networks with setups. In [1] the model allows for more flexible servers than in this chapter.

There has been a flurry of research efforts in pinpointing efficient dispatch policies, meaning policies that maximize throughput, given an achievable arrival rate. Examples of policies that work for any network can be found in Bramson [4, 6] and Dai and Li [16]. For policies that stabilize networks with special structure, see Kumar [28], Dai and Weiss [19], Kumar and Kumar [29], Bramson [3] and Chen and Yao [10]. The last of these examples is related to the multi-type network presented in Section 7.5.1.

The method of using Lyapunov functions for demonstrating stability of a fluid model is wide-spread and multifaceted. For example, Bramson [3, 4] uses entropy type Lyapunov functions, Down and Meyn [20] and Dai and Vande Vate [18] advocate piecewise linear Lyapunov functions, and Chen and Zhang [11] forward linear Lyapunov functions.

Acknowledgments

Jim Dai's research is supported in part by NSF grants DMI-9457336 and DMI-9813345 and by TLI-AP, a partnership between National University of Singapore and Georgia Institute of Technology.

7.9 References

[1] Andradóttir, S., Ayhan, H., and Down, D. G., Dynamic server allocation for queueing networks with flexible servers, Submitted to *Operations Research*, 2001.

[2] Bertsimas, D., Lecture notes on stability of multiclass queueing networks, 1996.

[3] Bramson, M., Convergence to equilibria for fluid models of FIFO queueing networks, *Queueing Systems: Theory and Applications*, **22** (1996), 5-45.

[4] Bramson, M., Convergence to equilibria for fluid models of head-of-the-line proportional processor sharing queueing networks, *Queueing Systems: Theory and Applications*, **23** (1997), 1-26.

[5] Bramson, M., Stability of two families of queueing networks and a discussion of fluid limits, *Queueing Systems: Theory and Applications*, **28** (1998), 7-31.

[6] Bramson, M., Stability of earliest-due-date, first-served queueing networks, *Queueing Systems: Theory and Applications*, **39** (2001), 79-102.

[7] Chen, H., Fluid approximations and stability of multiclass queueing networks I: Work-conserving disciplines, *Annals of Applied Probability*, **5** (1995), 637-665.

[8] Chen, H. and Mandelbaum, A., Discrete flow networks: Bottlenecks analysis and fluid approximations, *Mathematics of Operations Research*, **16** (1991), 408-446.

[9] Chen, H. and Shanthikumar, J. G., Fluid limits and diffusion approximations for networks of multi-server queues in heavy traffic, *Discrete Event Dynamic Systems*, **4** (1994), 269-291.

[10] CHEN, H. AND YAO, D. D., Stable priority disciplines for multiclass networks, *Lecture Notes in Statistics*, eds by P. Glasserman, K. Sigman and D.D. Yao, Springer-Verlag, (1996), 27-40.

[11] CHEN, H. AND ZHANG, H., Stability of multiclass queueing networks under priority service disciplines, *Operations Research*, **48** (2000), 26-37.

[12] DAI, J. G., On positive Harris recurrence of multiclass queueing networks: A unified approach via fluid limit models, *Annals of Applied Probability*, **5** (1995), 49-77.

[13] DAI, J. G., Stability of fluid and stochastic processing networks, *MaPhySto Miscellanea Publication, No. 9*, Centre for Mathematical Physics and Stochastics, 1999.

[14] DAI, J. G., HASENBEIN, J., AND VANDE VATE, J. H., Stability of a three-station fluid network, *Queueing Systems: Theory and Applications*, **33** (1999), 293-325.

[15] DAI, J. G. AND JENNINGS, O. B., Stabilizing queueing networks with setups, 2001, Preprint.

[16] DAI, J. G. AND LI, C., Discrete proportional processor sharing policy for multiclass queueing networks, 2001, Preprint.

[17] DAI, J. G. AND LI, C., Stabilizing batch processing networks, *Operations Research*, (2001), To Appear.

[18] DAI, J. G. AND VANDEVATE, J., The stability of two-station multitype fluid networks, *Operations Research*, **48** (2000), 721-744.

[19] DAI, J. G. AND WEISS, G., Stability and instability of fluid models for re-entrant lines, *Mathematics of Operations Research*, **21** (1996), 115-134.

[20] DOWN, D. AND MEYN, S. P., Piecewise linear test functions for stability and instability of queueing networks, *Queueing Systems: Theory and Applications*, **27** (1997), 205-226.

[21] EL-TAHA, M. AND STIDHAM JR., S., *Sample-Path Analysis of Queueing Systems*, Kluwer, 1999.

[22] GROSS, D. AND HARRIS, C. M., *Fundamentals of Queueing Theory*, Wiley, New York, 1985.

[23] HARRISON, J. M., Brownian models of queueing networks with heterogeneous customer populations, *Proceedings of the IMA Workshop on Stochastic Differential Systems*, Springer, 1988.

[24] HARRISON, J. M., Brownian models of open processing networks: canonical representation of workload, *Annals of Applied Probability*, **10** (2000), 75-103.

[25] HASENBEIN, J. J., Necessary conditions for global stability of multiclass queueing networks, *Operations Research Letters*, **21** (1997), 87-94.

[26] JENNINGS, O. B., *Multiclass Queueing Networks with Setup Delays: Stability Analysis and Heavy Traffic Approximation*, Ph.D. dissertation, School of ISyE, Georgia Institute of Technology, 2000.

[27] JENNINGS, O. B., On the stability of multiclass queueing networks with setups, 2000, Preprint.

[28] KUMAR, P. R., Re-entrant lines, *Queueing Systems, Theory and Applications*, **13** (1993), 87-110.

[29] KUMAR, S. AND KUMAR, P. R., Fluctuation smoothing policies are stable for stochastic reentrant lines, *Discrete Event Dynamical Systems*, **6** (1996), 361-370.

[30] KUMAR, S. AND ZHANG, H., Stability of reentrant lines with batch servers, 2000, Preprint.

[31] LU, S. H. AND KUMAR, P. R., Distributed scheduling based on due dates and buffer priorities, *IEEE Transactions on Automatic Control*, **36** (1991), 1406-1416.

[32] MAGLARAS, C. AND KUMAR, S., Capacity realization in stochastic batch-processing networks using discrete review policies, 1999, Preprint.

[33] RYBKO, A. N. AND STOLYAR, A. L., Ergodicity of stochastic processes describing the operation of open queueing networks, *Problems of Information Transmission*, **28** (1992), 199-220.

[34] TAKAGI, H., *Analysis of Polling Systems*, MIT Press, Cambridge, MA, 1986.

8

Large Deviations, Long-Range Dependence, and Queues

C.-S. Chang, David D. Yao and Tim Zajic

The objective of this chapter is to demonstrate the development of a large deviations theory for processes with long range dependence. Our starting point is the integral relationship between the standard Brownian motion, B, and the fractional Brownian motion, B_H (with H being the Hurst parameter). This integral is known to preserve self-similarity, and produce long-range dependence in B_H when $H \in (\frac{1}{2}, 1)$. To extend beyond the class of Gaussian processes, we replace the Brownian motion by other processes, in particular, a class of point processes that we call *sample-path processes*, denoted X. We shall use the integral as a filter that takes as input a process X that has short-range dependence and outputs a process with long-range dependence, denoted Y. This way, we show that a general theory follows for Y to satisfy a moderate deviations principle (MDP), based on on the MDP of X. Applying this approach to queueing systems, we can derive results, in terms of the MDP, regarding the asymptotic behavior of queues fed with the long-range dependent process Y.

Here is an outline of the rest of this chapter. We start with an overview in Section 8.1 to motivate our study as well as to introduce basic notions such as fractional Brownian motion, self-similarity, and long-range dependence. In Section 8.2, we introduce the sample-path process X; in particular, we develop a criterion for the sequence of scaled X processes to satisfy the moderate deviations principle (MDP). Based on the MDP of X, we establish the MDP of Y in Section 8.3. In Section 8.4, we feed the long-range dependent process Y into a single-server queue, and study both steady-state and transient behaviors of the system. A key ingredient in our approach is a condition on X that guarantees a certain exponential equivalence relation. In Section 8.5, we verify this condition, using martingale inequalities, through revisiting the sample-path processes of Section 8.2.

8.1 Fractional Brownian Motion and a Related Filter

It is well known, following Mandelbrot and van Ness [20], that the fractional Brownian motion (FBM), denoted B_H (with $B_H(0) = 0$), relates to the standard Brownian motion, B, through the following integral:

$$\begin{aligned} B_H(t) &= \frac{1}{\Gamma(H+\frac{1}{2})} \cdot \Big\{ \int_0^t (t-s)^{H-\frac{1}{2}} dB(s) \\ &\quad + \int_{-\infty}^0 [(t-s)^{H-\frac{1}{2}} - (-s)^{H-\frac{1}{2}}] dB(s) \Big\}, \end{aligned} \tag{8.1}$$

where $0 < H < 1$ denotes the Hurst parameter and $\Gamma(\cdot)$ is the gamma function.

Recall that a stochastic process $\{X(t)\}$ is termed self-similar if the increment process, $\{X(t+\tau) - X(t)\}$, is equal in distribution to $\{a^{-H}[X(t+a\tau) - X(t)]\}$, for any t and any positive a and τ, where $H \geq 0$ is the Hurst parameter. It is well known (and easy to verify from (8.1)) that the FBM is a *self-similar* process. Note that when $H = \frac{1}{2}$, the FBM is reduced to the standard Brownian motion. On the other hand, when $H \in (\frac{1}{2}, 1)$, the FBM is known to exhibit long-range dependence — for instance, the autocorrelation function is not summable.

The relationship in (8.1) is the predominant mechanism in the literature for generating self-similar processes with long-range dependence. Indeed, (8.1) can be made more general through replacing the Brownian motion by an α-stable process, and replacing the power, $H - \frac{1}{2}$, by $H - 1/\alpha$ (where α is the parameter that characterizes the stable process). For instance, suppose the Brownian motion is replaced by an α-stable Lévy motion, which retains the stationary independent increment property of the Brownian motion, but has a non-Gaussian marginal distribution — in particular, when $\alpha < 2$ the marginal distribution has an infinite variance. Then, (8.1) results in the so-called fractional Lévy motion. Refer to Samorodnitsky and Taqqu [26]. This generalization hence broadens the class of processes represented by (8.1) to processes outside of the Gaussian family.

To develop a large deviations theory for processes with long range dependence, our idea here is to use (8.1) as the basic model. To get beyond the class of Gaussian processes, we shall replace the Brownian motion in (8.1) by other random processes, in particular, a class of point processes that we call *sample-path processes* (see Section 8.2). On the other hand, we cannot allow the power $H - \frac{1}{2}$ in (8.1) to be replaced by the more general $H - 1/\alpha$ — in other words, we must have $\alpha = 2$ — as techniques of large deviations generally do not apply to processes with an infinite variance.

In an earlier work, [7], we have used an alternative FBM model, due to

Lévy ([19] (also refer to [20]):

$$B_H(t) = \frac{1}{\Gamma(H+\frac{1}{2})} \cdot \int_0^t (t-s)^{H-\frac{1}{2}} dB(s). \tag{8.2}$$

We used the integral relation in (8.2) as a "basic filter". Indeed, in (8.2), the FBM can be viewed as the output of a time-invariant linear filter with the impulse response $h(t) = t^{H-\frac{1}{2}}$ subject to an input that is the standard Brownian motion. Similar types of linear filter are widely used in modeling stochastic systems (see, e.g., Wong and Hajek [33]).

Specifically, with $H \in (\frac{1}{2}, 1)$ in (8.2), we replaced B by other stochastic processes with short-range dependence, denoted X; and we showed that the *moderate deviations principle* (MDP) of X can be carried over to the MDP of Y, the output of the filter.

The main advantage of working with (8.2) is that the integral is over a finite time interval, which leads to ready applications of the basic results and techniques in large deviations theory. The shortcoming, however, is that it does not preserve the stationary increment property; i.e., when X has stationary increments, Y need not have stationary increments. (Neither does the filter preserve self-similarity.) This creates certain difficulties in applications. For instance, the stationary increment property is the key that relates the workload process in queueing systems to the maximum of a random walk; and it is the monotonicity of the latter that leads to steady-state limiting results. Without this linkage, it is difficult to study the steady-state behavior of the workload process.

On the other hand, a filter based on (8.1), namely,

$$\begin{aligned} Y(t) &= \int_0^t (t-s)^{H-\frac{1}{2}} dX(s) \\ &\quad + \int_{-\infty}^0 [(t-s)^{H-\frac{1}{2}} - (-s)^{H-\frac{1}{2}}] dX(s), \end{aligned} \tag{8.3}$$

does preserve both stationary increments and self-similarity. That the integration takes place over an infinite interval, however, makes it more challenging to establish moderate deviations results.

In recent years, it has been demonstrated that queues with long-range dependent inputs have subexponential tails (Norros [23], and Duffield and O'Connell [11]). This phenomenon is qualitatively different from the exponential tail distribution in queues with renewal or Markov arrival processes — inputs that have short-range dependence. Our work is motivated in part by [11]. Assuming that the netput" — input minus (potential) output — to a queueing system satisfies a large deviations principle with a speed that is slower than n [cf. (8.4) below] or, more precisely, an MDP, (refer to Section 8.2), Duffield and O'Connell showed that the maximum of a corresponding random walk, which is equal in distribution to the workload process of a single-server queue, also satisfies the MDP. They give conditions for the netput to satisfy the MDP and show, in particular, that the MDP holds with speed $n^{2(1-H)}$ when the netput is an FBM with Hurst parameter $H \in (\frac{1}{2}, 1)$ (See Section 8.4 for more details). Consequently, with the FBM as the netput, the steady-state queueing distribution has a subexponential tail; i.e., the exponential decay rate is $n^{2(1-H)}$ rather than n.

Hence, the starting point of [11] is the MDP of the Y process, in the parlance here. In contrast, our starting point is to relate Y, through the FBM-like filter in (8.3), to more basic processes X, and establish the MDP of Y based on the MDP of X. Specifically, we will develop (in Section 8.2) a criterion for the sequence of scaled X processes, $\{X^{(n)}(t) = X(nt)/n\}$ to satisfy the MDP and give several examples that satisfy this criterion, including a stationary Markov chain, a moving average process, and a ϕ-mixing (or, uniform mixing) sequence. Based on the MDP of $\{X^{(n)}\}$, we establish the MDP of $\{Y^{(n)}(t) = Y(nt)/n\}$ (in Section 8.3). Feeding the long-range dependent process Y into a single-server queue, we study (in Section 8.4) both steady-state and transient behaviors of the queue.

What plays a key role in our approach is a condition in (8.5) below, which essentially makes it possible to work with a modification of the process Y obtained by truncating the second integral in (8.3). Using martingale inequalities, we show, in Section 8.5, that this condition is satisfied by the X processes highlighted as examples in Section 8.2.

8.2 Moderate Deviations for Sample-Path Processes

Let $\mathcal{D}$, $\mathcal{C}$, $\mathcal{BV}$ and $\mathcal{AC}$ denote the spaces of real-valued functions defined on $(-\infty, 1]$ that are, respectively, right continuous with left limits (*cad-lag*), continuous, of bounded variation on finite intervals and absolutely continuous. These function spaces are equipped with the metric of uniform convergence on compact sets,

$$d_\infty(\phi, \psi) = \sum_{n=1}^{\infty} \frac{2^{-n} \sup_{t\in[-n,1]} |\phi(t) - \psi(t)|}{1 + \sup_{t\in[-n,1]} |\phi(t) - \psi(t)|}.$$

With $\mathcal{K}$ referring to any one of these spaces, we let $\mathcal{K}_0 = \{\phi \in \mathcal{K} : \phi(0) = 0\}$ and denote by $\mathcal{K}[a, b]$ the space obtained by restricting the elements of $\mathcal{K}$ to the interval $[a, b]$, equipped with the norm of uniform convergence.

We next recall the definition of the large deviations principle on a metric space $\mathcal{S}$ equipped with the topology generated by the metric and the corresponding Borel σ-field.

Definition 8.1 A sequence of probability measures $\mu^{(n)}$ on a metric space $\mathcal{S}$ satisfies the large deviations principle, with speed $\zeta_n \uparrow \infty$ and good rate function $I(\cdot) : \mathcal{S} \to [0, \infty]$ if

(i) the level sets $\{x \in \mathcal{S} : I(x) \le \alpha\}$, $\alpha \in I\!R$, are compact,

(ii) for any closed set F,

$$\limsup_{n\to\infty} \frac{1}{\zeta_n} \log \mu^{(n)}(F) \le - \inf_{x\in F} I(x),$$

(iii) for any open set G,

$$\liminf_{n\to\infty} \frac{1}{\zeta_n} \log \mu^{(n)}(G) \ge - \inf_{x\in G} I(x).$$

A sequence of random variables is said to satisfy the large deviations principle if the corresponding sequence of distributions does. Furthermore, when the space $\mathcal{S}$ is realized as a space of functions, such as $\mathcal{D}$ or $\mathcal{D}[a, b]$, we refer to the large deviations principle as a *sample-path* large deviations principle.

We say that a sequence $\mu^{(n)}$ satisfies a *moderate* deviations principle (MDP) if $\mu^{(n)}$ satisfies the LDP with speed γ_n^2, where γ_n satisfies

$$\lim_{n\to\infty} \frac{\gamma_n}{\sqrt{n}} = 0 \quad \text{and} \quad \lim_{n\to\infty} \gamma_n = \infty. \tag{8.4}$$

(Refer to Mogulskii [22], and Wu [34].) In case $\mu^{(n)}$ satisfies a sample-path large deviation principle with speed γ_n^2 satisfying (8.4), we say $\mu^{(n)}$ satisfies a sample-path moderate deviations principle, denoted as MDP(sp). Allowing that the definition of $\mu^{(n)}$ may depend on γ_n, we say $\mu^{(n)}$ satisfies the *complete* MDP (resp. MDP(sp)) if $\mu^{(n)}$ satisfies the MDP (resp. MDP(sp)) for every sequence satisfying (8.4).

Our starting point in the following sections will be that a scaled version of some process X, which is the input to the filter in (8.3), satisfies a MDP(sp) in $\mathcal{D}$. In the remainder of this section we demonstrate that in fact the class of such processes is far from trivial.

With $I\!N$ denoting the set of all integers, $\{0, \pm 1, \pm 2, \ldots\}$, consider a sequence of real-valued random variables, $\{a(t), t \in I\!N\}$. Let $A(t_1, t_2) = \sum_{t=t_1+1}^{t_2} a(t)$; and define a sequence of stochastic processes (indexed by n) as follows:

$$A^{(n)}(t) = \begin{cases} \frac{1}{n} A(0, [nt]), & t \in [0, 1] \\ \\ -\frac{1}{n} A(\lfloor nt \rfloor, 0), & t \in (-\infty, 0). \end{cases}$$

Below, we shall refer to such processes as *sample-path processes.*

Theorem 8.2 Let $\{a(t), t \in \mathbb{N}\}$ be a stationary sequence of bounded, mean zero, real-valued random variables. For a fixed positive integer m and $t_0 < t_1 < \cdots < t_m \leq 1$, let

$$\begin{aligned} Z_n &= \frac{\sqrt{n}}{\gamma_n}\big(A^{(n)}(t_1) - A^{(n)}(t_0), A^{(n)}(t_2) - A^{(n)}(t_1), \\ &\qquad \ldots, A^{(n)}(t_m) - A^{(n)}(t_{m-1}) \big). \end{aligned}$$

Suppose that for each positive integer m and partition $\{t_i\}_{i=1}^m$, $\{Z_n\}$ satisfies the complete MDP in $\mathbb{R}^m$ with rate function

$$I_m(z) = \sum_{i=1}^m \frac{1}{2\sigma^2} \frac{z_i^2}{(t_i - t_{i-1})},$$

for some $\sigma^2 > 0$, where $z = (z_1, \ldots, z_m)$. In addition, assume

$$\limsup_{n\to\infty} \frac{1}{\gamma_n^2} \log \mathsf{E}[\exp(\frac{\gamma_n \sum_{i=1}^n a(i)}{\sqrt{n}}) + \exp(\frac{-\gamma_n \sum_{i=1}^n a(i)}{\sqrt{n}})] < \infty$$

for every sequence γ_n satisfying (8.4). Denote by $\nu_A^{(n)}$ the distribution of $\{\frac{\sqrt{n}}{\gamma_n} A^{(n)}(t),\ -\infty < t \leq 1\}$. Then the sequence $\nu_A^{(n)}$ satisfies the complete MDP(sp) in $\mathcal{D}$, with a good rate function $I(\cdot) : \mathcal{D} \to \mathbb{R}$,

$$I(\phi) = \begin{cases} \int_{-\infty}^1 \frac{1}{2\sigma^2}(\phi'(t))^2 dt & \text{if } \phi \in \mathcal{AC}_0 \\ \\ \infty & \text{otherwise}. \end{cases}$$

Proof. For any positive integer m, let $\nu_{A,m}^{(n)}$ denote the distribution of $\{\frac{\sqrt{n}}{\gamma_n} A^{(n)}(t),\ -m \leq t \leq 1\}$. A straightforward modification of the proof of [7, Theorem 2.2] yields that $\nu_{A,m}^{(n)}$ satisfies the complete MDP(sp) on $\mathcal{D}[-m, 1]$ with a good rate function

$$I_m(\phi) = \begin{cases} \int_{-m}^1 \frac{1}{2\sigma^2}(\phi'(t))^2 dt & \text{if } \phi \in \mathcal{AC}_0[-m, 1] \\ \\ \infty & \text{otherwise}. \end{cases}$$

Since we may view the space $\mathcal{D}(-\infty, 1]$ as the projective limit of the spaces $\mathcal{D}[-m, 1]$, $m = 1, 2, \ldots$, the proof now follows from an application of [10, Theorem 4.6.1]. □

Remark 8.3 The proof above extends readily to yield, under the assumptions of the theorem, the complete MDP(sp) for the sequence

$$\{\frac{\sqrt{n}}{\gamma_n}A^{(n)}(t), -\infty < t < \infty\},$$

on the space of real-valued RCLL (right continuous with left limit) functions on $I\!R$ equipped with the topology of uniform convergence on compact sets. Here, for $t > 1$, $A^{(n)}(t) = \frac{1}{n}A(0, [nt])$.

Next, we present three classes of stochastic processes that satisfy the hypotheses in Theorem 8.2. The verification that they do indeed satisfy the hypotheses is similar to the verification of their counterparts in [7], and hence omitted.

Example 8.4 Let $\{b(t), t \in I\!N\}$ be a stationary Markov chain taking values in a Polish space and satisfying the conditions in Theorem 2.1 of Wu [34]. Let

$$a(t) = f(b(t)) - \mathsf{E}_\mu[f(b(t))],$$

for any bounded real-valued function f, where μ is the stationary distribution of $\{b(t)\}$. Here,

$$\sigma^2 = \mathsf{E}_\mu[a^2(0)] + 2\sum_{k=1}^{\infty} \mathsf{E}_\mu[a(k)a(0)],$$

which is required to be positive.

Example 8.5 Let $\{\xi(t), t \in I\!N\}$ be a sequence of bounded, mean zero, i.i.d. random variables and $\{b(t), t \in I\!N\}$, a sequence of real numbers which is absolutely summable. Let the sequence $\{a(t), t \in I\!N\}$ be the infinite moving-average sequence defined as

$$a(t) = \sum_{i=-\infty}^{\infty} b(i+t)\xi(i).$$

Here,

$$\sigma^2 = (\sum_{i=-\infty}^{\infty} b_i)^2 \mathsf{E}[\xi^2(0)]$$

is assumed to be positive.

Example 8.6 Let $\{a(t), t \in I\!N\}$ be a stationary sequence of bounded, mean zero random variables. For n a positive integer, denote

$$\begin{aligned}\phi(n) &= \sup\{\, |P(A|B) - P(B)|;\ A \in \mathcal{F}_k \text{ with } P(A) > 0, \\ &\qquad\qquad B \in \mathcal{F}^{k+n}, k \in I\!N\,\},\end{aligned}$$

where, for $k \in \mathbb{N}$,

$$\mathcal{F}_k = \sigma\{a(i) : i \leq k\} \quad \text{and} \quad \mathcal{F}^k = \sigma\{a(i) : i \geq k\}.$$

Recall when $\lim_{n\to\infty} \phi(n) = 0$, the sequence is said to be ϕ-mixing or uniform-mixing (cf. Bradley [4]). We further assume that $\phi(n) = o(e^{-\alpha n})$ for some $\alpha > 0$, and that

$$\sigma^2 = \lim_{n\to\infty} \frac{1}{n} \mathsf{E}[\sum_{i=1}^{n} a(i)]^2$$

is positive.

8.3 MDP for the Filtered Process

To start with, we make precise the processes that we shall allow as possible input, X, to the filter in (8.3). First, we allow X to be a standard Brownian motion, so that our results will apply to FBM (output of the filter). Beyond this, we shall restrict X to be a sample-path process (i.e., X is the process $A^{(1)}$ of Section 8.2).

Letting

$$X^{(n)}(t) = X(nt)/n \quad \text{and} \quad Y^{(n)}(t) = Y(nt)/n,$$

where Y relates to X through (8.3), we have, for $t \in [0,1]$,

$$\begin{aligned} & Y^{(n)}(t) \\ = \; & \int_0^t (t-s)^{H-\frac{1}{2}} d(n^{H-\frac{1}{2}} X^{(n)}(s)) \\ & + \int_{-\infty}^{0} [(t-s)^{H-\frac{1}{2}} - (-s)^{H-\frac{1}{2}}] d(n^{H-\frac{1}{2}} X^{(n)}(s)) \,. \end{aligned}$$

Whereas $Y^{(n)}$ is the FBM when X is the standard Brownian motion, when X is a sample-path process, we need to *define* the above integral relationship. To this end, let $Y_a^{(n)}$ denote the process $Y^{(n)}$ modified so that the second integral above starts at a instead of $-\infty$. We assume the input process X is such that, for each $n = 1, 2, \ldots$ fixed, the sequence $\{Y^{(n)}_{-j/n}\}_{j=1}^{\infty}$ is a Cauchy sequence in $L^2[\mathcal{C}[0,1]]$. We then take $Y^{(n)}$ to be the limit of this sequence.

We shall also require the following condition, which essentially ensures that $Y_m^{(n)}$ is an exponentially good approximation of $Y^{(n)}$ (see [10, Definition 4.2.14]):

$$\begin{aligned} \lim_{m\to\infty} \limsup_{n\to\infty} \frac{1}{n^{2(1-H)}} \log \mathsf{P}\{ \sup_{t\in[0,1]} | \int_{-\infty}^{-m} [(t-s)^{H-\frac{1}{2}} - (-s)^{H-\frac{1}{2}}] \\ d(n^{H-\frac{1}{2}} X^{(n)}(s))| \geq \eta \} = -\infty, \quad \text{for any } \eta > 0. \end{aligned} \tag{8.5}$$

In Section 8.5 we shall verify the above assumptions. Specifically, we will show that,

- the condition in (8.5) holds when X is either the standard Brownian motion or the sample-path processes such as those in the examples of Section 8.2 (with additional assumptions);

- and, in the case of sample-path processes, for each n, the sequence $\{Y^{(n)}_{-j/n}\}_{j=1}^{\infty}$ is Cauchy in $L^2[\mathcal{C}[0,1]]$.

Finally, before stating the results of this section, we remark that by Schilder's theorem (cf. [10] and [27]), when X is the standard Brownian motion, the sequence $\{n^{H-\frac{1}{2}}X^{(n)}(t), t \in (-\infty,1]\}$ satisfies the MDP in $\mathcal{D}$ with speed $n^{2(1-H)}$ and a good rate function as in the statement of Theorem 8.2, with $\sigma^2 = 1$.

Theorem 8.7 Denote by $\mu_X^{(n)}$ the probability law of $\{n^{H-\frac{1}{2}}X^{(n)}(t), t \in (-\infty,1]\}$, for some $\frac{1}{2} < H < 1$. Suppose $\mu_X^{(n)}$ satisfies the MDP(sp) on $\mathcal{D}$ with speed $n^{2(1-H)}$ and a good rate function $I_X(\cdot) : \mathcal{D} \mapsto I\!R$,

$$I_X(\phi) = \begin{cases} \int_{-\infty}^{1} \frac{1}{2\sigma^2}(\phi'(t))^2 dt & \text{if } \phi \in \mathcal{AC}_0 \\ \\ \infty & \text{otherwise.} \end{cases}$$

Suppose the condition in (8.5) holds. Then, the sequence $\{Y^{(n)}(t), t \in [0,1]\}$ satisfies the MDP(sp) in $\mathcal{C}[0,1]$ with speed $n^{2(1-H)}$ and a good rate function $I_Y(\cdot) : \mathcal{C}[0,1] \mapsto I\!R$, given by

$$\begin{aligned} I_Y(\psi) &= \inf\{I_X(\phi) : \psi(t) = \int_0^t (t-s)^{H-\frac{1}{2}}\phi'(s)ds \\ &\quad + \int_{-\infty}^{0} [(t-s)^{H-\frac{1}{2}} - (-s)^{H-\frac{1}{2}}]\phi'(s)ds,\ t \in [0,1]\}. \end{aligned}$$

Corollary 8.8 Suppose the conditions in Theorem 8.7 hold. Then the sequence $\{Y(n)/n, n \geq 0\}$ satisfies the MDP(sp) with speed $n^{2(1-H)}$ and a good rate function $\Lambda_Y^*(\cdot) : I\!R \mapsto I\!R$,

$$\Lambda_Y^*(\gamma) = \frac{\tilde{H}}{\sigma^2}\gamma^2,$$

where

$$\tilde{H} = \frac{1}{2}\left[\frac{1}{2H} + \int_{-\infty}^{0} [(1-s)^{H-\frac{1}{2}} - (-s)^{H-\frac{1}{2}}]^2 ds\right]^{-1}. \tag{8.6}$$

Remark 8.9 It is known (e.g., [20]) that the variance of the FBM is $\mathsf{E}[B_H^2(t)] = t^{2H} V_H$, where $V_H = 1/(2\tilde{H})$ (ignoring the constant $\Gamma(H+\frac{1}{2})$ in (8.1)), with $\tilde{H}$ following (8.6) above.

Note that, here and below, δ_ϕ denotes the distribution defined as follows: for any Borel set A, $\delta_{\phi_\gamma}(A) = 1\{\phi_\gamma \in A\}$, with $1\{\cdot\}$ being the indicator function.

Theorem 8.10 Suppose the conditions in Theorem 8.7 are in force.

(i) Let $\tilde{\mu}_X^{(n)}$ denote the law of $\{n^{H-\frac{1}{2}} X^{(n)} | Y^{(n)}(1) \geq \gamma\}$ for some $\gamma \geq 0$. Then, $\tilde{\mu}_X^{(n)}$ converges in distribution on $\mathcal{D}$ to δ_{ϕ_γ}, where

$$\phi_\gamma(t) = \begin{cases} \gamma \frac{2\tilde{H}}{H+\frac{1}{2}} (1 - (1-t)^{H+\frac{1}{2}}) & t \in [0,1] \\ \\ \gamma \frac{2\tilde{H}}{H+\frac{1}{2}} (1 - (1-t)^{H+\frac{1}{2}} + (-t)^{H+\frac{1}{2}}) & t \in (-\infty, 0]. \end{cases}$$

(ii) Let $\tilde{\mu}_Y^{(n)}$ denote the law of $\{Y^{(n)}(\cdot) | Y^{(n)}(1) \geq \gamma\}$. Then $\tilde{\mu}_Y^{(n)}$ converges in distribution on $\mathcal{C}[0,1]$ to δ_{ψ_γ}, where

$$\begin{aligned} \psi_\gamma(t) &= \int_0^t (t-s)^{H-\frac{1}{2}} d\phi_\gamma(s) \\ &\quad + \int_{-\infty}^0 [(t-s)^{H-\frac{1}{2}} - (-s)^{H-\frac{1}{2}}] d\phi_\gamma(s), \quad t \in [0,1]. \end{aligned}$$

Proof. [Theorem 8.7]

Consider first the case when X is a sample-path process. Define the map $F^m : \mathcal{BV} \to \mathcal{C}[0,1]$ as follows:

$$\begin{aligned} F^m(\phi)(t) &= \int_0^t (t-s)^{H-\frac{1}{2}} d\phi(s) \\ &\quad + \int_{-m}^0 [(t-s)^{H-\frac{1}{2}} - (-s)^{H-\frac{1}{2}}] d\phi(s) \,, \end{aligned} \tag{8.7}$$

and a second map, $\tilde{F}^m : \mathcal{D} \to \mathcal{C}[0,1]$, as

$$\begin{aligned} &\tilde{F}^m(\phi)(t) \\ &= \int_0^t (H - \frac{1}{2})(t-s)^{H-\frac{3}{2}} \phi(s) ds \\ &\quad + \int_{-m}^0 (H - \frac{1}{2})[(t-s)^{H-\frac{3}{2}} - (-s)^{H-\frac{3}{2}}] \phi(s) ds \\ &\quad - [(t+m)^{(H-\frac{1}{2})} - m^{(H-\frac{1}{2})}] \phi(-m) \,. \end{aligned} \tag{8.8}$$

Clearly, $\tilde{F}^m$ is continuous. Also, integration by parts verifies that $\tilde{F}^m(\phi) = F^m(\phi)$ for all $\phi \in \mathcal{BV}$ with $\phi(0) = 0$.

Thus, it follows from the contraction principle (see, for example, [10, Theorem 4.2.1]) that the following sequence

$$\begin{aligned} & Y_m^{(n)}(t) \\ = & \int_0^t (t-s)^{H-\frac{1}{2}} d(n^{H-\frac{1}{2}} X^{(n)}(s)) \\ + & \int_{-m}^0 [(t-s)^{H-\frac{1}{2}} - (-s)^{H-\frac{1}{2}}] d(n^{H-\frac{1}{2}} X^{(n)}(s)), \quad t \in [0,1], \end{aligned}$$

satisfies the MDP(sp), with a good rate function given by

$$\begin{aligned} I_Y^m(\psi) \; = \; & \inf\{I_X(\phi) : \psi(t) = \int_0^t (t-s)^{H-\frac{1}{2}} \phi'(s) ds \\ & + \int_{-m}^0 [(t-s)^{H-\frac{1}{2}} - (-s)^{H-\frac{1}{2}}] \phi'(s) ds, \; t \in [0,1]\}. \end{aligned}$$

For $t \in [0,1]$ and $\phi \in \mathcal{BV}$,

$$\begin{aligned} & F^m(\phi)(t) - F^\infty(\phi)(t) \\ = & \int_{-\infty}^{-m} [(t-s)^{H-\frac{1}{2}} - (-s)^{H-\frac{1}{2}}] d\phi(s). \end{aligned} \tag{8.9}$$

Thus, by (8.5) $Y_m^{(n)}$ are exponentially good approximations of $Y^{(n)}$ (see [10, Definition 4.2.14]). Moreover, for any $\alpha < \infty$, making use of the Cauchy-Schwartz inequality and the definition of $I_X(\cdot)$ in Theorem 8.7, we have

$$\begin{aligned} & \sup_{\{\phi: I_X(\phi) \le \alpha\}} \sup_{0 \le t \le 1} |F^m(\phi)(t) - F^\infty(\phi)(t)| \\ = & \sup_{\{\phi: I_X(\phi) \le \alpha\}} \sup_{0 \le t \le 1} \left| \int_{-\infty}^{-m} [(t-s)^{H-\frac{1}{2}} - (-s)^{H-\frac{1}{2}}] \phi'(s) ds \right| \\ \le & \sup_{\{\phi: I_X(\phi) \le \alpha\}} \sup_{0 \le t \le 1} \Big(\int_{-\infty}^{-m} [(t-s)^{H-\frac{1}{2}} - (-s)^{H-\frac{1}{2}}]^2 ds \\ & \cdot \int_{-\infty}^{-m} [\phi'(s)]^2 ds \Big)^{\frac{1}{2}} \\ \le & \sqrt{2\alpha}\sigma \Big(\int_{-\infty}^{-m} [(1-s)^{H-\frac{1}{2}} - (-s)^{H-\frac{1}{2}}]^2 ds \Big)^{\frac{1}{2}}. \end{aligned}$$

Therefore,

$$\limsup_{m \to \infty} \sup_{\{\phi: I_X(\phi) \le \alpha\}} \sup_{0 \le t \le 1} |F^m(\phi)(t) - F^\infty((\phi)(t)| = 0. \tag{8.10}$$

It then follows from [10, Theorem 4.2.23] that $\{Y^{(n)}(t), t \in [0,1]\}$ satisfies the desired MDP(sp). □

Proof. [Corollary 8.8]
Recall that $Y^{(n)}(1) = Y(n)/n$. Since the mapping $f : \psi \mapsto \psi(1)$ from $\mathcal{C}[0,1]$ to $\mathbb{R}$ is continuous, it follows from the contraction principle and the MDP(sp) of Theorem 8.7 that the distribution of $Y^{(n)}(1)$ satisfies the MDP with speed $n^{2(1-H)}$ and a good rate function $\Lambda_Y^*(\cdot) : \mathbb{R} \to \mathbb{R}$,

$$\begin{aligned} & \Lambda_Y^*(\gamma) \\ = \ & \inf\{I_Y(\psi) : \psi(1) = \gamma\} \\ = \ & \inf\{I_X(\phi) : \int_0^1 (1-s)^{H-\frac{1}{2}} \phi'(s) ds \\ & + \int_{-\infty}^0 [(1-s)^{H-\frac{1}{2}} - (-s)^{H-\frac{1}{2}}] \phi'(s) ds = \gamma\} . \end{aligned}$$

From the definition of $I_X(\cdot)$, the infimum above amounts to solving the following minimization problem:

$$\begin{aligned} \min_{\phi \in \mathcal{AC}_0} \quad & \int_{-\infty}^1 \frac{1}{2\sigma^2} (\phi'(s))^2 ds \\ \text{s.t.} \quad & \int_0^1 (1-s)^{H-\frac{1}{2}} \phi'(s) ds \\ & + \int_{-\infty}^0 [(1-s)^{H-\frac{1}{2}} - (-s)^{H-\frac{1}{2}}] \phi'(s) ds = \gamma. \end{aligned}$$

It can be directly verified that

$$\phi'_\gamma(s) = \begin{cases} \gamma(2\tilde{H})(1-s)^{H-\frac{1}{2}} & s \in [0,1] \\ \\ \gamma(2\tilde{H})[(1-s)^{H-\frac{1}{2}} - (-s)^{H-\frac{1}{2}}] & s \in (-\infty, 0] \end{cases}$$

uniquely solves the above minimization problem. Furthermore, the achieved minimum value is $\tilde{H}\gamma^2/\sigma^2$, as desired.

When $X \equiv B$, the standard Brownian motion, simply replace ϕ by B and treat the integrals as Itô integrals. In particular, we still have $F^m(B) = \tilde{F}^m(B)$, and the above argument still applies. □

Proof. [Theorem 8.10]
(i) It suffices to show that, for $\epsilon > 0$ arbitrary, we have

$$\lim_{n\to\infty} \frac{\tilde{\mu}_X^{(n)}(B_\epsilon \cap \Gamma)}{\tilde{\mu}_X^{(n)}(\Gamma)} = 0 , \tag{8.11}$$

where

$$B_\epsilon = \{\phi \in \mathcal{D} : d_\infty(\phi, \phi_\gamma) \geq \epsilon\}$$

and

$$\begin{aligned}\Gamma &= \{\phi \in \mathcal{AC}_0 : \int_0^1 (1-s)^{H-\frac{1}{2}} \phi'(s) ds \\ &\quad + \int_{-\infty}^0 [(1-s)^{H-\frac{1}{2}} - (-s)^{H-\frac{1}{2}}] \phi'(s) ds \geq \gamma\}.\end{aligned}$$

We have from the proof of Corollary 8.8 that

$$\inf_{\phi\in\Gamma} \int_{-\infty}^1 \frac{1}{2\sigma^2} (\phi'(s))^2 ds = \int_{-\infty}^1 \frac{1}{2\sigma^2} (\phi'_\gamma(s))^2 ds = \Lambda_Y^*(\gamma) .$$

Furthermore, observe that the set $B_\epsilon \cap \Gamma \cap \{\phi : I_X(\phi) \leq 2\Lambda_Y^*(\gamma)\}$ is closed. An application of Lemma 2.8 of [28] yields

$$\inf_{\phi\in B_\epsilon\cap\Gamma} \int_{-\infty}^1 \frac{1}{2\sigma^2} (\phi'(s))^2 ds > \Lambda_Y^*(\gamma) .$$

Applying Theorem 8.7 yields (8.11).

(ii) Since the map $\tilde{F}^m : \mathcal{D} \to \mathcal{C}[0,1]$ defined in (8.8) is continuous, the probability law of

$$\{\tilde{F}^m(n^{H-\frac{1}{2}} X^{(n)}(\cdot)) | Y^{(n)}(1) \geq \gamma\}$$

converges in distribution to $\delta_{\tilde{F}^m(\phi_\gamma)}$. Also, since $\tilde{F}^m(\phi) = F^m(\phi)$ for all $\phi \in \mathcal{BV}$ and $\tilde{F}^m(B) = F^m(B)$, where B denotes Brownian motion and F^m is defined in (8.7), we conclude that the probability law of

$$\{F^m(n^{H-\frac{1}{2}} X^{(n)}(\cdot)) | Y^{(n)}(1) \geq \gamma\}$$

converges in distribution to $\delta_{F^m(\phi_\gamma)}$. Suppressing the argument $t \in [0,1]$, we have

$$\begin{aligned}&|Y^{(n)} - F^\infty(\phi_\gamma)| \\ = \ &|F^\infty(n^{H-\frac{1}{2}} X^{(n)}) - F^\infty(\phi_\gamma)| \\ \leq \ &|F^\infty(n^{H-\frac{1}{2}} X^{(n)}) - F^m(n^{H-\frac{1}{2}} X^{(n)})| \\ &+ |F^m(n^{H-\frac{1}{2}} X^{(n)}) - F^m(\phi_\gamma)| \\ &+ |F^m(\phi_\gamma) - F^\infty(\phi_\gamma)|.\end{aligned}$$

As in (8.10), it follows that

$$\lim_{m\to\infty} \sup_{0\leq t\leq 1} |F^m(\phi_\gamma)(t) - F^\infty(\phi_\gamma)(t)| = 0,$$

since $I_X(\phi_\gamma) < \infty$. Using (8.5) then allows us to conclude

$$\lim_{n\to\infty} \mathsf{P}\Big[\sup_{0\le t\le 1} |Y^{(n)}(t) - F^\infty(\phi_\gamma)(t)| > \epsilon \,\Big|\, Y^{(n)}(1) \ge \gamma\Big] = 0,$$

for all $\epsilon > 0$, which is the desired result, observing that $F^\infty(\phi_\gamma) = \psi_\gamma$. □

8.4 Queueing Applications: The Workload Process

Consider a single-server queue with an input process Y, which, in turn, is the output of the filter in (8.3) subject to an input process X. As noted before, for $H \in (\frac{1}{2}, 1)$ the input process to the queue, Y, typically has long-range dependence.

Denote

$$V(t) = \sup_{0\le s\le t} (Y(t) - Y(s) - c(t-s)). \tag{8.12}$$

Then, V is the *state* process of a queue with input Y and a deterministic service mechanism that depletes work at rate $c > 0$. (Alternatively, $Y(t)$ can be viewed as the mean-zero part of the "netput" — input minus output, with $-ct$ as the "deterministic negative drift" part of the netput. The negative drift is crucial to stability.) In the queueing literature, V is often referred to as the virtual waiting time process, or the workload/inventory process. See, for instance, [5] and [11]. Chang [5], Duffield and O'Connell [11], Harrison [17], and Sigman and Yao [29], among others. (That $c > 0$ and hence the netput has a negative drift is crucial for stability; see, e.g., [29].)

The steady-state limit, $V(\infty)$, is typically obtained through the limit of another process, $M(t) = \sup_{0\le s\le t}(-Y(-s) - cs)$, which has the interpretation of the maximum of a random walk. The key linkage between the two is the stationary increment property of the process Y, which guarantees that for each t,

$$\begin{aligned} V(t) &=_d \sup_{0\le s\le t} (Y(t-t) - Y(s-t) - c(t-s)) \\ &= \sup_{0\le s\le t} (-Y(s-t) - c(t-s)) \\ &= M(t). \end{aligned}$$

That is, for each t, $V(t)$ and $M(t)$ are equal in distribution, and hence so are $V(\infty)$ and $M(\infty)$. This fact plays a fundamental role in characterizing the steady-state limit, since whereas $M(t)$ is an increasing process (and hence approaches a limit), $V(t)$ is not.

Below, we begin by considering the tail behavior of $M(\infty)$ in Theorem 8.11, based on which the tail behavior of $V(\infty)$ readily follows (Theorem

8.15). Finally, in Theorem 8.17, we present the transient behavior of the workload process, in terms of conditional limits. Throughout, $H \in (\frac{1}{2}, 1)$ and $\tilde{H}$ follows (8.6).

Theorem 8.11 Let $M(t) = \sup_{0 \le s \le t}(-Y(-s) - cs)$ for some $c > 0$. Then, under the conditions of Theorem 8.7, $M(t)$ converges almost surely to a finite random variable $M(\infty)$, which has a subexponential tail distribution. Specifically,

$$\lim_{x \to \infty} \frac{1}{x^{2(1-H)}} \log \mathsf{P}[M(\infty) > x] = -\frac{\tilde{H}}{H}\theta^*, \tag{8.13}$$

where

$$\theta^* = \frac{H}{\sigma^2} \frac{c^{2H}}{H^{2H}(1-H)^{2(1-H)}}.$$

To prove Theorem 8.11, we require three lemmas, presented below. The proof of the first lemma follows closely the proofs of Theorem 8.7 and Corollary 8.8 (the only difference being $t \in [-1, 0]$ instead of $t \in [0, 1]$), and is hence omitted.

Lemma 8.12 Suppose the conditions in Theorem 8.7 hold. Then the distribution of $\{Y^{(n)}(t), t \in [-1, 0]\}$, denoted $\mu_Y^{(n)}$, satisfies the MDP(sp) in $\mathcal{C}[-1, 0]$ with speed $n^{2(1-H)}$ and a good rate function $\tilde{I}_Y(\cdot) : \mathcal{C}[-1, 0] \mapsto I\!R$,

$$\tilde{I}_Y(\psi) = \inf\{I_X(\phi) : \psi(t) = \int_{-\infty}^{t} (t-s)^{H-\frac{1}{2}} \phi'(s) ds$$
$$+ \int_{-\infty}^{0} -(-s)^{H-\frac{1}{2}} \phi'(s) ds, \ t \in [-1, 0]\}.$$

In addition, the sequence $\{Y(-n)/n, n \ge 0\}$ satisfies the MDP with speed $n^{2(1-H)}$ and a good rate function $\tilde{\Lambda}_Y^*(\cdot) : I\!R \mapsto I\!R$,

$$\tilde{\Lambda^*}_Y(\gamma) = \frac{\tilde{H}}{\sigma^2}\gamma^2 .$$

Lemma 8.13 For all $\gamma \ge 0$, under the conditions of Theorem 8.7,

$$\lim_{n \to \infty} \frac{1}{n^{2(1-H)}} \log \mathsf{P}[M(n)/n > \gamma] = -\tilde{g}(\gamma),$$

where

$$\tilde{g}(\gamma) = \begin{cases} \frac{\tilde{H}}{\sigma^2}(c+\gamma)^2 & \text{if } \gamma \ge \frac{c(1-H)}{H} \\ \frac{\tilde{H}}{\sigma^2} \frac{c^{2H}}{H^{2H}(1-H)^{2(1-H)}} \gamma^{2(1-H)} & \text{if } \gamma < \frac{c(1-H)}{H}. \end{cases}$$

Proof. By the contraction principle and Lemma 8.12, we know the sequence $\{M(n)/n\}$ satisfies the MDP(sp) with speed $n^{2(1-H)}$, and we may write

$$\limsup_{n\to\infty} \frac{1}{n^{2(1-H)}} \log \mathsf{P}[M(n)/n \geq \gamma]$$
$$\leq - \inf_{-1\leq s\leq 0} \{ \inf_{\tilde{F}_s} \int_{-\infty}^{0} \frac{1}{2\sigma^2} (\phi'(\tau))^2 d\tau \}, \qquad (8.14)$$

where

$$\begin{aligned}\tilde{F}_s &= \{ \phi \in \mathcal{AC}_0 : -[\int_{-\infty}^{s} (s-\tau)^{H-\frac{1}{2}} \phi'(\tau) d\tau \\ &\quad + \int_{-\infty}^{0} -(-\tau)^{H-\frac{1}{2}} \phi'(\tau) d\tau] + cs \geq \gamma \} \\ &= \{ \phi \in \mathcal{AC}_0 : -[\int_{-\infty}^{s} [(s-\tau)^{H-\frac{1}{2}} - (-\tau)^{H-\frac{1}{2}}] \phi'(\tau) d\tau \\ &\quad + \int_{s}^{0} -(-\tau)^{H-\frac{1}{2}} \phi'(\tau) d\tau] \geq (\gamma - cs) \} .\end{aligned}$$

It can be directly verified (using the Cauchy-Schwartz inequality) that the solution to the minimization problem within braces in (8.14), ϕ, satisfies the following:

$$\phi'(t) = \begin{cases} [2\tilde{H}(\gamma - cs)/|s|^{2H}] \cdot (-t)^{H-\frac{1}{2}} & t \in [s, 0] \\ -[2\tilde{H}(\gamma - cs)/|s|^{2H}] \cdot [(s-t)^{H-\frac{1}{2}} - (-t)^{H-\frac{1}{2}}] & t \in (-\infty, s]. \end{cases}$$

Furthermore,

$$\begin{aligned}&\inf_{\tilde{F}_s} \int_{-\infty}^{0} \frac{1}{2\sigma^2} (\phi'(\tau))^2 d\tau \\ &= (\gamma - cs)^2 / [2\sigma^2 (\int_{-\infty}^{s} [(s-\tau)^{(H-\frac{1}{2})} - (-\tau)^{(H-\frac{1}{2})}]^2 d\tau \\ &\quad + \int_{s}^{0} (-(-\tau)^{(H-\frac{1}{2})})^2 d\tau)].\end{aligned}$$

By a change of variables, $\tau \to |s|\tau$, the denominator above becomes

$$\begin{aligned}&2\sigma^2 |s|^{2H} (\int_{-\infty}^{-1} [(-1-\tau)^{(H-\frac{1}{2})} - (-\tau)^{(H-\frac{1}{2})}]^2 d\tau \\ &\quad + \int_{-1}^{0} (-(-\tau)^{(H-\frac{1}{2})})^2 d\tau)\end{aligned}$$

$$= 2\sigma^2|s|^{2H}\Big[\int_{-\infty}^{0}[(-\tau)^{(H-\frac{1}{2})}-(1-\tau)^{(H-\frac{1}{2})}]^2 d\tau+\frac{1}{2H}\Big]$$
$$= \sigma^2|s|^{2H}/\tilde{H}.$$

Therefore,

$$\inf_{\tilde{F}_s}\int_{-\infty}^{0}\frac{1}{2\sigma^2}(\phi'(\tau))^2 d\tau=\frac{\tilde{H}(\gamma-cs)^2}{\sigma^2|s|^{2H}}\ .$$

The right hand side above is minimized at

$$s^*=\begin{cases}-H\gamma/[c(1-H)] & \text{if } \gamma<c(1-H)/H\\ \\ -1 & \text{otherwise}\end{cases}$$

with the minimum following $\tilde{g}(\gamma)$ as stated. A lower bound is obtained similarly and the lemma follows by the continuity of $\tilde{g}$. □

Lemma 8.14 For $n=1,2,\ldots$, let $Z(n)=\sup_{n-1<s\le n}(-Y(-s)-cs)$. Then, for all $\gamma\ge 0$,

$$\lim_{n\to\infty}\frac{1}{n^{2(1-H)}}\log \mathsf{P}[Z(n)/n>\gamma]=-\frac{\tilde{H}}{\sigma^2}(c+\gamma)^2.$$

Proof. We establish the limit by showing that the liminf and the limsup are dominated from below and from above by the same bound. Note that

$$Z(n)\ge -Y(-n)-cn.$$

Hence, the lower bound follows from the liminf part of the MDP for $Y(-n)$ in Lemma 8.12.

To establish the upper bound, we have

$$\begin{aligned}&\limsup_{n\to\infty}\frac{1}{n^{2(1-H)}}\log \mathsf{P}[Z(n)/n\ge\gamma]\\ =\ &\limsup_{n\to\infty}\frac{1}{n^{2(1-H)}}\log \mathsf{P}[\sup_{1-1/n<s\le 1}(-Y^{(n)}(-s)-cs)\ge\gamma]\\ \le\ &\limsup_{n\to\infty}\frac{1}{n^{2(1-H)}}\log \mathsf{P}[\sup_{1-\delta\le s\le 1}(-Y^{(n)}(-s)-cs)\ge\gamma],\end{aligned}$$

for any $0<\delta<1$. Following an argument similar to the proof of Lemma 8.13, we have

$$\begin{aligned}&\limsup_{n\to\infty}\frac{1}{n^{2(1-H)}}\log \mathsf{P}[\sup_{1-\delta\le s\le 1}(-Y^{(n)}(-s)-cs)\ge\gamma]\\ \le\ &-\inf_{1-\delta\le s\le 1}\frac{\tilde{H}}{\sigma^2}\frac{(\gamma+cs)^2}{s^{2H}}\\ =\ &-g_\delta(1-\gamma),\end{aligned}$$

where

$$g_\delta(\gamma) = \begin{cases} \frac{\tilde H}{\sigma^2}(c+\gamma)^2 & \text{if } \gamma \geq \frac{c(1-H)}{H} \\ \frac{\tilde H}{\sigma^2}\frac{c^{2H}}{H^{2H}(1-H)^{2(1-H)}}\gamma^{2(1-H)} & \text{if } \frac{c(1-H)}{H}\delta < \gamma < \frac{c(1-H)}{H} \\ \frac{\tilde H}{\sigma^2}\frac{(c\delta+\gamma)^2}{\delta^{2H}} & \text{if } \frac{c(1-H)}{H}\delta \geq \gamma. \end{cases}$$

Since δ is arbitrary, letting $\delta \to 1$ yields the upper bound, noting that

$$\lim_{\delta\to 1} g_\delta(\gamma) = \frac{\tilde H}{\sigma^2}(c+\gamma)^2.$$

□

Proof. [Theorem 8.11]
Since $M(t)$ is an increasing sequence, it converges almost surely to a random variable $M(\infty)$, which is finite, due to the negative drift.

Again, we prove the limit (8.13) by arguing that the liminf and the limsup are dominated from below and from above by the same bound.

To establish the lower bound, we proceed as in Chang et al [7]. For all $\gamma > 0$, we have

$$\begin{aligned} & \liminf_{x\to\infty} \frac{1}{x^{2(1-H)}} \log \mathsf{P}[M(\infty) > x] \\ \geq \; & \liminf_{n\to\infty} \frac{1}{(\gamma n)^{2(1-H)}} \log \mathsf{P}[-Y(-n) - cn > \gamma n] \\ \geq \; & -\frac{\tilde H}{\sigma^2}\frac{(\gamma+c)^2}{\gamma^{2(1-H)}}, \end{aligned}$$

where we have used Lemma 8.12. Thus,

$$\begin{aligned} & \liminf_{x\to\infty} \frac{1}{x^{2(1-H)}} \log \mathsf{P}[M(\infty) > x] \\ \geq \; & -\inf_{\gamma>0} \frac{\tilde H}{\sigma^2}\frac{(\gamma+c)^2}{\gamma^{2(1-H)}} \\ = \; & -\frac{\tilde H}{H}\theta^*. \end{aligned}$$

To establish the upper bound, note that for all $\gamma > 0$

$$\begin{aligned} & \limsup_{x\to\infty} \frac{1}{x^{2(1-H)}} \log \mathsf{P}[M(\infty) \geq x] \\ = \; & \limsup_{n\to\infty} \frac{1}{(\gamma n)^{2(1-H)}} \log \mathsf{P}[M(\infty) \geq \gamma n], \end{aligned}$$

and write

$$\begin{aligned}
&\mathsf{P}[M(\infty) \geq \gamma n] \\
= \; &\mathsf{P}[M(n) \geq \gamma n] + \mathsf{P}[M(n) < \gamma n, \sup_{s>n}(-Y(-s) - cs) \geq \gamma n] \\
\leq \; &\mathsf{P}[M(n) \geq \gamma n] + \mathsf{P}[\sup_{m \geq n+1} Z(m) \geq 0],
\end{aligned} \tag{8.15}$$

where $Z(n)$ follows the definition in Lemma 8.14. If $\gamma < c(1-H)/H$ then, from Lemma 8.13,

$$\limsup_{n\to\infty} \frac{1}{(\gamma n)^{2(1-H)}} \log \mathsf{P}[M(n) \geq \gamma n] = -\frac{\tilde{H}}{H}\theta^*. \tag{8.16}$$

Next, in view of the upper bound for $\{Z(n)/n, n \geq 0\}$ in Lemma 8.14, for every $\epsilon > 0$ there exists an n_0 such that for all $n \geq n_0$ we have

$$\mathsf{P}(Z(n)/n \geq 0) \leq \exp\Big(- n^{2(1-H)}(\frac{\tilde{H}}{\sigma^2}c^2 - \epsilon)\Big).$$

Choosing ϵ such that $0 < \epsilon < \frac{\tilde{H}}{2\sigma^2}c^2$ and then γ small enough so that

$$\gamma^{2(1-H)}\theta^* \frac{\tilde{H}}{H} + \epsilon < \frac{\tilde{H}}{\sigma^2}c^2 - \epsilon \quad \text{and} \quad \gamma < \frac{c(1-H)}{H},$$

we therefore have, for $n \geq n_0$,

$$\begin{aligned}
&\mathsf{P}[\sup_{m \geq n+1} Z(m) \geq 0] \\
\leq \; &\sum_{m=n+1}^{\infty} \mathsf{P}[Z(m)/m \geq 0] \\
\leq \; &\sum_{m=n+1}^{\infty} \exp\Big(- n^{2(1-H)}(\gamma^{2(1-H)}\frac{\tilde{H}}{H}\theta^* + \epsilon)\Big).
\end{aligned} \tag{8.17}$$

Simple algebra yields

$$\begin{aligned}
&\limsup_{n\to\infty} \frac{1}{(\gamma n)^{2(1-H)}} \log \sum_{m=n+1}^{\infty} \exp\Big(- n^{2(1-H)}(\gamma^{2(1-H)}\frac{\tilde{H}}{H}\theta^* + \epsilon)\Big) \\
\leq \; & -\frac{\tilde{H}}{H}\theta^*.
\end{aligned} \tag{8.18}$$

Note that for $\alpha > 0, z > 0$, $n > (\frac{2}{\alpha z})^{1/\alpha}$,

$$ne^{-n^\alpha z} = \int_n^\infty (\alpha z s^\alpha - 1)e^{-s^\alpha z} ds \geq \int_n^\infty e^{-s^\alpha z} ds.$$

Choose $\alpha = 2(1-H)$, and

$$z = \gamma^{2(1-H)} \tilde{H} \theta^* / H + \epsilon$$

with $\epsilon > 0$ arbitrarily small. Therefore, combining (8.15), (8.16), (8.17) and (8.18), we have established the desired upper bound. □

As noted earlier, the stationary increment property of Y ensures that $V(t) =_d M(t)$, for each $t \in [0, \infty)$. This is certainly the case when $X \equiv B$, the standard Brownian motion. When X is a sample-path process, we can directly verify that $V(n) =_d M(n)$, for each integer n. To see this, it suffices to observe that, for any $u \in [0, 1]$, we have

$$\begin{aligned} & n^{\frac{3}{2}-H}[Y^{(n)}(1) - Y^{(n)}(u)] \\ = \ & \sum_{j=-\infty}^{n} (1 - j/n)^{H-\frac{1}{2}} a(j) - \sum_{j=-\infty}^{[nu]} (u - j/n)^{H-\frac{1}{2}} a(j) \\ =_d \ & \sum_{j=-\infty}^{n} (1 - j/n)^{H-\frac{1}{2}} a(j+n) - \sum_{j=-\infty}^{[nu]} (u - j/n)^{H-\frac{1}{2}} a(j+n) \\ = \ & \sum_{j=-\infty}^{0} (-j/n)^{H-\frac{1}{2}} a(j) - \sum_{j=-\infty}^{[n(u-1)]} [(u-1) - j/n]^{H-\frac{1}{2}} a(j) \\ = \ & n^{\frac{3}{2}-H}[Y^{(n)}(0) - Y^{(n)}(u-1)] \, . \end{aligned}$$

(The equality $=_d$ above follows from the stationarity of $\{a(t)\}$.)

The following theorem is a direct consequence of Lemma 8.13 and Theorem 8.11. (Given the above discussion, part (ii) of the Theorem should be interpreted in the discrete time setting when X is a sample-path process.)

Theorem 8.15 Suppose that the conditions in Theorem 8.11 hold.

(i) The sequence $\{V^{(n)}(1), n \geq 0\}$ satisfies the MDP with speed $n^{2(1-H)}$ and a good rate function $\Lambda_V^*(\cdot) : I\!R \mapsto I\!R$, where $\Lambda_V^*(\gamma) = \tilde{g}(\gamma)$, $\gamma \in I\!R$.

(ii) The random variable $V(t)$ converges in distribution to a finite random variable $V(\infty)$, which has a subexponential tail distribution. Specifically,

$$\lim_{x \to \infty} \frac{1}{x^{2(1-H)}} \log \mathsf{P}[V(\infty) > x] = -\frac{\tilde{H}}{H} \theta^* .$$

Remark 8.16 When Y is the FBM, the result in Theorem 8.15 is consistent with the result in Duffield and O'Connell [11] for FBM. To see this, observe that when $X \equiv \sqrt{2\tilde{H}}B$, with B being the standard Brownian motion, $\mathsf{E}[Y^2(t)] = t^{2H}$ for $t \in [0, \infty)$, which is consistent with relation (57) of

[11]. In addition, X satisfies the conditions in Theorem 8.7 with $\sigma^2 = 2\tilde{H}$. It then suffices to note that the minimization problem in (60) of [11] achieves the value θ^*.

Theorem 8.17 (i) Let $\hat{\mu}_X^{(n)}$ denote the law of $\{n^{H-\frac{1}{2}}X^{(n)}(\cdot)|V^{(n)}(1) \geq \gamma\}$ for some $\gamma \geq 0$. Then, $\hat{\mu}_X^{(n)}$ converges in distribution on $\mathcal{D}$ to $\delta_{\hat{\phi}_\gamma}$, where

$$\hat{\phi}_\gamma(t) = \phi_\gamma(t-1) - \phi_\gamma(-1),$$

for $t \in (-\infty, 1]$, with ϕ_γ given in the proof of Lemma 8.13.
(ii) Let $\hat{\mu}_V^{(n)}$ denote the law of $\{V^{(n)}(\cdot)|V^{(n)}(1) \geq \gamma\}$. Then $\hat{\mu}_V^{(n)}$ converges in distribution on $C[0,1]$ to $\delta_{\hat{\psi}_\gamma}$, where, for $t \in [0,1]$,

$$\begin{aligned}
\hat{\psi}_\gamma(t) &= \sup_{0 \leq s \leq t} \Big\{ \int_0^t (t-\tau)^{H-\frac{1}{2}} d\hat{\phi}_\gamma(\tau) \\
&\quad + \int_{-\infty}^0 [(t-\tau)^{H-\frac{1}{2}} - (-\tau)^{H-\frac{1}{2}}] d\hat{\phi}_\gamma(\tau) \\
&\quad - \int_0^s (s-\tau)^{H-\frac{1}{2}} d\hat{\phi}_\gamma(\tau) \\
&\quad - \int_{-\infty}^0 [(s-\tau)^{H-\frac{1}{2}} - (-\tau)^{H-\frac{1}{2}}] d\hat{\phi}_\gamma(\tau) \\
&\quad - c(t-s) \Big\}.
\end{aligned}$$

Remark 8.18 The most likely path to a large build up, $\hat{\psi}_\gamma$, has been studied in the case of input with short-range dependence by Anantharam [2], Chang [6] and Dembo and Zajic [9], where it has been found to be linear. Here, as in [7], the most likely path is *nonlinear*, as evident in Figure 8.1, where we have plotted the function $\hat{\psi}_\gamma(t)$ for different H values.

Proof. [Theorem 8.17] As in the proof of Theorem 3.3 (i), it suffices to show that

$$\inf_{\phi \in F \cap B_\epsilon} \int_{-\infty}^1 (\phi'(\tau))^2 d\tau > \Lambda_V^*(\gamma) = \tilde{g}(\gamma),$$

where

$$B_\epsilon = \{\phi : \sup_{t \in [0,1]} |\phi(t) - \hat{\phi}_\gamma(t)| \geq \epsilon\}, \qquad F = \cup_{s \in [0,1]} F_s,$$

and

$$\begin{aligned}
F_s &= \Big\{ \phi \in \mathcal{AC}_0 : \int_{-\infty}^1 (1-\tau)^{(H-\frac{1}{2})} \phi'(\tau) d\tau \\
&\quad - \int_{-\infty}^s (s-\tau)^{(H-\frac{1}{2})} \phi'(\tau) d\tau - c(1-s) \geq \gamma \Big\}.
\end{aligned}$$

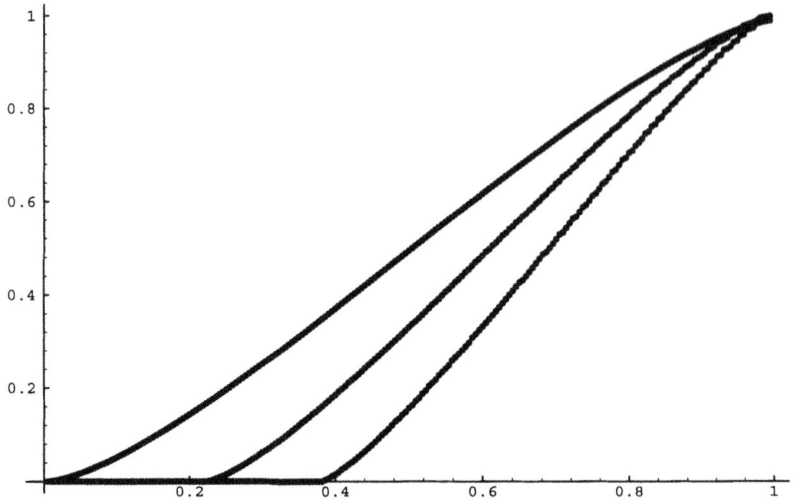

FIGURE 8.1. The most likely path: $\hat{\psi}_\gamma(t)$, with $c = 3$, $\gamma = 1$, and from left to right, $H = .75$, $H = .70$, $H = .65$.

Letting $\hat{\phi}(\cdot) = \phi(\cdot\ +1)$, we have

$$F_s = \{\hat{\phi} \in \mathcal{AC}_0 : \int_{-\infty}^{0} (-\tau)^{(H-\frac{1}{2})}\hat{\phi}'(\tau)d\tau - \int_{-\infty}^{s-1} (s-1-\tau)^{(H-\frac{1}{2})}\hat{\phi}'(\tau)d\tau - c(1-s) \geq \gamma\}\,.$$

If $\phi(\cdot) \in \tilde{F}_s$ (where $\tilde{F}_s$ was defined in the proof of Lemma 8.13), then

$$\hat{\phi}(\cdot) = \phi(\cdot\ -1) \in F_{1+s},$$

and

$$\int_{-\infty}^{0} (\phi'(\tau))^2 d\tau = \int_{-\infty}^{1} (\hat{\phi}'(\tau))^2 d\tau\,. \tag{8.19}$$

Likewise, if $\hat{\phi}(\cdot) \in F_s$, then

$$\phi(\cdot) = \hat{\phi}(\cdot\ +1) \in \tilde{F}_{s-1},$$

and (8.19) holds. From this it is clear that

$$\inf_{\phi\in F} \int_{-\infty}^{1} (\phi'(\tau))^2 d\tau = \int_{-\infty}^{1} (\hat{\phi}_\gamma'(\tau))^2 d\tau = \tilde{g}(\gamma)\,.$$

That

$$\inf_{\phi\in F\cap B_\epsilon} \int_{-\infty}^{1} (\phi'(\tau))^2 d\tau > \Lambda_V^*(\gamma)$$

holds true may be shown as in the proof of Theorem 8.10 (i).

For part (ii), make use of the continuity of the mapping as in the proof of Theorem 8.10 (ii). □

8.5 Verifying the Key Assumptions

Here we verify the assumptions outlined in Section 8.3 (preceding Theorem 8.7). First we show that the condition in (8.5) holds when X is the standard Brownian motion (Proposition 8.19). Next, when X is a sample-path process constructed from the sequence $\{a(t)\}$, we establish the condition, assuming that $\{a(t)\}$ admits an appropriate representation involving a martingale difference sequence (Lemma 8.20). With this representation, we also verify that for each n, $\{Y^{(n)}_{-j/n}\}_{j=1}^{\infty}$ is a Cauchy sequence in $L^2[\mathcal{C}[0,1]]$ (Proposition 8.22). Finally, in the last three propositions we revisit the examples of Section 8.2.

Proposition 8.19 When $X = B$, the standard Brownian motion, (8.5) holds for any $H \in (\frac{1}{2}, 1)$ and $\eta > 0$.

Proof. Since $B(nt)/\sqrt{n}$ is also a (standard) Brownian motion, we have

$$\begin{aligned} &\int_{-\infty}^{-m} [(t-s)^{H-\frac{1}{2}} - (-s)^{H-\frac{1}{2}}] n^{H-1} d\Big(\frac{B(ns)}{\sqrt{n}}\Big) \\ =_d\ &\int_{-\infty}^{-m} [(t-s)^{H-\frac{1}{2}} - (-s)^{H-\frac{1}{2}}] n^{H-1} dB(s)\,. \end{aligned}$$

By a result of Marcus and Shepp [21] (see also page 43 of Adler [1]) regarding the tail behavior of the distribution of the supremum of a Gaussian process, we have

$$\begin{aligned} \lim_{n\to\infty} \frac{1}{n^{2(1-H)}} \log \mathsf{P}\{ \sup_{t\in[0,1]} | \int_{-\infty}^{-m} &[(t-s)^{H-\frac{1}{2}} \\ -(-s)^{H-\frac{1}{2}}] n^{H-1} dB(s)| \geq \eta \} &= -(2\sigma_m^2)^{-1}\,, \end{aligned}$$

where

$$\begin{aligned} \sigma_m^2 &= \sup_{t\in[0,1]} \mathsf{E}[\int_{-\infty}^{-m} [(t-s)^{H-\frac{1}{2}} - (-s)^{H-\frac{1}{2}}] dB(s)]^2 \\ &= \sup_{t\in[0,1]} \int_{-\infty}^{-m} [(t-s)^{H-\frac{1}{2}} - (-s)^{H-\frac{1}{2}}]^2 ds\,. \end{aligned}$$

Since $\sigma_m \to 0$, the desired result follows. □

Next we consider the case of a sample-path process, i.e. $X = A^{(1)}$ in the notation of Section 8.2, which is constructed from a stationary sequence of bounded, mean zero, real-valued random variables, $\{a(t), t \in \mathbb{N}\}$.

By Taylor's theorem, we have, for any positive integers m and h,

$$\mathsf{P}(\sup_{t\in[0,1]} |\int_{-h}^{-m} [(t-s)^{H-\frac{1}{2}} - (-s)^{H-\frac{1}{2}}] d(n^{H-\frac{1}{2}} X^{(n)}(s))| \geq \eta)$$

$$\leq \mathsf{P}\Big(\sup_{t\in[0,1]} |\int_{-h}^{-m} (H-\frac{1}{2}) t(-s)^{H-\frac{3}{2}} d(n^{H-\frac{1}{2}} X^{(n)}(s))|$$

$$+ \sup_{t\in[0,1]} |\int_{-h}^{-m} \frac{1}{2}(H-\frac{1}{2})(H-\frac{3}{2})(t-s)^{H-\frac{5}{2}} d(n^{H-\frac{1}{2}} X^{(n)}(s))| \geq \eta\Big)$$

$$\leq \mathsf{P}\Big(\frac{1}{2} |\int_{-h}^{-m} (-s)^{H-\frac{3}{2}} d(n^{H-\frac{1}{2}} X^{(n)}(s))| \geq \eta/2\Big)$$

$$+ \mathsf{P}\Big(\frac{1}{8} \sup_{t\in[0,1]} |\int_{-h}^{-m} (t-s)^{H-\frac{5}{2}} d(n^{H-\frac{1}{2}} X^{(n)}(s))| \geq \eta/2\Big).$$

Hence, to establish the condition in (8.5), it suffices to show the following limits, for any $H \in (\frac{1}{2}, 1)$ and $\eta > 0$,

$$\lim_{m\to\infty} \limsup_{n\to\infty} \lim_{h\to\infty} \frac{1}{n^{2(1-H)}} \log \mathsf{P}\{ |\int_{-h}^{-m} (-s)^{H-\frac{3}{2}} d(n^{H-\frac{1}{2}} X^{(n)}(s))| \geq \eta\} = -\infty, \tag{8.20}$$

and

$$\lim_{m\to\infty} \limsup_{n\to\infty} \lim_{h\to\infty} \frac{1}{n^{2(1-H)}} \log \mathsf{P}\{ \sup_{t\in[0,1]} |\int_{-h}^{-m} (t-s)^{H-\frac{5}{2}} d(n^{H-\frac{1}{2}} X^{(n)}(s))| \geq \eta\} = -\infty. \tag{8.21}$$

To establish the limits in (8.20) and (8.21), we further assume that $a(-k)$ can be expressed in the following form:

$$a(-k) = D_k + Z_k - Z_{k+1}, \tag{8.22}$$

for $k = 0, 1, \ldots$, where $\{Z_k\}$ is a bounded sequence, and $\{D_k\}$ is a bounded martingale difference sequence. That is, D_k is $\mathcal{F}_j$-measurable for $k \leq j$, where $\{\mathcal{F}_j\}$ is some filtration, $\mathsf{E}(D_j|\mathcal{F}_{j-1}) = 0$ a.s., and $|D_j| \leq C$ for some constant $C < \infty$.

We shall find useful the following inequality for martingale differences, due originally to Azuma (following Rhee and Talagrand [24], where Stout [30] is referred to for a proof). In our context, the inequality takes the following form:

Lemma 8.20 Suppose $\{D_j;\, j \geq 0\}$ is a martingale difference sequence, and $|D_j| \leq d_j$ for each j, where d_j is a constant. Then, for any $z > 0$, we have

$$\mathsf{P}(|\sum_{j=0}^{n} D_j| \geq z) \leq 2\exp(-\frac{z^2}{2\sum_{j=0}^{n} d_j^2}).$$

Lemma 8.21 Suppose $a(-k) = D_k + Z_k - Z_{k+1}$, for $k = 0, 1, \ldots$, with $\{Z_k\}$ a bounded sequence, and $\{D_k\}$ a bounded martingale difference sequence. Then, (8.20) and (8.21) hold for any $H \in (\frac{1}{2}, 1)$ and $\eta > 0$.

Proof. We first establish the limit in (8.20). With $h = \ell/n + m$, $\ell \geq nm$ a positive integer, we have

$$\begin{aligned}
&\int_{-h}^{-m} (-s)^{H-\frac{3}{2}} d(n^{H-\frac{1}{2}} X^{(n)}(s)) \\
&= \sum_{j=nm}^{\ell+nm} j^{H-\frac{3}{2}} a(-j) \\
&= \sum_{j=nm}^{\ell+nm} j^{H-\frac{3}{2}} D_j + \sum_{j=nm}^{\ell+nm} [(j+1)^{H-\frac{3}{2}} - j^{H-\frac{3}{2}})] Z_{j+1} \\
&\qquad + (nm)^{H-\frac{3}{2}} Z_{nm} - (\ell + nm + 1)^{H-\frac{3}{2}} Z_{\ell+nm+1}\,.
\end{aligned}$$

Since Z_k are bounded, the second, third and fourth terms in the last expression above can be made less than $\eta/2$ by choosing m large enough. It suffices, therefore, to consider only the first term. Making use of Lemma 8.20, with $\{j^{H-\frac{3}{2}} D_j\}$ the martingale difference, and $d_j = j^{H-\frac{3}{2}} C$ (recall $C \geq |D_j|$), we have

$$\begin{aligned}
&\mathsf{P}(|\sum_{j=nm}^{l} j^{H-\frac{3}{2}} D_j| \geq \eta/2) \\
&\leq 2\exp\left[-\frac{\eta^2}{8C^2 \sum_{j=nm}^{\ell+nm} j^{2H-3}}\right] \\
&\leq 2\exp\left[-\frac{n^{2-2H}\eta^2}{C_1 m^{2H-2}}\right],
\end{aligned}$$

where C_1 is some positive constant. Therefore

$$\begin{aligned}
&\limsup_{n\to\infty} \lim_{h\to\infty} \frac{1}{n^{2(1-H)}} \log \mathsf{P}\{|\int_{-h}^{-m} (-s)^{H-\frac{3}{2}} d(n^{H-\frac{1}{2}} X^{(n)}(s))| \\
&\qquad \geq \eta\} \leq -\frac{\eta^2}{C_1} m^{2-2H}\,.
\end{aligned}$$

Letting $m \to \infty$ yields the desired limit in (8.20).

Next, we prove (8.21). As before, with $h = \ell/n + m$, we have

$$\begin{aligned}
&\int_{-h}^{-m} (t-s)^{H-5/2} d(n^{H-\frac{1}{2}} X^{(n)}(s)) \\
&= n \sum_{j=nm}^{\ell+nm} (j+tn)^{H-5/2} a(-j) \\
&= n \sum_{j=nm}^{\ell+nm} (j+tn)^{H-5/2} D_j \\
&\quad + n \sum_{j=nm}^{\ell+nm} [(j+tn+1)^{H-5/2} - (j+tn)^{H-5/2}] Z_{j+1} \\
&\quad + n(nm+tn)^{H-5/2} Z_{nm} - n(\ell+nm+1+tn)^{H-5/2} Z_{\ell+nm+1} \,.
\end{aligned}$$

Again, we only need to bound the first term above, which we denote as

$$S_n(t) = n \sum_{j=nm}^{\ell+nm} (j+tn)^{H-5/2} D_j.$$

Note that

$$\begin{aligned}
&\mathsf{P}(\sup_{0\le t\le 1} |S_n(t)| \ge \eta) \\
&\le \sum_{i=1}^{n} \mathsf{P}(\sup_{\frac{i-1}{n}\le t\le \frac{i}{n}} |S_n(t)| \ge \eta) \\
&\le n \max_{1\le i\le n} \mathsf{P}(\sup_{\frac{i-1}{n}\le t\le \frac{i}{n}} |S_n(t)| \ge \eta).
\end{aligned} \tag{8.23}$$

Furthermore,

$$\begin{aligned}
&\sup_{\frac{i-1}{n}\le t\le \frac{i}{n}} |S_n(t)| \\
&\le |S_n(\frac{i-1}{n})| + \sup_{\frac{i-1}{n}\le t\le \frac{i}{n}} |S_n(t) - S_n(\frac{i-1}{n})|,
\end{aligned} \tag{8.24}$$

while

$$\begin{aligned}
&|S_n(t) - S_n(\frac{i-1}{n})| \\
&= n \Big| \sum_{j=nm}^{\ell+nm} [(j+tn)^{H-5/2} - (j+i-1)^{H-5/2}) D_j \Big| \\
&\le nC \sum_{j=nm}^{\ell+nm} [(j+i-1)^{H-5/2} - (j+i)^{H-5/2}]
\end{aligned}$$

$$\begin{aligned}
&\leq\ nC\sum_{j=nm}^{\infty}(5/2-H)(j+i)^{H-7/2}\\
&\leq\ nC_2(nm)^{H-5/2},
\end{aligned}$$

for some positive constant C_2. Hence, choosing m sufficiently large, the second term on the right hand side of (8.24) can be bounded by $\eta/2$. Applying Lemma 8.20 once more, we have

$$\begin{aligned}
&\mathsf{P}(\sup_{\frac{i-1}{n}\leq t\leq\frac{i}{n}}|S_n(t)|\geq\eta)\\
\leq\ &\mathsf{P}[|S_n(\frac{i-1}{n})|\geq\eta/2]\\
=\ &\mathsf{P}[\sum_{j=nm}^{\ell+nm}(j+i-1)^{H-5/2}D_j\geq\eta/(2n)]\\
\leq\ &2\exp\Big[-\frac{\eta^2}{8n^2C^2\sum_{j=nm}^{\ell+nm}(j+i-1)^{2H-5}}\Big]\\
\leq\ &2\exp\Big[-\frac{\eta^2}{C_3n^2(nm)^{2H-4}}\Big].
\end{aligned}$$

for some positive constant C_3. Putting the above together with (8.23), we have

$$\begin{aligned}
&\limsup_{n\to\infty}\lim_{h\to\infty}\frac{1}{n^{2(1-H)}}\log\mathsf{P}(\sup_{0\leq t\leq 1}|S_n(t)|\geq\eta)\\
\leq\ &\limsup_{n\to\infty}[\frac{\log n}{n^{2(1-H)}}]-\frac{\eta^2}{C_3m^{2H-4}}\\
=\ &-\frac{\eta^2}{C_3}m^{4-2H}.
\end{aligned}$$

Letting $m\to\infty$ completes the argument. □

Proposition 8.22 For each positive integer n, the sequence $\{Y^{(n)}_{-j/n}\}_{j=1}^{\infty}$ is a Cauchy sequence in $L^2[\mathcal{C}[0,1]]$.

Proof. For each n, we must show the following:

$$\begin{aligned}
&\lim_{\ell,m\to\infty}\mathsf{E}\Big[\sup_{t\in[0,1]}|Y^{(n)}_{-\ell/n}(t)-Y^{(n)}_{-m/n}(t)|^2\Big]\\
=\ &\lim_{\ell,m\to\infty}\mathsf{E}\Big[\sup_{t\in[0,1]}\Big|\int_{-\ell/n}^{-m/n}[(t-s)^{H-\frac{1}{2}}\\
&\qquad -(-s)^{H-\frac{1}{2}}]d(n^{H-\frac{1}{2}}X^{(n)}(s))\Big|^2\Big]\\
=\ &0\,.
\end{aligned}$$

Applying Taylor's theorem as we did at the beginning of this section, we see that it is sufficient to show (keeping in mind that n is fixed):

$$\lim_{\ell,m\to\infty} \mathsf{E}(|\int_{-\ell/n}^{-m/n} (-s)^{H-\frac{3}{2}} d(X^{(n)}(s))|^2) = 0, \tag{8.25}$$

and

$$\lim_{\ell,m\to\infty} \mathsf{E}(\sup_{t\in[0,1]} |\int_{-\ell/n}^{-m/n} (t-s)^{H-5/2} d(X^{(n)}(s))|^2) = 0\,. \tag{8.26}$$

To see that (8.26) holds, observe that

$$\begin{aligned}
&\int_{-\ell/n}^{-m/n} (t-s)^{H-5/2} d(X^{(n)}(s)) \\
&= \frac{1}{n}\sum_{j=-\ell}^{-m} (t-j/n)^{H-5/2} a(j) \\
&= \frac{1}{n}\sum_{j=\ell}^{m} (t+j/n)^{H-5/2} a(-j).
\end{aligned}$$

Since $\{a(t), t\in \mathbb{N}\}$ is bounded and n is fixed, the limit in (8.26) is indeed zero.

To see that (8.25) holds, write

$$\begin{aligned}
&\mathsf{E}\Big[|\int_{-\ell/n}^{-m/n} (-s)^{H-\frac{3}{2}} d(X^{(n)}(s))|^2\Big] \\
&= \frac{1}{n}\mathsf{E}\Big[|\sum_{j=-\ell}^{-m} (-j/n)^{H-\frac{3}{2}} (D_j + Z_j - Z_{j+1})|^2\Big]\,.
\end{aligned}$$

As in the proof of Lemma 8.21, we may ignore the Z terms. For the remaining term, since $\{D_j\}$ is a martingale difference sequence, we have

$$\mathsf{E}(|\sum_{j=-\ell}^{-m} (-j/n)^{H-\frac{3}{2}} D_j|^2) = \sum_{j=-\ell}^{-m} (-j/n)^{2H-3} \mathsf{E}[D_j^2]\,.$$

From this, (8.25) readily follows (taking into account that D_j is bounded). □

Proposition 8.23 Let $\{a(t), t\in\mathbb{N}\}$ be the sequence defined in Example 8.4 of Section 8.2. Suppose that the reversed process, $\{b(-t),\ t = 0,1,2,\ldots\}$, corresponds to a Markov chain with kernel P such that the Poisson equation, $Ph - h = f$, admits a bounded solution h. (Without loss of generality, assume $\mathsf{E}[f(b(t))] = 0$.) Then, (8.5) holds for any $H \in (\frac{1}{2}, 1)$ and $\eta > 0$.

Proof. Write

$$\begin{aligned} & f(b(-t)) \\ = \ & Ph(b(-t)) - h(b(-t)) \\ = \ & [Ph(b(-t)) - h(b(-t-1))] - h(b(-t)) + h(b(-t-1)). \quad (8.27) \end{aligned}$$

Note that $\{Ph(b(-t)) - h(b(-t-1))\}$ is a martingale difference sequence with respect to the filtration $\mathcal{F}_t = \sigma\{b(-t), b(-t+1), \ldots\}$. We may now identify D_k and Z_k with $\{Ph(b(-k)) - h(b(-k-1))\}$ and $-h(b(-k))$, respectively. □

Proposition 8.24 Let $\{a(t), t \in \mathbb{N}\}$ be the moving-average sequence defined in Example 8.5 of Section 8.2. Assume that

$$\sum_{k=1}^{\infty}\{\sum_{j=-\infty}^{-k} |b(j)| + \sum_{j=k}^{\infty} |b(j)|\} < \infty\,. \tag{8.28}$$

Then, (8.5) holds for any $H \in (\frac{1}{2}, 1)$ and $\eta > 0$.

Proof. Let $\mathcal{F}_k = \sigma\{\xi(j),\ j = -k, -k+1, \ldots\}$. Then, it can be directly verified that the given condition on $b(j)$ leads to the following sequences

$$D_k = \sum_{i=-\infty}^{+\infty} \{\mathsf{E}[a(-i)|\mathcal{F}_k] - \mathsf{E}[a(-i)|\mathcal{F}_{k-1}]\},$$

and

$$Z_k = \sum_{i=0}^{+\infty} \mathsf{E}[a(-i)|\mathcal{F}_{k-1}] - \sum_{i=-\infty}^{-1} [a(-i-k) - \mathsf{E}(a(-i)|\mathcal{F}_{k-1})],$$

$(k = 0, 1, \ldots,)$ being well defined and bounded. Furthermore, it readily follows that D_k is a martingale difference sequence with respect to the filtration $\mathcal{F}_k$ and that the representation in (8.22) holds. The desired conclusion then follows from Lemma 8.21. □

Remark 8.25 Note that the condition in (8.28) also plays a key role in [16] to imply that the infinite moving-average sequence satisfies the functional central limit theorem and law of iterated logarithm.

Proposition 8.26 Let $\{a(t), t \in \mathbb{N}\}$ be as in Example 8.6 of Section 8.2. Assume that the corresponding reversed sequence is ϕ-mixing and $\sum_{k=1}^{\infty} \phi(k) < \infty$. We have that (8.5) holds for any $H \in (\frac{1}{2}, 1)$ and $\eta > 0$.

Proof. Let $\mathcal{F}_k = \sigma\{a(k),\ j = -k, -k+1, \ldots\}$. It can be directly verified (cf. [12]), under the assumed conditions, that

$$D_k = \sum_{i=0}^{+\infty} \{\mathsf{E}[a(-i-k)|\mathcal{F}_k] - \mathsf{E}[a(-i-k)|\mathcal{F}_{k-1}]\},$$

for $k = 0, 1, \ldots$, is a bounded martingale difference sequence, that

$$Z_k = \sum_{i=0}^{+\infty} \mathsf{E}[a(-i-k)|\mathcal{F}_{k-1}],$$

is bounded, and that the relation in (8.22) is satisfied. That (8.5) holds then follows from Lemma 8.21. □

8.6 Notes

Mandelbrot and van Ness [20] is a classical reference for fractional Brownian motion (FBM). Refer to Cox [8] and Beran [3] for preliminaries regarding processes with long-range dependence, and self-similarity. Samorodnitsky and Taqqu [26] provides an extensive treatment of stable processes, which extend FBM to outside of the Gaussian family.

Queues with renewal or Markov arrival processes — inputs that have short-range dependence — exhibit exponential tail behavior (in terms of work-load or delay). This characteristic is usually established via large deviations techniques; refer to, e.g., Chang [5, 6], Glynn and Whitt [14, 15], Whitt [31], and the references there.

In contrast, queues with long-range dependent inputs have subexponential tails, as noted in Norros [23], Duffield and O'Connell [11], and [7], among others. Interests in studying queues with long-range dependent inputs are largely motivated by studies of telecommunications traffic that reveal long-range dependence in the empirical data, typically accompanied by self-similarity. Refer to Leland *et al* [18] and Willinger [32], among others.

For further readings on large deviations theory, in particular materials that relate to Section 8.2, we refer to Dembo and Zeitouni [10], and Schwartz and Weiss [28].

Azuma's inequality in Lemma 8.20 is useful in handling martingale differences. For other applications of this inequality, in addition to Rhee and Talagrand [24] mentioned earlier, also refer to Ross [25] (Chapter 6) and Glasserman and Yao [13] (Chapter 7).

Acknowledgments

Thanks are due to Tom Kurtz for suggesting martingale techniques to verify (8.5) for sample-path processes. Cheng-Shang Chang's research is supported in part by the National Science Council, Taiwan, R.O.C., under the contract NSC 85-2121-M007-035. David Yao acknowledges support by NSF Grant ECS-97-05392. Part of Tim Zajic's work was supported by NSF Grant DMS-9508709 while he was a postdoctoral fellow at IBM T.J. Watson Research Center and the Department of Industrial Engineering and Operations Research, Columbia University.

8.7 References

[1] Adler, R., *An Introduction to Continuity, Extrema, and Related Topics for General Gaussian Processes.* Institute of Mathematical Statistics Lecture Notes-Monograph Series, **12** (1990), Hayward, California.

[2] Anantharam, V., How Large Delays Build up in a GI/G/1 Queue. *Queueing Systems,* **5** (1989), 345-368.

[3] Beran, J., *Statistics for Long-Memory Processes.* Chapman and Hall, New York, 1994.

[4] Bradley, R.C., Basic Properties of Strong Mixing Conditions. In: *Dependence in Probability and Statistics,* E. Eberlein and M. Taqqu (eds.), Birkhäuser, Basel, Switzerland, 1986, 165-192.

[5] Chang, C.S., Stability, Queue Length and Delay of Deterministic and Stochastic Queueing Networks. *IEEE Transactions on Automatic Control,* **39** (1994), 913-931.

[6] Chang, C.S., Sample Path Large Deviations and Intree Networks. *Queueing Systems,* **20** (1995), 7-36.

[7] Chang, C.S., Yao, D.D., and Zajic, T., Large Deviations, Moderate Deviations, and Queues with Long-Range Dependent Inputs. *Adv. Appl. Prob.,* **31** (1999), 254-277.

[8] Cox, D.R., Long-Range Dependence: a Review. In: *Statistics: An Appraisal,* H.A. David and H.T. David (eds.) Iowa State University Press, Ames, Iowa, 55-74, 1984.

[9] Dembo, A. and Zajic, T., Large Deviations from Empirical Mean and Measure to Partial Sums Processes. *Stoch. Proc. and Appl.,* **57** (1995), 191-224.

[10] DEMBO A. AND ZEITOUNI, O., *Large Deviations Techniques and Applications.* Jones and Barlett Publishers, Boston, 1992.

[11] DUFFIELD N.G. AND O'CONNELL, N., Large Deviations and Overflow Probabilities for the General Single-Server Queue, with Applications. *Mathematical Proc. of the Cambridge Phil. Soc.*, **118** (1995), 363-374.

[12] GAO, F.Q., Moderate Deviations for Martingales and Mixing Random Processes. *Stoch. Proc. and Appl.*, **61** (1996), 263-275.

[13] GLASSERMAN P. AND YAO, D.D., *Monotone Structure in Discrete-Event Systems.* Wiley, New York, 1994.

[14] GLYNN P.W. AND W. WHITT, Logarithmic Asymptotics for Steady-State Tail Probabilities in a Single-Server Queue. *J. Appl. Prob*, **31A** (1994), 131-156.

[15] GLYNN P.W. AND WHITT, W., Large Deviations Behavior of Counting Processes and Their Inverses. *Queueing Systems*, **17** (1994), 107-128.

[16] HALL P. AND HEYDE, C., *Martingale Limit Theory and Its Application.* Academic Press, New York, 1980.

[17] HARRISON, J.M., *Brownian Motion and Stochastic Flow Systems.* Wiley, New York, 1985.

[18] LELAND, W.E., TAQQU, M.S., WILLINGER, W., AND WILSON, D.V., On the Self-Similar Nature of Ethernet Traffic. *IEEE/ACM Trans. on Networking*, **2** (1953), 1-15.

[19] LÉVY, P., Random Functions: General Theory with Special Reference to Laplacian Random Functions. *Univ. California Publ. Statist.*, **1** (1953), 331-390.

[20] MANDELBROT, B.B. AND VAN NESS, J., Fractional Brownian Motions, Fractional Noises and Applications. *SIAM Review*, **10** (1968), 422-437.

[21] MARCUS, M.B. AND SHEPP, L.A., Sample Behaviour of Gaussian Processes. *Proc. Sixth Berkeley Symp. Math. Statist. Prob.*, **2** (1971), 423-442.

[22] MOGULSKII, A.A., Large Deviations for Trajectories of Multidimensional Random Walk. *Th. Prob. Appl.*, **21** (1976), 300-315.

[23] NORROS, I., A Storage Model with Self-Similar Input, *Queueing Systems.* **16** (1994), 387-396.

[24] RHEE W. AND TALAGRAND, M., Martingale Inequalities and NP-complete Problems. *Math. Oper. Res.*, **12** (1987), 177-181.

[25] ROSS, S.M., *Stochastic Processes.* Wiley, New York, 2nd ed, 1996.

[26] SAMORODNITSKY G., AND TAQQU, M.S., *Stable Non-Gaussian Random Processes.* Chapman & Hall, New York, 1994.

[27] SCHILDER, M., Some Asymptotic Formulae for Wiener Integrals. *Trans. Amer. Math. Soc.*, **125** (1966), 63-85.

[28] SHWARTZ, A. AND WEISS, A., *Large Deviations for Performance Analysis.* Chapman & Hall, London, 1995.

[29] SIGMAN, K. AND YAO, D.D., Finite Moments for Inventory Processes. *Ann. of Appl. Prob.*, **4** (1994), 765-778.

[30] STOUT, W.F., *Almost Sure Convergence.* Academic Press, New York, 1974.

[31] WHITT, W., Tail Probability with Statistical Multiplexing and Effective Bandwidths in Multi-Class Queues. *Telecommunication Systems*, **2** (1993), 71-107.

[32] Willinger, W., Traffic modeling for high-speed networks: theory versus practice. In: *Stochastic Networks,* F. Kelly and R. Williams (eds.), IMA Volume in Mathematics and Its Applications. Springer-Verlag, New York, 1994.

[33] WONG E. AND HAJEK, B., *Stochastic Processes in Engineering Systems.* Springer-Verlag, New York, 1984.

[34] WU, L., Moderate Deviations of Dependent Random Variables Related to CLT. *Ann. of Prob.*, **23** (1995), 420-445.

9

Markowitz's World in Continuous Time, and Beyond

Xun Yu Zhou

Mean–variance portfolio selection is concerned with the allocation of wealth among a basket of securities so as to achieve a prescribed trade-off between the return of the investment and the associated risk. The model was first proposed and solved in the single-period setting by Harry Markowitz in his Novel-Prize winning work [38] (see also [39]). The most important contribution of Markowitz's work is the introduction of quantitative and scientific approaches to risk analysis or risk management. This work has become the foundation of modern finance and has had tremendous impact on its further development. For example, in Markowitz's world (i.e., the world where all the investors believes in the mean–variance theory), one of the important consequences of Markowitz's results is the so-called mutual fund theorem, which asserts that two mutual funds (which are two efficient portfolios) can be established so that all investors will be indifferent in making portfolios from among the original set of assets or from these two funds. Moreover, if a risk-free asset (such as a back account) is available, then investors need only to make use of one specific risky fund – called the *tangent fund* – and the risk-free asset. Assuming that portfolios are not constrained, a logical consequence of this single-period mutual fund theorem is that the market portfolio turns out to be the tangent fund, which therefore must be efficient. This, in turn, leads to the elegant *capital asset pricing model* (CAPM).

The purpose of this chapter is to report the latest progress in studying the continuous-time mean–variance portfolio selection and some important related problems using the framework of stochastic linear–quadratic (LQ) control. In the LQ formulation, the dollar amounts in different assets, rather than the weights, are used to define a portfolio – or the control in the control terminology. This leads to a dynamic system that is linear in *both* the state (the wealth) and the control variables. Together with the quadratic form of the objective function, this formulation falls naturally into the realm of stochastic LQ control. Moreover, since there is no running cost in the objective function, the resulting problem is *inherently* an *indefinite stochastic LQ control problem*, the theory of which has been developed only very recently.

It is important to recognize that there are many advantages of using the framework of stochastic LQ control to study dynamic mean–variance portfolio selection. First, powerful results from the theory of stochastic control can be used to gain further insights into this problem. Recent results on the optimal control of indefinite/singular stochastic LQ problems mentioned above reveals the crucial role of uncertainty in stochastic control. This, in turn, has important implications in finance applications. Secondly, stochastic LQ control provides a uniform and general framework to deal with complicated situations such as those with additional constraints and random market parameters, as will be reported in this chapter. This is because constraints on the state and/or control variable and randomness in the parameters are part of the natural language of stochastic control; such considerations in turn have important interpretations in the setting of finance. Most importantly, this approach opens up new avenues in solving other mean–variance type problems.

In this chapter, we shall report results for five models which were studied in [60, 34, 61, 32, 28], respectively. The first four models are mean–variance portfolio selection problems with deterministic coefficients, with random coefficients that are adapted to the stocks, with Markov-modulated random coefficients that are independent of the stocks, and with short-selling prohibition, respectively. The last one is a mean–variance hedging problem of a given contingent claim. In all the models, explicit forms of efficient/optimal portfolios and efficient frontiers are presented. While many results in the continuous-time Markowitz world are analogous to their single-period counterparts, there are some results that are strikingly different. These will be highlighted and discussed. A number of examples are given to illustrate the similarities as well as differences between the continuous-time and single-period settings.

9.1 The Mean–Variance Portfolio Selection Model

Throughout this chapter, unless otherwise specified, T is a fixed terminal time and $(\Omega, \mathcal{F}, \mathsf{P}, \{\mathcal{F}_t\}_{t\geq 0})$ is a fixed filtered complete probability space on which defined a standard $\mathcal{F}_t$-adapted m-dimensional Brownian motion $W(t) \equiv (W^1(t), \cdots, W^m(t))'$ with $W(0) = 0$. Specific assumptions on this basic framework will be made in individual sections.

Notation. We use the following notation:

$\Re^m$	:	the space of m-dimensional column vectors;
M'	:	the transpose of a vector or matrix M;
$\lvert M \rvert$	:	$=\sqrt{\sum_{i,j} m_{ij}^2}$ for a matrix or vector $M = (m_{ij})$;
σ_ξ	:	the standard deviation of a random variable ξ;
$C(0,T;\Re^d)$	:	the Banach space of $\Re^d$-valued continuous function on $[0,T]$ endowed with the maximum norm;
$L^2_{\mathcal{F}}(0,T;\Re^d)$	:	the set of all $\Re^d$-valued, measurable stochastic processes $f(t)$ adapted to $\mathcal{F}_t$, such that $\mathsf{E}\int_0^T \lvert f(t)\rvert^2 dt < +\infty$;
$L^\infty_{\mathcal{F}}(\Omega, C(0,T;\Re^d))$	:	the set of all $\Re^d$-valued, continuous stochastic processes $f(t)$ adapted to $\mathcal{F}_t$, such that $\sup_{t\in[0,T],\omega\in\Omega} \lvert f(t,\omega)\rvert < +\infty$;
$L^2_{\mathcal{F}_T}(\Omega;\Re^d)$	:	the set of all $\Re^d$-valued, $\mathcal{F}_T$-measurable random variables η such that $\mathsf{E}\lvert\eta\rvert^2 < +\infty$.

Suppose there is a market in which $m+1$ *assets* (or *securities*) are traded continuously. One of the assets is a *bank account* whose value $P_0(t)$ is subject to the following (stochastic) ordinary differential equation (ODE):

$$\begin{cases} dP_0(t) = r(t)P_0(t)dt, \quad t \in [0,T], \\ P_0(0) = p_0 > 0, \end{cases} \tag{9.1}$$

where the *interest rate* $r(t) \geq 0$ is a uniformly bounded, $\mathcal{F}_t$-adapted, scalar-valued stochastic process. The other m assets are *stocks* whose price processes $P_1(t), \cdots, P_m(t)$ satisfy the following stochastic differential equation (SDE):

$$\begin{cases} dP_i(t) = P_i(t)\big[b_i(t)dt + \sum_{j=1}^m \sigma_{ij}(t)dW^j(t)\big], \quad t \in [0,T], \\ P_i(0) = p_i > 0, \end{cases} \tag{9.2}$$

where $b_i(t)$ and $\sigma_i(t) := (\sigma_{i1}(t), \cdots, \sigma_{im}(t))$ are the *appreciation rate* and *dispersion* (or *volatility*) *rate* of the ith stock, respectively, $i = 1, 2, \cdots, m$.

Once again, we assume that $b_i(t)$ and $\sigma_{ij}(t)$ are scalar-valued, $\mathcal{F}_t$-adapted, uniformly bounded stochastic processes.

Define the *volatility matrix* $\sigma(t) := (\sigma_{ij}(t))_{m\times m}$. The basic assumption throughout this chapter is

$$\sigma(t)\sigma(t)' \geq \delta I, \quad \forall t \in [0,T], \quad \mathsf{P}-\text{a.s.}, \tag{9.3}$$

for some $\delta > 0$.

Consider an agent whose total wealth at time $t \geq 0$ is denoted by $x(t)$. Assume that the trading of shares takes place continuously and transaction cost and consumptions are not considered. Then $x(\cdot)$ satisfies (see, e.g., Karatzas and Shreve [27] and Elliott and Kopp [15])

$$\begin{cases} dx(t) & = \Big\{r(t)x(t) + \sum_{i=1}^m \big[b_i(t) - r(t)\big]\pi_i(t)\Big\}dt \\ & \quad + \sum_{j=1}^m \sum_{i=1}^m \sigma_{ij}(t)\pi_i(t)dW^j(t), \\ x(0) & = x_0 \geq 0, \end{cases} \tag{9.4}$$

where $\pi_i(t), \;\; i = 0,1,2\cdots,m$, denotes the total market value of the agent's wealth in the ith asset at time t (in particular, $\pi_0(t)$ is the wealth in the bank account at t). We call $\pi(\cdot) := (\pi_1(\cdot),\cdots,\pi_m(\cdot))'$ a *portfolio* of the agent.

Definition 9.1 A portfolio $\pi(\cdot)$ is called *admissible* if $\pi(\cdot) \in L^2_{\mathcal{F}}(0,T;\Re^m)$, and the SDE (9.4) has a unique solution $x(\cdot)$ corresponding to $\pi(\cdot)$. In this case, we refer to $(x(\cdot),\, \pi(\cdot))$ as an *admissible (wealth–portfolio) pair.*

Note that with an admissible $\pi(\cdot)$ the corresponding wealth trajectory, $x(\cdot)$, is completely determined. As a result, $\pi_0(\cdot)$ is derived as $\pi_0(t) = x(t) - \sum_{i=1}^m \pi_i(t)$. This is the reason why we should not include $\pi_0(\cdot)$ in defining a portfolio $\pi(\cdot)$.

The agent's objective is to find an admissible portfolio $\pi(\cdot)$, among all such admissible portfolios that their expected terminal wealth $\mathsf{E}x(T) = d$, for some given $d \in \Re$, and any other prescribed constraints are satisfied (which may differ from case to case), so that the risk measured by the variance of the terminal wealth

$$\mathsf{Var}\; x(T) := \mathsf{E}[x(T) - \mathsf{E}x(T)]^2 = \mathsf{E}[x(T) - d]^2 \tag{9.5}$$

is minimized. The problem of finding such a portfolio $u(\cdot)$ is referred to as the *mean–variance portfolio selection problem.* In words, the solution of a mean–variance portfolio selection problem minimizes the terminal risk while satisfying the targeted mean payoff at the terminal time (along with other specific requirements such as short selling or bankruptcy prohibition). Mathematically, we have the following formulation.

Definition 9.2 The mean–variance portfolio selection problem is formulated as a constrained stochastic optimization problem, parameterized by $d \in \Re$:

$$\begin{cases} \text{minimize} & J_{\mathrm{MV}}(x_0, \pi(\cdot)) := \mathsf{E}[x(T) - d]^2, \\ \text{subject to} & \begin{cases} \mathsf{E}x(T) = d, \\ (x(\cdot), \pi(\cdot)) \ \text{admissible}, \\ \text{any other prescribed constraint(s)}. \end{cases} \end{cases} \tag{9.6}$$

Moreover, the problem is called *feasible* if there is at least one portfolio satisfying all the constraints. The problem is called *finite* if it is feasible and the infimum of $J_{\mathrm{MV}}(x_0, \pi(\cdot))$ is finite. Finally, an optimal portfolio to the above problem, if it ever exists, is called an *efficient portfolio* corresponding to d, the corresponding $(\mathsf{Var}\ x(T), d) \in \Re^2$ and $(\sigma_{x(T)}, d) \in \Re^2$ are interchangeably called an *efficient point*, where $\sigma_{x(T)}$ denotes the standard deviation of $x(T)$. The set of all the efficient points is called the *efficient frontier*.

Note that while in the above definition an efficient portfolio is broadly defined for any given $d \in \Re$, in the subsequent context we will see that it is practically sensible only for d greater or equal to a certain value. Also, the shape of the efficient frontier depends on whether it is plotted in the variance–return plane or standard-deviation–return plane. In the sequel, we will specify which one we are referring to only when ambiguity might arise. Finally, in the model (9.6) the "any other prescribed constraint(s)" can differ depending on the specified problem. For example, if short-selling of the stocks is not allowed, then the additional constraints are

$$\pi_i(t) \geq 0, \quad \forall t \in [0, T], \quad \mathsf{P} - \text{a.s.}, \quad i = 1, 2, \cdots, m;$$

see Section 9.6. If bankruptcy of the agent is prohibited, then the constraint that $x(t) \geq 0\ \forall t \in [0, T]$, P – a.s., should be imposed.

9.2 A Stochastic LQ Control Approach

The mean–variance problem formulated in the previous section will be solved via a stochastic LQ optimal control technique. In this section, we outline the LQ approach to solving (9.6). However, we shall do this for the model (9.6) where the generic "any other prescribed constraint(s)" are not present (these have to be dealt with using an *ad hoc* approach depending on the nature of the particular constraints; see Section 9.6 for an example).

First of all, equation (9.4) can be rewritten as

$$\begin{cases} dx(t) = [r(t)x(t) + B(t)\pi(t)]dt + \pi(t)'\sigma(t)dW(t), \\ x(0) = x_0, \end{cases} \tag{9.7}$$

where

$$B(t) := (b_1(t) - r(t), \ldots, b_m(t) - r(t)). \tag{9.8}$$

The mean–variance problem (9.6) under consideration is a dynamic optimization problem with a constraint $\mathsf{E}x(T) = d$. To handle this constraint, we apply the Lagrange multiplier technique. Define

$$\begin{aligned} J(x_0, \pi(\cdot), \lambda) := &\ \mathsf{E}\{|x(T) - d|^2 + 2\lambda[x(T) - d]\} \\ = &\ \mathsf{E}\big[x(T)^2 + 2(\lambda - d)x(T) + d^2 - 2\lambda d\big], \quad \lambda \in \Re. \end{aligned} \tag{9.9}$$

Since $J_{\mathrm{MV}}(x_0, \pi(\cdot))$ is strictly convex in $\pi(\cdot)$ and the constraint function $\mathsf{E}x(T) - d$ is affine in $\pi(\cdot)$, we can apply the well-known duality theorem (see, e.g., [36]) to conclude that if problem (9.6) is finite for *every* $d \in \Re$, then the optimal value is

$$\begin{aligned} J^*_{\mathrm{MV}}(x_0) := &\ \inf_{\text{all constraints}} J_{\mathrm{MV}}(x_0, \pi(\cdot)) \\ = &\ \sup_{\lambda \in \Re} \inf_{\pi(\cdot) \text{ admissible}} J(x_0, \pi(\cdot), \lambda). \end{aligned} \tag{9.10}$$

Based on this, we can solve the original problem (9.6) in three steps (for details see [34]):

Step 1. For each fixed λ, solve the problem $\inf_{\pi(\cdot) \text{ admissible}} J(x_0, \pi(\cdot), \lambda)$. This is an *unconstrained* stochastic optimal control problem where the dynamics (9.7) is linear and the cost functional (9.9) is quadratic. It is a special case of a more general class of problems, namely the indefinite stochastic LQ control problems[1]. Solving this problem leads to the optimal portfolio $\pi^*_\lambda(\cdot)$ and the optimal value $J(x_0, \pi^*_\lambda(\cdot), \lambda)$.

Step 2. Solve the problem $\sup_{\lambda \in \Re} J(x_0, \pi^*_\lambda(\cdot), \lambda)$ to get the *unique* optimal λ^* and the corresponding value $J(x_0, \pi^*_{\lambda^*}(\cdot), \lambda^*)$. This problem is easy since it is an unconstrained optimization problem in one-dimensional space.

---

[1]Solution to this problem depends crucially on the underlying filtration, $\{\mathcal{F}_t\}_{t\geq 0}$, when the market parameters are random. In particular, if $\mathcal{F}_t$ is generated by the Brownian motion (which means in financial terms that the market parameters are adapted to the stocks), then the problem can be solved via a *stochastic Riccati equation* (*SRE*), which is in general a nonlinear backward stochastic differential equation (BSDE); see Section 9.4 for such a situation. If $\mathcal{F}_t$ is strictly larger than the filtration generated by the Brownian motion (i.e., there is an additional dimension of uncertainty to that of the stocks), then the situation is more complicated. One of such cases is presented in Section 9.5.

Step 3. The unique optimal portfolio for (9.6) is then $\pi^*_{\lambda^*}(\cdot)$, and the optimal value is $J(x_0, \pi^*_{\lambda^*}(\cdot), \lambda^*)$.

9.3 Efficient Frontier: Deterministic Market Parameters

In this section, we assume that all the coefficients in the mean–variance problem (9.6) are deterministic and there is no "any other prescribed constraint(s)". In particular, shorting the stocks is allowed.

Denote

$$\theta(t) \equiv (\theta_1(\cdot), \cdots, \theta_m(\cdot)) := B(t)(\sigma(t)')^{-1}, \tag{9.11}$$

and

$$\rho(t) := \theta(t)\theta(t)', \tag{9.12}$$

where $B(t)$ is defined in (9.8). The process $\theta(t)$ is called a *risk premium process.*

The following result gives a complete solution to the mean–variance portfolio selection problem.

Theorem 9.3 (Efficient Portfolios and Frontier) Problem (9.6) is finite for every $d \in \Re$ if and only if $\int_0^T |B(t)|dt > 0$. In this case, the unique efficient portfolio corresponding to $d \in \Re$, as a function of the time t and the wealth x, is

$$\begin{aligned} \pi^*(t,x) \equiv & \ (\pi_1^*(t,x), \cdots, \pi_m^*(t,x)) \\ = & \ -[\sigma(t)\sigma(t)']^{-1}B(t)'(x - \gamma e^{-\int_t^T r(s)ds}), \end{aligned} \tag{9.13}$$

where

$$\gamma := \frac{d - x_0 e^{\int_0^T [r(t)-\rho(t)]dt}}{1 - e^{-\int_0^T \rho(t)dt}}. \tag{9.14}$$

Moreover, the efficient frontier is given by the following equation

$$\mathsf{Var}\, x^*(T) = \frac{1}{e^{\int_0^T \rho(t)dt} - 1}\left[d - x_0 e^{\int_0^T r(t)dt}\right]^2, \quad d \geq x_0 e^{\int_0^T r(t)dt}. \tag{9.15}$$

Proof. The assertion about the finiteness of the problem is an immediate consequence of [34, Corollary 5.1] (see also Proposition 9.25 in Section 9.4). The efficient portfolios and frontier are derived in [60, (5.12) and Theorem 6.1]. □

Remark 9.4 The first statement of the above theorem says that so long as $B(t) \neq 0$ for a non-zero subset (i.e., a set whose Lebesgue measure is non-zero) of $[0, T]$, there is at least one admissible portfolio that can achieve any given expected terminal payoff. Note that $B(t) \neq 0$ simply means that at least one of the stocks has an appreciation rate different than the risk-free interest rate at t. Clearly this is a reasonable and mild condition.

Remark 9.5 Each efficient strategy depends on the risk preference of an individual investor, which is described by the parameter γ. Note that γ is, in turn, determined by d, which is pre-set by an investor.

Remark 9.6 In the efficient frontier given above, the parameter d is restricted to be no less than $x_0 e^{\int_0^T r(t)dt}$, which is the terminal wealth if all the initial wealth is put into the bank account (or the future value of the initial wealth under the interest rate $r(\cdot)$). Clearly, this is a sensible restriction.

Corollary 9.7 (Reference Wealth Level) Let $\pi^*(\cdot)$ be an efficient strategy and $x^*(\cdot)$ the corresponding wealth trajectory. If $x^*(t_0, \omega_0) = \gamma e^{-\int_{t_0}^T r(s)ds}$ for some $t_0 \in [0, T]$ and $\omega_0 \in \Omega$, then

$$\pi^*(t, \omega_0) = 0, \quad \text{and} \quad x^*(t, \omega_0) = \gamma e^{-\int_t^T r(s)ds}, \quad \forall t \in [t_0, T]. \tag{9.16}$$

Moreover, for any efficient wealth trajectory $x^*(\cdot)$ it must hold that

$$x^*(T, \omega) \leq \gamma, \quad \forall \omega \in \Omega. \tag{9.17}$$

Proof. According to (9.13), $\pi^*(t_0, \omega_0) = 0$ if $x^*(t_0, \omega_0) = \gamma e^{-\int_{t_0}^T r(s)ds}$. This means that all the money should be put in the bank account at t_0. Suppose that we keep $\pi^*(t, \omega_0) = 0$ for $t \geq t_0$, then the corresponding wealth $x^*(t, \omega_0) = x^*(t_0, \omega_0)e^{\int_{t_0}^t r(s)ds} = \gamma e^{-\int_t^T r(s)ds}$. Therefore (9.13) stays true for $t \geq t_0$. The first assertion of the corollary thus follows from the uniqueness of the optimal feedback strategy (9.13). To prove the second assertion, suppose $x^*(T, \omega) > \gamma$ for some $\omega \in \Omega$. Set $\phi(t, \omega) := x^*(t, \omega)e^{\int_t^T r(s)ds}$. Then $\phi(0, \omega) \leq \gamma$ (due to $d \geq x_0 e^{\int_0^T r(t)dt}$) and $\phi(T, \omega) = x^*(T, \omega) > \gamma$. The continuity of $\phi(\cdot, \omega)$ ensures that there is $t_0 \in [0, T]$ so that $\phi(t_0, \omega) = \gamma$, or $x^*(t_0, \omega) = \gamma e^{-\int_{t_0}^T r(s)ds}$. From the first statement of the corollary it follows that $x^*(T, \omega) = \gamma$, which is a contradiction. □

Remark 9.8 The above corollary has a very interesting implication. It says that under an efficient strategy one needs to invest in the risky assets so long as his wealth level deviates from the present value of γ (i.e., $\gamma e^{-\int_t^T r(s)ds}$). On the other hand, one should stop investing in the risky assets as soon as this value is reached, and put all the money into the bank account thereafter. In that case his terminal wealth will exactly equal γ which, as can be shown easily, is no less than d under the assumption that $d \geq x_0 e^{\int_0^T r(t)dt}$. In this sense, the present value of γ becomes a reference point as to whether the agent should invest in the stocks, and once this value is reached the best strategy is to stay there. The corollary also implies that γ provides an upper bound for the efficient terminal wealth for any random scenario. It should be noted, certainly, that this upper bound may never be reached.

Let us now derive the continuous-time version of the *capital market line* (for the original single-period version see, e.g., [37]). Equation (9.15) gives rise to

$$d \equiv \mathsf{E}x^*(T) = x_0 e^{\int_0^T r(t)dt} + \sqrt{e^{\int_0^T \rho(t)dt} - 1}\,\sigma_{x^*(T)}. \tag{9.18}$$

Define $r^*(t) = \frac{x^*(t)-x_0}{x_0}$, the return rate of an efficient strategy at time t. Then we immediately have the following result.

Corollary 9.9 (Capital Market Line) We have

$$\mathsf{E}r^*(T) = r_f(T) + \sqrt{e^{\int_0^T \rho(t)dt} - 1}\,\sigma_{r^*(T)}, \tag{9.19}$$

where $r_f(T) := e^{\int_0^T r(t)dt} - 1$ is the risk-free return rate over the entire horizon $[0, T]$.

Hence the efficient frontier mapped onto the terminal mean–standard deviation plane (i.e., a plane where the vertical and horizontal axes represent the mean return rate and the standard deviation of the return rate, respectively, at the terminal time T) is still a straight line, as in the case of single period mean–variance portfolio selection [38, 39], which is also termed the capital market line. The slope of this line, $k = \sqrt{e^{\int_0^T \rho(t)dt} - 1}$, is called the *price of risk*.

Example 9.10 Take the same example as in [60] where a market has a bank account with $r(t) = 0.06$ and only one stock with $b(t) = 0.12$ and $\sigma(t) = 0.15$. In this case $\rho(t) = (\frac{0.12-0.06}{0.15})^2 = 0.16$. Assume an investment period $T = 1$ (year). It has been calculated in [60] that the capital market line (at the terminal time) is

$$\mathsf{E}r^*(1) = r_f(1) + 0.4165\sigma_{r^*(1)} = 0.0618 + 0.4165\sigma_{r^*(1)}. \tag{9.20}$$

Therefore, if an investor is expecting an annual return rate of 20%, the corresponding standard deviation is 33.17%. Moreover, his strategy at time 0 is $(\pi_0, \pi_1) = (-1.3468, 2.3468)$, meaning that he has to borrow \$1.3468 from the bank for every \$1 he has and invest the sum to the stock.

Let us see when the investor could refrain from investing in the stock. Suppose x_0 = \$1M (million) and d = \$1.2M. The reference wealth level $\gamma = \frac{1.2-e^{-0.1}}{1-e^{-0.16}}$ = \$1.9963M. This means that the investor will stop investing in the stock only when the present value of \$1.9963M is reached (this value is \$1.88M at the beginning, and \$1.9373M after half a year).

It is interesting to see what the results are if we treat this particular case as a single-period investment (but both the interest and appreciate rates are still continuously compounding), namely, buying at the initial time and holding for the entire period until the terminal time. Let us first calculate the expected return rate of the stock at the terminal time, which is

$$e^{\int_0^T b(s)ds} - 1 = e^{0.12} - 1 = 0.1275.$$

On the other hand, it is easy to show via the Ito formula that the standard deviation of the above rate is

$$e^{\int_0^T b(s)ds}\sqrt{e^{\int_0^T \sigma(s)^2 ds - 1}} = e^{0.12}\sqrt{e^{0.15^2} - 1} = 0.1701.$$

Now, since there is only one stock, the capital market line in this single-period case is the straight line joining the points representing the stock and the bank account, which is

$$\mathsf{E}r^*(1) = 0.0618 + \frac{0.1275 - 0.0618}{0.1701}\sigma_{r^*(1)} = 0.0618 + 0.3862\sigma_{r^*(1)}. \quad (9.21)$$

We see that the capital market line (9.20) in the continuous-time case moves upward from its single-period counterpart (9.21), meaning that with the same risk level the efficient portfolio for the continuous-time achieves higher expected return than that of the single period. Finally, in order to obtain an annual return rate of 20% in the single period, suppose one needs to put a fraction α of the wealth into the stock. Then we have

$$0.1275\alpha + 0.0618(1 - \alpha) = 0.2.$$

This gives $\alpha = 2.1035$, which means borrowing \$1.1035 for every \$1. The standard deviation of this strategy is

$$\frac{0.2 - 0.0618}{0.3862} = 0.3578 = 35.78\%.$$

Comparing this with the result for the continuous-time case, we find that the strategy in the continuous time is more aggressive initially, but ends up with a lower risk at the terminal time. This, however, makes perfect sense since the dynamic portfolio selection, being able to adjust the portfolio over time, clearly has an edge over the buy-and-hold strategy.

Example 9.11 Let us now consider an example where there are three stocks. The interest rate and the appreciation rates of the stocks are $r = 0.02$ and $(b_1, b_2, b_3)' = (0.01, 0.05, 0.06)'$, respectively, and the volatility matrix is $\sigma = \begin{bmatrix} 1 & 0 & \frac{2}{3} \\ 0 & 1 & 0 \\ 0 & 0 & \frac{2}{3} \end{bmatrix}$.

Then we have $\sigma^{-1} = \begin{bmatrix} 1 & 0 & -1 \\ 0 & 1 & 0 \\ 0 & 0 & \frac{3}{2} \end{bmatrix}$ and $B \equiv (b_1 - r, b_2 - r, b_3 - r)' = (-0.01, 0.03, 0.04)'$. Hence, $\theta \equiv B(\sigma')^{-1} = (-0.05, 0.03, 0.06)$, and $\rho = \theta\theta' = \frac{7}{1000}$. Suppose an agent wishes to achieve 10% return in one year with an initial endowment \$1M. With these data, it is easy to get $\gamma =$ \$12.46M. The initial portfolio, based on (9.13), is

$$(\pi_1^*, \pi_2^*, \pi_3^*) = (-0.5607, 0.3364, 1.5699).$$

This means that the agent should borrow \$0.3456M from the bank, short sell the first stock for \$0.5607M, and then buy the second and the third stocks for \$0.3364M and \$1.5699, respectively.

The capital market line in this case is

$$\mathsf{E}r^*(1) = 0.0202 + 0.0838\sigma_{r^*(1)}. \tag{9.22}$$

Thus, a 10% return will yield a standard deviation of 95.23%.

Recall that $\pi(\cdot)$ represents the allocation of wealth among only the risky assets. Theorem 9.3 shows that in an efficient portfolio the *fractions* of wealth among the risky assets are the same, irrespective of the wealth levels, the risk preferences, or even the investment horizons of different investors. The following mutual fund theorem, originally due to Tobin [52] for single-period investment and proved for the continuous-time case in [34] in the general setting of random market parameters, is one of the consequences of this observation.

Theorem 9.12 (Mutual Fund Theorem) Suppose an efficient portfolio $\pi_1^*(\cdot)$ is given by (9.13) corresponding to $d = d_1 > x_0 e^{\int_0^T r(t)dt}$. Then a portfolio $\pi^*(\cdot)$ is efficient if and only if there is $\mu \geq 0$ such that

$$\pi^*(t) = \mu\pi_1^*(t), \quad t \in [0, T]. \tag{9.23}$$

Proof. According to [34, Theorem 5.3], a portfolio $\pi^*(\cdot)$ is efficient if and only if there is $\mu \geq 0$ such that

$$\pi^*(t) = (1 - \mu)\pi_0^*(t) + \mu\pi_1^*(t), \quad t \in [0, T],$$

where $\pi_0^*(t)$ is the risk-free portfolio. Now, since the risk-free portfolio in the present case is $\pi_0^*(t) \equiv 0$, we immediately get the desired result. □

Remark 9.13 Theorem 9.12 yields that any efficient portfolio can be constructed as a combination of the give fund $\pi_1^*(\cdot)$ and the bank account. Therefore, all investors need only to use the mutual fund $\pi_1^*(\cdot)$ and the bank account in compiling their portfolios. The fact that the factor μ is non-negative seems to be counter-intuitive at first sight. But this simply means that one never short-sells this fund, for otherwise his portfolio would fall below the bank account in the mean–standard deviation plane, resulting in a terminal wealth less than the risk-free payoff.

A *market portfolio* describes the market capitalization of all the stocks in the whole market. Mathematically it is a time-varying vector with each component being the capitalization of each of the stocks at every time instant. Now, suppose that the entire market is mean–variance, i.e., every investor in the market agrees on the stock price processes with the same appreciation and volatility rates (termed *homogeneous expectation*; see Merton [43]), and everyone is a mean–variance investor. More importantly, it is assumed that the drift and diffusion coefficients in the stock dynamics (9.2) are already the result of dynamic trading and equilibrium adjustment. Under these assumptions, Theorem 9.3 becomes a rule of investment for everyone; that is everyone will follow (9.13) to form his/her portfolio. As discussed earlier, everyone has the same fractions of stocks in his/her portfolio across the market. As a consequence, collectively the market will have exactly the *same* portfolio (in terms of the proportions in different stocks) as every individual investor. In other words, the market portfolio has the same proportions as those of the efficient portfolio (9.13). To summarize, we have the following result.

Theorem 9.14 (Market Portfolio) Assume that

$$\eta(t) := \mathbf{e}'[\sigma(t)\sigma(t)']^{-1}B(t)' \neq 0, \quad \forall t \geq 0,$$

where $\mathbf{e} = (1, 1, \cdots, 1)'$. Then in a mean–variance market, the market portfolio, as a function of the time t and the total market capitalization x, is

$$\pi_M(t, x) = \frac{[\sigma(t)\sigma(t)']^{-1}B(t)'}{\eta(t)}x. \tag{9.24}$$

Based on the above, we can specify the dynamics of the total market capitalization, $x_M(\cdot)$.

Theorem 9.15 (Dynamics of Total Market Capitalization) Under the assumptions of Theorem 9.14, the total market capitalization $x_M(\cdot)$, with an initial value $x_{M0} > 0$, is determined by the following equation

$$\begin{cases} dx_M(t) = & [r(t) + \frac{\rho(s)}{\eta(t)}]x_M(t)dt + \frac{\theta(t)}{\eta(t)}x_M(t)dW(t), \quad t \geq 0, \\ x_M(0) = & x_{M0}. \end{cases} \tag{9.25}$$

Moreover,

$$\mathsf{E}x_M(t) = e^{\int_0^t [r(s) + \frac{\rho(t)}{\eta(s)}]ds} x_{M0}, \tag{9.26}$$

$$\mathsf{Var}\; x_M(t) = e^{\int_0^t 2[r(s) + \frac{\rho(s)}{\eta(s)}]ds}[e^{\int_0^t \frac{\rho(s)}{\eta(s)^2}ds} - 1]x_{M0}^2, \quad t \geq 0. \tag{9.27}$$

Proof. Putting the portfolio (9.24) into the market dynamics (9.7), we obtain (9.25). Expressions (9.26) and (9.27) are easily obtained by applying Ito's formula to (9.25) and solving the resulting ODEs. □

Define $r_M(t) = \frac{x_M(t)-x_{M0}}{x_{M0}}$ to be the return rate of the market portfolio at time t. Then the following results are straightforward by virtue of Theorem 9.15.

Corollary 9.16 We have

$$\begin{aligned} \mathsf{E}r_M(t) &= e^{\int_0^t [r(s)+\frac{\rho(s)}{\eta(s)}]ds} - 1, \\ \mathsf{Var}\ r_M(t) &= e^{\int_0^t 2[r(s)+\frac{\rho(s)}{\eta(s)}]ds}[e^{\int_0^t \frac{\rho(s)}{\eta^2(s)}ds} - 1]. \end{aligned} \tag{9.28}$$

In the single-period mean–variance portfolio selection, it is well known that the market portfolio itself is efficient, which is used conveniently as the one fund in the mutual fund theorem. Hence, every efficient portfolio can be constructed as a combination of the market portfolio and the risk-free account (in this case, the market portfolio is also termed *tangent fund*; see, e.g., Luenberger [37, Chapters 6 and 7]). However, in the continuous-time case, as will be evident in the following, the market portfolio is *inherently* inefficient for *any* investment horizon. This represents one of the major departures of the continuous-time world from the single-period one.

Theorem 9.17 (Inefficiency of Market Portfolio [8, Theorem 4.3]**)** Under the assumptions of Theorem 9.14, the market portfolio is inefficient for any investment horizon $T > 0$.

The following two examples demonstrate the inefficiency of the market portfolio.

Example 9.18 We continue with Example 9.10. With the given parameters, we get $\eta(t) = 2.667 \neq 0$. Therefore, by Corollary 9.16 the expected return rate of the market portfolio at the terminal time is

$$\mathsf{E}r_M(1) = e^{\int_0^1 [0.06+\frac{0.16}{2.667}]ds} - 1 = 0.1275$$

with the corresponding standard deviation

$$\sigma_{r_M(1)} = e^{\int_0^1 [0.06+\frac{0.16}{2.667}]ds}\sqrt{e^{\int_0^1 \frac{0.16}{2.667^2}ds} - 1} = 0.1701.$$

However, with an expected rate of return equal to 0.1275, by the capital market line (9.20), an efficient portfolio should produce a standard deviation equal to $\frac{0.1275-0.0618}{0.4165} = 0.1577 < 0.1701$. This means that the market portfolio is not efficient in this case. However, the market portfolio lies exactly on the single-period capital market line (9.21).

Example 9.19 We continue with Example 9.11. One can compute $\eta(t) = 0.12$. Thus, the expected return rate of the market portfolio at the terminal time is

$$\mathsf{E}r_M(1) = e^{\int_0^1 [0.02 + \frac{0.007}{0.12}]ds} - 1 = 0.0815,$$

with the corresponding standard deviation

$$\sigma_{r_M(1)} = e^{\int_0^1 [0.02 + \frac{0.007}{0.12}]ds} \sqrt{e^{\int_0^1 \frac{0.007}{0.12^2} ds} - 1} = 0.8557.$$

Again, the market portfolio falls below the capital market line (9.22).

Remark 9.20 In the single-period case, it is well known that the market portfolio itself is efficient (if there is no constraint on the portfolios), which is the prerequisite for the Sharpe–Lintner–Mossin capital asset pricing model (CAPM) [50, 35, 44]. Markowitz pointed out in [40, Chapter 12] that the market portfolio may not be efficient in the single period if the portfolios are constrained. Here we have shown that the market portfolio in continuous time *must* be inefficient, even if the portfolios are unconstrained. This seemingly surprising result can be explained intuitively by the following. The market portfolio is essentially a buy-and-hold policy, as the number of shares outstanding for any stock in the market never changes in our model. The market value of a stock in the market portfolio changes over time only because the price of the stock changes. As a result, the market portfolio is not likely to be efficient among all the portfolios that could continuously change number of shares (which was not a problem with the single-period model). This is further confirmed by Example 9.18 where the market portfolio lies on the single-period capital market line.

Remark 9.21 Merton [43] derived a continuous time CAPM under the assumption that the market portfolio is efficient. Since now the assumption has been proved to be inherently false, the CAPM will have to be revised to reflect the inefficiency of the market. Specifically there will be some additional terms in the continuous-time CAPM. This has been carried out in [8].

9.4 Efficient Frontier: Random Adaptive Market Parameters

In this section we study problem (9.6) where all the market coefficient processes, including the interest rate process, are random. We need to assume throughout this section that

$$\mathcal{F}_t = \sigma\{W(s) : s \le t\}, \tag{9.29}$$

which implies that all the random coefficients are adaptive to the underlying Brownian motion. Once again, there is no "any other prescribed constraint(s)" in (9.6) and, in particular, shorting the stocks is allowed.

Recall that $B(t)$, $\theta(t)$ and $\rho(t)$ are defined in (9.8), (9.11) and (9.12) respectively. The following stochastic Riccati equation (SRE), which is a nonlinear BSDE, plays an important role

$$\begin{cases} dp(t) = & -\Big\{[2r(t)-\rho(t)]p(t) - 2\theta(t)\Lambda(t) - \frac{1}{p(t)}\Lambda(t)'\Lambda(t)\Big\}dt \\ & +\Lambda(t)'dW(t), \quad t \in [0, T], \\ p(T) = & 1. \end{cases} \tag{9.30}$$

Proposition 9.22 ([34, Theorem 4.1]) There exists a unique solution $(p(\cdot), \Lambda(\cdot)) \in L^{\infty}_{\mathcal{F}}(\Omega, C(0, T; \Re)) \times L^2_{\mathcal{F}}(0, T; \Re^m)$ of the stochastic Riccati equation (9.30).

One of the major complications with the random market parameters is that the interest rate is also random, hence putting all the money in the bank account is no longer risk free. However, we will show that it is possible to use the stock market to *perfectly* hedge the risk arising from the random interest rate, thereby a risk-free portfolio is still available. To this end, consider the following BSDE:

$$\begin{cases} dg(t) = & [r(t)g(t) + \theta(t)\zeta(t)]dt + \zeta(t)'dW(t), \quad t \in [0, T], \\ g(T) = & 1. \end{cases} \tag{9.31}$$

This is a linear BSDE with bounded coefficients, whose unique solvability is well known (see, e.g., [58, Chapter 7]).

Theorem 9.23 (Risk-free Portfolio [34, Theorem 5.2]) A risk-free investment, with an initial endowment x_0, is given by the following portfolio

$$\pi_0^*(t) = \frac{x_0}{g(0)}(\sigma(t)^{-1})'\zeta(t) \tag{9.32}$$

with the deterministic terminal wealth

$$d = d_0 := \frac{x_0}{g(0)}. \tag{9.33}$$

Remark 9.24 If $r(t)$ is deterministic, then by the unique solvability of (9.31), $g(t) = e^{-\int_t^T r(s)ds}$ and $\zeta(t) = 0$. Thus $g(t)$ is nothing more than the discount factor between t and T. Moreover, $\pi_0^*(t) = 0$, meaning that the risk-free investment corresponds to putting all the wealth in the bank account and nothing into stocks. On the other hand, if $r(t)$ is random, then $\zeta(t)$ is generally non-zero and hence $\pi_0^*(t) \neq 0$. In this case, equation

(9.31) suggests that $g(t)$ is the price process of a contingent claim with a unit terminal payoff, while $(\sigma(t)^{-1})'\zeta(t)$ is the replicating portfolio of the claim; see Yong and Zhou [58, Proposition 6.1, p. 394]. Thus, $g(t)$ is the *risk-adjusted discount factor* between t and T; in particular, $g(0)$ is the risk-adjusted discount factor for the entire horizon $[0, T]$.

Quantitative properties of this discounting process $g(t)$ follow immediately from the comparison theorem for nonlinear BSDEs [14, Theorem 5.1, p. 199].

Proposition 9.1 If $r(t) \geq 0$ a.e. $t \in [0, T]$, P-a.s., then $0 < g(0) \leq 1$. If $r(t) \geq \delta$ for some $\delta > 0$ a.e. $t \in [0, T]$, P-a.s., then $0 < g(0) < 1$. Moreover, if (g, ζ) and $(\tilde{g}, \tilde{\zeta})$ are solutions of (9.31), corresponding to interest rate processes $r(\cdot)$ and $\tilde{r}(\cdot)$, respectively, then $r(t) \leq \tilde{r}(t)$ a.e. $t \in [0, T]$, P-a.s.implies $g(t) \leq \tilde{g}(t)$ a.e. $t \in [0, T]$, P-a.s..

We need to handle another issue – the feasibility of (9.6) – before presenting the efficient portfolios and frontier. Recall in the case of deterministic coefficients the problem is feasible (and finite) for *every* $d \in \Re$ if and only if $\int_0^T |B(t)|dt > 0$. There is a little twist on this in the random case.

Proposition 9.25 (Feasibility [34, Proposition 3.4]) Let $(\psi(\cdot), \xi(\cdot)) \in L^2_{\mathcal{F}}(\Omega; C(0, T; \Re)) \times L^2_{\mathcal{F}}(0, T; \Re^m)$ denote the unique solution of the BSDE

$$\begin{cases} d\psi(t) &= -r(t)\,\psi(t)\,dt + \xi(t)'\,dW(t), \quad t \in [0, T], \\ \psi(T) &= 1. \end{cases} \tag{9.34}$$

Then problem (9.6) is feasible for every $d \in \Re$ if and only if

$$\mathsf{E} \int_0^T \left| \psi(t)\,B(t) + \xi(t)'\,\sigma(t)' \right|^2 dt > 0. \tag{9.35}$$

Remark 9.26 From Proposition 9.25 and the condition (9.35), we obtain certain minimum requirements for the mean–variance problem (9.6) to be feasible for every $d \in \Re$. In particular, condition (9.35) implies that $B(\cdot)$ and $\sigma(\cdot)$ cannot be zero at the same time (this is certainly reasonable, for otherwise it is a pathological case). If $r(\cdot)$ is deterministic, then the unique solvability of the BSDE (9.34) implies that $\xi(t) \equiv 0$, reducing the condition (9.35) to $\mathsf{E}\int_0^T |B(t)|dt > 0$ (and $\int_0^T |B(t)|dt > 0$ if $B(\cdot)$ is also deterministic). On the other hand, when $r(\cdot)$ is random, then $\xi(\cdot)$ is in general non-zero. In this case, (9.35) holds even when $B(t) \equiv 0$. In conclusion, (9.35) is a very mild and reasonable condition.

Now we are ready to present the main results in this section.

Theorem 9.27 (Efficient Portfolios and Frontier [34, Theorem 5.1]) If (9.35) holds, then problem (9.6) is finite for every $d \in \Re$. The unique efficient portfolio corresponding to $d \in \Re$, as a function of the time t and the wealth x, is

$$\pi^*(t,x) = -[\sigma(t)\sigma(t)']^{-1}\Big[\Big(B(t)' + \sigma(t)\frac{\Lambda(t)}{p(t)}\Big)(x - \gamma g(t)) - \gamma\sigma(t)\zeta(t)\Big], \tag{9.36}$$

where

$$\gamma := \frac{d - p(0)g(0)x_0}{1 - p(0)g(0)^2}. \tag{9.37}$$

Moreover, the efficient frontier is given by

$$\mathsf{Var}\, x^*(T) = \frac{p(0)\,g(0)^2}{1 - p(0)\,g(0)^2}\Big[d - \frac{x_0}{g(0)}\Big]^2, \quad d \geq \frac{x_0}{g(0)}. \tag{9.38}$$

Remark 9.28 Theorem 9.23 shows that the agent can expect a terminal wealth of at least $d_0 \equiv \frac{x_0}{g(0)}$. This is why in the efficient frontier the value of d is restricted to be $d \geq \frac{x_0}{g(0)}$. Moreover, if $r(t) > \delta$ for some $\delta > 0$, then it follows from Proposition 9.1 that $g(0) < 1$ and hence in view of (9.33) an increase of initial wealth is guaranteed.

The following result is immediate.

Corollary 9.29 (Capital Market Line) The capital market line in the present case is

$$\mathsf{E}r^*(T) = r_f(T) + \sqrt{\frac{1 - p(0)\,g(0)^2}{p(0)}}\,\frac{1}{g(0)}\sigma_{r^*(T)}, \tag{9.39}$$

where $r_f(T) := \frac{1}{g(0)} - 1$ is the risk-free return rate over the entire horizon $[0,T]$.

In the case when only $r(t)$ is deterministic, we have the following result.

Corollary 9.2 Suppose that $r(t)$ is deterministic. Then the efficient frontier of the mean–variance problem (9.6) exists if and only if $\mathsf{E}\int_0^T |B(t)|dt > 0$. In this case

$$(g(t),\, \zeta(t)) = \Big(e^{-\int_t^T r(s)\,ds},\, 0\Big) \tag{9.40}$$

and the optimal variance associated with the return $\mathsf{E}x^*(T) = d$ is

$$\mathsf{Var}\, x^*(T) = \frac{p(0)\,e^{-2\int_0^T r(s)\,ds}}{1 - p(0)e^{-2\int_0^T r(s)\,ds}}\Big[d - x_0\,e^{\int_0^T r(s)\,ds}\Big]^2. \tag{9.41}$$

In addition, if $B(t)$ and $\sigma(t)$ are also deterministic, then

$$(p(t),\, \Lambda(t)) = \Big(e^{\int_t^T (2r(s) - \rho(s))\,ds},\, 0\Big). \tag{9.42}$$

In the case of the deterministic parameters, we have obtained the mutual fund theorem which implies that the allocation among the stocks in any efficient portfolio must have constant proportions independent of the total wealth of the agent. In the present case, however, the efficient portfolio obtained in (9.36) does not have this feature. This suggests that the previous version of the mutual fund theorem is no longer valid in the current setting. This is not surprising, though, for even the risk-free portfolio now has to include stocks (Theorem 9.23). However, we do have the following result.

Theorem 9.30 (**Mutual Fund Theorem** [34, Theorem 5.3]) Suppose an efficient portfolio $\pi_1^*(\cdot)$ is given by (9.36) corresponding to $d \equiv d_1 > \frac{x_0}{g(0)}$. Then a portfolio $\pi^*(\cdot)$ is efficient if and only if there is $\mu \geq 0$ such that

$$\pi^*(t) = (1-\mu)\pi_0^*(t) + \mu\pi_1^*(t), \quad t \in [0, T], \tag{9.43}$$

where $\pi_0^*(\cdot)$ is the risk-free portfolio defined in (9.32).

9.5 Efficient Frontier: Markov-Modulated Market Parameters

In the previous section the random market coefficients are assumed to be adapted to the Brownian motion used to define the stocks. As a result the market is essentially complete. A more sensible description of the reality, however, would be that the market parameters constitute an additional dimension of uncertainty. In this section we investigate such a case where there is *regime switching* in the market. More specifically, the interest rate and the appreciation and volatility rates of the stocks randomly depend on the market mode that switches between a finite number of states. The random regime switching is independent of the random sources that drive the stock prices. We will employ a Markov-chain modulated diffusion formulation to solve the problem.

To start, in addition to the complete probability space $(\Omega, \mathcal{F}, \mathsf{P})$ along with the Brownian motion $W(t) \equiv (W^1(t), \ldots, W^d(t))'$, there is a continuous-time stationary Markov chain $\alpha(t)$ taking value in a finite state space $\mathcal{M} = \{1, 2, \ldots, l\}$. Moreover, $\alpha(t)$ is independent of $W(t)$. The Markov chain $\alpha(t)$ has a generator $Q = (q_{ij})_{l\times l}$ and stationary transition probabilities

$$p_{ij}(t) = \mathsf{P}(\alpha(t) = j | \alpha(0) = i), \quad t \geq 0, \quad i, j = 1, 2, \ldots, l. \tag{9.44}$$

Finally, the filtration

$$\mathcal{F}_t = \sigma\{W(s), \alpha(s) : 0 \leq s \leq t\}. \tag{9.45}$$

Suppose that the initial market mode is

$$\alpha(0) = \alpha_0. \tag{9.46}$$

The market parameters in (9.1) and (9.2) take the form

$$r(t) = r(t, \alpha(t)), \quad b_i(t) = b_i(t, \alpha(t)), \quad \sigma_{ij}(t) = \sigma_{ij}(t, \alpha(t)), \tag{9.47}$$

where $r(t,\alpha), b_i(t,\alpha)$ and $\sigma_{ij}(t,\alpha)$, $\alpha = 1, 2, \ldots, l$, are given market data corresponding to the market mode α. Moreover, we can define $B(t,\alpha)$, $\theta(t,\alpha)$ and $\rho(t,\alpha)$ in the same way as that of (9.8), (9.11) and (9.12), respectively, for each $\alpha = 1, 2, \ldots, l$.

Proposition 9.31 (Feasibility [61, Theorem 3.1]) The problem (9.6) is feasible for every $d \in \Re$ if and only if

$$\mathsf{E} \int_0^T |B(t, \alpha(t))|^2 dt > 0. \tag{9.48}$$

Consider the following two systems of ODEs:

$$\begin{cases} \dot{P}(t,\alpha) = [\rho(t,\alpha) - 2r(t,\alpha)]\, P(t,\alpha) - \sum_{\beta=1}^{l} q_{\alpha\beta} P(t,\beta), \\ P(T,\alpha) = 1, \quad \alpha = 1, 2, \ldots, l, \end{cases} \tag{9.49}$$

and

$$\begin{cases} \dot{H}(t,\alpha) = r(t,\alpha) H(t,\alpha) - \frac{1}{P(t,\alpha)} \sum_{\beta=1}^{l} q_{\alpha\beta} P(t,\beta)[H(t,\beta) - H(t,\alpha)], \\ H(T,\alpha) = 1, \quad \alpha = 1, 2, \ldots, l. \end{cases} \tag{9.50}$$

The existence and uniqueness of solutions to the above two systems of equations are evident as both are linear with uniformly bounded coefficients.

Theorem 9.32 (Efficient Portfolios and Frontier [61, Theorem 5.1]) Assume that (9.48) holds. Then we must have

$$P(0,\alpha_0) H(0,\alpha_0)^2 + \theta - 1 < 0, \tag{9.51}$$

where

$$\begin{aligned} \theta := {} & \mathsf{E} \int_0^T \sum_{\beta=1}^{l} q_{\alpha(t)\beta} P(t,\beta)[H(t,\beta) - H(t,\alpha(t))]^2 dt \\ = {} & \sum_{\alpha=1}^{l} \sum_{\beta=1}^{l} \int_0^T P(t,\beta) p_{\alpha_0\alpha}(t) q_{\alpha\beta} [H(t,\beta) - H(t,\alpha)]^2 dt \geq 0. \end{aligned} \tag{9.52}$$

Moreover, the unique efficient portfolio corresponding to $d \in \Re$, as a function of the time t, the wealth level x, and the market mode α, is

$$\pi^*(t,x,\alpha) = -[\sigma(t,\alpha)\sigma(t,\alpha)']^{-1} B(t,\alpha)' [x - \gamma H(t,\alpha)], \tag{9.53}$$

where

$$\gamma := \frac{d - P(0,\alpha_0) H(0,\alpha_0) x_0}{1 - \theta - P(0,\alpha_0) H(0,\alpha_0)^2}. \tag{9.54}$$

Furthermore, the efficient frontier is given by

$$\begin{aligned}\mathsf{Var}\, x^*(T) = & \frac{P(0,\alpha_0)H(0,\alpha_0)^2+\theta}{1-\theta-P(0,\alpha_0)H(0,\alpha_0)^2}\left[d - \frac{P(0,\alpha_0)H(0,\alpha_0)}{P(0,\alpha_0)H(0,\alpha_0)^2+\theta}x_0\right]^2 \\ & + \frac{P(0,\alpha_0)\theta}{P(0,\alpha_0)H(0,\alpha_0)^2+\theta}x_0^2, \quad d \geq \frac{P(0,\alpha_0)H(0,\alpha_0)}{P(0,\alpha_0)H(0,\alpha_0)^2+\theta}x_0.\end{aligned} \tag{9.55}$$

Remark 9.33 Quite contrary to the cases without Markovian jumps studied in the previous two sections, the efficient frontier in the present case is no longer a perfect square (or, equivalently, the efficient frontier in the standard-deviation–return diagram is no more a straight line). As a consequence, one is not able to achieve a risk-free investment. This, certainly, is expected since now the interest rate process is modulated by the Markov chain, and the interest rate risk cannot be perfectly hedged by any portfolio consisting of the bank account and stocks, because the regime switching induces an additional dimension of uncertainty.

Nevertheless, the expression (9.55) does disclose the *minimum variance*, namely, the minimum possible terminal variance achievable by an admissible portfolio, along with the portfolio that attains this minimum variance.

Theorem 9.34 (Minimum Variance [61, Theorem 5.2]) The minimum terminal variance is

$$\mathsf{Var}\, x^*_{\min}(T) = \frac{P(0,\alpha_0)\theta}{P(0,\alpha_0)H(0,\alpha_0)^2+\theta}x_0^2 \geq 0 \tag{9.56}$$

with the corresponding expected terminal wealth

$$d_{\min} := \frac{P(0,\alpha_0)H(0,\alpha_0)}{P(0,\alpha_0)H(0,\alpha_0)^2+\theta}x_0, \tag{9.57}$$

and the corresponding Lagrange multiplier $\lambda^*_{\min} = 0$. Moreover, the portfolio that achieves the above minimum variance, as a function of the time t, the wealth level x and the market mode α, is

$$\pi^*_{\min}(t,x,\alpha) = -[\sigma(t,\alpha)\sigma(t,\alpha)']^{-1}B(t,\alpha)'[x - d_{\min}H(t,\alpha)]. \tag{9.58}$$

Remark 9.35 It is because of the above theorem that the parameter d can be restricted to $d \geq d_{\min}$ when one defines the efficient frontier for the present mean-variance problem.

The following is a version of the mutual fund theorem for the present model.

Theorem 9.36 (Mutual Fund Theorem [61, Theorem 5.3]) Suppose an efficient portfolio $\pi_1^*(\cdot)$ is given by (9.53) corresponding to $d = d_1 > d_{\min}$. Then a portfolio $\pi^*(\cdot)$ is efficient if and only if there is a $\mu \geq 0$ such that

$$\pi^*(t) = (1-\mu)\pi^*_{\min}(t) + \mu\pi_1^*(t), \quad t \in [0,T], \tag{9.59}$$

where $\pi^*_{\min}(\cdot)$ is the minimum variance portfolio defined in Theorem 9.34.

Remark 9.37 The above mutual fund theorem shows that in the present model, the feature that the fractions of wealth among the stocks are the same among all efficient portfolios is no longer available.

9.6 Efficient Frontier: No Short Selling

In the previous three sections we discussed mean–variance portfolio selection when short-selling of stocks was allowed, i.e., there is no sign constraint on the components of the portfolio vector $\pi(\cdot)$. In this section, we present results when short-selling stocks is prohibited (however, borrowing from the bank is still allowed) and all the market parameters are deterministic. Mathematically, this adds an extra constraint as "any other prescribed constraint(s)" in (9.6):

$$\pi_i(t) \geq 0, \ \ \forall t \in [0,T], \ \ \mathsf{P}-\text{a.s.}, \ \ i = 1, 2, \cdots, m. \tag{9.60}$$

Also, we assume that $b_i(t) \geq r(t)$ $\forall i$.

Define $B(t)$, $\theta(t)$ and $\rho(t)$ as in (9.8), (9.11) and (9.12) respectively. Set

$$\mu(t) := \text{argmin}_{\mu \in [0,\infty)^m} \|\sigma(t)^{-1}\mu + \theta(t)'\|^2, \tag{9.61}$$

and

$$\beta(t) := \sigma(t)^{-1}\mu(t) + \theta(t)'. \tag{9.62}$$

Note that $\mu(t)$ and $\beta(t)$ can be numerically computed off line by a standard quadratic programming solver.

Theorem 9.38 (Efficient Portfolios and Frontier [32, Theorem 4.2]) If problem (9.6) with short-selling prohibition is finite for every $d \geq x_0 e^{\int_0^T r(t)dt}$, then the efficient portfolio corresponding to $d \in \Re$, as a function of the time t and the wealth x, is

$$\pi^*(t,x) \equiv (\pi_1^*(t,x), \cdots, \pi_m^*(t,x))'$$
$$= \begin{cases} -[\sigma(t)\sigma(t)']^{-1}[\mu(t) + B(t)']\left[x - \gamma e^{-\int_t^T r(s)ds}\right], & \\ \qquad\qquad\qquad\qquad \text{if } x - \gamma e^{-\int_t^T r(s)ds} < 0, & \\ 0, \qquad\qquad\qquad\quad \text{if } x - \gamma e^{-\int_t^T r(s)ds} \geq 0, & \end{cases} \tag{9.63}$$

where

$$\gamma := \frac{d - x_0 e^{\int_0^T (r(t) - |\beta(t)|^2)dt}}{1 - e^{-\int_0^T |\beta(t)|^2 dt}}, \tag{9.64}$$

and $\mu(\cdot)$ and $\beta(\cdot)$ are defined in (9.61) and (9.62), respectively. The efficient frontier is given by

$$\mathsf{Var}\, x^*(T) = \frac{1}{e^{\int_0^T |\beta(t)|^2 dt} - 1}\left[d - x_0 e^{\int_0^T r(t)dt}\right]^2, \quad d \geq x_0 e^{\int_0^T r(t)dt}. \tag{9.65}$$

Remark 9.39 The form of the efficient strategy (9.63) suggests that one should invest in the stocks *only* when the wealth level is less than the present value of γ.

Remark 9.40 It follows from (9.63) that the mutual fund theorem does not hold under the short-selling prohibition, because the fraction of wealth in stocks in an efficient portfolio depends on the current wealth level. However, the failure of the mutual fund theorem is natural as the underlying market is essentially incomplete.

Corollary 9.41 (Capital Market Line) The capital market line in the present case is

$$\mathsf{E}r^*(T) = r_f(T) + \sqrt{e^{\int_0^T |\beta(s)|^2 ds} - 1}\,\sigma_{r^*(T)}. \tag{9.66}$$

9.7 Mean–Variance Hedging

In this section, we study a different problem (although in the broad sense it is still in the realm of mean–variance portfolio selection), which is mean–variance hedging. The classical hedging problem has to do with making a replicating portfolio to duplicate the payoff of a contingent claim at the maturity. The mean–variance hedging is to relax the requirement of *exactly* duplicating the payoff (which may not be possible anyway). Its goal is to minimize the expected squared error of the difference between the payoff and the value of the replicating portfolio at the maturity. To be precise, suppose an agent's wealth–portfolio process satisfies (9.4). The objective is, for a *given* initial wealth x_0 and a contingent claim ξ (which is an $\mathcal{F}_T$-measurable square integrable random variable), to choose an admissible (*mean–variance*) *hedging portfolio* $\pi(\cdot)$ so as to minimize

$$J(x_0, \pi(\cdot)) = \mathsf{E}|x(T) - \xi|^2. \tag{9.67}$$

This problem turns out to be also a special case of the LQ model presented in Section 9.2.

Theorem 9.42 (Optimal Portfolio [28, Theorem 5.1]**)** The optimal portfolio of the mean–variance hedging problem, as a function of time t and wealth x, is

$$\pi^*(t,x) = -[\sigma(t)\sigma(t)']^{-1}B(t)'[x - p(t)] + (\sigma(t)^{-1})'q(t), \tag{9.68}$$

where $(p(\cdot), q(\cdot)) \in L^2_{\mathcal{F}}(0,T;\Re) \times L^2_{\mathcal{F}}(0,T;\Re^m)$ is the unique $\mathcal{F}_t$-adapted solution of the BSDE

$$\begin{cases} dp(t) = [r(t)p(t) + \theta(t)q(t)]dt + q(t)'dW(t), \\ p(T) = \xi. \end{cases} \tag{9.69}$$

Moreover, the optimal value of (9.67) is

$$J^*(x_0) := \inf_{\pi(\cdot) \text{ admissible}} J(x_0, \pi(\cdot)) = e^{\int_0^T [2r(s)-\rho(s)]ds}[x_0 - p(0)]^2. \tag{9.70}$$

Remark 9.43 The formula (9.68) has a straightforward interpretation in financial terms. Indeed, the second term on the right hand side of (9.68), $(\sigma(t)^{-1})'q(t)$, is the replicating portfolio for the claim ξ when the initial wealth is the initial fair option price $p(0)$ (see [58, Proposition 6.1, p. 394]). The other term is similar in form to an efficient portfolio (say, (9.13)), where the present value of γ is replaced by the current claim price $p(t)$. In particular, if ξ has a deterministic payoff γ, then the first term of (9.68) is exactly the mean–variance efficient portfolio. Therefore, the optimal hedging policy (9.68) for our problem is the sum of the replicator for the claim and a certain mean–variance efficient portfolio. Consequently, if the initial endowment x_0 is different from the fair initial price $p(0)$ necessary to replicate the claim ξ, then the difference $x_0 - p(0)$ should be invested according to a mean–variance efficient portfolio.

In the above we investigated a model where only the terminal difference is to be minimized. It is more in line with the *European option* in the context of option theory where only the terminal situation is of interest. To motivate a modified model let us consider the pricing problem of an American contingent claim ξ where the holder has the right to exercise the claim at any time $\tau \in [0,T]$. This claim cannot be hedged by a self-financing portfolio and it is necessary to introduce a (nondecreasing) consumption process $C(\cdot)$ in the price process

$$\begin{cases} dp(t) = \big[r(t)p(t) + \theta(t)q(t)\big]dt + q(t)'dW(t) - dC(t), \\ p(T) = \xi, \end{cases} \tag{9.71}$$

in order to push upwards the process $p(t)$ to keep it above the expected value of the claim $E(\xi|\mathcal{F}_t)$ at *any* time, while the amount of pushing should be minimal in the sense that

$$\int_0^T [p(t) - \mathsf{E}(\xi|\mathcal{F}_t)]dC(t) = 0. \tag{9.72}$$

Motivated by this, we now introduce a modified mean–variance hedging problem where the wealth $x(t)$ is to be kept near the expected value of the claim, $\mathsf{E}(\xi|\mathcal{F}_t)$, at *any* time $t \geq 0$ (rather than just at the terminal time). Specifically, instead of cost functional (9.67), we consider the following

$$\tilde{J}(x_0, \pi(\cdot)) = \mathsf{E}\Big[\int_0^T |x(t) - \mathsf{E}(\xi|\mathcal{F}_t)|^2 dt + |x(T) - \xi|^2\Big], \tag{9.73}$$

while keeping the same dynamics (9.4). (One can also put different weights on the running cost and the terminal cost, but let us not bother to do it here.)

Theorem 9.44 (Optimal Portfolio for Modified Model [28, Theorem 6.3]) The optimal hedging portfolio for the modified model is

$$\pi^*(t,x) = -[\sigma(t)\sigma(t)']^{-1}B(t)'[x - p(t)] + (\sigma(t)^{-1})'q(t), \tag{9.74}$$

where $(p(\cdot), q(\cdot)) \in L^2_{\mathcal{F}}(0,T;\Re) \times L^2_{\mathcal{F}}(0,T;\Re^m)$ is the unique $\mathcal{F}_t$-adapted solution of the following BSDE:

$$\begin{cases} dp(t) = \big[r(t)p(t) + \theta(t)q(t) + \frac{p(t) - E(\xi|\mathcal{F}_t)}{P(t)}\big]dt + q(t)'dW(t), \\ p(T) = \xi, \end{cases} \tag{9.75}$$

with

$$P(t) = e^{\int_t^T [2r(s) - \rho(s)]ds} + \int_t^T e^{\int_t^s [2r(\tau) - \rho(\tau)]d\tau} ds > 0. \tag{9.76}$$

Remark 9.45 Equation (9.75) is exactly in the same form as (9.71) with

$$C(t) = \int_t^T \frac{p(s) - \mathsf{E}(\xi|\mathcal{F}_s)}{P(s)} ds. \tag{9.77}$$

This process, $C(\cdot)$, plays a similar role to the cumulative consumption process in the original American contingent claim problem. It is no longer a nondecreasing process due to the underlying mean–variance criterion, however the function of it is to correct any large deviation of the price $p(t)$ from the expected value of the claim, $\mathsf{E}(\xi|\mathcal{F}_t)$, at any time t. On the other hand, since in our model the price $p(t)$ is allowed to go either above or under $\mathsf{E}(\xi|\mathcal{F}_t)$, we do not have (9.72). Nevertheless, we do have the following analogous relations:

$$\begin{aligned} &\textstyle\int_0^T [p(t) - \mathsf{E}(\xi|\mathcal{F}_t)]^+ [dC(t)]^- = 0, \quad \text{and} \\ &\textstyle\int_0^T [p(t) - \mathsf{E}(\xi|\mathcal{F}_t)]^- [dC(t)]^+ = 0, \end{aligned} \tag{9.78}$$

where $a^+ = \max\{a, 0\}$ and $a^- = \max\{-a, 0\}$. Therefore, the interpretation of the portfolio (9.74) is that one should hedge (in the mean–variance sense) the claim using the auxiliary process $C(\cdot)$ and invest the rest of the wealth according to a mean–variance efficient portfolio.

9.8 Notes

Mean–variance portfolio selection model was proposed and solved in the single-period setting by Harry Markowitz [38]. In [38], the short selling of stocks is not allowed. This mathematically renders a quadratic programming (QP) problem which can be solved numerically by standard QP algorithms. When shorting is allowed and the stock covariance matrix is positive definite, Merton [42] obtained efficient portfolios and the efficient frontier analytically by employing the Lagrange multiplier. Later, Perold [46] developed a general technique to locate the efficient frontier for the case when the covariance matrix is merely positive semidefinite. On the other hand, single-period one-fund theorem can be found in Tobin [52], and CAPM in [50, 35, 44].

It is natural to extend Markowitz's work from the static setting to the *dynamic* ones, including those of multi-period and continuous-time portfolio selections. However, these extensions have, by and large, taken a somewhat different tack to Markowitz's original formulation; see, e.g., [4, 16, 22, 23, 44, 48] for the multi-period case, and [10, 12, 13, 20, 24, 26, 41, 47] for the continuous-time case. Specifically, rather than treating the $\mathsf{Var}\ x(T)$ and $\mathsf{E}x(T)$ of a portfolio as separate quantities and finding the relationship between them, a single quantity, the expected utility of terminal wealth $\mathsf{E}U(x(T))$, is considered instead. The conflicting "profit-seeking yet risk-averse" nature of the investor is captured by the *utility function* U, which is commonly a power, log, exponential, or quadratic form. It should be noted that mean-variance analysis and expected utility formulation are two *different* tools for dealing with portfolio selections. For comparisons of the performance of the mean-variance versus utility approaches, the reader is referred to [18, 19, 22, 52, 59]. In particular, in a recent paper [59], a simulation study was carried out to compare the performances of the mean–variance analysis and the growth optimal (logarithmic utility) return using data from the Dow Jones from March 1997 to February 2000. It was shown that the former will outperform the latter by more than an 80% chance if the targeted wealth in one year in the mean–variance model is set to be approximately double the initial wealth.

One major difficulty in extending Markowitz's model to the multi-period or continuous-time settings is that the variance $\mathsf{Var}\ x(T)$ involves a term $[\mathsf{E}x(T)]^2$ that is hard to analyze due to its non-separability in the dynamic programming sense. In a recent paper by Li and Ng [30], the multi-period mean–variance problem was studied in the framework of multi-objective optimization, where $\mathsf{Var}\,x(T)$ and $\mathsf{E}x(T)$ were viewed as competing objectives. An embedding technique was introduced by which dynamic programming can be applied and the mean–variance efficient frontier was explicitly obtained. Shortly after the work of [30], Zhou and Li [60] investigated the continuous-time mean–variance problem (with deterministic, time-varying coefficients) and formulated it, for the first time, as a stochastic LQ optimal control problem. The solution of this problem was then explicitly derived using the embedding technique introduced in [30] and results from stochastic LQ control. For recent development on general (indefinite) stochastic LQ control see [5, 7, 6, 33, 56, 58].

To the best of the author's knowledge there has not been much research on continuous-time mean–variance problem with random market parameters. Using a martingale approach, Sekine [49] studied a mean–variance hedging problem with stochastic interest rate. Lim and Zhou [34] solved the mean–variance portfolio selection problem where all the coefficients are random (adapted to the underlying Brownian motion) via LQ control and BSDE techniques. On the other hand, regime switching model has been investigated mainly to discuss options; see, e.g., [1, 11, 21, 3, 55]. Such a model in the context of mean–variance portfolio selection was studied in [61].

Portfolio selection with prohibition of short-selling is an example of a more general class of problems called constrained portfolio selection. In the past decade, the constrained portfolio selection problem has been extensively studied (see, e.g., [9, 25, 51, 53, 54]). However, again the expected utility model has been mainly adopted. Li, Zhou and Lim [32] solved the mean–variance portfolio selection without shorting by making use of the stochastic LQ control and viscosity verification theorem [62].

The classical hedging problem is to find an appropriate initial wealth and a replicating portfolio so that the terminal payoff of the claim is exactly duplicated. Now, one may decompose the classical hedging problem into several mean–variance hedging problems with different initial conditions in order to characterize the price as well as the hedging portfolio for the original problem. Gourerioux, Laurent, and Pham [17] recently introduced an "extended asset method" to carry out the decomposition. The method was later extended by Laurent and Pham [29]. On the other hand there is a growing interest in measuring risk when it is *a priori* known that with the initial capital the contingent claim cannot be reached. In this case, the mean–variance hedging seems to be a viable alternative model. Kohlmann and Zhou [28] solved this problem via the stochastic LQ control approach.

Finally, we remark that the power of the LQ approach is further demonstrated in some on-going research projects: A continuous-time CAPM is being developed in [8]; a finance index tracking problem is being investigated in [57]; a mean–variance selection model in an incomplete market (i.e., the assumption (9.3) is not imposed) is being tackled in [31]; and a mean–variance selection model where bankruptcy is not allowed (i.e., there is a wealth constraint $x(t) \geq 0\ \forall t$) is being studied in [2].

Acknowledgment

This work was supported by the RGC Earmarked Grants CUHK4435/99E and CUHK4175/00E. The author is grateful to Harry Markowitz for reading the chapter and sending the book [40] to the author alerting him to the results on the possible inefficiency of the market in the single-period setting. It should be acknowledged that the upper bound of efficient terminal wealth in Corollary 9.7 was first observed by Stanley Pliska. The author also wishes to thank the two anonymous referees for their very careful reading of an earlier version of the chapter, and for their insightful and constructive comments that have led to the current version.

9.9 References

[1] Barone-Adesi, G. and Whaley, R., Efficient analytic approximation of American option values, *J. Finance*, **42** (1987), 301-320.

[2] Bielecki, T.R., Pliska, S. and Zhou, X.Y., Mean–Variance portfolio selection with bankruptcy prohibition. Working paper, Department of Systems Engineering and Engineering Management, The Chinese University of Hong Kong, 2001.

[3] Buffington, J. and Elliott, R., American options with regime switching. Working paper, 2001.

[4] Campbell, J.Y., Lo, A.W. and MacKinlay, A.C., *The Econometrics of Financial Markets*, Princeton University Press, New Jersey, 1997.

[5] Chen, S., Li, X. and Zhou, X.Y., Stochastic linear quadratic regulators with indefinite control weight costs, *SIAM J. Contr. Optim.*, **36** (1998), 1685-1702.

[6] Chen, S. and Yong, J., Stochastic linear quadratic optimal control problems, *Appl. Math. Optim.* **43** (2001), 21-45.

[7] Chen, S. and Zhou, X.Y., Stochastic linear quadratic regulators with indefinite control weight costs. II, *SIAM J. Contr. Optim.*, **39** (2000), 1065-1081.

[8] Chiu, C.H. and Zhou, X.Y., Mean–variance market portfolio, its inefficiency, and CAPM in continuous time. Working paper, Department of Systems Engineering and Engineering Management, The Chinese University of Hong Kong, 2002.

[9] Cvitanic, J. and Karatzas, I., Convex duality in constrained portfolio optimization, *Ann. Appl. Probab.*, **2**, 767-818.

[10] Cox, J. and Huang, C.F., Optimal consumption and portfolio policies when asset prices follow a diffusion process, *J. Econ. Theory*, **49** (1989), 33-83.

[11] Di Masi, G.B., Kabanov, Y.M. and Runggaldier, W.J., Mean variance hedging of options on stocks with Markov volatility, *Theory Prob. Appl.*, **39** (1994), 173-181.

[12] Duffie, D. and Richardson, H., Mean-variance hedging in continuous time, *Ann. Appl. Prob.*, **14** (1991), 1-15.

[13] Dumas, B. and Luciano, E., An exact solution to a dynamic portfolio choice problem under transactions costs, *J. Finance*, **46** (1991), 577-595.

[14] El Karoui, N., Peng, S. and Quenez, M.C., Backward stochastic differential equations in finance, *Math. Finance*, **7** (1997), 1-71.

[15] Elliott, E.J. and Kopp, P.E., *Mathematics of Financial Markets*, Springer-Verlag, New York, 1999.

[16] Francis, J.C., *Investments: Analysis and Management*, McGraw-Hill, New York, 1976.

[17] Gourerioux, C., Laurent, J.P. and Pham, H., Mean variance hedging and numeraire, *Math. Finance*, **8** (1998), 179-200.

[18] Grauer, R.R., A comparison of growth optimal and mean-variance investment policies, *J. Financial Quant. Anal.*, **16** (1981), 1-21.

[19] Grauer, R.R. and Hakansson, N.H., On the use of mean-variance and quadratic approximations in implementing dynamic investment strategies: A comparison of the returns and investment policies, *Manag. Sci.*, **39** (1993), 856-871.

[20] Grossman, S.J. and Zhou, Z., Equilibrium analysis of portfolio insurance, *J. of Finance*, **51** (1996), 1379-1403.

[21] Guo, X., *Insider Information and Stock Fluctuations*, PhD Thesis, Rutgers University, 1999.

[22] Hakansson, N.H., Capital growth and the mean-variance approach to portfolio selection, *J. Financial Quant. Anal.*, **6** (1971), 517-557.

[23] Harrison, M. and Kreps, D., Martingale and multiperiod securities markets, *J. Econ. Theory*, **20** (1979), 382-408.

[24] Harrison, M. and Pliska, S., Martingale and stochastic integrals in the theory of continuous trading, *Stoch. Proc. Appl.*, **11** (1981), 215-260.

[25] Karatzas, I. and Kou, S.G., Hedging American contingent claims with constrained portfolios, *Finance and Stoch.*, **2** (1998), 215-258.

[26] KARATZAS, I., LEHOCZKY, J.P. AND SHREVE, S.E., Optimal portfolio and consumption decisions for a small investor on a finite horizon, *SIAM J. Contr. Optim.*, **25** (1987), 1157-1186.

[27] KARATZAS, I. AND SHREVE, S.E., *Methods of Mathematical Finance*, Springer-Verlag, New York, 1999.

[28] KOHLMANN, M. AND ZHOU, X.Y., Relationship between backward stochastic differential equations and stochastic controls: A linear-quadratic approach, *SIAM J. Contr. Optim.*, **38** (2000), 1392-1407.

[29] LAURENT, J.P. AND PHAM, H., Dynamic programming and mean variance hedging, *Finance Stoch.*, **3** (1999), 83-110.

[30] LI, D. AND NG, W.L., Optimal dynamic portfolio selection: Multiperiod mean-variance formulation, *Math. Finance*, **10** (2000), 387-406.

[31] LI, X. AND ZHOU, X.Y., Continuous-time Markowitz's model in an incomplete market. Working paper, Department of Systems Engineering and Engineering Management, The Chinese University of Hong Kong, 2001.

[32] LI, X., ZHOU, X.Y. AND LIM, A.E.B., Dynamic mean-variance portfolio selection with no-shorting constraints, *SIAM J. Contr. Optim.*, **40** (2001), 1540-1555.

[33] LIM, A.E.B. AND ZHOU, X.Y., Optimal stochastic LQR control with integral quadratic constraints and indefinite control weights, *IEEE Trans. Autom. Contr.*, **44** (1999), 1359-1369.

[34] LIM, A.E.B. AND ZHOU, X.Y., Mean-variance portfolio selection with random parameters in a complete market, *Math. Oper. Res.*, **27** (2002), 101-120.

[35] LINTNER, J., The valuation of risk assets and the selection of risky investment in stock portfolios and capital budgets, *Rev. Econ. Stat.*, **47** (1965), 13-37.

[36] LUENBERGER, D.G., *Optimization by Vector Space Methods*, John Wiley, New York, 1968.

[37] LUENBERGER, D.G., *Investment Science*, Oxford University Press, New York, 1998.

[38] MARKOWITZ, H., Portfolio selection, *J. of Finance*, **7** (1952), 77-91.

[39] MARKOWITZ, H., *Portfolio Selection: Efficient Diversification of Investment*, John Wiley & Sons, New York, 1959.

[40] MARKOWITZ, H., *Mean–Variance Analysis in Portfolio Choice and Capital Markets*, Frank J. Fabozzi Associates, New Hope, Pennsylvania, 1987.

[41] MERTON, R.C., Lifetime portfolio selection under uncertainty: the continuous-time case, *Rev. Econ. Stat.*, **51** (1989), 247-257.

[42] MERTON, R.C., An analytic derivation of the efficient frontier, *J. Financial Quant. Anal.*, **7** (1972), 1851-1872.

[43] MERTON, R.C., An intertemporal capital asset pricing model, *Econometrica*, **41** (1973), 867-887.

[44] MOSSIN, J., Equilibrium in a capital asset market, *Econometrica*, **34** (1966), 768-783.

[45] MOSSIN, J., Optimal multi-period portfolio policies, *J. of Business*, **41** (1968), 215-229.

[46] PEROLD, A.F., Large-scale portfolio optimization, *Manag. Sci.*, **30** (1984), 1143-1160.

[47] PLISKA, S., A stochastic calculus model for continuous trading: Optimal portfolio, *Math. Oper. Res.*, **11** (1986), 371-382.

[48] SAMUELSON, P.A., Lifetime portfolio selection by dynamic stochastic programming, *Rev. Econ. Stat.*, **51** (1969), 236-246.

[49] SEKINE, J., Mean-variance hedging in continuous-time with stochastic interest rate, *Stoch. Stoch. Rep.*, **67** (1999), 1-16.

[50] SHARPE, W.F., Capital asset prices: A theory of market equilibrium under conditions of risk, *J. of Finance* **19** (1964), 425-442.

[51] SHIRAKAWA, H., Optimal consumption and portfolio selection with incomplete markets and upper and lower bound constraints, *Math. Finance*, **4** (1994), 1-24.

[52] TOBIN, J., Liquidity preference as behavior toward risk, *Rev. Econ. Studies*, **26** (1958), 65-86.

[53] XU, G.L. AND SHREVE, S.E., A duality method for optimal consumption and investment under short-selling prohibition: I. general market coefficients, *Ann. Appl. Prob.*, **2** (1992), 87-112.

[54] XU, G.L. AND SHREVE, S.E., A duality method for optimal consumption and investment under short-selling prohibition: II. Constant market coefficients, *Ann. Appl. Prob.*, **2** (1992), 314-328.

[55] YAO, D.D., ZHANG, Q. AND ZHOU, X.Y., Option pricing with Markov-modulated volatility. Working paper, Department of Systems Engineering and Engineering Management, The Chinese University of Hong Kong, 2001.

[56] YAO, D.D., ZHANG, S. AND ZHOU, X.Y., Stochastic LQ control via semidefinite programming, *SIAM J. Contr. Optim.*, **40** (2001), 801-823.

[57] YAO, D.D., ZHANG, S. AND ZHOU, X.Y., Tracking a stock index with a few stocks. Working paper, Department of Systems Engineering and Engineering Management, The Chinese University of Hong Kong, 2001.

[58] YONG, J. AND ZHOU, X.Y., *Stochastic Controls: Hamiltonian Systems and HJB Equations*, Springer, New York, 1999.

[59] ZHAO, Y.G. AND ZIEMBA, W.T., Mean-variance versus expected utility in dynamic investment analysis. Working paper, Faculty of Commerce and Business Administration, University of British Columbia, 2000.

[60] ZHOU, X.Y. AND LI, D., Continuous time mean-variance portfolio selection: A stochastic LQ framework, *Appl. Math. Optim.*, **42** (2000), 19-33.

[61] ZHOU, X.Y. AND YIN, G., Dynamic mean-variance portfolio selection with regime switching: A Continuous-time model. Working paper, Department of Systems Engineering and Engineering Management, The Chinese University of Hong Kong, 2002.

[62] ZHOU, X.Y., YONG, J. AND LI, X., Stochastic verification theorems within the framework of viscosity solutions, *SIAM J. Contr. Optim.*, **35** (1997), 243-253.

10

Variance Minimization in Stochastic Systems

Duan Li, Fucai Qian and Peilin Fu

In portfolio selection, almost every investor would like to maximize his/her expected return while at the same time minimizing his/her risk that is often represented by a variance term. In dual control problems, the uncertainty, that can be characterized by a variance term, can be significantly reduced through active learning or probing. On the one hand, variance minimization problems are widely encountered in real-world applications. On the other hand, variance minimization is a notorious problem in optimization due to its associated properties of nonconvexity and nonseparability. The traditional optimal stochastic control theory concerns a sole objective of minimizing the expected value of a performance measure. There is a need to develop an efficient solution framework to deal with a general class of variance minimization problems. A novel solution approach is developed in this chapter to tackle variance minimization problems by exploring special features in variance minimization. Convexification and separation schemes are adopted to overcome the analytical and computational difficulties in variance minimization and to seek an analytical optimal feedback control law by a mathematically tractable setting.

10.1 Variance Minimization Problem

We consider the following general class of optimal stochastic control problems,

$$
\begin{aligned}
(P(c)) \quad & \min J = \mathsf{E}\left[J_1(x(T)) \mid I^0\right] + c\mathsf{Var}[J_2\left(x(T)\right) \mid I^0] && (10.1)\\
& \text{s.t. } x(t+1) = f\left(t, x(t), u(t), w(t)\right), \ t = 0, \ldots, T-1, \\
& \qquad y(t) = h\left(t, x(t), v(t)\right), \ t = 1, \ldots, T,
\end{aligned}
$$

where $x\left(t\right) \in R^n$ is the state, $u\left(t\right) \in R^m$ is the control, $y\left(t\right) \in R^p$ is the measured output, J_1: $R^n \rightarrow R$ and J_2: $R^n \rightarrow R$ are twice differentiable, c

is a nonnegative weighting coefficient, and I^0 is the initial information set that includes information about the distribution of the initial state $x(0)$ and the statistics of the random sequences $\{w(t)\}$ and $\{v(t)\}$ that are two independent Gaussian white noises: $w(t) \sim N(0, \sigma_w^2(t))$, $t = 0, 1, \ldots, T - 1$, $v(t) \sim N(0, \sigma_v^2(t))$, $t = 1, 2, \ldots, T$, $\mathsf{E}\{w(t)v'(l)\} = 0$, $t = 0, 1, \ldots, T - 1$, $l = 1, 2, \ldots, T$. The random initial state $x(0)$ is assumed to be Gaussian: $x(0) \sim N(\hat{x}(0), P(0))$ and it is independent of the process and observation noises.

Dynamic programming is the only universal solution scheme to obtain an optimal feedback policy for stochastic control problems. To apply dynamic programming requires certain crucial conditions, especially, the conditions of separability and monotonicity. Problem $(P(c))$ is difficult to be solved directly due to its nonseparability in the sense of dynamic programming. Let I^t be an information set available at time t and $I^{t-1} \subset I^t, \forall\, t$. A key observation is that while the expectation operator satisfies the smoothing property: $\mathsf{E}[\mathsf{E}(\cdot \mid I^j) \mid I^k] = \mathsf{E}(\cdot \mid I^k), \forall\, j > k$, the variance operation does not: $\mathsf{Var}[\mathsf{Var}(\cdot \mid I^j) \mid I^k] \neq \mathsf{Var}(\cdot \mid I^k), \forall\, j > k$, since $\mathsf{Var}(\cdot)$ includes a nonlinear term of $\mathsf{E}^2(\cdot)$. When the recursive equation of dynamic programming involves certain nonlinear terms of the expectation operator, there exists a nonseparability in the sense of dynamic programming.

In order to overcome this difficulty, a solution scheme is adopted to embed problem $(P(c))$ into a tractable auxiliary problem which is separable in the sense of dynamic programming. Solving the auxiliary problem and investigating the relationship between the solution sets of problem $(P(c))$ and the auxiliary problem, an optimal control of problem $(P(c))$ can be identified.

Let

$$J^{[1]} = \mathsf{E}\left\{J_1(x(T)) + c[J_2(x(T))]^2 \mid I^0\right\}, \tag{10.2}$$

$$J^{[2]} = \mathsf{E}\left\{J_2(x(T)) \mid I^0\right\}. \tag{10.3}$$

It is easy to see that the performance index in $(P(c))$, J, is a concave function of $J^{[1]}$ and $J^{[2]}$,

$$J = J^{[1]} - c[J^{[2]}]^2. \tag{10.4}$$

The following auxiliary problem is now constructed for problem $(P(c))$ with a fixed multiplier $\lambda \in R$,

$$\begin{aligned} (A(\lambda, c)) \qquad & \min J^{[1]} - 2\lambda J^{[2]} \\ & = \mathsf{E}\{J_1(x(T)) + c[J_2(x(T))]^2 - 2\lambda J_2(x(T)) \mid I^0\}, \quad (10.5) \\ & \text{s.t. } x(t+1) = f\,(t, x(t), u(t), w(t))\,,\ t = 0, \ldots, T-1, \\ & \qquad y(t) = h\,(t, x(t), v(t))\,,\ t = 1, \ldots, T. \end{aligned}$$

Theorem 10.1 Suppose u^* is an optimal control of problem $(P(c))$, then u^* is also an optimal control of the auxiliary problem $(A(\lambda^*, c))$ where λ^* satisfies

$$\lambda^* = -\frac{1}{2}\frac{\partial J(J^{[1]}, J^{[2]})}{\partial J^{[2]}} \mid_{u^*} = c\mathsf{E}(J_2(x(T)) \mid I^0) \mid_{u^*} . \tag{10.6}$$

Proof. The theorem can be proven by contradiction. Assume that u^* is not an optimal control of the auxiliary problem with parameter λ^*. Then there exists another admissible control $\hat{u}$ such that

$$\left[J^{[1]} - 2\lambda^* J^{[2]}\right] \mid_{\hat{u}} < \left[J^{[1]} - 2\lambda^* J^{[2]}\right] \mid_{u^*} . \tag{10.7}$$

Since J is a concave function of $J^{[1]}$ and $J^{[2]}$, we have

$$\begin{aligned} J \mid_{\hat{u}} &\leq J \mid_{u^*} + \frac{\partial J}{\partial J^{[1]}} \mid_{u^*} \left(J^{[1]} \mid_{\hat{u}} - J^{[1]} \mid_{u^*}\right) \\ &+ \frac{\partial J}{\partial J^{[2]}} \mid_{u^*} \left(J^{[2]} \mid_{\hat{u}} - J^{[2]} \mid_{u^*}\right). \end{aligned} \tag{10.8}$$

Notice $\frac{\partial J}{\partial J^{[1]}} \mid_{u^*} = 1$ and $\frac{\partial J}{\partial J^{[2]}} \mid_{u^*} = -2c\mathsf{E}(J_2(x(T)) \mid I^0) \mid_{u^*} = -2\lambda^*$. We have from (10.8)

$$\begin{aligned} J \mid_{\hat{u}} &\leq J \mid_{u^*} + \left[J^{[1]} - 2\lambda^* J^{[2]}\right] \mid_{\hat{u}} \\ &- \left[J^{[1]} - 2\lambda^* J^{[2]}\right] \mid_{u^*} . \end{aligned} \tag{10.9}$$

Combining (10.7) and (10.9) yields

$$J \mid_{\hat{u}} < J \mid_{u^*}, \tag{10.10}$$

which contradicts the assumption that u^* is an optimal control of the problem $(P(c))$. □

A feedback optimal control law is unachievable for the variance minimization problem $(P(c))$ when using dynamic programming directly since the nonseparability in $(P(c))$ blocks a direct application of dynamic programming. The implication of Theorem 10.1 is that any optimal solution to problem $(P(c))$ is in the set of solutions to auxiliary problem $(A(\lambda, c))$. Thus, a nonseparable variance minimization problem $(P(c))$ can be embedded into a family of parametric stochastic optimal control problems $(A(\lambda, c))$ which are separable in the sense of dynamic programming. Note that solution $\{u(k)\}$ to $(A(\lambda, c))$ is a function of λ. If $\mathsf{E}(J_2(x(T)) \mid I^0)$ under $\{u(k)\}$ can be expressed analytically in terms of λ, then solving (10.6) yields the optimal parameter λ^*.

Following a solution concept in [18], problem $(P(c))$ can be also tackled by considering the following equivalent formulation:

$$\begin{aligned}(EP(c)) \quad & \min \mathsf{E}\left\{J_1(x(T)) + c[J_2(x(T))]^2 - c\xi^2 \mid I^0\right\}, \qquad (10.11)\\ & \text{s.t. } x(t+1) = f\left(t, x(t), u(t), w(t)\right),\ t = 0, \ldots, T-1,\\ & y(t) = h\left(t, x(t), v(t)\right),\ t = 1, \ldots, T,\\ & \mathsf{E}[J_2(x(T)) \mid I^0] = \xi.\end{aligned}$$

A primal-dual method can be used to solve $(EP(c))$ where the following Lagrangian problem is constructed at the lower level for a given Lagrangian multiplier $\rho \in R$:

$$\begin{aligned}(L(c,\rho)) \quad & \min \mathsf{E}\{J_1(x(T)) + c[J_2(x(T))]^2 - c\xi^2\\ & + \rho(J_2(x(T)) - \xi) \mid I^0\}, \qquad (10.12)\\ & \text{s.t. } x(t+1) = f\left(t, x(t), u(t), w(t)\right),\ t = 0, \ldots, T-1,\\ & y(t) = h\left(t, x(t), v(t)\right),\ t = 1, \ldots, T.\end{aligned}$$

Notice that problem $(L(c, \rho))$ is separable in the sense of dynamic programming. The minimum value of $(L(c, \rho))$ is defined as the dual function $H(\rho; c)$. Maximizing $H(\rho; c)$ with respect to ρ yields the optimal value of the Lagrangian multiplier ρ. The success of the primal-dual method, however, can be guaranteed only when problem $(EP(c))$ possesses a convexity. When problem $(EP(c))$ is nonconvex, the primal-dual method often fails to identify an optimal control for $(P(c))$ due to the existence of a duality gap. Compared to the primal-dual method, the solution scheme proposed in this chapter does not depend on the convexity of problem $(P(c))$.

10.2 General Variance Minimization Problem

In problem $(P(c))$ the overall performance index J is a linear function of $\mathsf{E}[J_1(x(T)) \mid I^0]$ and $\mathsf{Var}[J_2(x(T)) \mid I^0]$. In this section, we consider a more general situation where J can be a nonlinear function of $\mathsf{E}[J_1(x(T)) \mid I^0]$ and $\mathsf{Var}[J_2(x(T)) \mid I^0]$,

$$\begin{aligned}(GP) \quad & \min J = J(\mathsf{E}[J_1(x(T)) \mid I^0], \mathsf{Var}[J_2(x(T)) \mid I^0]), \qquad (10.13)\\ & \text{s.t. } x(t+1) = f\left(t, x(t), u(t), w(t)\right),\ t = 0, \ldots, T-1,\\ & y(t) = h\left(t, x(t), v(t)\right),\ t = 1, \ldots, T,\end{aligned}$$

where all the assumptions in $(P(c))$ also apply in (GP). In addition, J is assumed to be strictly increasing with respect to both $\mathsf{E}[J_1(x(T))]$ and $\mathsf{Var}[J_2(x(T))]$, i.e.,

$$\frac{\partial J}{\partial \mathsf{E}[J_1(x(T))]} > 0, \qquad (10.14)$$

$$\frac{\partial J}{\partial \mathsf{Var}[J_2(x(T))]} > 0. \tag{10.15}$$

A solution u^* of (GP) is said to be efficient if there exists no other admissible $\tilde{u}$ such that $\mathsf{E}[J_1(x(T))] \mid_{\tilde{u}} \le \mathsf{E}[J_1(x(T))] \mid_{u^*}$ and $\mathsf{Var}[J_2(x(T))] \mid_{\tilde{u}} \le \mathsf{Var}[J_2(x(T))] \mid_{u^*}$ with at least one strict inequality. Since J is a strictly increasing function of $\mathsf{E}[J_1(x(T))]$ and $\mathsf{Var}[J_2(x(T))]$, it is easy to prove that any solution to (GP) must be an efficient solution. In other words, any solution to (GP) must be an efficient solution to the following two-objective stochastic optimal control problem,

$$\begin{aligned}(M)\quad & \min\{\mathsf{E}[J_1(x(T)) \mid I^0], \mathsf{Var}[J_2(x(T)) \mid I^0]\}' \\ & \text{s.t. } x(t+1) = f(t, x(t), u(t), w(t)),\ t = 0, \ldots, T-1, \\ & \quad\ y(t) = h(t, x(t), v(t)),\ t = 1, \ldots, T.\end{aligned} \tag{10.16}$$

If problem (M) is R_+^2-convex, each efficient solution of (M) can be generated by solving $(P(c))$, the Lagrangian formulation of (M), for a certain value of c. The next question to be answered is which efficient solution of (M) gives an optimal solution to (GP).

Define the following

$$J_E(u) = \frac{\partial J\left(\mathsf{E}\left[J_1(x(T))\right], \mathsf{Var}\left[J_2(x(T))\right]\right)}{\partial \mathsf{E}\left[J_1(x(T))\right]}\Big|_u, \tag{10.17}$$

$$J_V(u) = \frac{\partial J\left(\mathsf{E}\left[J_1(x(T))\right], \mathsf{Var}\left[J_2(x(T))\right]\right)}{\partial \mathsf{Var}\left[J_2(x(T))\right]}\Big|_u. \tag{10.18}$$

Theorem 10.2 Assume that problem (M) is R_+^2-convex and u^* solves $(P(c^*))$. A necessary condition for u^* being a solution to (GP) is $c^* = \frac{J_V(u^*)}{J_E(u^*)}$.

Proof. The efficient frontier in the $\{\mathsf{E}\left[J_1(x(T))\right], \mathsf{Var}\left[J_2(x(T))\right]\}$ space can be parameterized by the coefficient c. In other words, each point on the efficient frontier can be represented by $(\mathsf{E}\left[J_1(x(T))\right] \mid_c, \mathsf{Var}\left[J_2(x(T))\right] \mid_c)$. Since each solution to (GP) is also a solution to $(P(c))$ for certain c, problem (GP) can be reduced in abstract to the following equivalent form,

$$\min_{c \ge 0} \quad J\left(\mathsf{E}\left[J_1(x(T)) \mid I^0\right] \mid_c, \mathsf{Var}\left[J_2(x(T)) \mid I^0\right] \mid_c\right). \tag{10.19}$$

A first-order necessary condition for optimum $c^* > 0$ is

$$J_E(u^*) \frac{\partial \mathsf{E}\left[J_1(x(T))\right] \mid_{c^*}}{\partial c} + J_V(u^*) \frac{\partial \mathsf{Var}\left[J_2(x(T))\right] \mid_{c^*}}{\partial c} = 0, \tag{10.20}$$

where u^* solves $(P(c^*))$. On the other hand, for a u^* that solves $(P(c^*))$, we have from [5],

$$\frac{\partial \mathsf{E}\left[J_1(x(T))\right] \mid_{c^*}}{\partial c} + c^* \frac{\partial \mathsf{Var}\left[J_2(x(T))\right] \mid_{c^*}}{\partial c} = 0. \tag{10.21}$$

Thus vector $[J_E(u^*), J_V(u^*)]$ is proportional to $[1, c^*]$. We must have $c^* = \frac{J_V(u^*)}{J_E(u^*)}$. □

Problem (GP) can be embedded into problem $(P(c))$. Theorem 10.2 gives a necessary condition for a solution of $(P(c))$ to attain an optimum of (GP). Problem $(P(c))$ can be further embedded into the auxiliary problem $(A(\lambda, c))$ as we know from the previous section. Thus we can conclude that problem (GP) can be also embedded into $(A(\lambda, c))$. The following theorem gives the condition for optimal parameters with which the solution of $(A(\lambda, c))$ attains the optimal point of (GP).

Theorem 10.3 Assume u^* solves problem $(A(\lambda^*, c^*))$. Necessary conditions under which u^* also solves (GP) are

$$c^* = \frac{J_V(u^*)}{J_E(u^*)}, \tag{10.22}$$

$$\lambda^* = \frac{J_V(u^*)}{J_E(u^*)} \mathsf{E}(J_2(x(T)))\,|_{u^*}. \tag{10.23}$$

Proof. The theorem can be easily proven by combining Theorems 10.1 and 10.2. □

The general variance minimization problem discussed in this section possesses a nonseparability in the sense of dynamic programming, thus prohibiting derivation of a feedback optimal policy when applying dynamic programming directly to (GP). Theorem 10.3 extends the reach of dynamic programming to a general class of variance minimization problems. The general variance minimization problem (GP) can be solved by a parametric dynamic programming. More specifically, an optimal solution to problem (GP) can be generated by solving a parametric dynamic programming problem formulation $(A(\lambda, c))$ while the optimality conditions for optimal parameters are specified in Theorem 10.3.

10.3 Variance Minimization in Dynamic Portfolio Selection

Portfolio selection is to seek a best allocation of wealth among a basket of securities. The mean-variance formulation by Markowitz [12] provides a fundamental basis for single-period portfolio selection. Analytical expression of the mean-variance efficient frontier in single-period portfolio selection was derived by Merton [13]. The literature in multi-period portfolio selection has been dominated by the results of maximizing expected utility functions of the terminal wealth and/or multi-period consumption. There are two major problems when adopting a utility function approach. First, it is not practical for investors to construct their utility functions. Second, the trade-off between the expected return and the risk, that is clearly shown in the mean-variance formulation, is hidden to investors when using a utility formulation. In this sense, the concept of the Markowitz's mean-variance formulation has not been fully utilized in the multi-period portfolio selection until recently.

Let $x(T)$ be the wealth at the final stage. One key difficulty in extending Markowitz's idea to a multi-period setting is that the variance $\mathsf{Var}(x(T))$ involves a term $[\mathsf{E}(x(T))]^2$ that is hard to analyze due to its nonseparability in the sense of dynamic programming. Only recently have Li and Ng [10] faithfully extended the Markowitz's mean-variance model to the multi-period setting by using an idea of embedding.

Consider a capital market with $(n+1)$ risky securities, with random rates of returns. An investor joins the market at time 0 with an initial wealth $x(0)$. The investor can allocate his wealth among the $(n+1)$ assets. The wealth can be reallocated among the $(n+1)$ assets at the beginning of each of the following $(T-1)$ consecutive time periods. The rates of return of the risky securities at time period t within the planning horizon are denoted by a vector $e(t) = \left[e^0(t), e^1(t), ..., e^n(t)\right]'$, where $e^i(t)$ is the random return for security i at time period t. It is assumed that vectors $e(t)$, $t = 0, 1, \ldots, T - 1$, are statistically independent and return $e(t)$ has a known mean $\mathsf{E}\left(e(t)\right)$ and a known covariance $\mathsf{Cov}(e(t))$. Let $x(t)$ be the wealth of the investor at the beginning of the t-th period, $u^i(t)$, $i = 1, 2, \ldots, n$, be the amount invested in the i-th risky asset at the beginning of the t-th time period. The amount investigated in the 0-th risky asset at the beginning of the t-th time period is equal to $x(t) - \sum_{i=1}^n u^i(t)$. Using security 0 as the reference, the wealth dynamics can be described using the following formula,

$$x(t+1) = \sum_{i=1}^{n} e^i(t)u^i(t) + \left(x(t) - \sum_{i=1}^{n} u^i(t)\right) e^0(t). \tag{10.24}$$

Define $u(t) = [u^1(t), u^2(t), ..., u^n(t)]'$ and $P(t) = [P^1(t), P^2(t), \ldots, P^n(t)]' = [(e^1(t) - e^0(t)),\ (e^2(t) - e^0(t)),\ \ldots,\ (e^n(t) - e^0(t))]'$. Equation (10.24) can be rewritten in a more compact form:

$$x(t+1) = e^0(t)x(t) + P'(t)u(t), \quad t = 0, \ldots, T-1. \tag{10.25}$$

If $\mathsf{E}(e(t)(e(t))')$ is positive definite for all time periods that is reasonable to assume, we have for $t = 0, \ldots, T-1$,

$$\mathsf{E}(P(t)P'(t)) > 0, \tag{10.26}$$

$$\mathsf{E}((e^0(t))^2) - \mathsf{E}(e^0(t)P'(t))\mathsf{E}^{-1}(P(t)P'(t))\mathsf{E}(e^0(t)P(t)) > 0. \tag{10.27}$$

An investor is seeking a best investment strategy, $u(t) = [u^1(t), u^2(t), \ldots, u^n(t)]'$ for $t = 0, 1, \ldots, T - 1$, such that the expected value of the terminal wealth $x(T)$, $\mathsf{E}(x(T))$, is maximized while the risk represented by the variance of the terminal wealth, $\mathsf{Var}(x(T))$, is minimized. A multi-period portfolio policy is an investment sequence,

$$\begin{aligned} \pi &= \{\mu(0), \mu(1), \ldots, \mu(T-1)\} \\ &= \left\{ \begin{bmatrix} \mu_0^1 \\ \mu_0^2 \\ \vdots \\ \mu_0^n \end{bmatrix}, \begin{bmatrix} \mu_1^1 \\ \mu_1^2 \\ \vdots \\ \mu_1^n \end{bmatrix}, \ldots, \begin{bmatrix} \mu_{T-1}^1 \\ \mu_{T-1}^2 \\ \vdots \\ \mu_{T-1}^n \end{bmatrix} \right\}. \end{aligned} \quad (10.28)$$

More specifically, π is a feedback policy and $\mu(t)$ maps the wealth at the beginning of the t-th period, $x(t)$, into a portfolio decision in the t-th period,

$$\begin{bmatrix} u^1(t) \\ u^2(t) \\ \vdots \\ u^n(t) \end{bmatrix} = \begin{bmatrix} \mu_t^1(x(t)) \\ \mu_t^2(x(t)) \\ \vdots \\ \mu_t^n(x(t)) \end{bmatrix}. \quad (10.29)$$

A multi-period portfolio policy, π^*, is said to be efficient if there exists no other multi-period portfolio policy, π, such that $\mathsf{E}(x(T))\,|_\pi \geq \mathsf{E}(x(T))\,|_{\pi^*}$ and $\mathsf{Var}(x(T))\,|_\pi \leq \mathsf{Var}(x(T))\,|_{\pi^*}$ with at least one strict inequality. A mathematical formulation to generate efficient multi-period portfolio policies is given as follows,

$$\begin{aligned} (E(w)) \qquad & \max \mathsf{E}(x(T)) - w\mathsf{Var}(x(T)) \qquad (10.30) \\ & \text{s.t. } x(t+1) = e^0(t)x(t) + P'(t)u(t), \quad t = 0, 1, ..., T-1, \end{aligned}$$

where $w \in [0, \infty)$ represents the trade-off between the expected return and the risk. By varying w, the set of efficient solutions can be generated.

Problem $(E(w))$ is difficult to be solved directly due to its nonseparability in the sense of dynamic programming. The solution scheme in [10] is to embed problem $(E(w))$ into a tractable auxiliary problem that is separable, to investigate the relationship between the solution sets of problem $(E(w))$ and the auxiliary problem, and to search for the solution of the auxiliary problem that attains the optimum point of problem $(E(w))$.

Define $\Pi_E(w)$ to be the set of optimal solutions of problem $(E(w))$ with given w,

$$\Pi_E(w) = \{\pi \mid \pi \text{ is a maximizer of } (E(w))\}. \quad (10.31)$$

Define

$$\begin{aligned}
&\widetilde{U}\left(\mathsf{E}\left((x(T))^2\right), \mathsf{E}\left(x(T)\right)\right) \\
= \; & \mathsf{E}\left(x(T)\right) - w\mathsf{Var}\left(x(T)\right) \\
= \; & -w\mathsf{E}\left((x(T))^2\right) + \left[w\mathsf{E}^2\left(x(T)\right) + \mathsf{E}\left(x(T)\right)\right]. \qquad (10.32)
\end{aligned}$$

It is obvious that $\widetilde{U}$ is a convex function of $\mathsf{E}\left((x(T))^2\right)$ and $\mathsf{E}\left(x(T)\right)$. The following auxiliary problem is now constructed for $(E\left(w\right))$,

$$\begin{aligned}
(A_{DPS}\left(\lambda, w\right)) \qquad & \max \mathsf{E}\left\{-w(x(T))^2 + \lambda x(T)\right\} \qquad (10.33) \\
& \text{s.t. } x(t+1) = e^0(t)x(t) + P'(t)u(t), \quad t = 0, ..., T-1.
\end{aligned}$$

Prominent features of problem $(A_{DPS}\left(\lambda, w\right))$ include its separable structure in the sense of dynamic programming, its quadratic objective function, and its linear system dynamics. Define $\Pi_{A_{DPS}}\left(\lambda, w\right)$ to be the set of the optimal solutions of problem $(A_{DPS}\left(\lambda, w\right))$ for given λ and w,

$$\Pi_{A_{DPS}}\left(\lambda, w\right) = \left\{\pi \mid \pi \text{ is a maximizer of } \left(A_{DPS}\left(\lambda, w\right)\right)\right\}. \qquad (10.34)$$

Denote

$$\begin{aligned}
d(\pi, w) &= \frac{\partial \widetilde{U}\left(\mathsf{E}\left((x(T))^2\right), \mathsf{E}\left(x(T)\right)\right)}{\partial \mathsf{E}\left(x(T)\right)} \Big|_{\pi} \\
&= 1 + 2w\mathsf{E}\left(x(T)\right)\big|_{\pi}. \qquad (10.35)
\end{aligned}$$

Notice that $(E(w))$ is a special case of $(P(c))$ with $J_1(x(T)) = -x(T)$, $J_2(x(T)) = x(T)$ and $c = w$. From Theorem 10.1, we have the following Theorem.

Theorem 10.4 For any $\pi^* \in \Pi_E\left(w\right)$, $\pi^* \in \Pi_{A_{DPS}}\left(d\left(\pi^*, w\right), w\right)$.

The optimal solution of the auxiliary problem $(A_{DPS}\left(\lambda, w\right))$ can be derived analytically using dynamic programming. Define $\gamma = \frac{\lambda}{w}$. Note that the solution of $(A_{DPS}\left(\lambda, w\right))$ only depends on γ. The optimal portfolio policy for auxiliary problem $(A_{DPS}\left(\lambda, w\right))$ at each time period t is of the following form,

$$u_\gamma^*\left(t; x(t)\right) = -K_t x(t) + v_\gamma\left(t\right) \quad t = 0, 1, \ldots, T-1, \qquad (10.36)$$

where

$$\begin{aligned}
K_t &= \mathsf{E}^{-1}\left(P(t)P'(t)\right)\mathsf{E}\left(e^0(t)P(t)\right), \qquad (10.37) \\
v_\gamma\left(t\right) &= \frac{\gamma}{2}\left(\prod_{k=t+1}^{T-1}\frac{A_k^1}{A_k^2}\right)\mathsf{E}^{-1}\left(P(t)P'(t)\right)\mathsf{E}\left(P(t)\right), \\
& \qquad t = 0, 1, .., T-2 \qquad (10.38)
\end{aligned}$$

with the following boundary condition,

$$v_\gamma(T-1) = \frac{\gamma}{2}\mathsf{E}^{-1}(P(T-1)P'(T-1))\,\mathsf{E}(P(T-1)), \tag{10.39}$$

where

$$\begin{aligned} A_t^1 &= \mathsf{E}(e^0(t)) - \mathsf{E}(P'(t))\,\mathsf{E}^{-1}(P(t)P'(t))\,\mathsf{E}(e^0(t)P(t)), \quad (10.40)\\ A_t^2 &= \mathsf{E}((e^0(t))^2)\\ &\quad -\mathsf{E}(e^0(t)P'(t))\,\mathsf{E}^{-1}(P(t)P'(t))\,\mathsf{E}(e^0(t)P(t)). \quad (10.41)\end{aligned}$$

Substituting (10.36) into the equation of wealth dynamics (10.25) yields the dynamics of the wealth under portfolio policy $u_\gamma^*(t;x(t))$,

$$x_\gamma(t+1) = (e^0(t) - P'(t)K_t)\,x_\gamma(t) + P'(t)v_\gamma(t). \tag{10.42}$$

The explicit expressions for the expected value and the variance of the terminal wealth under portfolio policy $u_\gamma^*(t;x(t))$ can be derived as follows,

$$\mathsf{E}(x_\gamma(T) \mid I^0) = \mu x(0) + \nu\gamma, \tag{10.43}$$

$$\mathsf{Var}(x_\gamma(T) \mid I^0) = a(\gamma - bx(0))^2 + c(x(0))^2, \tag{10.44}$$

where $\mu = \prod_{t=0}^{T-1} A_t^1$, $\nu = \sum_{t=0}^{T-1}\left(\prod_{k=t+1}^{T-1} A_k^1\right) \mathsf{E}(P'(t))\ \mathsf{E}^{-1}(P(t)P'(t))$ $\mathsf{E}(P(t))\,\frac{\prod_{k=t+1}^{T-1} A_k^1}{2\prod_{k=t+1}^{T-1} A_k^2}$, $\tau = \prod_{t=0}^{T-1} A_t^2$, $a = \frac{\nu}{2} - \nu^2$, $b = \frac{\mu\nu}{a}$, and $c = \tau - \mu^2 - ab^2$. Here we define $\prod_{k=T}^{T-1} A_k^i = 1$, $i = 1, 2$.

From Theorem 10.4, (10.35) and (10.43), we have the following equation for optimal λ^*,

$$\lambda^* = 1 + 2w(\mu x(0) + \nu\frac{\lambda^*}{w}), \tag{10.45}$$

i. e.,

$$\gamma^* = bx(0) + \frac{\nu}{2wa}. \tag{10.46}$$

Substituting the optimal γ^* in (10.46) into (10.36) yields the following analytical form of the optimal multi-period portfolio policy for $(E(w))$:

$$\begin{aligned} u^*(t) &= -\mathsf{E}^{-1}(P(t)P'(t))\,\mathsf{E}(e^0(t)P(t))\,x(t)\\ &\quad +\frac{1}{2}(bx(0)+\frac{\nu}{2wa})\left(\prod_{k=t+1}^{T-1}\frac{A_k^1}{A_k^2}\right)\mathsf{E}^{-1}(P(t)P'(t))\,\mathsf{E}(P(t)),\\ &\qquad \forall\, t = 0,1,\ldots,T-2, \quad (10.47)\\ u^*(T-1) &= \\ &-\mathsf{E}^{-1}(P(T-1)P'(T-1))\,\mathsf{E}(e^0(T-1)P(T-1))\,x(T-1)\\ &+\frac{1}{2}(bx(0)+\frac{\nu}{2wa})\mathsf{E}^{-1}(P(T-1)P'(T-1))\,\mathsf{E}(P(T-1)). \quad (10.48)\end{aligned}$$

Substituting (10.46) into (10.43) and (10.44) yields the expression for the expected value and the variance of the terminal wealth on the efficient frontier in terms of w,

$$\mathsf{E}\left(x_w(T) \mid I^0\right) = (\mu + b\nu)x(0) + \frac{\nu^2}{2wa}, \tag{10.49}$$

$$\mathsf{Var}\left(x_w(T) \mid I^0\right) = \frac{\nu^2}{4aw^2} + c(x(0))^2. \tag{10.50}$$

Eliminating the parameter w in (10.49) and (10.50) yields the mean-variance efficient frontier for $(E(w))$:

$$\begin{aligned} \mathsf{Var}\left(x(T) \mid I^0\right) &= \frac{a}{\nu^2}[\mathsf{E}\left(x(T) \mid I^0\right) - (\mu + b\nu)x(0)]^2 + c(x(0))^2, \\ &\quad \text{for } \mathsf{E}\left(x(T) \mid I^0\right) \geq (\mu + b\nu)x(0). \end{aligned} \tag{10.51}$$

With the analytical solution derived above, the implementation of an optimal multi-period portfolio policy for the dynamic portfolio selection problem with a mean-variance formulation, $(E(w))$, is straightforward. The continuous-time dynamic portfolio selection with a mean-variance formulation is derived in [21].

While the mean-variance approach provides a balance between the expected return and the risk, the safety-first approach proposed by Roy [14] represents another school of thinking. Roy [14] stated that most investors are primarily concerned with avoiding trapping themselves into a possible "disaster" and that the principle of safety-first plays a crucial role in the decision making process in investment. The objective of an investor with such consideration is to minimize the probability that his/her terminal wealth $x(T)$ is below a preselected threshold d, $\mathsf{P}(x(T) \leq d)$. The value d can be viewed as a "disaster" level from the investor's point of view. The value of d is assumed to be smaller than $\mathsf{E}(x(T))$ that is associated with portfolio policies under consideration. From Bienaymé-Tchebycheff inequality, we have $\mathsf{P}(x(T) \leq d) \leq \frac{\mathsf{Var}(x(T))}{(\mathsf{E}(x(T))-d)^2}$. Thus, minimizing $\mathsf{P}(x(T) \leq d)$ can be achieved by maximizing $\frac{\mathsf{E}(x(T))-d}{\sqrt{\mathsf{Var}(x(T))}}$ as suggested in Roy [14]. The safety-first approach of Roy [14] has been extended to multi-period portfolio selection recently in [7]. The key contribution in [7] is to overcome the nonseparability in the dynamic safety-first formulation, thus deriving an optimal feedback investment policy.

A multi-period safety-first portfolio selection problem can be formulated as:

$$\begin{aligned} (SF) \quad & \max\ U\{\mathsf{E}(x(T) \mid I^0), \mathsf{Var}(x(T) \mid I^0)\} \\ & = \frac{\mathsf{E}\left(x(T) \mid I^0\right) - d}{\sqrt{\mathsf{Var}\left(x(T) \mid I^0\right)}} \qquad (10.52) \\ & \text{s.t.}\ \ x(t+1) = e^0(t)x(t) + P'(t)u(t), \quad t = 0, ..., T-1. \end{aligned}$$

It is clear that U satisfies $\frac{\partial U}{\partial \mathsf{E}(x(T))} > 0$ for all $\mathsf{E}(x(T))$ and $\mathsf{Var}(x(T))$ and $\frac{\partial U}{\partial \mathsf{Var}(x(T))} < 0$ for all $\mathsf{Var}(x(T))$ and $\mathsf{E}(x(T)) > d$.

Define Π_{SF} to be the set of the optimal solutions of problem (SF),

$$\Pi_{SF} = \{\pi \mid \pi \text{ is the maximizer of } (SF)\}. \tag{10.53}$$

Problem (SF) is not separable in the sense of dynamic programming and can not be directly solved by dynamic programming.

Denote by U_E and U_V the partial derivatives of U with respect to $\mathsf{E}(x(T))$ and $\mathsf{Var}(x(T))$, respectively,

$$U_E(\pi) = \left.\frac{\partial U}{\partial \mathsf{E}(x(T))}\right|_{\pi} = \left.\frac{1}{\sqrt{\mathsf{Var}(x(T))}}\right|_{\pi}, \tag{10.54}$$

$$U_V(\pi) = \left.\frac{\partial U}{\partial \mathsf{Var}(x(T))}\right|_{\pi} = \left.\frac{-1}{2}\frac{\mathsf{E}(x(T)) - d}{\mathsf{Var}(x(T))^{\frac{3}{2}}}\right|_{\pi}. \tag{10.55}$$

From the results in Section 10.2, we can conclude any solution in Π_{SF} can be generated by solving $(A_{DPS}(\lambda, w))$ for certain λ and w. Also from Theorem 10.3, we have the following Theorem.

Theorem 10.5 Assume π^* solves $(A_{DPS}(\lambda^*, w^*))$ with $\frac{\lambda^*}{w^*} > 0$. A necessary condition for $\pi^* \in \Pi_{SF}$ is

$$\begin{aligned}\lambda^*/w^* &= [-U_E(\pi^*)/U_V(\pi^*) + 2\mathsf{E}(x(T))] \mid_{\pi^*} \\ &= 2\frac{\mathsf{Var}(x(T))}{\mathsf{E}(x(T)) - d} \mid_{\pi^*} + 2\mathsf{E}(x(T)) \mid_{\pi^*}.\end{aligned} \tag{10.56}$$

The analytical solution to $(A_{DPS}(\lambda, w))$ is obtained in (10.36) for a given $\gamma = \frac{\lambda}{w}$. The corresponding expected value and the variance of the terminal wealth are given in (10.43) and (10.44), respectively. It is clear from (10.43) and (10.44) that γ varies in the range of $[bx(0), \infty)$ on the mean-variance efficient frontier. Recall that the proposed solution scheme requires the condition of $\mathsf{E}(x(T)) > d$. The range of d, upon which the proposed solution schemes can be applied, can be determined by

$$d < (\mu + \nu b)\, x(0). \tag{10.57}$$

Theorem 10.5 specifies the following optimal condition for parameter γ:

$$\gamma^* = \frac{2[a(\gamma^* - bx(0))^2 + c(x(0))^2]}{\mu x(0) + \nu\gamma^* - d} + 2[\mu x(0) + \nu\gamma^*], \tag{10.58}$$

i.e.,

$$\gamma^* = \frac{(ab^2 + c)\,\nu(x(0))^2 + ab\,(\mu x(0) - d)\, x(0)}{ab\nu x(0) + a\,(\mu x(0) - d)}. \tag{10.59}$$

Notice that both $[(ab^2 + c)\,\nu(x(0))^2 + ab\,(\mu x(0) - d)\,x(0)]$ and $[ab\nu x(0) + a(\mu x(0) - d)]$ are positive when $d \le \mu x(0) + \nu b x(0)$.

For a given multi-period safety-first portfolio selection problem, finding the optimal policy becomes a straightforward two-step task. The optimal value of γ is found first by using (10.59). Substituting the optimal value of γ into (10.36) yields the optimal portfolio policy for (SF).

10.4 Variance Minimization in Dual Control

Except for a few ideal situations, an optimal control for a stochastic system usually pursues two often conflicting objectives: to drive the system toward a desired state, and to perform active learning to reduce the systems uncertainty, as pointed by Feldbaum in his seminal work on dual control [3]. Prominent features and fundamental properties of dual control have been extensively studied in the literature. The dual roles of an optimal control, optimization and estimation, in general situations, cannot be separated. This coupling between optimization and estimation makes an analytical form of optimal control, in most situations, unattainable. Previous efforts in dual control have thus mainly been devoting to developing certain suboptimal solution schemes, such as certainty equivalence scheme and open-loop feedback control, by bypassing this essential feature of coupling in dual control.

To power a control law with a property of active learning, which is indispensable in reducing reducible system's uncertainty, is a key to developing an optimal control law in dual control. Solution proposals have been seen in the literature to attach certain variance terms of the state or of the innovation process into the objective function or use it as a constraint in order to force the control to perform active learning. These solution schemes, however, truncate the time horizon into shorter time periods of one stage, prompting a concern of possible myopic behaviors.

Consider in this section the following class of stochastic optimal control problems where there exists a parameter uncertainty in the observation equation,

$$
\begin{aligned}
(DC) \qquad & \min \mathsf{E}\{x'(N)Q(N)x(N) + \sum_{k=0}^{N-1}[x'(k)Q(k)x(k) \\
& \qquad + u'(k)R(k)u(k)] \mid I^0\}, \qquad (10.60)\\
& \text{s.t. } x(k+1) = A(k)x(k) + B(k)u(k) + w(k),\\
& \qquad k = 0, \ldots, N-1,\\
& \qquad y(k) = C(k,\theta)x(k) + v(k),\ k = 1, \ldots, N,
\end{aligned}
$$

where the state $x(k)$, control $u(k)$, measured output $y(k)$, process noise $w(k)$, observation noise $v(k)$ and information state I^k are defined as the same in problem ($P(c)$) in Section 10.1. The quantities $A(k), B(k)$ and $C(\theta, k)$ are matrices of appropriate dimensions, where $\{A(k)\}$ and $\{B(k)\}$ are completely known while the value of $\{C(\theta, k)\}$ depends on an unknown parameter θ. It is assumed that θ belongs to a finite set $\Theta = \{\theta_1, \theta_2, \ldots, \theta_s\}$ and is a constant over the whole control horizon. The a-priori probabilities of parameter θ are

$$q_i(0) = \mathsf{P}(\theta = \theta_i \mid I^0), \; i = 1, 2, \ldots, s.$$

Further, $\{Q(k)\}$ and $\{R(k)\}$ are sequences of positive semidefinite and positive definite symmetric matrices of appropriate dimensions, respectively. The dual control problem for this system is to find a feedback control law,

$$u(k) = f_k\left(I^k\right), \quad k = 0, 1, \ldots, N-1$$

such that the performance index in (DC) is minimized.

Optimal solution to problem (DC) has not been achieved in the literature until recently due to an analytical complexity in the cost-to-go function when applying dynamic programming. Essentially, this analytical complexity is caused by certain nonlinear terms of the state estimation which are nonseparable in the sense of dynamic programming. A variance minimization approach has been recently proposed in [11] to obtain an optimal feedback control for problem (DC).

Define $\hat{x}_i(k \mid k)$ to be the state estimation or the conditional mean of the state at stage k when assuming the i-th model ($\theta = \theta_i$):

$$\hat{x}_i(k|k) = \mathsf{E}\left\{x(k)|\theta = \theta_i, I^k\right\}. \tag{10.61}$$

The conditional mean of the state when assuming model i can be obtained by the Kalman filter. Define $q_i(k)$ to be the posterior probability of model i at stage k,

$$q_i(k) = \mathsf{P}(\theta = \theta_i \mid I^k), \quad k = 0, 1, \ldots, N-1. \tag{10.62}$$

The posterior probabilities, $q_i(k)$, $i = 1, 2, \ldots, s$, satisfy the following recursive equation:

$$q_i(k) = \frac{L_i(k)}{\sum_{j=1}^{s} q_j(k-1)L_j(k)} q_i(k-1), \quad k = 1, 2, \ldots, N \tag{10.63}$$

with initial condition $q_i(0)$, where

$$\begin{aligned}
L_i(k) &= \mid P_y(k \mid k-1, \theta_i) \mid^{-\frac{1}{2}} \exp[-\frac{1}{2}\tilde{y}(k \mid k-1, \theta_i)' \\
&\quad \times P_y(k \mid k-1, \theta_i)^{-1}\tilde{y}(k \mid k-1, \theta_i)], \\
\tilde{y}(k \mid k-1, \theta_i) &= y(k) - C(k, \theta_i)\hat{x}_i(k \mid k-1), \\
P_y(k \mid k-1, \theta_i) &= C(k, \theta_i)P_i(k \mid k-1)C'(k, \theta_i) + \sigma_v^2(k).
\end{aligned}$$

Define $\hat{x}(k)$ to be the conditional mean at stage k,

$$\hat{x}(k) = \mathsf{E}\left\{x(k) \mid I^k\right\}, \tag{10.64}$$

and it can be obtained by

$$\hat{x}(k) = \sum_{i=1}^{s} q_i(k)\hat{x}_i(k \mid k). \tag{10.65}$$

A passive learning algorithm (DUL Algorithm) has been proposed by Deshpande *et al.* [2] for (DC):

$$u(k) = -\Gamma(k)\hat{x}(k), \quad k = 0, 1, \ldots, N-1 \tag{10.66}$$

with

$$\Gamma(k) = D(k)B'(k)S(k+1)A(k), \tag{10.67}$$

where

$$D(k) = \left[B'(k)S(k+1)B(k) + R(k)\right]^{-1}, \tag{10.68}$$

$$S(k) = A'(k)S(k+1)A(k) + Q(k) - A'(k)S(k+1)B(k)D(k)B'(k)S(k+1)A(k) \tag{10.69}$$

with the boundary condition,

$$S(N) = Q(N). \tag{10.70}$$

Assigning the current posterior probabilities as the weighting coefficients, the control law in the DUL Algorithm is a weighting sum of the optimal control laws for s different models. It is a passive learning algorithm because the control does not probe.

Casiello and Loparo [1] later proved that this passive control law in DUL Algorithm is optimal to the following modified problem of (DC),

$$\begin{aligned}(M_p) \quad & \min \mathsf{E}\{x'(N)Q(N)x(N) + \sum_{k=0}^{N-1}[x'(k)Q(k)x(k) + u'(k)R(k)u(k)] \\ & - \sum_{k=0}^{N-1}[x(k) - \hat{x}(k)]'T(k)[x(k) - \hat{x}(k)] \mid I^0\}, \qquad (10.71) \\ & \text{s.t. } x(k+1) = A(k)x(k) + B(k)u(k) + w(k), \\ & \qquad k = 0, \ldots, N-1, \\ & y(k) = C(k,\theta)x(k) + v(k), \; k = 1, \ldots, N,\end{aligned}$$

where

$$T(k) = A'(k)S(k+1)B(k)D(k)B'(k)S(k+1)A(k). \tag{10.72}$$

The only difference between problems (DC) and (M_p) is that a summation of weighted conditional covariances is subtracted in (M_p) from the original performance index in (DC). As pointed by Casiello and Loparo[1], subtracting the conditional covariance weighted by the sequence $\{T(k)\}$ balances out the function of active learning such that the passive learning controller becomes optimal in this context.

Notice that the smoothing property does not hold for $J_k(\hat{x}(k))$ when function J_k is nonlinear, i. e. , $\mathsf{E}\{J_k(E[x(k+1) \mid I^{k+1}]) \mid I^k\} \neq J_k(\mathsf{E}[x(k+1) \mid I^k])$. In other words, a nonlinear term of the state estimation is nonseparable in the sense of dynamic programming. Investigating further on (M_p), we can also conclude that subtracting the conditional covariance weighted by the sequence $\{T(k)\}$ removes these nonlinear terms involving the conditional mean of the state in the recursive equation of dynamic programming, thus making the problem tractable.

A degree of success for an active learning can be measured by the variance of the final state. Therefore minimizing a variance term of the final state will add a feature of active learning to the derived control law. Li et al. [11] consider a modified problem of (M_p) in which a variance term at the final stage is attached to the objective function of (M_p),

$$
\begin{aligned}
(M_a(\mu)) \quad & \min \mathsf{E}\{x'(N)Q(N)x(N) \\
& + \sum_{k=0}^{N-1} [x'(k)Q(k)x(k) + u'(k)R(k)u(k)] \\
& - \sum_{k=0}^{N-1} [x(k) - \hat{x}(k)]' T(k) [x(k) - \hat{x}(k)] \mid I^0\} \\
& + \mu \mathrm{Tr}[\mathsf{Cov}(x(N) \mid I^0)] \qquad (10.73) \\
= & \min \mathsf{E}\{x'(N)Q(N)x(N) \\
& + \sum_{k=0}^{N-1} [x'(k)Q(k)x(k) + u'(k)R(k)u(k)] \\
& - \sum_{k=0}^{N-1} [x(k) - \hat{x}(k)]' T(k) [x(k) - \hat{x}(k)] \mid I^0\} \\
& + \mu \mathsf{E}\left\{[x(N) - \mathsf{E}(x(N) \mid I^0)]' [x(N) - \mathsf{E}(x(N) \mid I^0)] \mid I^0\right\}, \\
\text{s.t. } & x(k+1) = A(k)x(k) + B(k)u(k) + w(k), \\
& \qquad k = 0, \ldots, N-1, \\
& y(k) = C(k, \theta)x(k) + v(k), \ k = 1, \ldots, N,
\end{aligned}
$$

where

$$T(k) = A'(k)S(k+1)B(k)D(k)B'(k)S(k+1)A(k), \quad (10.74)$$

$$S(k) = A'(k)S(k+1)A(k) + Q(k) - T(k), \tag{10.75}$$
$$D(k) = \left[B'(k)S(k+1)B(k) + R(k)\right]^{-1}, \tag{10.76}$$
$$S(N) = Q(N) + \mu I. \tag{10.77}$$

Parameter $\mu \in [0, \infty)$ is a weighting coefficient of active learning and Tr denotes the trace of a square matrix. A larger μ implies that the more importance has been placed on active learning and uncertainty reduction.

Problem $(M_a(\mu))$ is difficult to be solved directly, since the recursive equation of dynamic programming involves certain nonlinear terms of the state estimation (conditional mean) which introduces a nonseparability in the sense of dynamic programming. In order to overcome this difficulty, a solution scheme described in Section 10.1 is adopted to embed problem $(M_a(\mu))$ into a tractable auxiliary problem which is separable in the sense of dynamic programming. Solving the auxiliary problem and investigating the relationship between the solution sets of problem $(M_a(\mu))$ and the auxiliary problem, the optimal control of problem $(M_a(\mu))$ can be identified.

The performance index of $(M_a(\mu))$ can be written as

$$\begin{aligned} J &= \mathsf{E}\{x'(N)S(N)x(N) + \sum_{k=0}^{N-1}[x'(k)Q(k)x(k) + u'(k)R(k)u(k)] \\ &- \sum_{k=0}^{N-1}[x(k) - \hat{x}(k)]'T(k)[x(k) - \hat{x}(k)] \mid I^0\} \\ &- \mu\mathsf{E}(x(N) \mid I^0)'\mathsf{E}(x(N) \mid I^0). \end{aligned} \tag{10.78}$$

Let

$$\begin{aligned} J_1 &= \mathsf{E}\{x'(N)S(N)x(N) + \sum_{k=0}^{N-1}[x'(k)Q(k)x(k) + u'(k)R(k)u(k)] \\ &\quad - \sum_{k=0}^{N-1}[x(k) - \hat{x}(k)]'T(k)[x(k) - \hat{x}(k)] \mid I^0\}, \\ J_2 &= \mathsf{E}(x(N) \mid I^0). \end{aligned}$$

It is easy to see that the performance index in $(M_a(\mu))$, J, is a concave function of J_1 and J_2,

$$J(J_1, J_2) = J_1 - \mu J_2' J_2. \tag{10.79}$$

The following auxiliary problem is now constructed for problem $(M_a(\mu))$ with a fixed multiplier vector $\lambda \in R^n$,

$$(A_{DC}(\lambda, \mu)) \qquad \min \mathsf{E}\{x'(N)S(N)x(N) + \sum_{k=0}^{N-1}[x'(k)Q(k)x(k)$$

$$+ u'(k)R(k)u(k)] - \sum_{k=0}^{N-1} [x(k) - \hat{x}(k)]'T(k)[x(k) - \hat{x}(k)]$$
$$- 2\lambda' x(N) \mid I^0\}, \tag{10.80}$$
$$\text{s.t. } x(k+1) = A(k)x(k) + B(k)u(k) + w(k),$$
$$k = 0, \ldots, N-1,$$
$$y(k) = C(k, \theta)x(k) + v(k), \ k = 1, \ldots, N.$$

Although there is a difference in the dimension of the variable in the variance term between the problem addressed in Section 10.1 and the problem addressed here, one is of one-dimension and one is of n-dimension, these two variance minimization problems are of a similar nature. Similar to the result in Theorem 10.1, the following theorem can be derived.

Theorem 10.6 Suppose u^* is an optimal control of problem $(M_a(\mu))$, then u^* is also an optimal control of the auxiliary problem $(A_{DC}(\lambda^*, \mu))$ where λ^* satisfies

$$\lambda^* = -\frac{1}{2}\frac{\partial J(J_1, J_2)}{\partial J_2} \mid_{u^*} = \mu \mathsf{E}(x(N) \mid I^0) \mid_{u^*} . \tag{10.81}$$

The implication of Theorem 10.6 is that any optimal solution to problem $(M_a(\mu))$ is in the set of solutions to auxiliary problem $(A_{DC}(\lambda, \mu))$. Note that the auxiliary problem is strictly convex with respect to $\{u(k)\}$. Thus the following solution to $(A_{DC}(\lambda, \mu))$ is unique for a given λ:

$$u^*(k) = -\Gamma_1(k)\hat{x}(k) + \Gamma_2(k)\lambda, \quad k = 0, 1, \ldots, N-1, \tag{10.82}$$

where for $k = N-1, N-2, \ldots, 1, 0$,

$$\Gamma_1(k) = D(k)B'(k)S(k+1)A(k), \tag{10.83}$$
$$\Gamma_2(k) = D(k)B'(k)L'(k+1), \tag{10.84}$$
$$L(k) = L(k+1)\,[A(k) - B(k)\Gamma_1(k)] \tag{10.85}$$

with the boundary conditions $S(N) = Q(N) + \mu I$ and $L(N) = I$.

Define

$$\Phi = I - \mu\{\sum_{s=1}^{N-1} \prod_{i=s}^{N-1} [A(i) - B(i)\Gamma_1(i)]B(s-1)\Gamma_2(s-1)$$
$$+ B(N-1)\Gamma_2(N-1)\}, \tag{10.86}$$
$$\Psi = \mu \prod_{i=0}^{N-1} [A(i) - B(i)\Gamma_1(i)]. \tag{10.87}$$

Theorem 10.7 Assume that Φ is invertible. Then the optimal λ^* with which the optimal solution to $(A(\lambda^*, \mu))$ solves $(M_a(\mu))$ is equal to

$$\lambda^* = \Phi^{-1}\Psi\hat{x}(0). \tag{10.88}$$

Proof. Substituting the optimal control in (10.82) into the system state equation and taking the expected value, we get

$$\begin{aligned}\mathsf{E}(x(k+1) \mid I^0) &= [A(k) - B(k)\Gamma_1(k)]\mathsf{E}(x(k) \mid I^0) + B(k)\Gamma_2(k)\lambda, \\ & \qquad k = 0, 1, \ldots, N-1. \end{aligned} \tag{10.89}$$

Solving the above equation, we have

$$\begin{aligned}\mathsf{E}(x(N) \mid I^0) &= \prod_{i=0}^{N-1}[A(i) - B(i)\Gamma_1(i)]\hat{x}(0) \\ &+ \{\sum_{s=1}^{N-1}\prod_{i=s}^{N-1}[A(i) - B(i)\Gamma_1(i)]B(s-1)\Gamma_2(s-1) \\ &+ B(N-1)\Gamma_2(N-1)\}\lambda. \end{aligned} \tag{10.90}$$

Based on the optimal condition $\lambda^* = \mu\mathsf{E}(x(N) \mid I^0)$ in (10.81), we have the following from (10.90),

$$\begin{aligned}\lambda^* &= \mu\prod_{i=0}^{N-1}[A(i) - B(i)\Gamma_1(i)]\hat{x}(0) \\ &+ \mu\{\sum_{s=1}^{N-1}\prod_{i=s}^{N-1}[A(i) - B(i)\Gamma_1(i)]B(s-1)\Gamma_2(s-1) \\ &+ B(N-1)\Gamma_2(N-1)\}\lambda^*, \end{aligned} \tag{10.91}$$

i. e.,

$$\Phi\lambda^* = \Psi\hat{x}(0). \tag{10.92}$$

If Φ^{-1} exists, then

$$\lambda^* = \Phi^{-1}\Psi\hat{x}(0). \tag{10.93}$$

□

In summary, the optimal control of problem $(M_a(\mu))$ can be expressed as

$$u^*(k) = -\Gamma_1(k)\hat{x}(k) + \Gamma_2(k)\Phi^{-1}\Psi\hat{x}(0). \tag{10.94}$$

An optimal feedback control is derived in the above discussion for problem $(M_a(\mu))$ with a fixed value of μ. The ultimate goal is to derive an optimal control law for the primal problem (DC). It is evident that the solution to problem $(M_a(\mu))$ will become an optimal solution to problem (DC) if we select a parameter μ such that the last two terms in the performance index of $(M_a(\mu))$ are cancelled out each other in average,

$$\begin{aligned}&\mu\mathsf{E}\{[x(N) - \mathsf{E}(x(N) \mid I^0)]'[x(N) - \mathsf{E}(x(N) \mid I^0)] \mid I^0\} \\ &= E\{\sum_{k=0}^{N-1}[x(k) - \hat{x}(k)]'T(k)[x(k) - \hat{x}(k)] \mid I^0\}, \end{aligned} \tag{10.95}$$

and this cancellation should be independent of control u. The issue of how to determine an optimal degree μ, thus solving (DC), has been further addressed in [11].

10.5 Notes

Variance minimization is a kind of stochastic optimization problems. The celebrated dynamic programming is the only universal solution scheme for stochastic optimization problems. Dynamic programming, however, is only applicable to problems that satisfy the property of separability and monotonicity. Variance minimization problems are not directly solvable by dynamic programming due to its nonseparability property. The difficulty to involve a variance term in Markov decision process is studied in [17]. The solution schemes for variance minimization and variance constrained problems are suggested in [18] [19] [20] [15] [16]. The previous research results in [4] [6] [8] [9] extended the applicability of dynamic programming to a general class of nonseparable optimization problems by developing a parametric iterative dynamic programming.

Using an embedding scheme, a feedback optimal policy for a general class of variance minimization problems can be obtained using parametric dynamic programming method, while the optimal condition for the parameter is derived in this chapter. The optimal value of the parameter, in general, depends on the initial state of the system. Thus, adopting a rolling horizon concept in the solution process is necessary to update the optimal value of the parameter at every realized current state.

Acknowledgments

This research was partially supported by the grants from the Research Grants Council of the Hong Kong Special Administrative Region, China (REG Ref. No. CUHK4130/97E and CUHK4392/99E). The authors appreciate the comments from Shushang Zhu.

10.6 References

[1] Casiello, F. and Loparo, K. A., Optimal policies for passive learning controllers, *Automatica*, **25** (1989), 757-763.

[2] Deshpande, J. G., Upadhyay, T. N. and Lainiotis, D. G., Adaptive control of linear stochastic systems, *Automatica*, **9** (1973), 107-115.

[3] Feldbaum, A. A., *Optimal Control Systems*, New York: Academic Press, 1965.

[4] Li, D., Multiple objectives and nonseparability in stochastic dynamic programming, *International Journal of Systems Sciences*, **21** (1990), 933-950.

[5] Li, D., On general linear multiple linear-quadratic control problems, *IEEE Transactions on Automatic Control*, **38** (1993), 1722-1727.

[6] Li, D., Iterative parametric dynamic programming and its application in reliability optimization, *Journal of Mathematical Analysis and Applications*, **191** (1995), 589-607.

[7] Li, D. Chan, T. F. and Ng, W. L., Safety-first dynamic portfolio selection, *Dynamics of continuous, Discrete and Impulsive Systems*, **4** (1998), 585-600.

[8] Li, D. and Haimes, Y. Y., New approach for nonseparable dynamic programming, *Journal of Optimization Theory and Applications*, **64** (1990), 311-330.

[9] Li, D. and Haimes, Y. Y., Extension of dynamic programming to nonseparable problems, *Computers & Mathematics with Applications*, **21** (1991), No. 11/12, 51-56.

[10] Li, D. and Ng, W. L., Optimal dynamic portfolio selection: Multiperiod mean variance formulation, *Mathematical Finance*, **10** (2000), 387-406.

[11] Li, D., Qian, F. C. and Fu, P. L., Variance minimization approach for a class of dual control problems, Accepted for publication in *IEEE Transactions on Automatic Control*, 2002.

[12] Markowitz, H. M., *Portfolio Selection: Efficient Diversification of Investment*, New York: John Wiley & Sons, 1959.

[13] Merton, R. C., An analytical derivation of the efficient portfolio frontier, *Journal of Financial and Quantitative Analysis*, **7** (1972), 1851-1872.

[14] Roy, A. D., Safety first and the holding of assets, *Econometrica*, **20** (1952) 431-449.

[15] Sniedovich, M., A class of variance-constrained problems, *Operations Research*, **31** (1983), 338-353.

[16] Sniedovich, M., Solution strategies for variance minimization problems, *Computers & Mathematics with Applications*, **21** (1991), No. 2-3, 49-56.

[17] Sobel, M. J., The variance of discounted Markov decision processes, *J. Appl. Prob.*, **19** (1982), 794-802.

[18] WHITE, D. J., Dynamic programming and probabilistic constraints, *Operations Research*, **22** (1974), 654-664.

[19] WHITE, D. J., Utility, probabilistic constraints, mean and variance of discounted rewards in Markov decision process, *OR Spektrum*, **9** (1987), 13-22.

[20] WHITE, D. J., Mean, variance, and probabilistic criteria in finite Markov decision processes: A review, *Journal of Optimization Theory and Applications*, **56** (1988), 1-29.

[21] ZHOU, X. Y. AND LI, D., Continuous time mean-variance portfolio selection: A stochastic LQ framework, *Applied Mathematics and Optimization*, **42** (2000), 19-33.

11

A Markov Chain Method for Pricing Contingent Claims

Jin-Chuan Duan, Geneviève Gauthier and Jean-Guy Simonato

This chapter introduces the use of a time-homogenous Markov chain for the valuation of options. In the computational finance literature, three categories of numerical techniques have been explored extensively for the valuation of contingent claims. The first category involves the use of a lattice structure to approximate the price movement of the underlying asset under the risk neutral probability measure and then computes the price of a contingent claim as a discounted expected payoff. The lattice approach essentially discretizes both time and state in a particular way. The second category is the finite difference/element approach. This technique numerically solves the partial differential equation that the value function of a contingent claim must obey under the no-arbitrage condition. The third category is the Monte Carlo method, which simulates the system under the risk-neutral probability measure so that the expectation of a contingent payoff can be approximated by the sample average.

In Duan and Simonato (2001), a Markov chain method was developed as an alternative to the existing numerical valuation methods. The method uses a time homogenous Markov chain to approximate the price system under the risk neutral probability measure. The approximating model is a discrete-time finite-state Markov chain but the target model can be either a discrete-time continuous-state or continuous-time continuous-state process. This approach, which can be viewed as a more general lattice method, was developed with the specific purpose of valuing American options in a time-varying volatility framework. However, as shown in Duan, Gauthier and Simonato (1999) and Lebel (2001), this approach can be easily adapted to many other problems encountered in financial engineering. The Markov chain method has one distinct feature. Unlike the traditional lattice and finite difference methods, the Markov chain approach allows one to decouple the partitioning of time and state. In other words, one can use time steps suitable for a particular contingent claim without being unduly constrained to have a particular set of state values. This feature proves to be extremely useful when early exercise or path dependency is present and/or when the postulated underlying price dynamic is a discrete-time stochastic process. This design feature provides three useful operational properties. First, the approach yields a simple recursive matrix formula for valuing options (European, American and exotic). Second, the transition probability matrix associated with the Markov chain is highly sparse, which makes it possible to significantly reduce storage requirement and computation time with the use of sparse matrix techniques. Third, the method can be easily adapted to different theoretical models. The Markov chain design may also be modified to deal with a d-dimensional valuation framework where d is greater than two. We refer readers to Duan, Gauthier and Simonato (2001) for details.

The organization of this chapter is as follows. We first describe in general terms the basic idea of the Markov chain method. Detailed discussions are provided for two special cases - the Black-Scholes (1973) and Duan's (1995) GARCH option pricing models. The Black-Scholes model, which uses a geometric Brownian motion for the asset price dynamic, constitutes the simplest valuation problem, and we use it to show that the commonly used binomial tree method is actually a special case of the Markov chain method. We then use the GARCH option pricing model to illustrate the Markov chain approach to a bivariate valuation problem for which both asset price and its volatility are random variables. Numerical results presented in this chapter include standard options of American and European styles as well as examples of exotic options of the barrier type.

11.1 The Markov Chain Pricing Method

Consider the filtered probability space $(\Omega, \mathcal{F}, \{\mathcal{F}_t : t \in \mathcal{T}\}, P)$ where $\mathcal{T}$ is a set of indices such as $[0,\infty)$, $[0,T]$, $\{0,1,...,T\}$, etc. Denote the time-t value of the contingent claim by $v(\mathbf{S}_t, t)$ where $\{\mathbf{S}_t = (S_1(t), ..., S_d(t)) : t \in \mathcal{T}\}$ is a d-dimensional stochastic process that captures all pricing-relevant information such as underlying asset, interest rate, exchange rate, convenience yield, volatility, etc. The typical derivative pricing theory implies that, under the hypothesis of no-arbitrage, there is an equivalent pricing measure Q with which all discounted values of tradable assets are martingales. The logical consequence is that replicable contingent claims can be priced uniquely by simply taking an expectation under Q of their associated payoffs and then discount the averages using the discount factor determined by the risk-free interest rate.

In principle, one can invoke the dynamic programming technique to compute (or approximate) the value of the Bermudan (American) contingent claim recursively:

$$\begin{aligned} v(\mathbf{S}_t, t) &= \max\left\{f(\mathbf{S}_t, t), e^{-r_t\tau}\mathsf{E}_t^Q\left[v(\mathbf{S}_{t+\tau}, t+\tau)\right]\right\}, \\ v(\mathbf{S}_T, T) &= f(\mathbf{S}_T, T), \end{aligned} \tag{11.1}$$

for $t \in \{0, \tau, 2\tau, ..., (T^* - 1)\tau\}$ where r_t is the risk-free interest rate (continuously compounded); $f(\mathbf{S}_t, t)$ is the time-t payoff of the contingent claim if exercised; T is the maturity date of the contingent claim; τ is the length of a time period dictated by the model or chosen for approximation purposes[1]; $T^* = T/\tau$ is the number of time periods with the maturity measured in terms of the basic valuation time interval; and $\mathsf{E}_t^Q[\bullet]$ is a short-hand notation for the conditional expectation $\mathsf{E}^Q[\bullet|\mathcal{F}_t]$.

The theory aside, the early exercise possibility related to Bermudan (American) style derivatives creates numerical complications. The major difficulty in implementing (11.1) is related to computing the conditional expectation and that is where the Markov chain method comes in. The Markov chain approach replaces the stochastic process $\mathbf{S}$ by a discrete-state process for which the conditional expectation can be easily computed. Since the methodology relies on time-homogenous Markov chains, we need to ensure that there is a time-homogeneous Markovian process $\mathbf{X}$ behind $\mathbf{S}$. More formally, the Markov chain approach assumes existence of a function $g : \mathcal{R}^d \times \mathcal{T} \to \mathcal{R}^d$ such that 1) the d-dimensional process $\mathbf{X} = \{\mathbf{X}_t : t \in \mathcal{T}\}$ where $\mathbf{X}_t = (X_1(t), ..., X_d(t)) = g(\mathbf{S}_t; t)$ is a time-homogeneous Markovian process under Q and 2) for any fixed t, $g(\cdot; t)$ is invertible ensuring that $\mathbf{S}_t = g^{-1}(\mathbf{X}_t; t)$. An illustration of this transformation is given in Section 11.2. The time-homogenous Markov process $\mathbf{X}$ is then approximated by an m-state Markov chain

$$\left\{\mathbf{X}_{t^*}^{(m)} : t^* \in \{0, 1, 2, ..., T^*\}\right\}$$

so that $\mathbf{X}_{t^*}^{(m)}$ converges to $\mathbf{X}_{t^*\tau}$ weakly for any $t^* \in \{0, 1, 2, ..., T^*\}$ as m tends to infinity. To ensure that the price of a contingent claim obtained from applying the Markov chain method converges to the right theoretical value (given by the model $\mathbf{S}$), we also need that $\phi\left(\mathbf{X}_{t^*}^{(m)}\right)$ converges to $\phi(\mathbf{X}_{t^*\tau})$ in mean (as $m \to \infty$) for a reasonable class of payoff functions ϕ.

Under the time-homogenous Markov chain approximation, the valuation problem described in (11.1) can be written as

$$\overrightarrow{v}_{t^*}^{(m)} = \max\left[f\left(g^{-1}(\mathbf{X}^{(m)}; t^*\tau), t^*\tau\right), e^{-r_{t^*}\tau}\mathbf{\Pi}^{(m)}\overrightarrow{v}_{t^*+1}^{(m)}\right],$$

[1]In the following, the number of time periods is marked by a star while time is represented without such superscript. This distinction is needed since in some cases we use a discrete-time structure to approximate a continuous-time model. We need the liberty of choosing τ to suit the characteristics of the contingent claim in question.

$$\overrightarrow{v}_{T^*}^{(m)} = f\left(g^{-1}(\mathbf{X}^{(m)}; T^*\tau), T^*\tau\right), \tag{11.2}$$

for $t^* \in \{0, 1, ..., T^* - 1\}$ where $\mathbf{X}^{(m)}$ is an $m \times d$ matrix that contains each of the m possible states of the Markov chain (each state belongs to $\mathcal{R}^d$); $f\left(g^{-1}(\mathbf{X}^{(m)}; t^*), t^*\tau\right)$ is also a $m \times 1$ vector filled with payoffs of the contingent claim, if exercised at time $t = t^*\tau$, corresponding to m states of the chain; $\mathbf{\Pi}^{(m)}$ is the time-homogeneous $m \times m$ transition probability matrix over a time period of length τ, and the $m \times 1$ vector $\overrightarrow{v}_{t^*}^{(m)}$ contains, for each of the possible states, the time $t = t^*\tau$ values of the contingent claim. Note that $\mathsf{E}_t^Q\left[v\left(\mathbf{S}_{t+\tau}, t+\tau\right)\right]$ is being approximated by $\mathbf{\Pi}^{(m)}\ \overrightarrow{v}_{t^*+1}^{(m)}$, a simple matrix-vector multiplication. After the recursive operation is completed, the element of $\overrightarrow{v}_0^{(m)}$ corresponding to the initial state of the Markov system is the Markov chain value for the contingent claim.

The implementation of the above Markov chain method requires that we have available a suitable set of state values and a transition probability matrix associated with these values. The next section discusses how we can find these quantities in such a way that the approximating Markov chain and the target stochastic process share identical probabilistic behavior in the limit. The discussion will first take place in the Black-Scholes context which has been the workhorse of financial engineering over the past twenty-thirty years, and then move on to the GARCH setting.

11.2 The Black-Scholes (1973) Pricing Model

In the Black-Scholes model, only the asset price is random and is modeled as a geometric Brownian motion process. We thus have $d = 1$ and a risk-neutralized asset price dynamic given by:

$$dS_t = rS_t dt + \sqrt{h} S_t dW_t, \tag{11.3}$$

where r is a constant risk-free rate of interest, h is the instantaneous variance of the asset return and W is a standard Brownian motion with respect to the risk-neutralized probability measure Q. Using the strong solution of this stochastic differential equation, one can rewrite the asset return under the risk-neutralized probability measure Q in a form that is convenient for the purpose of constructing an approximating Markov chain:

$$\ln \frac{S_{t+\tau}}{S_t} = \left(r - \frac{1}{2}h\right)\tau + \sqrt{h}\left(W_{t+\tau} - W_t\right). \tag{11.4}$$

The Black-Scholes model, due to its constant volatility, has a temporal aggregation property, i.e., τ can be an arbitrary positive number. This thus enables one to freely adjust the time step of the Markov chain to match the specific needs of the pricing problem.

A straightforward application of a finite-state Markov chain approximation is ill advised if there exists a trend in the target process and particularly if the valuation problem is of long term in nature. Allowing for the target process to have an upward drift requires of setting a large value for the maximum state so as to have a proper coverage. If the number of states is fixed, the resulting partition will naturally be coarse. Such a problem can be easily addressed by targeting an adjusted asset price instead; that is,

$$X_t = \ln S_t - (r - \frac{1}{2}h)t = \ln S_0 + \sqrt{h}W_t. \tag{11.5}$$

Such a transformation from S_t to X_t (i.e., $g(s,t) = \ln s - (r - \frac{1}{2}h)t$) removes the inherent growth trend in the asset price. Since this transformation is invertible, the value of the adjusted asset price can be easily converted back later to recover the asset price.

To construct a Markov chain that will, in the limit, share the same probabilistic behavior as the adjusted asset price process X, we need to ascertain that the set of state values for the Markov chain provides a reasonable coverage of the possible asset values at the maturity of the option. Obviously there are many reasonable ways to construct such a set. A simple procedure is adopted here. First, we find an interval that extends for δ standard deviations in each direction from X_0, the starting point of the target process. This interval can then be partitioned into equal-size cells. The middle points of these cells are used as the state values of the Markov chain. These state values in conjunction with the cell boundaries allow us to compute the entries of the transition probability matrix in accordance with equation (11.5). Such a Markov chain will thus mimic the probabilistic behavior of the target geometric Brownian motion process over the set of discrete time points. The next two subsections formalize the procedure.

11.2.1 Choosing the Set of Asset Prices

The target Markov process is

$$\left\{\mathbf{X}_t = (X_t) = \ln S_t - (r - \frac{1}{2}h)t : t \geq 0\right\}.$$

The overall interval covering the set of representative adjusted asset prices is written as $[X_0 - I_x\ ,\ X_0 + I_x]$ where $X_0 = \ln S_0$ is the logarithm of the known asset price at time 0. The quantity I_x is computed using the conditional standard deviation of the asset return over the life of the option contract multiplied by a scaling factor. Formally,

$$I_x = \delta(m)\sqrt{hT} \tag{11.6}$$

is one half of the length of the interval over which we choose our representative asset prices. The quantity, $\sqrt{hT}$, is the standard deviation of the asset return at time T. The scaling factor $\delta(m)$ is an increasing function of m, satisfying the partition condition: 1) $\delta(m) \to \infty$ as $m \to \infty$ and 2) $\frac{\delta(m)}{m} \to 0$ as $m \to \infty$. The first item of the partition condition ensures that the interval is eventually extended to cover arbitrarily large and small values. The second item ensures that the increase (decrease) in the upper (lower) bound does not outpace the increase in the number of subintervals. In other words, an increase in refinement is achieved as m goes to infinity. This partition condition guarantees that the approximating Markov chain converges to its target process. Examples of the scaling factor satisfying the partition condition will be given later.

Once the overall interval for the adjusted asset price is determined, the interval is divided equally into $m-1$ parts to yield m discrete values

$$x(i) = X_0 + \frac{2i-m-1}{m-1} I_x, \quad i \in \{1, \cdots, m\},$$

for the adjusted asset price. Note that $x(1) = X_0 - I_x$ and $x(m) = X_0 + I_x$ are the minimum and maximum values, respectively. In order to have X_0 among the state prices, m needs to be an odd integer, and in which case, $x\left(\frac{m+1}{2}\right) = X_0$. The m cells are constructed as:

$$C(1) = (c(1), c(2)), C(i) = [c(i), c(i+1)) \text{ for } i = \{2, \cdots, m\}, \quad (11.7)$$

where $c(1) = -\infty$, $c(i) = \frac{x(i)+x(i-1)}{2}$ for $i = 2$ to m and $c(m+1) = +\infty$. The computation of the transition probabilities is the next crucial issue.

11.2.2 Computing Transition Probabilities and Option Prices

The transition probability from the adjusted asset price $x(i)$ to $x(j)$ over the time period $[t, t+\tau]$ is defined as:

$$\pi(i, j; \tau) = \mathsf{P}^Q [X_{t+\tau} \in C(j) | X_t = x(i)], \quad \text{for } i, j \in \{1, \cdots, m\} \quad (11.8)$$

where $C(j)$ has been defined in (11.7). Note that time-homogeneity allows one to ignore the time index t. Since equation (11.5) implies that

$$X_{t+\tau} = X_t + \sqrt{h}\,(W_{t+\tau} - W_t),$$

the transition probability can be computed by

$$\begin{aligned}
\pi(i, j; \tau) &= \mathsf{P}^Q [c(j) \le X_{t+\tau} < c(j+1) | X_t = x(i)] \\
&= \mathsf{P}^Q \left[c(j) \le x(i) + \sqrt{h}\,(W_{t+\tau} - W_t) < c(j+1)\right] \\
&= \mathsf{P}^Q \left[\frac{c(j) - x(i)}{\sqrt{h\tau}} \le \frac{W_{t+\tau} - W_t}{\sqrt{\tau}} < \frac{c(j+1) - x(i)}{\sqrt{h\tau}}\right] \\
&= \Phi\left(\frac{c(j+1) - x(i)}{\sqrt{h\tau}}\right) - \Phi\left(\frac{c(j) - x(i)}{\sqrt{h\tau}}\right), \quad (11.9)
\end{aligned}$$

where $\Phi(\cdot)$ stands for the cumulative standard normal distribution function.

Using these transition probabilities and the discretized asset prices, an option price can be computed using equation (11.2) with the early exercise consideration performed for every period of length τ. For plain vanilla European call and put options, the payoff function can be written as

$$f\left(g^{-1}(\mathbf{X}^{(m)};t),t\right) \tag{11.10}$$

$$= \max\left\{w\left[\exp\left[(r-\frac{1}{2}h)t\mathbf{1}+\mathbf{X}^{(m)}\right]-K\mathbf{1}\right],\mathbf{0}\right\}, \tag{11.11}$$

where w indicates a call ($w=1$) or a put ($w=-1$), K is the strike price, $\mathbf{0}$ and $\mathbf{1}$ denote $m\times 1$ vectors of zeros and ones, respectively. The functions exp and max are vector-valued functions applied element by element. For options without early exercise privilege (European style), the early exercise feature embedded in the recursion can be ignored. In fact, even for American call options, the early exercise consideration is unnecessary if there is no cash dividends. For these cases, the recursive system can be simplified to:

$$\vec{v}_0^{(m)} = e^{-rT^*\tau}\left(\Pi^{(m)}\right)^{T^*} f\left(g^{-1}(\mathbf{X}^{(m)};T^*\tau),T^*\tau\right). \tag{11.12}$$

It is worth noting that the above expression is solely conceptual in nature. A direct computation of the quantity $\left(\Pi^{(m)}\right)^{T^*}$ will take more computing time than performing the matrix-vector multiplication repeatedly unless diagonalization of $\Pi^{(m)}$ is performed first.

The Markov chain method described here has been shown in Duan and Simonato (2001) to converge to the right theoretical values in the GARCH setting for typical payoff functions. Since the GARCH model contains the Black-Scholes model as a special case, their result immediately applies here. There are two essential elements in this convergence result. First, the approximating Markov chain converges weakly to the target asset price process. Second, the expected value of any "regular" payoff function, computed using the approximating scheme, converges to the theoretical expected value.

11.2.3 An Illustrative Example

A simple example is presented here to demonstrate the steps in implementing the Markov chain method. In this example, a Markov chain with 9 states and 4 time steps is used to price a European call with maturity of one year and strike price of 45 in the Black-Scholes pricing framework. The parameter values are thus $T=1$, $\tau=0.25$, $T^*=4$. We use the initial asset price $S_0=50$, the risk free interest rate (annualized) $r=0.05$ and the annual variance $h=0.04$. The scaling factor chosen is $\delta(m)=\sqrt{(m-1)/2}$.

This particular choice of scaling factor will become clear later when the link to binomial tree is examined. The implementation steps are laid out as follows.

Step 1 comes up with the overall adjusted asset price interval to be partitioned. First, compute

$$I_x = \sqrt{(9-1)/2} \times \sqrt{0.04 \times 1} = 0.4.$$

The initial value of the adjusted asset price is $X_0 = \ln(S_0) - (r - \frac{1}{2}h) \times 0 = 3.912$. This then yields the overall interval for the adjusted asset price: $[X_0 - I_x\ ,\ X_0 + I_x] = [3.512, 4.312]$.

Step 2 sets up 9 equally spaced adjusted asset prices and 10 boundary values for cells. The first adjusted asset price is thus $x(i) = 3.512$ and the 8 others are built recursively with an increment of $(4.312 - 3.512)/8 = 0.1$ for the subsequent values. This yields the following quantities for the adjusted asset prices, $x(i)$, and the boundary values for cells, $c(i)$:

i	$x(i)$	$c(i)$
1	3.512	$-\infty$
2	3.612	$\frac{3.612+3.512}{2} = 3.562$
3	3.712	3.662
4	3.812	3.762
5	3.912	3.862
6	4.012	3.962
7	4.112	4.062
8	4.212	4.162
9	4.312	4.262
10	-	∞

Step 3 computes the Markov transition probability matrix. The matrix for transition over three months (0.25 years) can be constructed using equation (11.9). For example, element $(1, 2)$ of this transition probability matrix can be computed as:

$$\pi(1,2;0.25) = \Phi\left(\frac{3.662 - 3.512}{\sqrt{0.04 \times 0.25}}\right) - \Phi\left(\frac{3.562 - 3.512}{\sqrt{0.04 \times 0.25}}\right) = 0.2417.$$

Table 11.1 provides values for the entire transition probability matrix. Notice here that probabilities with values smaller than 1e-6 are set to zero. The first row gives the 9 adjusted asset prices.

Step 4 computes the option price. For example, the fourth element of the payoff vector for the European call option with strike price of 45 at time $T = T^*\tau$ is computed as

$$\max\left(\exp\left[\left(0.05 - \frac{1}{2}0.04\right) \times 4 \times 0.25 + 3.8120\right] - 45, 0\right) = 1.6186.$$

The adjusted asset prices

3.5120	3.6120	3.7120	3.8120	3.9120	4.0120	4.1120	4.2120	4.3120

Transition probability matrix (3 months)

0.6915	0.2417	0.0606	0.0060	0.0002	0.0000	0.0000	0.0000	0.0000
0.3085	0.3829	0.2417	0.0606	0.0060	0.0002	0.0000	0.0000	0.0000
0.0668	0.2417	0.3829	0.2417	0.0606	0.0060	0.0002	0.0000	0.0000
0.0062	0.0606	0.2417	0.3829	0.2417	0.0606	0.0060	0.0002	00000
0.0002	0.0060	0.0606	0.2417	0.3829	0.21417	0.0606	0.0060	0.0002
0.0000	0.0002	0.006	0.0606	0.2417	0.3829	0.2417	0.0606	0.0062
0.0000	0.0000	0.0002	0.0060	0.0606	0.2417	0.3829	0.2417	0.0668
0.0000	0.0000	0.0000	0.0002	0.0060	0.0606	0.2417	0.3829	0.3085
0.0000	0.0000	0.0000	0.0000	0.0002	0.0060	0.0606	0.2417	0.6915

TABLE 11.1. Transition probability matrix for the Black and Scholes model (9 states)

Repeating the matrix-vector multiplication four times and then applying the scalar discount factor given by $\exp(-0.05 \times 0.25 \times 4)$ will yield the following vector of option prices:

$$\vec{v}_0^{(m)} = \begin{bmatrix} 0.57345 \\ 1.2542 \\ 2.6702 \\ 5.0306 \\ 8.3789 \\ 12.5020 \\ 16.8710 \\ 20.7020 \\ 23.1900 \end{bmatrix}.$$

Because the current asset price corresponds to the state value $X_0 = 3.912$ which is the fifth element of state vector, the computed option price is 8.3789, the fifth element in the final value vector.

11.2.4 A Markov Chain Interpretation of Binomial Tree

We now show that the binomial tree method involves the use of a specific Markov chain design. This link not only helps one better understand the Markov chain method but also provides insights into the lattice method. We use the binomial tree of Jarrow and Rudd (1983) to demonstrate this link. Similar links can be easily established for different binomial and trinomial tree schemes. The Jarrow and Rudd (1983) n-step binomial lattice deals with the logarithmic asset price and uses both up and down probabilities

The adjusted asset prices

3.5120	3.6120	3.7120	3.8120	3.9120	4.0120	4.1120	4.2120	4.3120

Transition probability matrix (3 months)

1	0	0	0	0	0	0	0	0
1/2	0	1/2	0	0	0	0	0	0
0	1/2	0	1/2	0	0	0	0	0
0	0	1/2	0	1/2	0	0	0	0
0	0	0	1/2	0	1/2	0	0	0
0	0	0	0	1/2	0	1/2	0	0
0	0	0	0	0	1/2	0	1/2	0
0	0	0	0	0	0	1/2	0	1/2
0	0	0	0	0	0	0	0	1

TABLE 11.2. Transition probability matrix for the binomial-tree method (4 steps)

of 1/2. If the deterministic growth component, $\left(r - \frac{1}{2}h\right) T/n$, is removed, one effectively deals with the adjusted asset price as described earlier. The up and down sizes for the adjusted asset price are given by $u = \sqrt{h\frac{T}{n}}$ and $d = -\sqrt{h\frac{T}{n}}$.

Note that an n-step binomial tree yields $2n+1$ different asset prices (the prices on the nodes of the last two time steps). A 4-step binomial tree for the option described in the preceding subsection generates 9 different asset prices with $u = \sqrt{0.04 \times 1/4} = 0.1$ and $d = -0.1$. Since the initial adjusted asset price is 3.912, the nine adjusted asset prices according to this binomial tree are therefore $\{3.512, 3.612, 3.712, 3.812, 3.912, 4.012, 4.112, 4.212, 4.312\}$. Every step of the binomial tree can thus be described as a Markov chain with the transition probability matrix shown in Table 11.2. For this transition probability matrix, both the top-left and bottom-right corners have a value of 1 because they correspond to the maximum and minimum values possible in the binomial tree. In other words, these two values cannot go anywhere in the binomial tree and thus they constitute absorbing states.

The normal backward recursive operation according to the binomial tree gives rise to the option value of 8.4841, which is equivalent to going through the recursive system of equation (11.2) with the matrix given in Table 11.2. The resulting option price vector obtained with this operation is $[0, 0.99854, 2.6168, 4.8762, 8.4841, 12.811, 18.246, 24.086, 30.307]'$. The option price is again the center value of the final value vector, which agrees with the direct calculation from a tree.

Comparing this Markov chain corresponding to a binomial tree with the direct Markov chain constructed in the preceding subsection, it is clear that the specific scaling factor $\delta(m) = \sqrt{(m-1)/2}$ is used to generate the exact same set of state prices corresponding to the binomial tree. The real difference is in assigning values to the transition probability matrix. The binomial tree gives rise to a very specific transition probability matrix with 1/2 as the up and down probabilities. In other words, even though the binomial tree method can be interpreted as a specific Markov chain design, its assignment of transition probabilities has nothing to do with the Markov chain method described in this article. Assigning 1/2 probability to the up and down moves leads to a poor approximation of the normal distribution over one time step implied by the theoretical model. In contrast, the direct Markov chain method offers a far better approximation of the conditional distribution. For the particular option in question, for example, the Black-Scholes formula value equals 8.3497, which is better approximated by the direct Markov chain method.

The lattice method is highly rigid in its geometry. For example, in order for a binomial tree to have 9 possible asset prices, one must use a 4-step tree. In contrast, the direct Markov chain method allows for a complete separation of time and state, which in turn offers much of the needed flexibility for pricing exotic derivatives such as barrier options. This aspect will be explored further later.

In general, one may want to consider a scaling factor different from $\delta(m) = \sqrt{(m-1)/2}$ as long as it still satisfies the partition condition. An alternative scaling factor may be desirable because letting the maximum (minimum) value tend to infinity (minus infinity) at a rate of square root of m may not give the best numerical performance. The choice of scaling factor adds additional flexibility to the Markov chain method. In the numerical examples in the next subsection, a different scaling factor is employed.

11.2.5 Numerical Examples

Table 11.3 presents the numerical results for European put options using the binomial tree and Markov chain methods. We consider three maturities (1, 3 and 9 months) and three strike-to-asset price ratios (1.1, 1.0 and 0.9). In the numerical analyses, we assume 365 days per year for annualizing. The current asset price is set at 50. The annualized risk-free rate (continuously compounded) and the asset return standard deviation are 5% and 20%, respectively. Notice also that in the transition probability matrix, probabilities with values smaller than 1e-6 are set to zero. The first row of this table presents the Black-Scholes formula prices for these European puts.

The first panel of this table presents the European put option prices using the Jarrow and Rudd (1983) version of the binomial-tree method with various numbers of steps. As discussed earlier, a $\frac{m-1}{2}$-step binomial tree yields m different asset prices (the prices in the last 2 binomial steps) so

	Maturity = 1 month			**Maturity = 3 months**			**Maturity = 9 months**		
K/S$_0$	1.10	1.00	0.90	1.10	1.00	0.90	1.10	1.00	0.90
				Theoretical prices					
	4.8457	1.0416	0.0293	4.9098	1.6769	0.2706	5.2318	2.5343	0.9150
				Binomial tree prices ($\frac{m-1}{2}$ steps)					
$m = 11$	4.8361	1.0987	0.0283	4.9635	1.7715	0.2846	5.0865	2.6773	0.9718
$m = 51$	4.8461	1.0520	0.0274	4.9160	1.6912	0.2735	5.2614	2.5403	0.9291
$m = 101$	4.8454	1.0418	0.0291	4.9066	1.6821	0.2646	5.2469	2.5503	0.9237
$m = 201$	4.8446	1.0426	0.0292	4.9122	1.6811	0.2715	5.2298	2.5411	0.9181
$m = 301$	4.8450	1.0427	0.0292	4.9114	1.6801	0.2690	5.2364	2.5361	0.9182
$m = 401$	4.8457	1.0426	0.0292	4.9102	1.6793	0.2692	5.2274	2.5328	0.9158
$m = 501$	4.8455	1.0425	0.0291	4.9079	1.6787	0.2695	5.2342	2.5315	0.9146
				Markov chain prices (time step=1 month), $I_x = [2 + \ln(\ln(m))]\sqrt{hT}$)					
$m = 11$	4.8478	1.0363	0.0304	4.9716	1.7270	0.2832	5.5170	2.8340	1.1487
$m = 51$	4.8452	1.0427	0.0290	4.9129	1.6819	0.2729	5.2466	2.5484	0.9297
$m = 101$	4.8455	1.0419	0.0292	4.9106	1.6772	0.2708	5.2348	2.5389	0.9189
$m = 201$	4.8457	1.0416	0.0292	4.9098	1.6769	0.2706	5.2330	2.5357	0.9158
$m = 301$	4.8457	1.0416	0.0292	4.9098	1.6768	0.2706	5.2321	2.5348	0.9154
$m = 401$	4.8457	1.0416	0.0292	4.9098	1.6768	0.2706	5.2320	2.5345	0.9151
$m = 501$	4.8457	1.0416	0.0293	4.9098	1.6769	0.2706	5.2319	2.5345	0.9151

Parameters: $S_0 = 50$, $r = 0.05$ (annualized), $h = 0.20$ (annualized) and 1 day = 1/365.

TABLE 11.3. European put options in the Black and Scholes framework

	Maturity = 1 month			Maturity = 3 months			Maturity = 9 months		
K/S_0	1.10	1.00	0.90	1.10	1.00	0.90	1.10	1.00	0.90
				Binomial tree prices (10,000 steps)					
	5.0001	1.0567	0.0295	5.1608	1.7295	0.2758	5.7473	2.7182	0.9637
				Binomial tree prices ($\frac{m-1}{2}$ steps)					
$m = 11$	5.0000	1.1166	0.0283	5.2043	1.8249	0.2915	5.7029	2.8353	0.9919
$m = 51$	5.0000	1.0682	0.0276	5.1649	1.7459	0.2783	5.7628	2.7301	0.9738
$m = 101$	5.0000	1.0571	0.0293	5.1611	1.7344	0.2706	5.7565	2.7305	0.9719
$m = 201$	5.0000	1.0578	0.0294	5.1625	1.7336	0.2770	5.7475	2.7245	0.9669
$m = 301$	5.0000	1.0579	0.0295	5.1620	1.7327	0.2747	5.7499	2.7207	0.9665
$m = 401$	5.0000	1.0578	0.0294	5.1614	1.7320	0.2747	5.7459	2.7182	0.9642
$m = 501$	5.0000	1.0577	0.0293	5.1602	1.7315	0.2750	5.7490	2.7168	0.9637
				Markov chain prices (time step = 1 day, $I_x = [2 + \ln(\ln(m))]\sqrt{hT}$)					
$m = 11$	5.0000	1.0991	0.0465	5.0000	0.5763	0.0082	5.0000	0.0013	0.0000
$m = 51$	5.0000	1.0803	0.0333	5.2230	1.8465	0.3370	6.1077	3.1438	1.2918
$m = 101$	5.0000	1.0627	0.0306	5.1765	1.7611	0.2921	5.8734	2.8720	1.0791
$m = 201$	5.0000	1.0576	0.0297	5.1637	1.7375	0.2801	5.7809	2.7603	0.9948
$m = 301$	5.0000	1.0567	0.0296	5.1612	1.7327	0.2778	5.7620	2.7372	0.9779
$m = 401$	5.0000	1.0563	0.0295	5.1603	1.7310	0.2768	5.7551	2.7288	0.9716
$m = 501$	5.0000	1.0561	0.0295	5.1598	1.7301	0.2764	5.7518	2.7248	0.9687

Parameters: $S_0 = 50$, $r = 0.05$ (annualized), $h = 0.20$ (annualized) and 1 day = 1/365.

TABLE 11.4. American put options in the Black and Scholes framework

that the number of asset prices already tells us about the number of steps in the binomial tree. The results indicate that the binomial-tree prices exhibit a typical jagged convergence pattern.[2] The Markov chain option prices for various numbers of states are presented in the second panel of this table. The values in this panel are obtained with one month as the basic transition time step. (Recall the earlier discussion on the time-state separation.) The value for I_x is computed using the scaling factor $\delta(m) = 2+\ln(\ln(m))$. Note that this scaling factor, in comparison with $\delta(m) = \sqrt{(m-1)/2}$, tends to infinity much slower. As a result, the possible asset prices are more densely scattered in a relevant target region. It is clear that the computed value converges to the corresponding theoretical price as the number of states increases. Moreover, the Markov chain method seems to converge faster than does the binomial tree method.[3]

The results for American put options are presented in Table 11.4. Because there is no formula value for American options, we benchmark the computed prices to the binomial-tree prices with 10,000 steps, which, for all practical purposes, can be regarded as the theoretical values on the basis of continuous early exercise. For the Markov chain method, we again employ the double logarithmic scaling factor to determine the asset price range. The values in the first panel are based on the $\frac{m-1}{2}$-step binomial tree, where m is the number of discrete asset prices. American options thus have $\frac{m-1}{2}$ early exercise opportunities. In contrast to the binomial-tree method, an increase in the number of asset prices under the Markov chain method does not need to increase the number of early exercise opportunities, again due to the time-state separation. The results in 11.4 are obtained by fixing the number of time steps equal to maturity (in terms of number of days) of the option contract. Because early exercise is only allowed on a daily basis, the theoretical value that the Markov chain method converges to should be lower than the theoretical value by allowing for early exercise on a continuous-time basis. Indeed, the results are in agreement with the prediction.

For the constant-volatility case (the Black-Scholes pricing framework), a fairly accurate price can be obtained in fractions of a second on a standard desktop computer. For example, computing a standard European call option with a maturity of 90 days using 201 states and daily time steps takes 0.016 second of computer time. For the same option contract, a crude

[2] As discussed in Broadie and Detemple (1996), it is possible to modify the binomial-tree method so that this oscillatory pattern is removed and the convergence speed is improved. The modification calls for replacing the calculation of the option values in the last time step with the value from the Black-Scholes formula.

[3] Although not reported here, other functional forms for $\delta(m)$ were also considered. We checked the performance of the Markov chain using $\ln(m)$ and $\sqrt{(m-1)/2}$ as the factors for determining the asset price range. These alternative specifications were found to have slower convergence.

Monte Carlo simulation with 50,000 sample paths required 0.047 seconds. The computing time of the Markov chain of course depends on the number of time steps employed. A smaller transition time step will make the transition probability matrix more sparse. This will result in a shorter computation time in building the transition probability matrix and in the subsequent matrix-vector multiplications. A smaller transition time step, however, results in more steps and consequently needs more matrix multiplications in computing option prices.

11.3 The GARCH Pricing Model

One advantage of the Markov chain method is its flexibility in dealing with different theoretical models. It is rather straightforward, for example, to apply it to the model with stochastic volatility. In this section, we show how the Markov chain method can be applied to option pricing in the setting of the non-linear asymmetric GARCH(1,1) process, NGARCH(1,1) for short. Assume that the asset price is governed by the following stochastic process under the data generating probability measure P. For any $t^* \in \{0, 1, 2, ...\}$,

$$\ln \frac{S_{t^*+1}}{S_{t^*}} = r + \lambda\sqrt{h_{t^*+1}} - \frac{1}{2}h_{t^*+1} + \sqrt{h_{t^*+1}}\varepsilon_{t^*+1}, \tag{11.13}$$

$$h_{t^*+1} = \beta_0 + \beta_1 h_{t^*} + \beta_2 h_{t^*}(\varepsilon_{t^*} - \theta)^2, \tag{11.14}$$

$$\varepsilon_{t^*+1}|\mathcal{F}_{t^*} \sim N(0,1), \tag{11.15}$$

where $\{\varepsilon_{t^*} : t^* \in \mathcal{N}\}$ is a sequence of independent and identically distributed standard normal random variables, $\mathcal{F}_{t^*}$ is the σ-field generated by $\{S_0, h_0, \varepsilon_{u^*}\}$ for $u^* = 0, 1, 2, \cdots, t^*$; r is the one-period, continuously compounded risk-free rate of interest; λ is the constant unit risk premium; h_{t^*+1} is the conditional variance of the asset return; θ determines the leverage effect. The conditional variance follows the NGARCH(1,1) process with the typical parameter restrictions: $\beta_0 > 0, \beta_1 \geq 0, \beta_2 \geq 0$, and $\beta_1 + \beta_2(1 + \theta^2) < 1$. According to the valuation theory developed by Duan (1995), the derivative contract contingent on S_{t^*} can be valued by using a locally risk-neutralized price dynamic. Specifically, the asset price dynamic under the locally risk-neutralized probability measure Q is, for any $t^* \in \{0, 1, 2, ...\}$,

$$\ln \frac{S_{t^*+1}}{S_{t^*}} = r - \frac{1}{2}h_{t^*+1} + \sqrt{h_{t^*+1}}\epsilon_{t^*+1}, \tag{11.16}$$

$$h_{t^*+1} = \beta_0 + \beta_1 h_{t^*} + \beta_2 h_{t^*}\left[\epsilon_{t^*} - (\theta + \lambda)\right]^2, \tag{11.17}$$

$$\epsilon_{t^*+1}|\mathcal{F}_{t^*} \sim N(0,1), \tag{11.18}$$

where $\{\epsilon_{t^*} : t^* \in \mathcal{N}\}$ is a sequence of independent and identically distributed standard normal random variables under the measure Q. Note that the above system is Markovian when expressed in a vector form of (S_{t^*}, h_{t^*+1})

for $t^* \in \{0, 1, 2, ...\}$. Using the general notation described earlier, we have $d = 2$ and $\{\mathbf{S}_{t^*} = (S_{t^*}, h_{t^*}) : t^* \in \mathcal{N}\}$.

We will again choose to work with the adjusted asset price. For volatility, we will use the logarithm of variance. The logarithmic transformation is expected to make the variance distribution more symmetrical. By the previously defined notation, we thus have for any $t^* \in \{0, 1, 2, ...\}$,

$$\mathbf{X}_{t^*} = (X_1(t^*), X_2(t^*)) = \left(\ln(S_{t^*}) - (r - \frac{1}{2}h^*)t^*, \ln(h_{t^*+1})\right),$$

where

$$h^* = \beta_0\{1 - \beta_1 - \beta_2[1 + (\theta + \lambda)^2]\}^{-1}$$

is the stationary variance under measure Q. Indeed, as in the Black-Scholes case, the asset price process has an upward drift. However, since volatility is stochastic, we have used the expected value of variance in defining the adjusted asset return; that is,

$$X_1(t^*) = \ln S_{t^*} - (r - \frac{1}{2}h^*)t^*. \tag{11.19}$$

It is convenient to note that under measure Q, we have for any $t^* \in \{0, 1, 2, ...\}$,

$$\begin{aligned} & X_1(t^* + 1) \\ = \; & X_1(t^*) + \frac{1}{2}(h^* - \exp(X_2(t^*))) + \sqrt{\exp(X_2(t^*))}\epsilon_{t^*+1}, \end{aligned} \tag{11.20}$$

$$\begin{aligned} & \exp(X_2(t^* + 1)) \\ = \; & \beta_0 + \beta_1 \exp(X_2(t^*)) + \beta_2 \exp(X_2(t^*)) \left[\epsilon_{t^*+1} - (\theta + \lambda)\right]^2, \end{aligned} \tag{11.21}$$

To obtain a set of discrete values and transition probability, we adopt a strategy similar to the one used earlier for the Black Scholes model. Additional complexity occurs because of the need for variance partitioning. We again construct an overall interval that extends δ standard deviations away in each direction from the starting point of the process. However, there will now be one interval for the adjusted asset price and the other for the logarithmic variance. The set of states of the Markov chain thus consists of all possible pairs of discrete values for the two variables. In other words, if m_1 values are used for the adjusted asset price and m_2 values for the logarithmic variance, the dimension of the chain will be $m = m_1 \times m_2$. Using these state values in conjunction with the cell boundaries and Equations (11.20) and (11.21) enables us to come up with an approximating Markov chain. The procedure described in the remainder of this section is due to Duan and Simonato (2001).

11.3.1 Choosing the Set of Discrete Prices and Volatilities

Similar to the Black-Scholes case, the interval for the adjusted asset return is $[X_1(0) - I_{x_1}, X_1(0) + I_{x_1}]$ where $X_1(0) = \ln(S_0)$,

$$I_{x_1} = \delta_{x_1}(m_1)\sqrt{\sum_{t^*=1}^{T^*} \mathsf{E}^Q[h_{t^*}|\mathcal{F}_0]},$$

and m_1 is the number of discrete values for the adjusted asset return. In this last expression, $\sqrt{\sum_{t^*=1}^{T^*} \mathsf{E}^Q[h_{t^*}|\mathcal{F}_0]}$ serves as an approximation for the variance of the asset return at period T^*. This sum can be evaluated analytically and the formula for $\mathsf{E}^Q[h_{t^*}|\mathcal{F}_0]$ is available in Appendix 11.5.

For the conditional variance, it might not be desirable to center the interval around $X_2(0) = \ln h_1$. The option valuation is performed in accordance with the locally risk-neutralized system and this system's conditional volatility reverts to its stationary level, h^*, when maturity becomes longer. Centering a symmetric interval around $X_2(0)$ might result in a bad coverage of the possible values for the variance. To allow for a better coverage, we define a symmetric interval around some value denoted by $X_2^*(0)$. This interval is written as $[X_2^*(0) - I_{x_2}, X_2^*(0) + I_{x_2}]$ where the determination of I_{x_2} is based on the dispersion of the logarithmic conditional variance. A simple choice for $X_2^*(0)$ is the logarithm of a weighted sum between h_1 and h^* with the weights determined by the maturity of the option. For a short-maturity option, a case for which the volatility process has less time to revert to its long-run mean, more weight is given to h_1. For a long-maturity option, a case where the process has more time to revert to its long-run mean, more weight is loaded on h^*. Specifically, $X_2^*(0)$ is defined as:

$$X_2^*(0) = \ln\left(\frac{\kappa - \min(T,\kappa)}{\kappa} h_1 + \frac{\min(T,\kappa)}{\kappa} h^*\right), \tag{11.22}$$

where $T = T^*\tau$ is the maturity of the option to be valued and $\kappa > 0$ is some preset value for determining weighting. For example, if κ equals three months, the volatility partitioning will be centered at the logarithm of h^* for all options with a maturity longer than or equal to three months. For shorter term options, the weighting is tilted more toward the initial conditional volatility, h_1. It is important to ensure that $X_2(0) \in [X_2^*(0) - I_{x_2}, X_2^*(0) + I_{x_2}]$. This can always be achieved by increasing κ. This strategy is also intuitively appealing, because if the current conditional volatility is far away from its stationary value, it is better to move the center of the interval closer to the current level of conditional volatility.

The value of I_{x_2} also needs to be determined to compute the overall interval for the logarithmic variance. Note that we do not have a formula for the standard deviation of the logarithmic variance. We thus resort to an expression based on the standard deviation of the variance. Specifically,

$$I_{x_2} = \ln[h_1 + \delta_{x_2}(m_2)\sigma_h] - \ln[h_1], \tag{11.23}$$

where

$$\sigma_h = \sqrt{\mathsf{E}^Q\left[h_{T^*}^2|\mathcal{F}_0\right] - \left(\mathsf{E}^Q\left[h_{T^*}|\mathcal{F}_0\right]\right)^2}. \tag{11.24}$$

and m_2 is the number of discrete values for the logarithmic variance. The analytical expression for $\mathsf{E}^Q\left[h_{T^*}^2|\mathcal{F}_0\right]$ is available in Appendix 11.5.

We divide these intervals equally into $m_1 - 1$ and $m_2 - 1$ cells respectively to yield m_1 adjusted asset prices and m_2 values for the logarithmic variance. These values are constructed as follows:

$$\left\{x_1(i) = X_1(0) + \frac{2i - m_1 - 1}{m_1 - 1} I_{x_1}, i = 1, \cdots, m_1\right\},$$

and

$$\left\{x_2(j) = X_2^*(0) + \frac{2j - m_2 - 1}{m_2 - 1} I_{x_2}, j = 1, \cdots, m_2\right\},$$

respectively. Note that m_1 and m_2 need not be equal. The relevant cells for the adjusted asset price are defined as:

$$C(1) = (c(1), c(2)) \text{ and } C(i) = [c(i), c(i+1)),\ i = 2, \cdots, m_1, \tag{11.25}$$

where $c(1) = -\infty$, $c(i) = \frac{x_1(i)+x_1(i-1)}{2}$ for $i = 2$ to m_1 and $c(m_1+1) = +\infty$. For the logarithmic variance, the cells are defined as:

$$D(1) = (d(1), d(2)) \text{ and } D(j) = [d(j), d(j+1)),\ j = 2, \cdots, m_2, \tag{11.26}$$

where $d(1) = -\infty$, $d(j) = \frac{x_2(i)+x_2(i-1)}{2}$ for $j = 2$ to m_2 and $d(m_2 + 1) = +\infty$. Finally, the Cartesian product of these two sets of cells constitutes a partition of the target region.

11.3.2 Computing Transition Probabilities and Option Prices

First note that the nature of the GARCH model makes $X_2(t+1)$ a deterministic function of $X_1(t)$, $X_1(t+1)$ and $X_2(t)$. Indeed, isolating ϵ_{t+1} in Equation (11.20) and placing it in Equation (11.21) gives rise to

$$X_2(t+1) = \Gamma(X_1(t), X_1(t+1), X_2(t)), \tag{11.27}$$

where

$$\Gamma(x, y, z) = \ln\left[\beta_0 + \beta_1 e^z + \beta_2\left(y - x + \frac{1}{2}[e^z - h^*] - (\theta + \lambda)e^{\frac{z}{2}}\right)^2\right].$$

The transition probability under measure Q is

$$\pi(i,j;k,l) = \mathsf{P}^Q\{X_1(t+1) \in C(k) | X_1(t) = x_1(i), X_2(t) = x_2(j)\}\mathbf{1}_\Gamma,$$

where the indicator function $\mathbf{1}_\Gamma$ equals 1 if the condition $\Gamma(x_1(i), x_1(k), x_2(j)) \in D(l)$ is satisfied and 0 otherwise. The conditional probability can be analytically computed as follows:

$$\begin{aligned} & \mathsf{P}^Q\left[X_1(t+1) \in C(k)\, |X_1(t) = x_1(i), X_2(t) = x_2(j)\right] \\ = \; & \mathsf{P}^Q\left[L_{ij}(k) \leq Z < L_{ij}(k+1)\right] \end{aligned}$$

where

$$L_{ij}(k) = \frac{c(k) - x_1(i) + \frac{1}{2}[\exp(x_2(j)) - h^*]}{\sqrt{\exp(x_2(j))}},$$

and Z is a standard normal random variable under measure Q. To obtain the option price, we need to define the payoff vector which is a $m_1 \times m_2$ dimensional vector. This vector repeats m_2 times the m_1 dimensional vector corresponding to the m_1 different asset prices.

After applying the recursive system (11.2), one needs to linearly interpolate the two option values corresponding to the two volatilities surrounding the current volatility. This interpolation is required since equation (11.22) is used to set the centering point for the volatilities.

11.3.3 Numerical Examples

To implement the GARCH option pricing model, we need to know the current asset price and conditional volatility. For the numerical analyses of this section, the initial conditional variance is set equal to the stationary variance under the data generating probability measure P; that is, $h_1 = \beta_0\left[1 - \beta_1 - \beta_2(1+\theta^2)\right]^{-1}$. Recall that due to the notational convention of the GARCH literature, h_1 is the conditional variance for the first asset return innovation, and it is known at time 0. We use the stationary variance under measure P instead of measure Q because it represents the true average variance from observing the data series. For the centering scheme in equation (11.22), κ is chosen to be 3 months. Recall that $h^* = \beta_0\left[1 - \beta_1 - \beta_2[1+(\theta+\lambda)^2]\right]^{-1}$. Since the typical GARCH parameter estimates are combined to yield a value of $\beta_1 + \beta_2[1+(\theta+\lambda)^2]$ close to one, reversion to h^* takes a longer time and κ should not be set at too small a value. We assume a NGARCH model for daily frequency, and use the following set of GARCH parameter values: $\beta_0 = 0.00001, \beta_1 = 0.8, \beta_2 = 0.1, \theta = 0.3$ and $\lambda = 0.2$ in the numerical analyses.

For each option contract, we consider three maturities (1, 3 and 9 months) and three strike-to-asset price ratios (1.1, 1.0 and 0.9). In the numerical analyses, we assume 365 days per year for annualizing. The current asset price is set at 50. The annualized risk-free rate (continuously compounded) is fixed at 5%. The scaling factors used to determine the asset price and volatility range are: $\delta_{x_1} = 2 + \ln(\ln(m_1))$ and $\delta_{x_2} = 2 + \ln(\ln(m_2))$. Recall that the double logarithmic transformation was also used in the Black-Scholes case. Since the GARCH model is assumed for a daily frequency, the Markov transition probability matrix over one day can be constructed as described in the two preceding subsections. It should be noted that one is not restricted to using one day as the length of a time step because the Markov transition probability matrix over any number of days can be obtained by raising the daily Markov transition probability matrix to an appropriate power. For the values in Tables 11.5 and 11.6, the length of the time step is always kept at one day, however.

	Maturity = 1 month			Maturity = 3 months			Maturity = 9 months		
K/S$_0$	1.10	1.00	0.90	1.10	1.00	0.90	1.10	1.00	0.90
				Monte Carlo with control variate					
price	4.8388	1.0880	0.0778	4.9546	1.8197	0.4158	5.4773	2.8416	1.1945
std.	(0.0012)	(0.0009)	(0.0007)	(0.0021)	(0.0018)	(0.0014)	(0.0033)	(0.0028)	(0.0022)
				Markov chain prices (time step = 1 day) :					
$m_2 = 25$ $m_1 = 25$	4.8756	1.2502	0.1023	5.2628	2.2142	0.6334	4.0344	0.6721	0.0878
$m_2 = 31$ $m_1 = 31$	4.8599	1.1937	0.0902	5.3264	2.3070	0.6884	4.8435	2.0628	0.6681
$m_2 = 35$ $m_1 = 35$	4.8535	1.1707	0.0855	5.2752	2.2371	0.6396	5.3373	2.6470	1.0516
$m_2 = 41$ $m_1 = 41$	4.8482	1.1469	0.0813	5.2053	2.1544	0.5877	5.9755	3.3301	1.5539
$m_2 = 45$ $m_1 = 45$	4.8472	1.1427	0.0802	5.1771	2.1117	0.5618	6.1733	3.5370	1.7238
$m_2 = 51$ $m_1 = 51$	4.8443	1.1271	0.0773	5.1301	2.0542	0.5293	6.2313	3.6008	1.7689
$m_2 = 25$ $m_1 = 75$	4.8417	1.1132	0.0738	5.0498	1.9456	0.4660	6.0560	3.4061	1.6048
$m_2 = 31$ $m_1 = 93$	4.8412	1.1066	0.0739	5.0073	1.8915	0.4416	5.8631	3.2223	1.4687
$m_2 = 35$ $m_1 = 105$	4.8394	1.0995	0.0723	4.9957	1.8743	0.4312	5.7944	3.1522	1.4124
$m_2 = 41$ $m_1 = 123$	4.8376	1.0914	0.0713	4.9823	1.8549	0.4206	5.7015	3.0596	1.3433
$m_2 = 45$ $m_1 = 135$	4.8394	1.0978	0.0725	4.9789	1.8524	0.4206	5.6749	3.0314	1.3229
$m_2 = 51$ $m_1 = 153$	4.8379	1.0918	0.0715	4.9753	1.8469	0.4171	5.6135	2.9704	1.2773
$m_2 = 25$ $m_1 = 125$	4.8425	1.1080	0.0729	4.9931	1.8678	0.4262	5.6722	3.0298	1.3201
$m_2 = 31$ $m_1 = 155$	4.8386	1.0951	0.0724	4.9694	1.8399	0.4134	5.6268	2.9829	1.2857
$m_2 = 35$ $m_1 = 175$	4.8389	1.0940	0.0714	4.9651	1.8321	0.4084	5.5760	2.9323	1.2478
$m_2 = 41$ $m_1 = 205$	4.8379	1.0893	0.0712	4.9645	1.8316	0.4087	5.5325	2.8889	1.2162
$m_2 = 45$ $m_1 = 225$	4.8386	1.0921	0.0716	4.9569	1.8218	0.4041	5.5377	2.8944	1.2209
$m_2 = 51$ $m_1 = 255$	4.8373	1.0879	0.0713	4.9606	1.8268	0.4073	5.5202	2.8774	1.2086
$m_2 = 25$ $m_1 = 175$	4.8392	1.0940	0.0713	4.9760	1.8453	0.4140	5.6101	2.9649	1.2711
$m_2 = 31$ $m_1 = 217$	4.8385	1.0919	0.0719	4.9607	1.8284	0.4081	5.5338	2.8913	1.2193
$m_2 = 35$ $m_1 = 245$	4.8387	1.0929	0.0720	4.9604	1.8269	0.4067	5.5222	2.8787	1.2090
$m_2 = 41$ $m_1 = 287$	4.8380	1.0893	0.0713	4.9555	1.8207	0.4042	5.5014	2.8582	1.1946
$m_2 = 45$ $m_1 = 315$	4.8381	1.0899	0.0715	4.9524	1.8170	0.4025	5.4916	2.8477	1.1860
$m_2 = 51$ $m_1 = 357$	4.8377	1.0884	0.0715	4.9550	1.8197	0.4036	5.4899	2.8471	1.1867

Parameters: $S_0 = 50$, $r = 0.05$ (annualized), $\beta_0 = 0.00001$, $\beta_1 = 0.80$, $\beta_2 = 0.10$, $\theta = 0.3$ and $\lambda = 0.2$.

TABLE 11.5. European put option in the GARCH framework

	Maturity = 1 month			Maturity = 3 months			Maturity = 9 months		
K/S_0	1.10	1.00	0.90	1.10	1.00	0.90	1.10	1.00	0.90
				Markov chain prices (time step = 1 day) :					
$m_2 = 25$ $m_1 = 25$	5.0099	1.2772	0.1163	5.4688	2.2950	0.6736	5.0000	0.7594	0.0970
$m_2 = 31$ $m_1 = 31$	5.0021	1.2170	0.1011	5.5373	2.3923	0.7309	5.2492	2.1886	0.7026
$m_2 = 35$ $m_1 = 35$	5.0004	1.1921	0.0948	5.4843	2.3132	0.6715	5.7483	2.8044	1.1040
$m_2 = 41$ $m_1 = 41$	5.0000	1.1667	0.0892	5.4169	2.2244	0.6135	6.4207	3.5394	1.6443
$m_2 = 45$ $m_1 = 45$	5.0000	1.1620	0.0876	5.3877	2.1788	0.5849	6.6331	3.7644	1.8273
$m_2 = 51$ $m_1 = 51$	5.0000	1.1454	0.0839	5.3434	2.1181	0.5491	6.7039	3.8351	1.8763
$m_2 = 25$ $m_1 = 75$	5.0000	1.1300	0.0788	5.2688	2.0043	0.4802	6.5222	3.6239	1.6935
$m_2 = 31$ $m_1 = 93$	5.0000	1.1228	0.0784	5.2319	1.9489	0.4545	6.3353	3.4338	1.5497
$m_2 = 35$ $m_1 = 105$	5.0000	1.1153	0.0766	5.2215	1.9309	0.4433	6.2677	3.3598	1.4894
$m_2 = 41$ $m_1 = 123$	5.0000	1.1069	0.0752	5.2101	1.9110	0.4319	6.1795	3.2648	1.4169
$m_2 = 45$ $m_1 = 135$	5.0000	1.1132	0.0764	5.2064	1.9081	0.4318	6.1528	3.2355	1.3953
$m_2 = 51$ $m_1 = 153$	5.0000	1.1070	0.0752	5.2035	1.9023	0.4280	6.0959	3.1733	1.3477
$m_2 = 25$ $m_1 = 125$	5.0000	1.1236	0.0767	5.2190	1.9240	0.4377	6.1539	3.2350	1.3926
$m_2 = 31$ $m_1 = 155$	5.0000	1.1102	0.0760	5.1993	1.8955	0.4242	6.1078	3.1856	1.3564
$m_2 = 35$ $m_1 = 175$	5.0000	1.1089	0.0748	5.1954	1.8872	0.4189	6.0607	3.1337	1.3165
$m_2 = 41$ $m_1 = 205$	5.0000	1.1041	0.0745	5.1945	1.8863	0.4190	6.0206	3.0895	1.2835
$m_2 = 45$ $m_1 = 225$	5.0000	1.1068	0.0747	5.1881	1.8764	0.4142	6.0246	3.0949	1.2884
$m_2 = 51$ $m_1 = 255$	5.0000	1.1024	0.0743	5.1907	1.8812	0.4173	6.0081	3.0773	1.2754
$m_2 = 25$ $m_1 = 175$	5.0000	1.1090	0.0746	5.2051	1.9007	0.4246	6.0924	3.1669	1.3406
$m_2 = 31$ $m_1 = 217$	5.0000	1.1066	0.0751	5.1914	1.8832	0.4183	6.0219	3.0923	1.2868
$m_2 = 35$ $m_1 = 245$	5.0000	1.1074	0.0750	5.1907	1.8814	0.4167	6.0101	3.0787	1.2758
$m_2 = 41$ $m_1 = 287$	5.0000	1.1037	0.0742	5.1869	1.8750	0.4140	5.9913	3.0580	1.2609
$m_2 = 45$ $m_1 = 315$	5.0000	1.1042	0.0742	5.1845	1.8713	0.4122	5.9821	3.0470	1.2516
$m_2 = 51$ $m_1 = 357$	5.0000	1.1026	0.0742	5.1861	1.8737	0.4132	5.9800	3.0463	1.2524

Parameters: $S_0 = 50$, $r = 0.05$ (annualized), $\beta_0 = 0.00001$, $\beta_1 = 0.80$, $\beta_2 = 0.10$, $\theta = 0.3$ and $\lambda = 0.2$.

TABLE 11.6. American put option in the GARCH framework

Table 11.5 presents the results for European call options. Since it is possible to compute the option prices using the Monte Carlo method, we have them as benchmarks. The benchmark European option prices are computed using a 200,000 sample path control-variate Monte Carlo simulation. The control variable is the Black-Scholes formula price using the stationary variance, under measure P, as the volatility. The Monte Carlo prices and their corresponding standard errors are presented in the top two rows of Table 11.5. The rest of the table reports the Markov chain prices under different partitioning of the state space.

The results indicate that the Markov chain method with a moderate number of states can in many cases yield prices with penny accuracy. Typically, the convergence is slower for longer-term options. This is, of course, not surprising. The results also suggest that a finer price partitioning seems to perform better in relation to a finer volatility partitioning.

American put option prices in the GARCH setting are reported in Table 11.6. Although there is no benchmark available for checking the pricing accuracy, the comparison of Tables 11.5 and 11.6 clearly indicates consistency. The computed American option for any strike price and maturity has a higher value than its European counterpart computed using the same m_1 and m_2. The convergence pattern for American options is similar to that for European ones.

The computing time required depends on the number of time steps and the length of a time step. With the GARCH model, the transition probability matrix is usually highly sparse with only 1 or 2% of the elements being non-zero. Although pricing in the GARCH setting requires more computing time than in the Black-Scholes setting, a reasonably accurate price estimate for a 90 days option with daily transition steps can be obtained with a standard desktop computer in a few seconds. For example, computing a standard European call option with a maturity of 90 days using $m_1 = 201$, $m_2 = 31$ and daily time steps takes 2.266 seconds on a standard desktop computer. For the same option contract, a crude Monte Carlo simulation with 50,000 sample paths requires 4.968 seconds.

11.4 Valuing Exotic Options

There are many exotic option contracts traded in the over-the-counter fashion. Typically, exotic options have some path dependency. Path dependency in a computational sense increases the relevant dimension of the valuation problem. For example, the Asian option whose payoff depends on the average asset price of the path has the underlying asset price and the average price of the path as the two relevant information dimensions in the sense of a Markov system even if one employs the simplest valuation setting like the Black-Scholes model. If the GARCH model is used, there will then be three relevant information dimensions - price, volatility and average price. In this section, we use the barrier option to demonstrate the use of the Markov chain method. The Markov chain method has previously been applied to barrier options in the Black-Scholes and GARCH settings. We refer readers to Duan, Dudley, Gauthier and Simonato (1999) for more details.

A knock-out barrier option differs from the standard option in that the option gets knocked out and becomes worthless whenever the underlying asset price has touched or crossed a deterministic barrier H_{t^*} at the monitoring time points t^*. In our setup, the monitoring of the barrier is assumed to take place at time $0, \tau, 2\tau ..., T^*\tau$. For a double barrier option, there are two barriers - lower barrier H_{t^*} and upper barrier $H^*_{t^*}$ - between which the underlying asset price at the monitoring times must remain or the option will be knocked out. As before, the maturity date of the option is $T = T^*\tau$.

The construction of the approximating Markov chain in the Black-Scholes setting has already been discussed in Section 11.2. Here we focus on the treatment of barrier. To value barrier options, it is convenient to augment the system by an auxiliary variable B_t to accommodate the added information dimension. This auxiliary variable takes on two possible values: $B_{t^*} = 1$ if the barrier condition is triggered before or at time $t = t^*\tau$ and $B_{t^*} = 0$ if otherwise. We also define the set $\mathcal{S}_{t^*}$ of the states for which the option is knocked out for time t^*.

Let $v(x(i), t^*; B_{t^*})$ and $f\left(g^{-1}(x(i), t^*\tau), t^*\tau\right)$ denote respectively the barrier option value and the immediate exercise value at time $t = t^*\tau$ corresponding to the state $x(i)$. It is obvious that the knock-out option's value at time $t = t^*\tau$ is 0 when $B_{t^*} = 1$ regardless of the prevailing underlying asset price. Therefore, we only need to consider the option's value for the case of $B_{t^*} = 0$, which yields the following recursive system for $t^* \in \{0, 1, ..., T^* - 1\}$,

$$v(x(i), t^*; B_{t^*} = 0) = \max \begin{bmatrix} f\left(g^{-1}(x(i), t^*\tau), t^*\tau\right), \\ e^{-r\tau}\sum_{j=1}^{m}\pi(i,j;\tau)v(x(j), t^*+1; B_{t^*+1} = 0) \end{bmatrix}.$$

Note that certain values of $x(i)$ are inconsistent with $B_{t^*} = 0$. In those cases, we always set $v(x(i), t^*; B_{t^*} = 0) = 0$ without loss of generality.

Operationally, it is more convenient to consider a quasi transition probability matrix $\widehat{\mathbf{\Pi}}^{(m)}_{t^*} = (\widehat{\pi}(i,j,t^*))_{i,j\in\{1,2.,,,.m\}}$ of passing from state $x(i)$ to state $x(j)$ without crossing the barrier:

$$\widehat{\pi}(i,j,t^*) = \begin{cases} \pi(i,j,\tau) & \text{if } i \notin \mathcal{S}_{t^*} \text{ and } j \notin \mathcal{S}_{t^*+1} \\ 0 & \text{otherwise.} \end{cases} \tag{11.28}$$

With $\widehat{\pi}(i,j,t^*)$ the valuation system can be stated in the following matrix form and ignore B_{t^*}:

$$\begin{aligned}\vec{v}(\mathbf{X}^{(m)}, t^*) &= \max \begin{bmatrix} f\left(g^{-1}\left(\mathbf{X}^{(m)}, t^*\tau\right), t^*\tau\right), \\ e^{-r\tau}\widehat{\mathbf{\Pi}}_{t^*}^{(m)}\vec{v}(\mathbf{X}^{(m)}, t^*+1) \end{bmatrix}, \\ \vec{v}(\mathbf{X}^{(m)}, T^*) &= f\left(g^{-1}\left(\mathbf{X}^{(m)}, T^*\tau\right), T^*\tau\right),\end{aligned}$$

for $t^* \in \{0, 1, ..., T^* - 1\}$ where $\vec{v}(\mathbf{X}^{(m)}, t^*)$ is the $m \times 1$ vector containing the value of the option at time $t = t^*\tau$ for each of the possible states of the Markov chain and $f\left(g^{-1}\left(\mathbf{X}^{(m)}, t^*\tau\right), t^*\tau\right)$ is a vector containing the immediate exercised value of the option. Again, the time-0 option price is the $\left(\frac{m+1}{2}\right)$-th element of $\vec{v}(\mathbf{X}^{(m)}, 0)$.

As pointed out by Boyle and Lau (1994) and Boyle and Tian (1998), the position of discrete prices in relation to the barrier is important in obtaining accurate barrier option prices. It is particularly sensitive in the cases where the barrier is located near the current asset price. The Markov chain method is flexible, due to the time-state separation, to allow for a suitable placement of asset prices and cell boundaries. Our earlier exposition sets $x(i)$ at the center of a cell. In the case of a constant barrier, the cells can be placed in such a way that the barrier corresponds exactly to one particular cell's border. Doing so ensures that the probability below or above the barrier is precisely equal to the value prescribed by the theory.

The pricing of a down-and-out option in the Black-Scholes setting is analyzed for three different barriers which are monitored daily and weekly. The results are summarized in Table 11.7. The parameter values employed in the analysis are provided at the bottom of the table. The benchmark values at the top of the table are based on the conditional Monte Carlo simulation method of Boyle, *et al.* (1997). The values corresponding to "Cheuk" are taken from Cheuk and Vorst (1996), which were obtained by using a trinomial tree with 100 steps between any two consecutive monitoring time points. The values corresponding to "Zvan" at the bottom of the table were taken from Zvan, *et al.* (1998), which are accurate up to 0.01. We have also included the cases where the barrier is close to the current asset price. These cases are known to be difficult to obtain accurate barrier option prices. We have assumed in these calculations 5 and 250 days for a week and a year, respectively. The convergence of the Markov chain method appears to be fairly fast. In most cases, using 701 discrete asset prices, i.e., $m = 701$, is enough to obtain penny accuracy. Similar to Cheuk and Vorst (1996), the Markov chain method can tackle the difficult cases where the barrier is close to the current asset price. Yet the Markov chain method requires a far smaller number of discrete asset prices to achieve the same level of accuracy. Note that for weekly monitoring, the difference between the price of Cheuk and Vorst (1996) and that of the Markov chain in part arises from the difference in the definition of a week.

	Daily			Weekly		
Barrier	95	99.5	99.9	95	99.5	99.9
			Monte Carlo			
Price	6.1662	1.9580	1.5104	6.6370	3.3494	3.0118
Std.	0.0045	0.0051	0.0047	0.0040	0.0058	0.0058
			Markov chain			
$m = 51$	6.8367	1.8935	1.3928	6.8145	3.3494	2.9713
$m = 101$	6.1887	2.0468	1.4804	6.6347	3.3825	2.9980
$m = 201$	6.2304	2.0906	1.5046	6.6496	3.3922	3.0059
$m = 301$	6.1779	1.9663	1.5096	6.6334	3.3563	3.0077
$m = 401$	6.1688	1.9853	1.5115	6.6305	3.3622	3.0085
$m = 501$	6.1735	1.9610	1.5124	6.6323	3.3547	3.0088
$m = 601$	6.1683	1.9686	1.5129	6.6306	3.3572	3.0091
$m = 701$	6.1709	1.9602	1.5133	6.6316	3.3546	3.0093
$m = 801$	6.1680	1.9636	1.5135	6.6306	3.3557	3.0094
$m = 901$	6.1694	1.9602	1.5137	6.6312	3.3546	3.0095
$m = 1001$	6.1679	1.9618	1.5138	6.6307	3.3552	3.0095
Ceuk	6.1692	1.9624	1.5116	6.6181	3.3122	2.9626
Zvan	NA	NA	1.506	NA	NA	2.997

"Daily" and "Weekly" are discretely monitored barrier options with daily and weekly frequency based on the assumption that 1 day = 1/250 years and 1 week = 1/50 years; "Monte Carlo" are prices computed with the conditional Monte Carlo simulation method described in Boyle, *et al.* (1997) using 200 000 sample paths; "Std." are standard deviations of the Monte Carlo prices; "Markov Chain" are prices computed with the Markov chain method; "Ceuk" and "Zvan" are prices taken from Ceuk and Vorst (1996) and Zvan, *et al.* (1998), respectively; Parameters: $S_0 = 100$, $r = 0.10$, $K = 100$, $T = 0.5$ (annualized), $\sqrt{h} = 0.20$ (annualized).

TABLE 11.7. European down-and-out call options in the Black-Scholes framework

	Daily			Weekly		
Barrier	95	99.5	99.9	95	99.5	99.9
			Monte Carlo			
Price	6.1614	1.9406	1.3906	6.5784	3.3368	2.9245
Std.	0.0064	0.0112	0.0116	0.0057	0.0010	0.0103
			Markov Chain			
$m_2 = 25$ $m_1 = 25$	5.8890	0.8355	0.6071	6.2758	0.9449	0.6639
$m_2 = 31$ $m_1 = 31$	8.5691	1.2925	0.9198	9.4729	1.5341	1.0449
$m_2 = 35$ $m_1 = 35$	6.2506	1.4944	1.0558	6.3921	1.8347	1.2338
$m_2 = 41$ $m_1 = 41$	6.4074	1.6712	1.1714	6.8282	2.1629	1.4390
$m_2 = 45$ $m_1 = 45$	6.7323	1.7550	1.2252	7.4531	2.3506	1.5532
$m_2 = 51$ $m_1 = 51$	7.5332	1.8384	1.2765	8.5268	2.5816	1.6971
$m_2 = 25$ $m_1 = 75$	6.2454	1.9871	1.3617	6.6131	3.2009	2.1485
$m_2 = 31$ $m_1 = 93$	6.2200	2.0316	1.3828	6.6027	3.4419	2.3594
$m_2 = 35$ $m_1 = 105$	6.3577	2.0525	1.3917	6.9296	3.5334	2.4458
$m_2 = 41$ $m_1 = 123$	6.2809	2.0703	1.3979	6.8024	3.6223	2.5626
$m_2 = 45$ $m_1 = 135$	6.1987	2.0802	1.4012	6.6360	3.6547	2.6124
$m_2 = 51$ $m_1 = 153$	6.1845	2.0914	1.4066	6.6125	3.6895	2.6821
$m_2 = 25$ $m_1 = 125$	6.3003	2.0729	1.4048	6.8369	3.6279	2.5764
$m_2 = 31$ $m_1 = 155$	6.1925	2.0935	1.4119	6.6374	3.6954	2.6979
$m_2 = 35$ $m_1 = 175$	6.1834	2.1000	1.4139	6.6282	3.7072	2.7507
$m_2 = 41$ $m_1 = 205$	6.1594	1.9205	1.4159	6.5624	3.1628	2.8043
$m_2 = 45$ $m_1 = 225$	6.1599	1.9222	1.4161	6.5694	3.2128	2.8216
$m_2 = 51$ $m_1 = 255$	6.1674	1.9308	1.4183	6.6079	3.2768	2.8497
$m_2 = 25$ $m_1 = 175$	6.1805	2.0984	1.4165	6.6238	3.7042	2.7519
$m_2 = 31$ $m_1 = 217$	6.1808	1.9257	1.4201	6.6336	3.2038	2.8268
$m_2 = 35$ $m_1 = 245$	6.1568	1.9288	1.4207	6.5709	3.2639	2.8490
$m_2 = 41$ $m_1 = 287$	6.1564	1.9391	1.4205	6.5810	3.3290	2.8752
$m_2 = 45$ $m_1 = 315$	6.1614	1.9474	1.4209	6.5994	3.3577	2.8850
$m_2 = 51$ $m_1 = 357$	6.1629	1.9585	1.4207	6.6032	3.3917	2.8996

"Daily" and "Weekly" are discretely monitored barrier options with daily and weekly frequency based on the assumption that 1 day = 1/250 years and 1 week = 1/50 years; "Monte Carlo" are prices computed with Monte Carlo simulation using 500 000 sample paths and the Black-Scholes price as the control variate; "Std." are standard deviations of the Monte Carlo prices; "Markov Chain" are prices computed with the Markov chain method; Parameters: $S_0 = 100$, $r = 0.10$, $K = 100$, $T = 0.5$ (annualized), $\beta_0 = 0.00001$ $\beta_1 = 0.80$ $\beta_2 = 0.10$, $\theta = 0.3$, $\lambda = 0.2$, $\sqrt{h_1} = 0.010483$.

TABLE 11.8. European down-and-out call options in the GARCH framework

The results of using the NGARCH(1,1) model are summarized in Table 11.8. The GARCH methodology, due to time-varying volatilities, requires of setting an initial conditional volatility. Here, we again consider an average situation; that is, we assume its conditional volatility equal to the stationary volatility under the data generating probability measure P. The specific parameter values used in the analysis are the same as in Duan and Simonato (2001) and are provided in the table. All price estimates in Table 11.8 tend to their respective Monte Carlo benchmark values as the number of states increases. For a given precision level, the numerical scheme for the NGARCH model requires more computing time because the transition probability matrix under the GARCH model is much larger in dimension than that under the Black-Scholes model. Unlike the constant volatility case, it is only possible to match the time step of the Markov chain with the monitoring frequency if the underlying GARCH model is defined exactly over the monitoring frequency. To be specific, if the GARCH model is defined on a daily basis but monitoring only takes place weekly, then weekly monitoring of a constant barrier is equivalent to monitoring an intermittent barrier on a daily basis. Alternatively, one can first obtain the relevant transition probability matrix over one week by raising the daily transition probability matrix to an appropriate power so that option valuation can be conducted on a weekly basis.

11.5 Appendix: The Conditional Expected Value of h_{T^*} and $h_{T^*}^2$

We will first justify the use of $\sqrt{\sum_{t^*=1}^{T^*} \mathsf{E}^Q\left[h_{t^*}|\mathcal{F}_0\right]}$ as an approximation for the variance of the asset return at period T^* (see Section 11.3.1). Duan, Gauthier and Simonato (1999) have shown that

$$\begin{aligned}
\mathsf{Var}_0^Q\left[\ln S_{t^*}\right] &= \sum_{u^*=1}^{t^*} \mathsf{E}_0^Q\left[h_{u^*}\right] \\
&\quad +\frac{1}{4}\sum_{u^*=1}^{t^*} \mathsf{E}_0^Q\left[h_{u^*}^2\right] - \frac{1}{4}\left(\sum_{u^*=1}^{t^*} \mathsf{E}_0^Q\left[h_{u^*}\right]\right)^2 \\
&\quad +\frac{1}{2}\sum_{u^*=1}^{t^*}\sum_{v^*=1}^{t^*-u^*} \mathsf{E}_0^Q\left[h_{u^*}h_{u^*+v^*}\right] \\
&\quad -\sum_{u^*=1}^{t^*}\sum_{v^*=1}^{t^*-u^*} \mathsf{E}_0^Q\left[\sqrt{h_{u^*}}\epsilon_{u^*}h_{u^*+v^*}\right].
\end{aligned}$$

The last four terms are generally small compared to the first one. Using induction on t^*, one can show that

$$\mathsf{E}^Q\left[h_{t^*}|\mathcal{F}_0\right] = h_1 v^{t^*-1} + \frac{\beta_0(1-v^{t^*-1})}{1-v}, \tag{11.29}$$

where $v = \beta_1 + \beta_2[1+(\theta+\lambda)^2]$. Indeed, the result is obvious when $t^* = 1$. If we assume that the result is verified for $t^* - 1$, then

$$\begin{aligned}
\mathsf{E}^Q\left[h_{t^*}|\mathcal{F}_0\right] &= \mathsf{E}^Q\left[\mathsf{E}^Q\left[h_{t^*}|\mathcal{F}_1\right]|\mathcal{F}_0\right] \\
&= \mathsf{E}^Q\left[h_2 v^{t^*-2} + \frac{\beta_0(1-v^{t^*-2})}{1-v}\right] \\
&\quad \text{(from the induction hypothesis)} \\
&= v^{t^*-2}\mathsf{E}^Q\left[h_2|\,\mathcal{F}_0\right] + \frac{\beta_0(1-v^{t^*-2})}{1-v} \\
&= v^{t^*-2}\mathsf{E}^Q\left[\beta_0 + \beta_1 h_1 + \beta_2 h_1(\epsilon_{t^*} - \theta - \lambda)^2|\,\mathcal{F}_0\right] \\
&\quad + \frac{\beta_0(1-v^{t^*-2})}{1-v} \\
&= v^{t^*-2}\left[\beta_0 + h_1\left(\beta_1 + \beta_2\left[1+(\theta+\lambda)^2\right]\right)\right] + \frac{\beta_0(1-v^{t^*-2})}{1-v} \\
&= h_1 v^{t^*-1} + \frac{\beta_0(1-v^{t^*-1})}{1-v}.
\end{aligned}$$

Notice that, since $h^* = \beta_0\{1-\beta_1-\beta_2[1+(\theta+\lambda)^2]\}^{-1} = \beta_0/(1-v)$, then

$$\mathsf{E}^Q\left[h_{t^*}|\mathcal{F}_0\right] = h^* + (h_1 - h^*)\left(\frac{h^*-\beta_0}{h^*}\right)^{t^*-1}. \tag{11.30}$$

A similar argument is used to prove that

$$\begin{aligned}
\mathsf{E}^Q\left[h_{t^*}^2|\mathcal{F}_0\right] &= h_1^2 u^{t^*-1} + 2\beta_0 h_1 \frac{v(u^{t^*-1} - v^{t^*-1})}{u-v} \\
&\quad + \beta_0^2\left[\frac{1-u^{t^*-1}}{1-u} + 2\frac{v}{u-v}\left(\frac{1-u^{t^*-1}}{1-u} - \frac{1-v^{t^*-1}}{1-v}\right)\right]
\end{aligned} \tag{11.31}$$

where $v = \beta_1 + \beta_2[1+(\theta+\lambda)^2]$ and $u = \beta_2^2[3+6(\theta+\lambda)^2+(\theta+\lambda)^4] + 2\beta_1\beta_2[1+(\theta+\lambda)^2] + \beta_1^2$.

11.6 References

[1] Black, F. and Scholes, M., The pricing of options and corporate liabilities, *Journal of Political Economy*, **81** (1973), 637-659.

[2] Boyle, P., Broadie, M. and Glasserman, P., Monte Carlo methods for security pricing, *Journal of Economic Dynamics and Control*, **21** (1997), 1263-1321.

[3] BOYLE, P., AND LAU, S., Bumping up against the barrier with the binomial method, *Journal of Derivatives*, **1** (1994), 6-14.

[4] BOYLE, P. AND TIAN, Y., An explicit finite difference approach to the pricing of barrier options, *Applied Mathematical Finance*, **5** (1998), 17-43.

[5] CHEUK, T. AND VORST, T., Complex barrier options, *Journal of Derivatives*, **3** (1996), 8-22.

[6] DUAN, J.-C., The GARCH option pricing model, *Mathematical Finance*, **5** (1995), 13-32.

[7] DUAN, J.-C., DUDLEY, E., GAUTHIER, G. AND SIMONATO, J.G., Pricing discretely monitored barrier options by a Markov Chain, *cahier de recherche du CIRANO*, (1999), Montréal, Canada.

[8] DUAN, J.-C., GAUTHIER, G. AND SIMONATO, J.G., An analytical approximation for the GARCH option pricing model *Journal of Computational Finance*, **2** (1999), 75-116.

[9] DUAN, J.-C., GAUTHIER, G. AND SIMONATO, J.G., Numerical pricing of contingent claims on multiple assets and/or factor - a low-discrepancy Markov Chain approach, working paper, École des Hautes Études Commerciales, Montréal, 2001.

[10] DUAN, J.-C. AND SIMONATO, J.G., American option pricing under GARCH by a Markov Chain approximation, *Journal of Economic Dynamics and Control*, **25** (2001), 1689-1718.

[11] JARROW, R. RUDD, A., Option pricing, R.D. Irwin, Homewood, Illinois, 1983.

[12] LEBEL, M., L'évaluation des options sur extremum à l'aide de chaînes de Markov, M.Sc. Thesis, École des Hautes Études Commerciales, Montréal, 2001.

[13] ZVAN, R., VETZAL, K. AND FORSYTH, P., PDE methods for pricing barrier options, working paper, University of Waterloo, 1998.

12

Stochastic Network Models and Optimization of a Hospital System

Xiuli Chao, Liming Liu and Shaohui Zheng

This chapter studies the design and operations of a hospital system based on the practice of Hong Kong's public health care system.

In order to provide convenient and adequate service to patients and to operate with an economy of scale, a health care organization often operates a number of hospitals and clinics that offer identical or similar services and are geographically distributed to serve the needs of different regions and local communities. The Hong Kong Hospital Authority (HKHA), for example, operates several clusters of hospitals and clinics to provide general and emergency services to different regions throughout Hong Kong. It is expected that patients will receive services from their local hospitals and clinics. However, with continuing economic and urban development, the population base and the demographics of a region may change significantly so as to create a mismatch between demands and the hospital capacity of the region. As a result of this mismatch, the service quality at different locations may become drastically different and the overall service quality and resource utilization of the entire system will then deteriorate. In such an unbalanced situation, a physician at one clinic may refer a patient to another clinic where a shorter waiting time is expected. This situation leads to inter-hospital patient flow, which, in general, can partially ease the burden of those hospitals with heavy work loads. On the other hand, the cost and inconvenience associated with the switching of hospitals, which is likely to involve repeated hospital visits, may become socially unacceptable. A longer term solution is to solve the mismatch problem by (new) resource allocation.

Obviously, to improve service quality and resource utilization, a health care administrator, such as the CEO of the HKHA, must consider two related issues. At an operational level, what should be the policy and guidelines to manage patient flows among hospitals and clinics? At a strategic level, how should the limited resources be allocated to hospitals in different regions, being aware of the possible reactions to the capacity allocation from the patients, so that the overall system service level is maximized? The study presented in this chapter was motivated by the need for analytical tools that can model the behavior of a hospital system and can be used to obtain the optimal decision rules regarding these two issues, using the hospital system run by the HKHA as a practical background.

We now briefly outline the organization of this chapter. We peel layers off the complex hospital system and propose a multi-site service network model in Section 12.1. Section 12.2 is devoted to patient flow management while Section 12.3 discusses the capacity design issue. Section 12.4 presents an informal attempt to put the two issues and the corresponding models together and discusses a number of modeling and managerial issues using some special cases and examples. Section 12.5 summarizes the results, highlights the managerial insights, and provides some future research directions. Section 12.6 provides a brief summary of the literature on related research, sources of information, and relevant assumptions and results used in the chapter.

12.1 A Multi-Site Service Network Model

Consider a health care delivery system providing services to an extended urban area. The operations of such a system are very complex involving multiple hospitals, clinics, specialists, nurses, technicians, treatments equipment, testing facilities, operation rooms, recovery rooms, and inpatient wards for overnight pre and post treatment stays. To simplify the analysis, we focus on a single specialty, for example, cardiology, thus reducing the system to a group of specialty clinics. Patients diagnosed with the need for a specialty treatment originate randomly from different regions. While the specific treatment required by a patient is known, exactly how long this patient will take to complete his or her treatment is difficult to predict accurately, making treatment times random in general. Because the resources required for specialty treatments are extremely costly, the total available capacity in the system for the specialty treatments is limited and must be highly utilized to justify the cost. This may cause long waiting times for the patients. Thus, one of the most prominent issues for such a system is the pre-treatment waiting time. Obviously, for an organization that provides a crucial service to the public, this waiting time is the key service quality indicator.

Generally, a patient would prefer a clinic that can provide earlier treatment. When even the shortest expected waiting time for a service within the system is deemed too long, a patient often has the option of receiving treatments from a clinic outside of the system. Thus, patients originating from one region may receive service from a clinic in another region (*switching*) or even outside the system (*rejection*) in order to reduce their waiting times. This creates inter-clinic patient flows, which interconnect all clinics, and flows to the outside of the system. So we need to treat this clinic system as a stochastic service network instead of a number of parallel servers. However, this system is different from a standard queueing network in that each node has an original demand stream and while patients may travel from one node to another, they only receive the service from the last node they travel to and then leave the system. We call this system *the multi-site service network system.* Figure 12.1 provides a sketch of a three-clinic service network model.

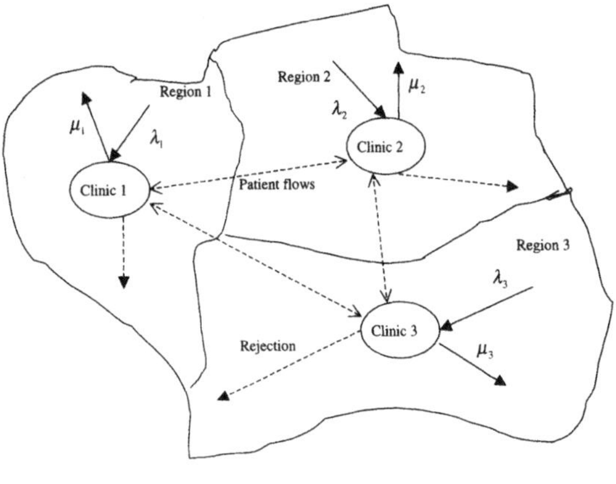

FIGURE 12.1. A Three-Clinic Service Network Model

Specifically, the multi-site service model consists of N clinics. Clinic j is associated with a population Region j and provides a set of services primarily to patients from this region. The services provided at every clinic are identical, thus it is technically feasible for a patient from Region j to receive services from Clinic k, $k \neq j$.

To model demands, we assume that patients originating from Region j follow a Poisson process with rate $\lambda_j, j = 1, \ldots, N$. With the inter-clinic flow, the exact rate of the demand arrival stream to a clinic is usually unknown beforehand. In Section 12.2 and Section 12.3, we will provide more details on the demand models appropriate to the specific modeling issues.

There are different ways to model the capacity of a clinic. Here, we consider the simple case of exponential service times and use the service rate to represent the capacity. Let the total available capacity (resource) be U and the service rate of clinic j be μ_j, $j = 1, \ldots, N$ and $\sum_{j=1}^{N} \mu_j = U$. Here μ_j is the maximum rate at which patients can be released from Clinic j. This assumption simplifies our modeling efforts and enables us to obtain analytical results that can be used to develop insights and guidelines for decision making in a complex hospital system.

12.2 Patient Flow Management

Upon arrival to the multi-site hospital system, patients may be given three options: staying at their home clinic, switching to another clinic with a shorter expected waiting time, or leaving the system to obtain services from another system. They will make a choice according to the available information on waiting times, i.e., the system state, and costs associated with the decision. For management, the routing recommendations that they present to a patient should be based on the overall system objective of minimizing the system-wide total cost. The service quality consideration is implied in the patient's waiting cost to be defined below. In this section, we formulate a stochastic dynamic programming model to determine, from the system's point of view, the optimal routing recommendation for an arriving patient.

We call patients originating from region j type-j patients. The following costs are considered for the optimization model: the unit waiting cost w; the switching cost C_j if a type-j patient is routed to Clinic k, $k \neq j$; and the penalty cost R_j for rejecting a type-j patient. For the HKHA, this penalty cost could be the compensation for a patient to pay (partially) for the treatment from a private clinic. For example, a patient pays around US$7 for a visit to a government clinic. This covers the physician's consultation and the prescription. A similar visit to a private clinic may cost anywhere from US$20 to US$60. The cost differences for hospitalization and surgery are much more significant. Obviously, a policy maker will be interested to know the impact of a rejection-and-compensation policy on the quality of service and the effectiveness of resource utilization of his or her hospital system.

For tractability, we construct a stochastic dynamic programming model for a system with two clinics and two patient arrival streams. Let $\alpha > 0$ be a discount factor and $(x_1,\ x_2)$ denote the state that there are x_1 patients in Clinic 1 and x_2 patients in Clinic 2. Without loss of generality, we assume

$$\alpha + \lambda_1 + \lambda_2 + \mu_1 + \mu_2 = 1.$$

Let $V(x_1, x_2)$ denote the optimal long-run discounted cost, starting from state (x_1, x_2). From the theory of Markov decision processes (MDP) (refer to, for instance, Puterman [17]), it satisfies the following optimality equation:

$$
\begin{aligned}
& V(x_1, x_2) \\
= & \sum_{j=1}^{2} w x_j \\
& +\lambda_1 \min\{V(x_1+1, x_2), C_1 + V(x_1, x_2+1), R_1 + V(x_1, x_2)\} \\
& +\lambda_2 \min\{C_2 + V(x_1+1, x_2), V(x_1, x_2+1), R_2 + V(x_1, x_2)\} \\
& +\mu_1 V(x_1-1, x_2) 1[x_1 > 0] + \mu_1 V(0, x_2) 1[x_1 = 0] \\
& +\mu_2 V(x_1, x_2-1) 1[x_2 > 0] + \mu_2 V(x_1, 0) 1[x_2 = 0]. \qquad (12.1)
\end{aligned}
$$

Let $V_0(x_1, x_2) \equiv 0$, and define $V_n(x_1, x_2)$ recursively by

$$
\begin{aligned}
& V_{n+1}(x_1, x_2) \\
= & \sum_{j=1}^{2} w x_j \\
& +\lambda_1 \min\{V_n(x_1+1, x_2), C_1 + V_n(x_1, x_2+1), R_1 + V_n(x_1, x_2)\} \\
& +\lambda_2 \min\{C_2 + V_n(x_1+1, x_2), V_n(x_1, x_2+1), R_2 + V_n(x_1, x_2)\} \\
& +\mu_1 V_n(x_1-1, x_2) 1[x_1 > 0] + \mu_1 V_n(0, x_2) 1[x_1 = 0] \\
& +\mu_2 V_n(x_1, x_2-1) 1[x_2 > 0] + \mu_2 V_n(x_1, 0) 1[x_2 = 0], \qquad (12.2)
\end{aligned}
$$

for $n \geq 0$ and $x_1, x_2 \geq 0$. Then, $V_n(x_1, x_2)$ is the discounted cost function for the first n transitions, starting from state (x_1, x_2), and we have

$$V_n(x_1, x_2) \rightarrow V(x_1, x_2) \qquad \text{as} \quad n \rightarrow \infty,$$

for all x_1 and x_2.

Lemma 12.1 The following properties are true for $V_n(x_1, x_2)$:

(i) $V_n(x_1+1, x_2) - V_n(x_1, x_2+1)$ is increasing in x_1 and decreasing in x_2, i.e.,

$$V_n(x_1+1, x_2) - V_n(x_1, x_2+1) \geq V_n(x_1, x_2) - V_n(x_1-1, x_2+1), \qquad (12.3)$$

and

$$V_n(x_1+1, x_2) - V_n(x_1, x_2+1) \leq V_n(x_1+1, x_2-1) - V_n(x_1, x_2); \qquad (12.4)$$

(ii) $V_n(x_1, x_2)$ is supermodular, i.e.,

$$V_n(x_1+1, x_2+1) - V_n(x_1+1, x_2) \geq V_n(x_1, x_2+1) - V_n(x_1, x_2); \qquad (12.5)$$

(iii) $V_n(x_1, x_2)$ is component-wise convex, i.e.,

$$V_n(x_1 + 1, x_2) - V_n(x_1, x_2) \geq V_n(x_1, x_2) - V_n(x_1 - 1, x_2), \quad (12.6)$$

and

$$V_n(x_1, x_2 + 1) - V_n(x_1, x_2) \geq V_n(x_1, x_2) - V_n(x_1, x_2 - 1). \quad (12.7)$$

Noticing that $V_0(x_1, x_2) \equiv 0$, all properties in the theorem are satisfied when $n = 0$. Suppose the same hold for n, i.e., $V_n(x_1, x_2)$ satisfies (12.3)-(12.7). The properties then can be proved by induction. The detail will be omitted here. The readers are referred to [3] for this and other omitted proofs.

Given Lemma 12.1, we have the following result.

Theorem 12.2 For type-1 patients, there exist $f_{1n}(x_2)$ and $g_{1n}(x_2)$ for each x_2, such that

(1) $f_{1n}(x_2)$ is decreasing in x_2;

(2) $g_{1n}(x_2)$ is increasing in x_2; and

(3) to minimize $V_{n+1}(x_1, x_2)$, the following control policy is optimal at (x_1, x_2):

(i) reject the patient if $x_1 > f_{1n}(x_2)$; otherwise,

(ii) refer the patient to Clinic 2 if $g_{1n}(x_2) \leq x_1 \leq f_{1n}(x_2)$; or otherwise,

(iii) accept the patient to Clinic 1.

Proof. From the results of discrete time MDP (refer to [19] or [17]), it is optimal to reject a type-1 patient at (x_1, x_2) if and only if

$$\min[V_n(x_1 + 1, x_2), C_1 + V_n(x_1, x_2 + 1)] > R_1 + V_n(x_1, x_2). \quad (12.8)$$

Let

$$\begin{aligned} f_{1n}(x_2) \quad =: \quad & \max\{x_1 \geq 0 : \min[V_n(x_1 + 1, x_2), C_1 + V_n(x_1, x_2 + 1)] \\ & -[R_1 + V_n(x_1, x_2)] \leq 0\}, \end{aligned}$$

and assume $f_{1n}(x_2) = -1$ if the set in the right-hand side of the above equation is empty. Since $V_n(x_1, x_2)$ is component-wise convex,

$$\min[V_n(x_1 + 1, x_2), C_1 + V_n(x_1, x_2 + 1)] - [R_1 + V_n(x_1, x_2)]$$

is increasing in both x_1 and x_2. Hence, $f_{1n}(x_2)$ above is well defined, and it is also decreasing in x_2 due to the supermodularity of $V_n(x_1, x_2)$. Note that (12.8) is equivalent to $x_1 > f_{1n}(x_2)$. Therefore, $f_{1n}(x_2)$ is the threshold value for rejecting type-1 patients.

Suppose that a type-1 patient should not be rejected at (x_1, x_2), i.e., $x_1 \leq f_{1n}(x_2)$. Then it is optimal to refer the patient to Clinic 2 if

$$V_n(x_1 + 1, x_2) > C_1 + V_n(x_1, x_2 + 1), \tag{12.9}$$

or accept the patient to Clinic 1 otherwise.

Let

$$g_{1n}(x_2) =: \min[x_1 \geq 0 : V_n(x_1 + 1, x_2) - C_1 - V_n(x_1, x_2 + 1) > 0].$$

From the induction hypothesis of (12.3) and (12.4),

$$V_n(x_1 + 1, x_2) - C_1 - V_n(x_1, x_2 + 1)$$

is increasing in x_1 and decreasing in x_2. Hence, $g_{1n}(x_2)$ above is well defined and is increasing in x_2. Furthermore, (12.9) is equivalent to $x_1 \geq g_{1n}(x_2)$. Therefore, it is optimal to refer the patient to clinic 2 if

$$g_{1n}(x_2) \leq x_1 \leq f_{1n}(x_2),$$

and accept him or her to Clinic 1 if $x_1 < \min\{g_{1n}(x_2), f_{1n}(x_2)\}$. □

Clearly, the above theorem presents a set of optimal switching curves for type-1 patients in a finite horizon. These curves divide the (x_1, x_2) plane into three regions corresponding to the three options that the patients may have upon arrival. Detailed discussion is left to two examples later in this section. With similar arguments, results parallel to those present in Theorem 12.2 are also true for type-2 patients and they are not repeated here.

Since $V_n(x_1, x_2) \to V(x_1, x_2)$ as $n \to \infty$, we can easily see that $V(x_1, x_2)$ also has the same properties as those in Lemma 12.1. With arguments similar to those in Theorem 12.2, we then have the following main theorem for the patient flow management.

Theorem 12.3 Define

$$\begin{aligned}\hat{y}_1(x_2) = \max\{x_1 \geq 0 : \min[&V(x_1 + 1, x_2), C_1 + V(x_1, x_2 + 1)] \\ &-[R_1 + V(x_1, x_2)] \leq 0\},\end{aligned}$$

and

$$y_1^*(x_2) = \min[x_1 \geq 0 : V(x_1 + 1, x_2) - C_1 - V(x_1, x_2 + 1) > 0].$$

Then $\hat{y}_1(x_2)$ is decreasing in x_2, $y_1^*(x_2)$ is increasing in x_2, and the following control policy for type-1 patients is optimal at (x_1, x_2):

X_2 \ X_1	0	1	2	3	4	5	6	7	>7
0	A	S	S	S	S	S	S	S	S
1	A	A	S	S	S	S	S	S	S
2	A	A	A	S	S	S	S	S	S
3	A	A	A	A	S	S	S	S	S
4	A	A	A	A	A	S	R	R	R
5	A	A	A	A	A	R	R	R	R
6	A	A	A	A	A	R	R	R	R
7	A	A	A	A	A	R	R	R	R
8	A	A	A	A	A	R	R	R	R
9	A	A	A	A	A	R	R	R	R
10	A	A	A	A	R	R	R	R	R
>10	A	A	A	A	R	R	R	R	R

TABLE 12.1. Optimal Control For Type-1 Patient Flow

(1) reject the patient if $x_1 > \hat{y}_1(x_2)$; otherwise,

(2) refer the patient to Clinic 2 if $y_1^*(x_2) \le x_1 \le \hat{y}_1(x_2)$; or otherwise,

(3) accept the patient to Clinic 1.

Similar conclusions hold for type-2 patients.

Example 12.4 Let the arrival rates from Regions 1 and 2 be $\lambda_1 = 0.85$ and $\lambda_2 = 3.5$ per week, respectively; the service rates at Clinics 1 and 2 be $\mu_1 = 1.0$ and $\mu_2 = 3.0$ per week, respectively; the unit waiting cost be $w = 10.0$; the switching costs be $C_1 = C_2 = 20.0$; the rejection penalty costs be $R_1 = R_2 = 200.0$; and the discount factor be $\alpha = 0.1$.

The optimal control policy for type-1 patients is described by: $\hat{y}_1(x_2) = \infty$ for $x_2 = 0, \ldots, 3$; $\hat{y}_1(4) = 5$; $\hat{y}_1(x_2) = 4$ for $x_2 = 5, \ldots, 9$; and $\hat{y}_1(x_2) = 3$ for $x_2 \ge 10$; and $y_1^*(0) = 1$; $y_1^*(1) = 2$; $y_1^*(2) = 3$; $y_1^*(3) = 4$; $y_1^*(4) = 5$; and $y_1^*(5) = 7$. The corresponding control actions at each state (x_1, x_2) are in Table 12.1, where "**A**" represents "accept"; "**S**" represents "switch"; and "**R**" represents "reject". This can be referred to as the decision table that should be generated for the decision makers. Similarly, the optimal control policy for type-2 patients is described by $\hat{y}_2(0) = \hat{y}_2(1) = \infty$; $\hat{y}_2(2) = 9$; $\hat{y}_2(3) = 8$; $\hat{y}_2(x_1) = 7$ for $x_1 = 4, \ldots, 7$; and $\hat{y}_2(x_1) = 6$ for $x_1 \ge 8$; and $y_2^*(0) = 2$; $y_2^*(1) = 5$, and $y_2^*(2) = 12$. Table 12.2 provides the decision table for type-2 patients. Figure 12.2 and 12.3 show how the state space is divided into three regions by the switching curves for type-1 and type-2 patients, respectively.

With simulation, we can easily find that, with the above optimal policies, about 4.06% of type-1 patients and 10.85% of type-2 patients are rejected; about 8.6% of type-1 patients are referred to Clinic 2, while about 6.66% of type-2 patients are referred to Clinic 1. These translate to the effective

X_2 \ X_1	0	1	2	3	4	5	6	7	8	>8
0	A	A	A	A	A	A	A	A	A	A
1	A	A	A	A	A	A	A	A	A	A
2	S	A	A	A	A	A	A	A	A	A
3	S	A	A	A	A	A	A	A	A	A
4	S	A	A	A	A	A	A	A	A	A
5	S	S	A	A	A	A	A	A	A	A
6	S	S	A	A	A	A	A	A	R	R
7	S	S	A	A	R	R	R	R	R	R
8	S	S	A	R	R	R	R	R	R	R
9	S	S	R	R	R	R	R	R	R	R
>9	S	S	R	R	R	R	R	R	R	R

TABLE 12.2. Optimal Control For Type-2 Patient Flow

workload intensities of $\rho_1 = 0.975$ and $\rho_2 = 0.987$ at the two clinics, respectively. The corresponding system performance are: 3.6862 and 5.0929 weeks for the average waiting (queueing) times; and 5.0 and 10.9 for the average queue lengths at Clinic 1 and Clinic 2, respectively.

Example 12.5 Consider another example with the following parameters: $w = 10.0$, $\alpha = 0.1$, $\lambda_1 = 0.9$/week, $\lambda_2 = 1$/week, $\mu_1 = 1$/week, $\mu_2 = 3$/week, $R_1 = R_2 = 200.0$, and $C_1 = C_2 = 20.0$. Figure 12.4 and Figure 12.5 show the switching curves for type-1 and type-2 patients, respectively.

In Example 12.4, the total arrival rate is greater than the total service rate. Without rejection, the system will be unstable. Naturally, the rejection rates are high, especially for type-2 patients. In the second example, we have a reverse situation: the total capacity is much higher than the demand and clinic 2 is over capacitated. Figure 12.4 and Figure 12.5 show that a patient, either type-1 or type-2, may be rejected only when the system is very crowded, and hence we can expect that the rejection rates for both types of patients are very low. In fact our simulation shows that rejection from the system and switching from Clinic 2 to Clinic 1 are negligible. The percentage of patients switched from Clinic 1 to 2 is very high at 27%. Obviously, the impact of system capacity and its distribution is significant. In the next section, we investigate how system performance is affected by capacity and identify the optimal structure of the system capacity design.

12.3 Capacity Design

In the previous section, we developed a dynamic programming model for patient flow management, in which system-wide costs are considered ex-

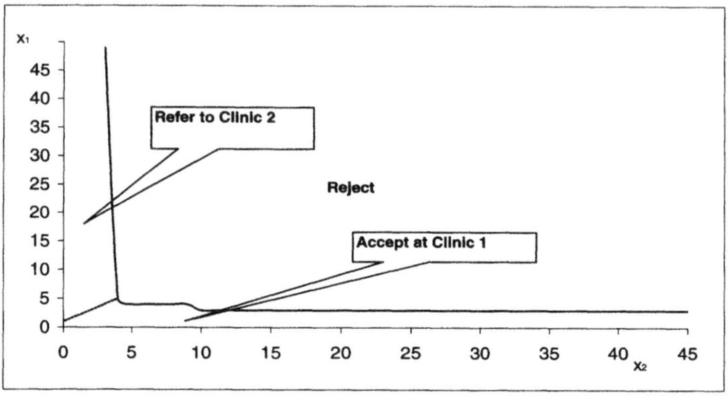

FIGURE 12.2. Switching Curves for Type 1 Patients, Example 12.4

plicitly. In this section, we examine the impact of service capacity on performance, i.e, how to allocate the capacity so as to optimize the overall system performance. Costs are not considered explicitly here. Instead, we focus on a single performance measure, the expected total waiting time. If we ignore the cost differences of setting up one unit of capacity at different locations, then the total cost of deploying the capacity U for the system is fixed regardless of how it is allocated. So the objective for the capacity allocation should be the minimization of the system-wide total average waiting cost.

To construct an appropriate allocation model, patients' response towards an allocation policy should be considered, i.e., we need a model of the effective demand process to each clinic. Since it is extremely difficult to model the patients' dynamic response in the capacity allocation model, we propose a static characterization of patients' response as follows: Clinic j receives a Poisson demand process with rate $\nu_j = \gamma_j \lambda_j$ from its home region. There is another Poisson arrival process to the system with rate $\lambda = \sum_{j=1}^{N}(1 - \gamma_j)\lambda_j$ and each patient from this arrival process chooses to join a clinic so that the total average waiting time in the system is minimized for a given capacity allocation. This demand model conforms to the following facts or considerations: (1) The proportion of patients who will leave their home clinics is ultimately related to the capacity of the clinics, provided that various cost parameters are identical for different clinics; (2) We often need to allocate resources among clinics in an existing system in which some minimum capacity must be maintained for each existing service region; (3) We expect the system to fulfill all the internal demands

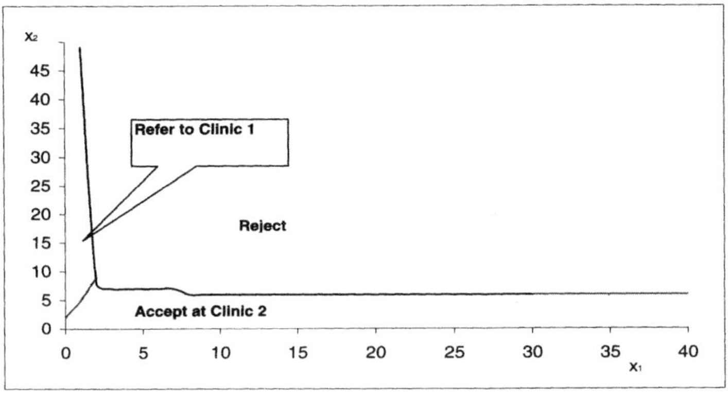

FIGURE 12.3. Switching Curves for Type 2 Patients, Example 12.4

to the best of its ability, thus rejections should only be made during the operation on a case-by-case basis; (4) Patients have no knowledge of the exact waiting time at each station and they are given choices based on the objective of minimizing the system-wide total average waiting time.

Given the demand model, we construct a two-layer optimization model: first, for a given resource allocation $(\mu_1, \ldots, \mu_N)$, our model establishes α_j, the optimal rate out of λ at equilibrium that joins clinic j; α_j's are then used as parameters in the next step allocation model for optimal $(\mu_1, \ldots, \mu_N)$. We now present the models and solutions.

First, for the system to be stable we need

$$U > \lambda + V, \tag{12.10}$$

where

$$V = \sum_{j=1}^{N} \nu_j.$$

According to the above discussion, the total arrival rate at Region j is $\nu_j + \alpha_j$. By standard queueing results, the total expected number of customers in the system is

$$\sum_{j=1}^{N} \frac{\alpha_j + \nu_j}{\mu_j - \alpha_j - \nu_j} = \sum_{j=1}^{N} \frac{\mu_j}{\mu_j - \alpha_j - \nu_j} - N.$$

Since the system-wide total arrival rate is fixed and waiting costs at different clinics are identical, minimizing the total average waiting cost rate

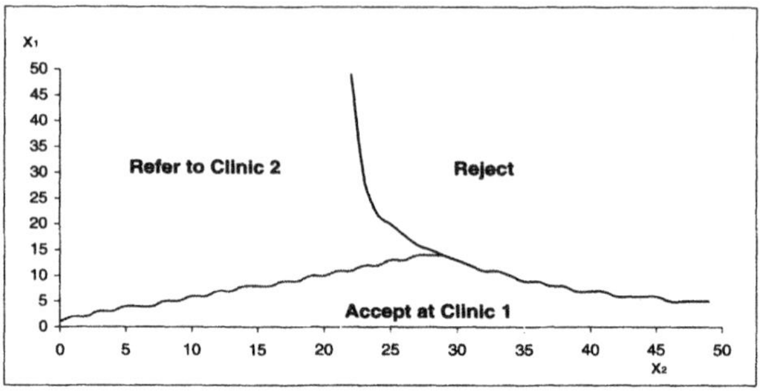

FIGURE 12.4. Switching Curves for Type 1 Patients, Example 12.5

is equivalent to minimizing the total average number of patients in the system. Thus, our first model is the following optimization problem.

$$\begin{aligned} \min \quad & \sum_{j=1}^{N} \frac{\mu_j}{\mu_j - \alpha_j - \nu_j} \\ \text{s.t.} \quad & \sum_{j=1}^{N} \alpha_j = \lambda, \\ & \alpha_j \geq 0, \quad j = 1, 2, \ldots, N. \end{aligned}$$

Note that the objective function above is a convex function.

Lemma 12.6 The optimal solution to the convex programming (12.11) is

$$\alpha_j = \max\left(0, \mu_j - \nu_j - \sqrt{\frac{\mu_j}{\theta}}\right), \quad j = 1, \ldots, N, \tag{12.11}$$

where θ is the unique solution of

$$\sum_{j=1}^{N} \max\left(0, \mu_j - \nu_j - \sqrt{\frac{\mu_j}{\theta}}\right) = \lambda. \tag{12.12}$$

Proof. Since the problem is a convex programming, Kuhn-Tucker conditions yield the optimal solution. It follows that $(\alpha_1, \alpha_2, \ldots, \alpha_N)$ is an optimal solution if and only if

$$\frac{\mu_j}{(\mu_j - \alpha_j - \nu_j)^2} - \theta \geq 0, \tag{12.13}$$

$$\alpha_j \left(\frac{\mu_j}{(\mu_j - \alpha_j - \nu_j)^2} - \theta \right) = 0, \tag{12.14}$$

$$\alpha_j \geq 0, \text{for } j = 1, 2, \ldots, N. \tag{12.15}$$

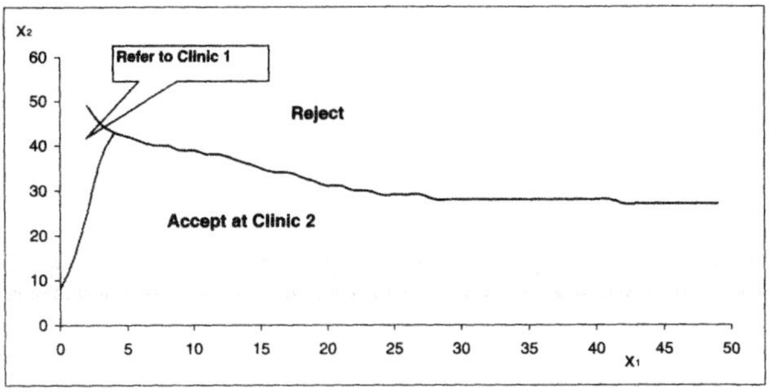

FIGURE 12.5. Switching Curves for Type 2 Patients, Example 12.5

Note that

$$\sum_{j=1}^{N} \max\left(0, \mu_j - \nu_j - \sqrt{\frac{\mu_j}{\theta}}\right)$$

is a continuous function of θ. It increases from 0^+ to $U - V$ as θ goes from 0^+ to ∞. Since $\lambda < U - V$, there exists a unique θ in $(0, \infty)$ satisfying

$$\sum_{j=1}^{N} \max\left(0, \mu_j - \nu_j - \sqrt{\frac{\mu_j}{\theta}}\right) = \lambda.$$

For this θ, we define

$$I_0 = \{j : \mu_j - \nu_j - \sqrt{\frac{\mu_j}{\theta}} \leq 0\},$$

and

$$I_1 = \{j : \mu_j - \nu_j - \sqrt{\frac{\mu_j}{\theta}} > 0\}.$$

Let

$$\alpha_j = \max\left(0, \mu_j - \nu_j - \sqrt{\frac{\mu_j}{\theta}}\right). \tag{12.16}$$

Then, for $j \in I_0$, $\alpha_j = 0$, $\mu_j - \nu_j - \sqrt{\mu_j/\theta} \leq 0$ and hence, all conditions in (12.13) - (12.15) are satisfied. For $j \in I_1$,

$$\alpha_j = \mu_j - \nu_j - \sqrt{\frac{\mu_j}{\theta}} > 0.$$

In this case, (12.13) becomes an equality, and (12.14) and (12.15) are also satisfied. Therefore, α_j as defined in (12.16) is the optimal solution of (12.11). □

As mentioned earlier, the optimal α_j's obtained become parameters in the next optimization procedure. However, the "max" format of α as given in equation (12.11) is inconvenient for us to derive the next optimal solution. If we remove the constraint $\alpha_j \geq 0$ in (12.11), the solution becomes

$$\alpha_j = \mu_j - \nu_j - \frac{\sqrt{\mu_j}}{\sum_{k=1}^N \sqrt{\mu_k}}(U - V - \lambda), \qquad j = 1, \ldots, N. \tag{12.17}$$

Here, α_j may be negative. If this is the case, it implies that the current fixed demand traffic is already more than the optimal level corresponding to the capacity allocated at Clinic j. In other words, it should be optimal to reduce the arrival rate at Clinic j to $\nu_j + \alpha_j$. However, this is not possible because the minimume arrival rate at Clinic j is set to ν_j. Thus, in the following, we will only consider the class of allocation policies $(\mu_1, \ldots, \mu_N)$ such that the resulting solution (12.17) is nonnegative, i.e.,

$$\alpha_j = \mu_j - \nu_j - \frac{\sqrt{\mu_j}}{\sum_{k=1}^N \sqrt{\mu_k}}(U - V - \lambda) \geq 0, \quad j = 1, \ldots, N. \tag{12.18}$$

Substituting (12.18) into the objective function in (12.11), the resources allocation problem can be stated as

$$\begin{aligned} \min \ & (\sum_{j=1}^N \sqrt{\mu_j})^2/(U - V - \lambda) \\ \text{s.t.} \ & \\ & \alpha_j = \mu_j - \nu_j - \frac{\sqrt{\mu_j}}{\sum_{k=1}^N \sqrt{\mu_k}}(U - V - \lambda), \\ & \sum_{j=1}^N \mu_j = U, \\ & \mu_j \geq 0, \quad \alpha_j \geq 0, \quad j = 1, \ldots, N. \end{aligned} \tag{12.19}$$

Let us denote $\Delta = U - V - \lambda$. Since minimizing

$$(\sum_{j=1}^N \sqrt{\mu_j})^2/(U - V - \lambda)$$

is equivalent to minimizing $\sum_{j=1}^N \sqrt{\mu_j}$, the non-linear programming problem (12.19) becomes

$$\min \sum_{j=1}^N \sqrt{\mu_j}$$

s.t.

$$\mu_j \geq \nu_j + \frac{\sqrt{\mu_j}}{\sum_{k=1}^{N} \sqrt{\mu_k}} \Delta, \qquad j = 1, \ldots, N, \tag{12.20}$$

$$\sum_{j=1}^{N} \mu_j = U, \qquad \mu_j \geq 0, \quad j = 1, \ldots, N.$$

This is a non-convex programming. Indeed, the feasible solution region of (12.20) is not a convex set. In general, such a problem is extremely difficult to solve analytically if not impossible. Fortunately, we are able to construct the analytical solution through a complicated procedure to be discussed below.

Without loss of generality, suppose $\nu_1 \leq \nu_2 \leq \cdots \leq \nu_N$. The following main theorem of this section gives the optimal allocation rules.

Theorem 12.7 The optimal solution for the resource allocation problem, i.e. the non-convex programming (12.20), is

$$\mu_j = (a\Delta + \sqrt{\nu_j + a^2\Delta^2})^2, \qquad j = 1, \ldots, N-1, \tag{12.21}$$

$$\mu_N = (a\Delta + \sqrt{\lambda + \nu_N + a^2\Delta^2})^2, \tag{12.22}$$

where a is the unique solution to

$$2a^2 N\Delta + 2a \sum_{j=1}^{N-1} \sqrt{\nu_j + a^2\Delta^2} + 2a\sqrt{\lambda + \nu_N + a^2\Delta^2} = 1. \tag{12.23}$$

We present a few remarks before giving the proof of Theorem 12.7.

Remark 12.8 Theorem 12.7 states that the optimal resource allocation rule has the structure of "one large and many small": there are one large facility and $N-1$ small ones. With the optimal solution in (12.21) and (12.22), the first $N-1$ constraints in (12.20) are tight, making $\mu_1, \ldots, \mu_{N-2}$ and μ_{N-1} as small as possible, and only the last inequality is strict, giving the last facility, which is initially the largest (because ν_N is the largest), the maximum allocation of resource μ_N.

Remark 12.9 Under the optimal resource allocation, $\alpha_j = 0, j = 1, \ldots, N-1$ and $\alpha_N = \lambda$, i.e., all the "switching customers" join the largest facility, Clinic N.

Remark 12.10 Under the optimal resource allocation rule the average queue length at station j, $j = 1, \ldots, N-1$, is

$$\begin{aligned} L_j &= \frac{\nu_j}{\mu_j - \nu_j} \\ &= \frac{\nu_j}{(a\Delta + \sqrt{\nu_j + a^2\Delta^2})^2 - \nu_j} \end{aligned}$$

$$= \frac{\sqrt{\nu_j + a^2\Delta^2}}{2a\Delta} - \frac{1}{2},$$

and similarly the average queue length at station N is

$$L_N = \frac{\sqrt{\nu_N + \lambda + a^2\Delta^2}}{2a\Delta} - \frac{1}{2}.$$

Clearly, the average queue lengths at the N stations satisfy

$$L_1 \le L_2 \le \cdots \le L_N.$$

By Little's formula $W_j = L_j/(\alpha_j + \nu_j)$, and the average waiting time at the N stations satisfy

$$W_1 \ge W_2 \ge \cdots \ge W_N.$$

The average waiting time for the entire system is

$$W = \sum_{j=1}^{N} \frac{\alpha_j + \nu_j}{V + \lambda} W_j.$$

Therefore, the optimal resource allocation rule minimizes the average waiting time W but gives rise to different average waiting times at the N stations.

Remark 12.11 One interesting observation is that the optimal resource allocation satisfies

$$\mu_j = \nu_j + \frac{\sqrt{\mu_j}}{\sum_{k=1}^{N-1} \sqrt{\mu_k} + \sqrt{U - \sum_{k=1}^{N-1} \mu_k}}(U - V - \lambda), \tag{12.24}$$

for $j = 1, \ldots, N-1$. Thus if $\mu_1, \ldots, \mu_{N-1}$ can be determined through these $N-1$ equations, then μ_N can be calculated through $\mu_N = U - \sum_{j=1}^{N-1} \mu_j$. It is easily verified that if $\mu_1, \ldots, \mu_{N-1}$ solves (12.24), then $(\mu_1, \ldots, \mu_N)$ obtained from this procedure is a feasible solution to the non-linear program (12.20) which, by Theorem 12.7, is the optimal resource allocation solution. Solving equations (12.24) is a fixed point problem for $\mu_1, \ldots, \mu_{N-1}$.

Proof of Theorem 12.7. Recalling the analysis following Lemma 12.6, we may assume $\nu_j > 0$ for all j. Let

$$\sum_{j=1}^{N} \sqrt{\mu_j} = \frac{1}{2a}. \tag{12.25}$$

The first set of constraints of (12.20) becomes

$$(\sqrt{\mu_j} - a\Delta)^2 \geq \nu_j + a^2\Delta^2,$$

after some manipulation. This inequality is equivalent to

$$\sqrt{\mu_j} \geq a\Delta + \sqrt{\nu_j + a^2\Delta^2} \tag{12.26}$$

or

$$\sqrt{\mu_j} \leq a\Delta - \sqrt{\nu_j + a^2\Delta^2}.$$

Because $\nu_j > 0$, the second inequality has no solution. Thus we only need to focus on (12.26). Our problem can now be stated as finding the maximum possible a such that there exists a nonnegative solution to (12.25), (12.26) and $\sum_{j=1}^N \mu_j = U$. Let

$$\sqrt{\mu_j} = a\Delta + \sqrt{\nu_j + a^2\Delta^2} + x_j, \qquad j = 1, \ldots, N.$$

We must have $x_j \geq 0$ due to (12.26). It follows from (12.25) that

$$2a^2 N\Delta + 2a\sum_{j=1}^{N} \sqrt{\nu_j + a^2\Delta^2} + 2a\sum_{j=1}^{N} x_j = 1. \tag{12.27}$$

Also, since

$$\begin{aligned} \mu_j = \nu_j &+ 2a^2\Delta^2 + 2a\Delta\sqrt{\nu_j + a^2\Delta^2} + x_j^2 \\ &+ 2x_j(a\Delta + \sqrt{\nu_j + a^2\Delta^2}), \qquad \text{for } j = 1, \ldots, N, \end{aligned}$$

and $\sum_{j=1}^N \mu_j = U$, we have

$$\begin{aligned} 2a^2 N\Delta &+ 2a\sum_{j=1}^{N} \sqrt{\nu_j + a^2\Delta^2} \\ &+ \frac{1}{\Delta}\left(\sum_{j=1}^{2}(x_j^2 + 2x_j a\Delta + 2x_j\sqrt{\nu_j + a^2\Delta^2})\right) \\ &= 1 + \frac{\lambda}{\Delta}. \end{aligned} \tag{12.28}$$

Subtracting (12.27) from (12.28) yields

$$\sum_{j=1}^{N}(x_j^2 + 2x_j\sqrt{\nu_j + a^2\Delta^2}) = \lambda. \tag{12.29}$$

We call a solution $(x_1, \ldots, x_N)$ feasible if it is nonnegative and satisfies (12.27) and (12.29). Our problem becomes that of finding a feasible solution $(x_1, \ldots, x_N)$ that maximizes a. We show that the optimal solution $(x_1, \ldots, x_N)$ satisfies $x_j = 0$ for $j = 1, \ldots, N-1$. This then implies that the solution claimed in the theorem is indeed optimal. We prove it by contradiction. Suppose the optimal solution is $(y_1^{(0)}, y_2^{(0)}, \ldots, y_N^{(0)})$ and it has some positive elements y_j for $j \neq N$. Let the corresponding a value be $a^{(0)}$.

We start with another solution $(0, 0, \ldots, y_N^{(1)})$, where $y_N^{(1)}$ is the solution of

$$y_N^2 + 2y_N\sqrt{\nu_N + (a^{(0)})^2\Delta^2} = \lambda. \tag{12.30}$$

If y_N equals $\sum_{j=1}^N y_j^{(0)}$ then the left-hand side of (12.30) becomes

$$\begin{aligned}&\sum_{j=1}^N \left[(y_j^{(0)})^2 + 2y_j^{(0)}\sqrt{\nu_N + (a^{(0)})^2\Delta^2)}\right] + \sum_{i\neq j} y_i^{(0)} y_j^{(0)} \\ &\geq \sum_{j=1}^N \left[(y_j^{(0)})^2 + 2y_j^{(0)}\sqrt{\nu_j + (a^{(0)})^2\Delta^2}\right] = \lambda,\end{aligned}$$

where the inequality follows from $\nu_N \geq \nu_j$ and $y_i^{(0)} y_j^{(0)} \geq 0$, and the equality follows from the feasibility of $(y_1^{(0)}, \ldots, y_N^{(0)})$. Because the left-hand side of (12.30) is continuous and increasing in y_N, there exists an $y_N^{(1)} \in (0, \sum_{j=1}^N y_j^{(0)})$ that satisfies (12.30), i.e.,

$$(y_N^{(1)})^2 + 2y_N^{(1)}\sqrt{\nu_N + (a^{(0)})^2\Delta^2} = \lambda. \tag{12.31}$$

It follows from $y_N^{(1)} \leq \sum_{j=1}^N y_j^{(0)}$ and

$$2(a^{(0)})^2 N\Delta + 2a^{(0)}\sum_{j=1}^N \sqrt{\nu_j + (a^{(0)})^2\Delta^2} + 2a^{(0)}\sum_{j=1}^N y_j^{(0)} = 1$$

that

$$2(a^{(0)})^2 N\Delta + 2a^{(0)}\sum_{j=1}^N \sqrt{\nu_j + (a^{(0)})^2\Delta^2} + 2a^{(0)} y_N^{(1)} \leq 1$$

must be satisfied. Noting that

$$2a^2 N\Delta + 2a\sum_{j=1}^N \sqrt{\nu_j + a^2\Delta^2} + 2a y_N^{(1)}$$

is increasing in a and tends to infinity as a goes to infinity, there exists an $a^{(1)} > a^{(0)}$, such that

$$2(a^{(1)})^2 N\Delta + 2a^{(1)} \sum_{j=1}^{N} \sqrt{\nu_j + (a^{(1)})^2\Delta^2} + 2a^{(1)} y_N^{(1)} = 1. \qquad (12.32)$$

Similarly, we must also have

$$(y_N^{(1)})^2 + 2y_N^{(1)} \sqrt{\nu_N + (a^{(1)})^2\Delta^2} \geq \lambda. \qquad (12.33)$$

The left-hand side of (12.33) is 0 when $y_N^{(1)} = 0$ and is increasing in $y_N^{(1)}$. Thus there must exist a $y_N^{(2)}$ with $0 < y_N^{(2)} < y_N^{(1)}$ such that

$$(y_N^{(2)})^2 + 2y_N^{(2)} \sqrt{\nu_N + (a^{(1)})^2\Delta^2} = \lambda. \qquad (12.34)$$

Because $y_N^{(2)} \leq y_N^{(1)}$, the value of left-hand side of (12.32) with $y_N^{(1)}$ being replaced by $y_N^{(2)}$ is less than or equal to 1, and we have to increase $a^{(1)}$ to $a^{(2)}$ to make it an equality. This process continues and we generate a sequence of $\{y_N^{(n)}; n \geq 0\}$ and $\{a^{(n)}; n \geq 0\}$, where $y_N^{(n)}$ is decreasing in n and is bounded from below by 0. Hence it converges to some $\hat{y}_N > 0$, and $a^{(n)}$ also converges to some finite number $\hat{a}$ that satisfies

$$\hat{y}_N^2 + 2\hat{y}_N \sqrt{\nu_N + \hat{a}^2\Delta^2} = \lambda.$$

Thus we obtain another feasible solution for (12.27) and (12.29), $(0, \ldots, \hat{y}_N)$ with its corresponding $\hat{a} \geq a^{(0)}$. This solution is optimal and thus the claim of optimal solution in this theorem is true. □

The optimal solution can be considered as a result of an interactive decision process. First, for a given resource allocation $(\mu_1^{(1)}, \ldots, \mu_N^{(1)})$, patients react to it which leads to the corresponding demand allocation. This is equivalent to solving an optimization problem for given $(\mu_1, \ldots, \mu_N) = (\mu_1^{(1)}, \ldots, \mu_N^{(1)})$. Let the optimal solution be $(\alpha_1^{(1)}, \ldots, \alpha_N^{(1)})$. Then, for this demand allocation the decision maker re-optimizes and seeks the corresponding resource allocation rule. This step is equivalent to solving an optimization problem for given $(\alpha_1, \ldots, \alpha_N) = (\alpha_1^{(1)}, \ldots, \alpha_N^{(1)})$. Let the optimal solution be $(\mu_1^{(2)}, \ldots, \mu_N^{(2)})$, and the patients react again to the new resource allocation by solving an optimization problem for given $(\mu_1, \ldots, \mu_N) = (\mu_1^{(2)}, \ldots, \mu_N^{(2)})$, and again the decision maker re-optimizes, and so on and so forth. This process continues until an equilibrium is reached. Indeed, as shown in the next theorem, the optimal resource allocation solution obtained in Theorem 12.7 is a Nash equilibrium

Theorem 12.12 The solution obtained in Theorem 12.7 is a Nash equilibrium.

Proof. We need to show that the demand allocation corresponding to the optimal resource allocation satisfies $\alpha_j = 0, j = 1, \ldots, N-1$, $\alpha_N = \lambda$. We use Lemma 12.6. For that, we only need to find θ such that

$$\mu_N - \nu_N - \sqrt{\frac{\mu_N}{\theta}} = \lambda, \tag{12.35}$$

$$\mu_j - \nu_j - \sqrt{\frac{\mu_j}{\theta}} \leq 0, \qquad j = 1, \ldots, N-1. \tag{12.36}$$

Apparently, it follows from (12.35) that θ must be given by

$$\begin{aligned}\theta &= \frac{\mu_N}{(\mu_N - \lambda - \nu_N)^2} \\ &= (\sum_{j=1}^{N-1} \sqrt{\nu_j} + \sqrt{\nu_N + \lambda})^2 + \frac{(\sum_{j=1}^{N-1} \sqrt{\nu_j} + \sqrt{\nu_N + \lambda})}{\sqrt{\nu_N + \lambda}},\end{aligned}$$

and (12.36) is equivalent to, for $j = 1, \ldots, N-1$,

$$\theta \leq (\sum_{j=1}^{N-1} \sqrt{\nu_j} + \sqrt{\nu_N + \lambda})^2 + \frac{(\sum_{j=1}^{N-1} \sqrt{\nu_j} + \sqrt{\nu_N + \lambda})}{\sqrt{\nu_j}}.$$

This is clearly satisfied. □

12.4 Switching Costs and Quality of Service

In Sections 12.2 and 12.3, we developed models and optimal solutions for the two challenging issues that health care system administrators often have to face. These models and solutions are given under some general conditions and are only the first step in providing decision rules and implementation guidelines for practical applications. In this chapter, we make no attempt to provide a complete solution package. On the other hand, there are many issues and implications that are related to the models and results from Sections 12.2 and 12.3. In this section, we will discuss some of them with a focus on the quality of service and switching costs.

In developing the optimal resource allocation policy in the last section, our sole objective is to minimize the system-wide total average waiting time. For a service system, this means that we aim at providing the best overall quality of service with the available resources. One may ask: is this enough? Should other costs be considered in making the allocation decision? These include the switching costs and rejection costs considered in the flow management model. Our argument for the allocation model is the following. As described earlier, capacity design and flow management can be seen as two related issues at different levels: capacity design decision at the strategic level and flow control decision at the operational level. We must first develop a robust system structure with which one can utilize his or her available resources cost-effectively. Once such a structure is in place, we can then deal with various costs and individual patients' needs dynamically based on the current operating conditions. This is the role played by the flow management model. To make the point that a good capacity design is crucial for the quality of service, we consider the following simple example.

Consider a system with three clinics, i.e., $N = 3$. The patient demand streams are Poisson with rates $\lambda_1 = \lambda_2 = \lambda$ and $\lambda_3 = 2\lambda$. This is a case with one larger region and two smaller regions. Let U be the total available service capacity. Suppose the capacity allocation is given as the following.

$$\mu_1 = \mu_2 = \lambda + \frac{U - 4\lambda}{2 + \sqrt{2}}, \quad \mu_3 = 2\lambda + \sqrt{2}\frac{U - 4\lambda}{2 + \sqrt{2}}. \tag{12.37}$$

From queueing theory, the average sojourn times (waiting time plus service time) at the three clinics are

$$S_1 = S_2 = \frac{2 + \sqrt{2}}{U - 4\lambda}, \quad S_3 = \frac{1}{\sqrt{2}}\frac{2 + \sqrt{2}}{U - 4\lambda}.$$

It is easy to see that the proportion of resources allocated to Clinic 3 is below the proportion of the patient traffic received at the clinic, but the average sojourn time is about 30% lower than those of the two other stations. This shows that resources can be more efficiently utilized at a facility with a higher demand rate, i.e., consolidation is preferred from the customer service point of view. This is consistent with the third remark in the previous section. Thus, it is important to allocate resources correctly at the beginning.

A natural question one will then ask is that if resources can be more effectively used when concentrated, why not allocate more to Clinic 3 which has the largest patient stream? Actually, (12.37) is the optimal allocation when there is no inter-clinic patient flow. This is one extreme scenario. At the opposite extreme when all patients are allowed to switch, we can show that the optimal capacity allocation is to have only one clinic with all the capacity. Obviously, it is unlikely that a health care system can be reduced to a single clinic or hospital in reality. A more likely scenario is when a proportion of patients may make a switch. In fact, the optimal allocation policy developed in Section 12.3 is for this realistic scenario and it dictates that after satisfying the basic capacity requirements of all the clinics, all the remaining capacity should be given to the clinic that has the largest patient flow from the home region.

This phenomenon points to the fact that we must allow the flexibility of choice in a health care system. This has been the practice among HKHA's clinics and it is obviously a good practice. The question is whether it is optimal or how can it be optimal from a system point of view.

The flow management model in Section 12.2 tries to identify the best switching options for the patients in terms of system-wide optimization. This is what may be called social optimization. The practice in Hong Kong is believed to be what is called individual optimization in that each patient will decide on their own, based on some waiting time information, where to obtain service. We want to point out through an example that, in general, given all the waiting time and cost information, everyone will be better off on average if the choice is made based on social optimization. Before showing the example, we present the theorem on individual optimal switching policies without proof.

Consider the same two-clinic system as defined in Section 12.1. Suppose that each patient is given the information on the number of patients in each clinic and will make a choice to join one clinic to minimize his or her waiting time. Then the following theorem gives the optimal individual decision rules for type-1 patients (those for type-2 patients are parallel).

Theorem 12.13 Let (x_1, x_2) be the state of a two-clinic system. For a type-1 patient, it is optimal to reject the service if

$$x_1 > \frac{\log(1 - \frac{\alpha R_1}{w})}{\log(\frac{\mu_1}{\alpha+\mu_1})} \qquad \text{and} \qquad x_2 > \frac{\log[1 - \frac{\alpha(R_1 - C_1)}{w}]}{\log(\frac{\mu_2}{\alpha+\mu_2})}.$$

Otherwise, it is optimal for a type-1 customer to seek service from Clinic 1 if

$$x_2 \geq \frac{\log(\frac{\alpha}{w} C_1)}{\log(\frac{\mu_2}{\alpha+\mu_2})}, \tag{12.38}$$

or

$$x_1 \leq \frac{\log[(\frac{\mu_2}{\alpha+\mu_2})^{x_2} - \frac{\alpha}{w} C_1]}{\log(\frac{\mu_1}{\alpha+\mu_1})}. \tag{12.39}$$

Example 12.14 Consider Example 12.4 discussed in Section 12.2, but assume that the decisions are made entirely by each individual patient. For this problem, the switching curves and optimal decisions for type-1 patients are given in Figure 12.6 and Table 12.3 respectively. The decisions for types-2 patients are similar and will not be presented here. Various results on the system performance are obtained through simulation. In particular, there are 9.64% of type-1 customers and 8.46% of type-2 customers rejecting the internal service; 4.75% of type-1 customers choose to switch to Clinic 2, while 7.64% of type-2 customers choose to switch to Clinic 1; the average waiting (queueing) time and queue length at Clinic 1 are 19.304 and 19.3, respectively; and the average waiting time and queue length at Clinic 2 are 19.889 and 59.3, respectively. Comparing the two sets of results, we observe some interesting behaviors. Because the cost of switching to Clinic 1 is much smaller than that of rejecting the internal service, more type-2 patients than in example 12.4 switch to Clinic 1 and few of them reject in-system service. As a result of the in-flow of type-2 patients, more type-1

X_2 \ X_1	0	1	2	3	4	5	6	7	8	9	10	11	12	13	14	15	16	17	18	19	20	21	22	23	>23
0	A	A	A	S	S	S	S	S	S	S	S	S	S	S	S	S	S	S	S	S	S	S	S	S	S
1	A	A	A	S	S	S	S	S	S	S	S	S	S	S	S	S	S	S	S	S	S	S	S	S	S
2	A	A	A	S	S	S	S	S	S	S	S	S	S	S	S	S	S	S	S	S	S	S	S	S	S
3	A	A	A	A	S	S	S	S	S	S	S	S	S	S	S	S	S	S	S	S	S	S	S	S	S
4	A	A	A	A	S	S	S	S	S	S	S	S	S	S	S	S	S	S	S	S	S	S	S	S	S
5	A	A	A	A	S	S	S	S	S	S	S	S	S	S	S	S	S	S	S	S	S	S	S	S	S
6	A	A	A	A	A	S	S	S	S	S	S	S	S	S	S	S	S	S	S	S	S	S	S	S	S
7	A	A	A	A	A	S	S	S	S	S	S	S	S	S	S	S	S	S	S	S	S	S	S	S	S
8	A	A	A	A	A	S	S	S	S	S	S	S	S	S	S	S	S	S	S	S	S	S	S	S	S
9	A	A	A	A	A	A	S	S	S	S	S	S	S	S	S	S	S	S	S	S	S	S	S	S	S
10	A	A	A	A	A	A	S	S	S	S	S	S	S	S	S	S	S	S	S	S	S	S	S	S	S
11	A	A	A	A	A	A	S	S	S	S	S	S	S	S	S	S	S	S	S	S	S	S	S	S	S
12	A	A	A	A	A	A	A	S	S	S	S	S	S	S	S	S	S	S	S	S	S	S	S	S	S
13	A	A	A	A	A	A	A	S	S	S	S	S	S	S	S	S	S	S	S	S	S	S	S	S	S
14	A	A	A	A	A	A	A	S	S	S	S	S	S	S	S	S	S	S	S	S	S	S	S	S	S
15	A	A	A	A	A	A	A	A	S	S	S	S	S	S	S	S	S	S	S	S	S	S	S	S	S
16	A	A	A	A	A	A	A	A	S	S	S	S	S	S	S	S	S	S	S	S	S	S	S	S	S
17	A	A	A	A	A	A	A	A	S	S	S	S	S	S	S	S	S	S	S	S	S	S	S	S	S
18	A	A	A	A	A	A	A	A	A	S	S	S	S	S	S	S	S	S	S	S	S	S	S	S	S
19	A	A	A	A	A	A	A	A	A	S	S	S	S	S	S	S	S	S	S	S	S	S	S	S	S
20	A	A	A	A	A	A	A	A	A	S	S	S	S	S	S	S	S	S	S	S	S	S	S	S	S
21	A	A	A	A	A	A	A	A	A	A	S	S	S	S	S	S	S	S	S	S	S	S	S	S	S
22	A	A	A	A	A	A	A	A	A	A	S	S	S	S	S	S	S	S	S	S	S	S	S	S	S
23	A	A	A	A	A	A	A	A	A	A	S	S	S	S	S	S	S	S	S	S	S	S	S	S	S
24	A	A	A	A	A	A	A	A	A	A	A	S	S	S	S	S	S	S	S	S	S	S	S	S	S
25	A	A	A	A	A	A	A	A	A	A	A	S	S	S	S	S	S	S	S	S	S	S	S	S	S
26	A	A	A	A	A	A	A	A	A	A	A	S	S	S	S	S	S	S	S	S	S	S	S	S	S
27	A	A	A	A	A	A	A	A	A	A	A	A	S	S	S	S	S	S	S	S	S	S	S	S	S
28	A	A	A	A	A	A	A	A	A	A	A	A	S	S	S	S	S	S	S	S	S	S	S	S	S
29	A	A	A	A	A	A	A	A	A	A	A	A	S	S	S	S	S	S	S	S	S	S	S	S	S
30	A	A	A	A	A	A	A	A	A	A	A	A	A	S	S	S	S	S	S	S	S	S	S	S	S
31	A	A	A	A	A	A	A	A	A	A	A	A	A	S	S	S	S	S	S	S	S	S	S	S	S
32	A	A	A	A	A	A	A	A	A	A	A	A	A	S	S	S	S	S	S	S	S	S	S	S	S
33	A	A	A	A	A	A	A	A	A	A	A	A	A	A	S	S	S	S	S	S	S	S	S	S	S
34	A	A	A	A	A	A	A	A	A	A	A	A	A	A	S	S	S	S	S	S	S	S	S	S	S
35	A	A	A	A	A	A	A	A	A	A	A	A	A	A	S	S	S	S	S	S	S	S	S	S	S
36	A	A	A	A	A	A	A	A	A	A	A	A	A	A	A	S	S	S	S	S	S	S	S	S	S
37	A	A	A	A	A	A	A	A	A	A	A	A	A	A	A	S	S	S	S	S	S	S	S	S	S
38	A	A	A	A	A	A	A	A	A	A	A	A	A	A	A	A	S	S	S	S	S	S	S	S	S
39	A	A	A	A	A	A	A	A	A	A	A	A	A	A	A	A	S	S	S	S	S	S	S	S	S
40	A	A	A	A	A	A	A	A	A	A	A	A	A	A	A	A	S	S	S	S	S	S	S	S	S
41	A	A	A	A	A	A	A	A	A	A	A	A	A	A	A	A	A	S	S	S	S	S	S	S	S
42	A	A	A	A	A	A	A	A	A	A	A	A	A	A	A	A	A	S	S	S	S	S	S	S	S
43	A	A	A	A	A	A	A	A	A	A	A	A	A	A	A	A	A	S	S	S	S	S	S	S	S
44	A	A	A	A	A	A	A	A	A	A	A	A	A	A	A	A	A	A	S	S	S	S	S	S	S
45	A	A	A	A	A	A	A	A	A	A	A	A	A	A	A	A	A	A	S	S	S	S	S	S	S
46	A	A	A	A	A	A	A	A	A	A	A	A	A	A	A	A	A	A	S	S	S	S	S	S	S
47	A	A	A	A	A	A	A	A	A	A	A	A	A	A	A	A	A	A	A	S	S	S	S	S	S
48	A	A	A	A	A	A	A	A	A	A	A	A	A	A	A	A	A	A	A	S	S	S	S	S	S
49	A	A	A	A	A	A	A	A	A	A	A	A	A	A	A	A	A	A	A	S	S	S	S	S	S
50	A	A	A	A	A	A	A	A	A	A	A	A	A	A	A	A	A	A	A	A	S	S	S	S	S
51	A	A	A	A	A	A	A	A	A	A	A	A	A	A	A	A	A	A	A	A	S	S	S	S	S
52	A	A	A	A	A	A	A	A	A	A	A	A	A	A	A	A	A	A	A	A	S	S	S	S	S
53	A	A	A	A	A	A	A	A	A	A	A	A	A	A	A	A	A	A	A	A	A	S	S	S	S
54	A	A	A	A	A	A	A	A	A	A	A	A	A	A	A	A	A	A	A	A	A	S	S	S	S
55	A	A	A	A	A	A	A	A	A	A	A	A	A	A	A	A	A	A	A	A	A	S	S	S	S
56	A	A	A	A	A	A	A	A	A	A	A	A	A	A	A	A	A	A	A	A	A	A	S	S	S
57	A	A	A	A	A	A	A	A	A	A	A	A	A	A	A	A	A	A	A	A	A	A	S	S	S
58	A	A	A	A	A	A	A	A	A	A	A	A	A	A	A	A	A	A	A	A	A	A	S	S	S
59	A	A	A	A	A	A	A	A	A	A	A	A	A	A	A	A	A	A	A	A	A	A	A	S	S
>59	A	A	A	A	A	A	A	A	A	A	A	A	A	A	A	A	A	A	A	A	A	A	A	R	R

TABLE 12.3. Optimal Individual Decision For Type-1 Patients

patients are forced to seek services from outside of the system, even though their rejection cost is the same as the type-2 patients. The overall system performance is much worse than when the social optimization is applied, i.e, waiting times of 19.304 and 19.889 from the individual optimization versus waiting times of 3.6862 and 5.0929 from the social optimization at the two clinics, respectively. This example supports social optimization in the hospital system.

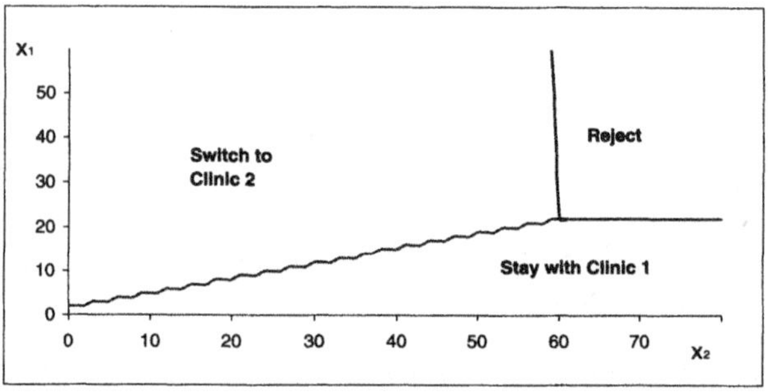

FIGURE 12.6. Individual Optimal Decisions for Type-1 Patients

The examples presented so far are for flow control with a fixed capacity allocation which is not necessarily optimal. In the following example, we integrate the dynamic flow control problem with the capacity design problem. Our objective is not to propose a formal approach. Instead, we want to demonstrate the need to tackle the two problems together.

Example 12.15 Suppose that there are two clinics in the system with $\lambda_1 = 1.2$ and $\lambda_2 = 1.8$. The total available capacity is $U = 3.4$. The waiting, switching, and rejection costs are $w = 20.0$, $C_1 = C_2 = 20.0$, and $R_1 = R_2 = 200.0$, respectively. We solve the flow control and capacity design problem in the following four steps:

Step 1: Assume proportional capacity allocation, i.e., $\mu_1 = 1.36$ and $\mu_2 = 2.04$. Solve a flow control problem according to Theorem 12.3. Simulate the performance of the system based on the optimal flow control policy.

Step 2: Let $1 - \gamma_1 =$ switching rate + rejection rate obtained for type-1 patients from the simulation and $1 - \gamma_2 =$ switching rate + rejection rate for type-2 patients. Solve a capacity allocation problem according to Theorem 12.7.

Step 3: Given the optimal μ_1 and μ_2 from the last step, solve the flow control problem again. Simulate the performance based on the new optimal control policy.

Step 4: Compute $1 - \gamma$ again based on results from step 3. Solve the capacity allocation problem again. If the resulting allocation is close to those obtained in step 2 or if $\gamma_1 \leq 0.5$, stop; otherwise, repeat from step 2.

After the first iteration, the flow control optimization results are as follows: the proportions of type-1 and type-2 patients rejected are 4.6% and 4.5%, respectively; the proportions of type-1 and type-2 patients switching to the other clinic are 18.6% and 12.2%, respectively; and the optimal total expected number of patients in the system (after optimizing the capacity design based on the above rates) is 14.43.

The computation should stop at iteration 4 when γ_1 drops below 0.5. From Table 12.4, we can see that after each new capacity allocation, the dynamic flow control leads to more switching or rejection at clinic 1 and the system performance continues to improve. So to stop, we need to have a minimum proportion of type-1 (type-2) patients who must stay. The switching curves of type-1 patients in iteration 1 and iteration 4 are shown in Figure 12.7 and Figure 12.8 respectively, The curves for type-2 patients are similar. The simulated performance for iteration 4 are as follows: the proportions of type-1 and type-2 patients rejected are 6.65% and 2.96%, respectively; the proportions of type-1 and type-2 patients switching to the other clinic are 36.2% and 4.54%, respectively; and the optimal total expected number of patients in the system (after optimizing the capacity design based on the above rates) is 13.61.

In actual computation, we allow the iteration to continue until iteration 8. Notice that the total cost fluctuates, but seems to be on a decreasing trend. The cost reductions are perhaps mainly due to the reduction of the waiting cost. This is because the switching cost for this example is relatively small, the same as the waiting cost. Since rejection is a case-by-case tactical decision, the main trade-off is between the quality of service and the switching cost. Lowering the switching cost, such as having better transportation infrastructure, will enable the system to consolidate resources more and achieve shorter waiting time for all patients.

12.5 Insights and Future Research Directions

This chapter introduces some novel modeling approaches to tackle two challenging and inter-connected issues in hospital system management: al-

Iteration	μ_1^*	μ_2^*	α_1	α_2	γ_1	Waiting cost	Total cost
1	1.084	2.316	0.000814	0.580189	0.662	275.999	309.311
2	0.947834	2.452404	0.000091	0.607347	0.613	274.856	309.643
3	0.884578	2.515634	0.000079	0.604934	0.5715	274.494	311.073
4	0.830918	2.569387	0.000111	0.649394	0.4561	240.894	278.982
5	0.681374	2.720544	0.000640	0.781398	0.396	230.752	271.347
6	0.602324	2.799349	0.000530	0.809456	0.353	231.580	273.788
7	0.545065	2.855066	0.000040	0.864073	0.33	226.238	268.483
8	0.514385	2.885694	0.000024	0.859556	0.318	229.841	272.679

TABLE 12.4. Results in Each Iteration

locating limited resources and managing patient flows among hospitals and clinics within the system.

For patient flow management, we construct a stochastic dynamic programming model for a two-clinic system. The optimal switching curves and decision tables can be generated from the model. The two-clinic system is a simplification of the systems in real applications. We expect the general control policy structure for each type of patients to be similar when there are more than two clinics in the system. However, we cannot extend the results in Section 12.2 directly to multi-clinic systems. There are different ways to construct heuristic control policies for multi-clinic systems. Consider a three-clinic system for instance. Suppose the capacity of each clinic is fixed. We can construct one dynamic programming model for each two-clinic pair, and obtain the corresponding decision tables. For type-j patients, there are two decision tables. Whenever a patient arrives, the two table will be checked in sequence to determine the best course of action.

For the capacity design, we construct a two-layer model. The first model assumes a fixed capacity allocation. Also a fixed proportion of patients will not make a switch while the rest are allocated to minimize system-wide waiting time. Given the patient allocation, we can then allocate the capacity to minimize the waiting time. The primary managerial insight from this study is significant and very clear: simple proportional resource allocation rules do not contribute to better customer services; instead, after providing "adequate" resources to each clinic, the extra resources should be concentrated in the area with the highest demand intensity so that the system-wide service quality can be optimized. We note that the implemen-

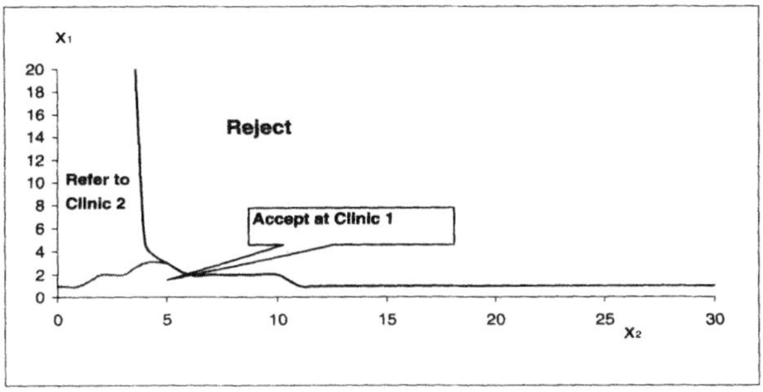

FIGURE 12.7. Switching Curves for Type-1 Patients in Iteration 1

tation of the "one large and many small" rule will not be simple. But the key is to identify the demand patterns and a central region in the system. In addition, we also need to analyze and determine the minimal service requirements of each region so that it will receive "adequate" resources.

There are many opportunities for future research related to the two issues discussed in this chapter. Here we discuss a few of them. First, we want to point out that it should not be too difficult to extend our results to systems with more general service times and switching costs (e.g., dependent on the destination clinics). A more interesting extension is to consider some non-financial constraints on switching. In some situations, for example, the travel time from one region to another can no longer be ignored and has to be added in the analysis explicitly. Also, a patient originating from one region may only be willing to search from a subset of clinics for services and join the shortest queue among those stations. This may require us to treat the whole problem hierarchically. In fact, some practical applications will likely require us to find more than one center, each of which is clustered with a number of smaller facilities. Another possible extension is to impose a limit on the number of patients waiting at each clinic, i.e., models with finite queues. Such models can be used to study other aspects of the service system, such as achieved throughput rate and appropriate maximum waiting time. Finally, the switching cost and service quality trade-off discussed in the last section needs further investigation, in terms of the trade-off patterns and key factors affecting the trade-off. In particular, how to provide the right incentive to individual patients so that they will follow the social optimal rules is an interesting, but difficult issue and needs to be studied carefully. Any policy solely based on the analysis of the trade-off between the cost of the incentive and the saving from social optimization may not be implementable since waiting for treatment can sometimes have

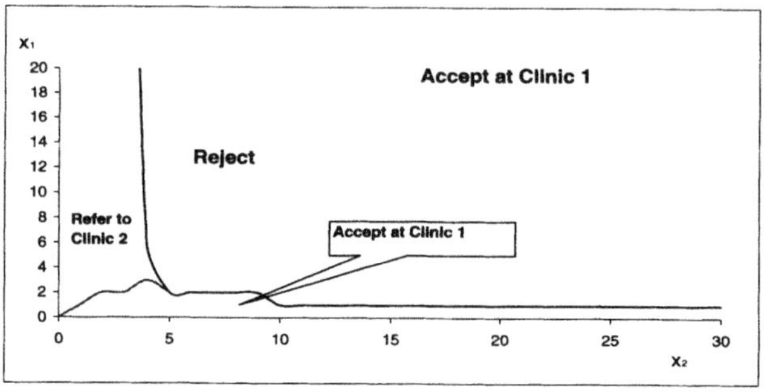

FIGURE 12.8. Switching Curves for Type-1 Patients in Iteration 4

a crucial impact on the patient's health and the individual should have the final choice on whether to switch, to stay, or to seek prompt service from outside of the system.

12.6 Notes

Health care has always been a rich and fruitful area for operations research. Many researchers have been attracted to a variety of health care related issues. Recent examples include Ahn and Hornberer [1] and Zenios, Chertow, and Wein [31] on organ allocation, Ho and Lau [10] and Liu and Liu [12] on outpatient appointments, So and Tang [23] and Fuloria and Zenios [7] on reimbursement policy, to name only a few. Resource management is of particular importance to health care administrators and has motivated many recent publications, such as Murray and Kreuser [14], McKillop [13], Zon and Kommer [32]. The issues and models presented in this chapter are different from those in the existing literature. We draw much of the materials from two recent works of the authors, [2] and [3] where readers will find more technical details and other related results.

Literature reviews on flow control and resources allocation can also be found from [2] and [3]. Here, we only provide a few key references for this chapter. Capacity allocation has been an important issue in queueing networks and manufacturing systems. Representative publications on this issue include, but are not limited to, [6], [16], [20], and [24]. There are many interesting works on admission and routing control. Among them, we would refer to [9], [11], [15], [4], [21], and [27].

In this chapter, we have used some standard results. Readers can find the corresponding references on stochastic dynamic programming from Puterman [17] or Ross [19], on convexity and submodularity from Topkis [25] and Glasserman and Yao [8], and on queueing theory from Walrand [26]. Besides, the switching curve structures and some related arguments in S12.2 are similar to those in Chen, Yao and Zheng [5], and Yao and Zheng [28], [29], [30].

Acknowledgments

The authors have benefited tremendously from discussions with management and staff of the HKHA over the past several years. In particular, Dr. Dickson Chang, who had until recently been the deputy director for planning at the HKHA for many years, has been a rare source of inspiration through his unlimited knowledge and insights on health care and his acute understanding of operations research methodologies and philosophy. This research is partially supported by Hong Kong Research Grant Council through a grant HKUST6089/00E; the first author is supported in part by NSF through a grant DMI-9908294, and the third author is also supported in part by Hong Kong Research Grant Council through a grant HKUST6012/00E. The authors would also like to acknowledge the valuable assistance provided by Xiang Zhu in performing the numerical computations and simulations.

12.7 References

[1] Ahn, J-H and Hornberger, J. C., Involving patients in the cadaveri kidney transplant allocation process: A decision-theoretic perspective, *Management Science*, **42** (1996), 629-854.

[2] Chao, X., Liu, L., and Zheng, S., Customer flow control in a multiclass network: individual and social optimization, working paper, 2001.

[3] Chao, X., Liu, L., and Zheng, S., Resources allocation in multi-site service systems with inter-site customer flow, working paper, 2001.

[4] Chen, H., Yang, P. and Yao, D.D., Control and Scheduling in a Two-Station Network: Optimal Policies and Heuristics, *Queueing System*, **18** (1994), 301-332.

[5] Chen, J., Yao, D.D. and Zheng, S., Quality Control for Products Supplied with Warranty, *Operations Research,* **46** (1998), 107-115.

[6] Dallery, Y. and Stecke, K., On the optimal allocation of servers and workloads in closed queueing networks, *Operations Research*, **38** (1990), 694-703.

[7] Fuloria, P.C. and Zenios, S.A., Outcomes-adjusted reimbursement in a health-care delivery system, *Management Science*, **47** (2001), 735-751.

[8] GLASSERMAN, P. AND YAO, D.D., *Monotone Structure in Discrete-Event Systems*, Wiley, 1994.

[9] HAJEK, B., Optimal Control of Two Interacting Service Stations, *IEEE Trans. Auto. Control*, **AC 29** (1984), 491-499.

[10] HO, C. AND LAU, H.S., Minimizing total cost in scheduling outpatient appointments, *Management Science*, **38** (1992), 1750-1764.

[11] LIPPMAN, S.A. AND STIDHAM, S. JR. , Individual versus Social Optimization in Exponential Congestion Systems, *Operations Research*, **25**(1977), 233-247.

[12] LIU, L. AND LIU, X., Dynamic and static job allocation for multi-server systems, *IIE Transactions*, **30** (1998), 845-854.

[13] MCKILLOP, I., *The Patient Flow Model: A Resource-Based Approach to Optimizing Patient Flows*, University of Waterloo, Ontario, Canada, 1997.

[14] MURRAY, C.J. AND KREUSER, J., *A Cost-Effectiveness Model for Allocating Health Sector Resources*, Harvard School of Public Health, 1994.

[15] NAOR, P., On the regulation of queue size by levying tolls, *Econometrica*, **37** (1969), 15-24.

[16] PINEDO, M. AND SESHADRI, S., Optimal allocation of resources in a job shop environment, *IIE Transactions*, **31** (1999), 195-206.

[17] PUTERMAN, M.L., *Markov Decision Processes*, John Wiley & Sons, New York, 1994.

[18] ROSS, S., *Stochastic Processes*, John Wiley & Sons, New York, 1983.

[19] ROSS, S., *Introduction to Stochastic Dynamic Programming*, Academic Press, 1983.

[20] SHAKED, M. AND SHANTHIKUMAR, J.G., Optimal allocation of resources to nodes of parallel and series systems, *Adv. Appl. Prob.*, **24** (1992), 894-914.

[21] SHANTHIKUMAR, J G. AND XU, S.H., Asymptotically optimal routing and service rate allocation in a multiserver queueing system, *Operations Research*, **45** (1997), 464-469.

[22] SHORTELL, S. M., BAZZOLI, G.J., DUBBS, N. L., AND KRALOVEC, P., Clarifying health networks and systems: managerial and policy implications, *Health Care Management Review*, **25** (2000), 9-17.

[23] So, K.C. and Tang, C.S., Modeling the impact of an outcome-oriented reimbursement policy on clinic, patients, and pharmaceutical firms, *Management Science*, **46** (2000), 875-892.

[24] Stecke, K.E. and Solberg, J.J., The optimality of unbalancing both workloads and machine group sizes in closed queueing networks of multiserver queues, *Operations Research*, **33** (1985), 882-910.

[25] Topkis, D.M., Minimizing a submodular function on a lattice, *Operations Research,* **26** (1978), 305-321.

[26] Walrand, J., *An Introduction to Queueing Networks*, Prentice Hall, New Jersey, 1988.

[27] Xu, S.H. and Shanthikumar, J.G., Optimal expulsion control–A dual approach to admission control of an ordered-entry system, *Operations Research*, **41** (1993), 1137-1152.

[28] Yao, D.D. and Zheng, S., *Dynamic Control of Quality in Production-Inventory Systems–Coordination and Optimization*, Springer-Verlag, New York, to be published.

[29] Yao, D.D. and Zheng S., Sequential inspection under capacity constraints, *Operations Research*, **47** (1999), 410-421.

[30] Yao, D.D. and Zheng S., Coordinated quality control in a two-stage system, *IEEE Transactions on Automatic Control*, **44** (1999), 1166-1179.

[31] Zenios, S. A., Chertow, G. M., and Wein, L. M., Dynamic allocation of kidneys to candidates on the transplant waiting list, *Operations Research*, **48** (2000), 549-569.

[32] Zon, A. H. and Kommer, G. J., Patient flows and optimal health-care resource allocation at the macro-level: a dynamic linear programming approach, *Health Care Management Science*, **2** (1999), 87-96.

13

Optimal Airline Booking Control with Cancellations

Youyi Feng, Ping Lin and Baichun Xiao

In airline revenue management, cancellations refer to returns or changes of booked seats prior to flight departures, and no-shows refer to passengers that do not check in without notifying the airline in advance. Cancellations and no-shows present a major barrier to effective seat inventory control systems. While cancelled seats can be rebooked in the future, no-shows lead to ultimate vacancies. According to Smith et al. (1992), customers cancel more than 50 percent of their booking requests before or at departures. To curtail potential revenue losses resulting from cancellations and no-shows, airlines commonly adopt overbooking policies. Smith et al. (1992) reported that without overbooking, 15 percent of seats would be spoiled. However, overbooking also runs a risk of over-sold if seat capacity is exceeded at departure. Hence, a good overbooking policy should strike a balance between the opportunity cost of vacant seats and the penalty of not honoring passengers with reservations.

The main difficulty for cancellations lies in the fact that booking requests and cancellations generate two concurrent stochastic processes which are interdependent. It is thus difficult to apply the recursive solution procedure in the continuous-time revenue management models without cancellations. Chatwin (1999) avoids this obstacle by using the net revenue rate. Because of the continuous nature of reservation processes, the discrete-time models are somehow always limited into approximations. The proximity of approaching continuous-time models by discretizing time scale is at most linear with the step size. To seek higher precision, we are motivated to study continuous-time overbooking models with cancellations directly.

This chapter proposes a new approach to tackle the optimal booking policy for single-leg models with cancellations and no-shows. We assume that demand of each fare class obeys a non-homogeneous Poisson process. Passengers may cancel their reservations at any time prior to departure. The cancelled tickets are partially refunded. The rate of cancellations is time-dependent. We derive an optimal seat control policy. This control policy suggests that a fare be accepted if it is above the *minimum acceptable fare* (MAF) and be rejected otherwise. With a newly developed technique, we explore monotone properties of MAF and develop a method for computing its values. Because MAF is a decreasing function of unbooked seats, the optimal seat control policy is of threshold. Each threshold point corresponds to a combined state of fare class and seat inventory, and marks an occurrence of price change for the given inventory. The existence of time thresholds makes the implementation of optimal control policy remarkably simple. Finally, although numerical examples indicate that MAF may not be a decreasing function of time because of cancellations, the property holds true when demand intensities decrease in time and cancellation rates remain constant.

The rest of the chapter is organized in the following orders. Section 13.1 presents a mathematical formulation for the problem and discusses the optimality conditions for the value function. Section 13.2 focuses on two issues around MAF. We show that MAF decreases in the unbooked seats but not in the remaining time. We also provide the solution procedure for computing MAF and the optimal booking control policy. Two extensions of the basic model are presented in Section 13.3 . In Section 13.4 we explain through numerical experiments why MAF may not decrease over time. A simple review for the literature is given in Section 13.5.

13.1 Preliminaries

13.1.1 Model Description

We consider a booking control model on a single-leg flight. The flight has M seats available at time 0 and is scheduled to depart at $T(>0)$. There are K fare classes, denoted by p_k, $k = 1, \cdots, K$. Without loss of generality, assume $p_1 > p_2 > \cdots > p_K$. Flight reservations for fare class k obey a Poisson process whose intensity $\lambda_k(t)$, $0 \le t \le T$, is time-dependent. We call $\lambda_k(t)$ the *booking curve* of class k. The booked seats can be canceled prior to the departure with a refund $c > 0$. We assume that the seat control policy has no influence on the cancellation process. Furthermore, passengers cancel their booked seats independently of one another. Let $\beta(t)$ be the average cancellation rate at t, $0 \le t \le T$, for each seat booked. The function $\beta(t)$ can be viewed as the probability density function for a reservation cancelled at t. Hence, if m seats have been booked at t, the cancellation rate is $m\beta(t)$. We assume that the cancellation probability is class independent. This assumption may not be observed in reality. For example, refundable fares usually have higher cancellation rates than nonrefundable fares.

For ease of analysis, we assume that $\lambda_k(t), k = 1, \cdots, K$ and $\beta(t)$ are all piecewise continuous and bounded over $[0, T]$. This assumption is for technical purpose only and innocuous in general. For instance, Van Slyke and Young (2000) assume this condition for dealing with Knapsacks problems.

Usually some of the passengers with confirmed tickets may not show up in the last minute, and thus it is uncertain how much the seat capacity is to be filled at the departure. Different from cancellations which permit rebooking, no-shows result in irrecoverable vacancies. To overcome revenue losses resulting from no-shows, airlines rationally adopt the overbooking policies.

We call an integer m, $m \geq M$, the *aggregate overbooking upper bound* if no more than m seats can be booked during the entire sales season. In this paper we only consider such overbooking policies that have been designated to the aggregate overbooking upper bounds. Given the overbooking upper bound m, we denote function $H(m-n)$ by the expected revenue at the departure when there are $m-n$ seats reserved. In Feng and Xiao (1997), $H(x)$ is called *no-show payoff function* while actually $-H(x)$ is the loss caused by those no-shows at the departure. To evaluate this function, given n vacant seats, or $m-n$ reservations, let Y_j be a binary variable representing the status of passenger j, $j = 1, \cdots, m-n$, at the departure time. Assume that all Y_j are i.i.d. random variables. Thus,

$$Y_j = \begin{cases} 1 & \text{w.p. } \xi, \\ 0 & \text{w.p. } 1-\xi, \end{cases}$$

where ξ, $0 < \xi < 1$, is the probability of each passenger to show up, and "w.p" reads "with probability". The total number of passengers presenting at $t = T$ then is a binomial variable $\sum_{j=1}^{m-n} Y_j$. As a result, the expected revenue at $t = T$ can be expressed as

$$H(m-n) = \mathsf{E}\Big[-\bar{p}\Big(\sum_{j=1}^{m-n} Y_j - M \Big)^+ - \bar{c}\Big(m - n - \sum_{j=1}^{m-n} Y_j \Big) \Big],$$

where $\bar{p}(x)$ is the total reimbursement for the exceeding x passengers who relinquish their boarding and $\bar{c}$ is the partial refund for each no-show passenger. From now on, $\bar{p}(x)$ is assumed to be an increasing convex function of x. In practice this assumption makes sense since marginal cost will become higher when more extra passengers are needed to pay. Apparently,

$$H(x) = \begin{cases} -\bar{c}(1-\xi)x - \sum_{j=1}^{x-M} \bar{p}(j) q_j^x, & M < x, \\ -\bar{c}(1-\xi)x, & M \geq x, \end{cases}$$

where

$$q_j^x = \binom{x}{M+j} \xi^{M+j}(1-\xi)^{x-M-j}, \qquad j = 1, 2, \cdots,$$

is the probability that out of x passengers, there are $M+j$ people showing up at departure. A fact is worth noting; that is, since $\bar{p}(x)$ is convex, $H(x)$ is concave.

A booking control policy is *non-anticipative* if it solely relies on the past booking history. A policy is *Markovian* if it depends only on the elapsed time and remaining seats. It is apparent that a Markovian booking control policy is non-anticipative because the elapsed time and unbooked seats are part of the history. Let

$$u(t) = (S_1(t), S_2(t), \cdots, S_K(t)),$$

where

$$S_i(t) = \begin{cases} 1 & \text{if fare class } i \text{ is open,} \\ 0 & \text{otherwise,} \end{cases}$$

for $i = 1, \cdots, K$ and $0 \leq t \leq T$. It is clear that $u(t)$ represents a seat control rule at t. A non-anticipative policy u then is defined to be

$$u = \{u(t) : 0 \leq t \leq T\}.$$

By $\mathcal{U}$ it is meant the collection of all non-anticipative control policies.

If policy u is chosen, then the expected revenue over the interval $[t, T]$ with n remaining seats at t is denoted by $\Psi^u(t, n)$. The optimal value of $\Psi^u(t, n)$ is represented by

$$V(t, n) = \sup_{u \in \mathcal{U}} \Psi^u(t, n).$$

We call $V(t, n)$ the *value function.*

13.1.2 Optimality Conditions and the Value Function

Booking and cancellation processes can be viewed as *point processes.* Hence, optimization techniques of the point process, such as intensity control, can be applied to the single-leg flight model formulated in the preceding section. The analysis starts with the *basic model* where $m = M$.

The following lemma presents the optimality conditions for the basic model.

Lemma 13.1 Suppose that $m = M$ and there exists a $\bar{V}(t, n)$ which is continuous and piecewise continuously differentiable in the interval $[0, T]$ and satisfies $\bar{V}(T, n) = 0$ for $0 \leq n \leq M$. In addition, for $0 < n \leq M$,

$$\frac{\partial \bar{V}(t, n)}{\partial t} + \sum_{i=1}^{K} \lambda_i(t)[\bar{V}(t, n-1) - \bar{V}(t, n) + p_i]^+$$
$$+(M-n)\beta(t)[\bar{V}(t, n+1) - \bar{V}(t, n) - c] = 0, \qquad (13.1)$$

and for $n = 0$,

$$\frac{\partial \bar{V}(t,0)}{\partial t} + M\beta(t)[\bar{V}(t,1) - \bar{V}(t,0) - c] = 0. \tag{13.2}$$

Then $\bar{V}(t,n) = V(t,n)$.

Proof. In view of the Markovian intensity control theory (see, for instance, Brémaud (1980)), a function $\bar{V}(t,n)$ is equivalent to $V(t,n)$ if $\bar{V}(t,n)$ satisfies the following Hamilton-Jacobi equation

$$\frac{\partial \bar{V}(t,n)}{\partial t} + \max_{\alpha_i \in \{0, \lambda_i(t)\}} \sum_{i=1}^{K} \alpha_i [\bar{V}(t,n-1) - \bar{V}(t,n) + p_i]$$
$$+(M-n)\beta(t)[\bar{V}(t,n+1) - \bar{V}(t,n) - c] = 0,$$

for $0 < n \leq M$. This relation is tantamount to (13.1). When $n = 0$, all booking requests are rejected, as (13.2) ratifies. □

When cancellations are not permitted, Feng and Xiao (1997) define a kind of overbooking policies corresponding to aggregate overbooking upper bounds. Since the aggregate overbooking upper bound is a constraint, it is always better off to relax this constraint by setting a higher m. However, assuming the increasing margin of the oversale penalty, $\bar{p}(x)$, there exists an effective overbooking upper bound m^* such that the number of reservations admitted can never exceed m^*. This effective upper bound is set as the maximum m which satisfies $H(m-1) - H(m) \leq p_1$. The latest version of Feng and Xiao (1997) shows that the overbooking policies with higher than m^* is identical with that of m^* which is the optimal overbooking policy. For details we refer the reader to Feng and Xiao (1997). Because of their work, we set m to be the aggregate overbooking upper bound in this paper.

Taking into account no-shows, we re-address the optimality condition modified in the following Lemma 13.2.

Lemma 13.2 Suppose that $m > M$ and there exists a $\bar{V}(t,n)$ which is continuous and piecewise continuously differentiable over the interval $[0,T]$ for fixed n and satisfies $\bar{V}(T,n) = H(m-n)$, where $m \geq M$ and $0 \leq n \leq m$. In addition, for $0 < n \leq m$,

$$\frac{\partial \bar{V}(t,n)}{\partial t} + \sum_{i=1}^{K} \lambda_i(t)[\bar{V}(t,n-1) - \bar{V}(t,n) + p_i]^+$$
$$+(m-n)\beta(t)[\bar{V}(t,n+1) - \bar{V}(t,n) - c] = 0; \tag{13.3}$$

and for $n = 0$,

$$\frac{\partial \bar{V}(t,0)}{\partial t} + m\beta(t)[\bar{V}(t,1) - \bar{V}(t,0) - c] = 0. \tag{13.4}$$

Then $\bar{V}(t,n) = V(t,n)$.

Replacing M and $V(T, n)$ by m and $H(m-n)$, respectively in the basic model, we secure Lemma 13.2.

According to Lemma 13.2, if there exists a function $\bar{V}(t,n)$ satisfying Equations (13.3) and (13.4), and $\bar{V}(T,n) = H(m-n), n = 0, 1, \cdots, m$, then the optimization solution has been found. Thus, our goal is to construct such a $\bar{V}(t,n)$ to fit Lemma 13.2. Because of the equivalence of $\bar{V}(t,n)$ and $V(t,n)$, interchange of these two notations will not give rive to ambiguous. To conceptualize the approach, we call $V(t,n) - V(t,n-1)$ the *minimum acceptable fare* (MAF). This term has lasted years in the literature and air travel industry and is referred to as *expected marginal seat revenue*, or EMSR in short, in Feng and Xiao (1996, 1997). EMSR was originated in Belobaba (1987, 1989).

Our primary interest is to investigate the structure of the optimal booking policy,. Once the optimization solution is sought, this policy is of the *nesting fare controls.*

Corollary 13.3 Assume that there exists a function $\bar{V}(t,n)$ that satisfies Lemma 13.1. Then optimal booking control at any time $0 \le t \le T$ has a nested-fare structure; i.e., there exists an index k, $1 \le k \le K$, such that $S_i(t) = 1 \; \forall \; 1 \le i \le k$ and $S_i(t) = 0 \; \forall \; k < i \le K$.

Proof. In view of Lemma 13.1, $V(t,n) = \bar{V}(t,n)$. Moreover, fare p_i is effective at time t if and only if

$$V(t,n) - V(t,n-1) < p_i.$$

Hence, there exists an index k, $1 \le k \le K$, such that

$$V(t,n) - V(t,n-1) < p_i, \quad i = 1, \cdots, k,$$

and

$$V(t,n) - V(t,n-1) \ge p_i, \quad i = k+1, \cdots, K.$$

□

13.2 The Minimum Acceptable Fare and Threshold Control

The optimal booking control can be implemented either by MAF or by the thresholds.

13.2.1 The Minimum Acceptable Fare

Equation (13.3) shows that $V(t,n)$ can be solved at time t if $V(t,n-1)$, $V(t,n+1)$, and the set of active fares are all known. However, this assumption makes the system a closed loop because $V(t, n-1)$ in turn depends on $V(t,n)$. Without cancellations, $V(t, n-1)$ does not rely on $V(t,n)$ and as a result, the value function can be solved recursively provided that the set of active fares are known at t. This technique fails to work for the present model because of the inter-dependence between $V(t, n-1)$ and $V(t,n)$.

Let n, $n = 1, \cdots, M$, be the number of remaining seats at time t and u^* the optimal policy. Assuming $p_1 > p_2 \cdots > p_K$, we denote by

$$k_n(t) = \max\{k : p_k > V(t,n) - V(t,n-1)\},$$

the highest index of the optimal active fares. Hence, under u^*, the optimal active fare mix can be expressed as $\{p_1, \cdots p_{k_n(t)}\}$. For simplicity, we may replace $k_n(t)$ with k_n if the substitution will not result in a confusion.

According to Corollary 13.3, all booking requests for fare class i, $1 \leq i \leq k_n$, should be accepted at t. Hence, conditions (13.3) and (13.4) generate the following system of linear differential equations.

$$\left\{\begin{array}{l}
\frac{\partial \bar{V}(t,0)}{\partial t} = M\beta(t)\bar{V}(t,0) - M\beta(t)\bar{V}(t,1) + M\beta(t)c, \\
\frac{\partial \bar{V}(t,1)}{\partial t} = -\sum_{i=1}^{k_1} \lambda_i(t)\bar{V}(t,0) + [(M-1)\beta(t) + \sum_{i=1}^{k_1} \lambda_i(t)]\bar{V}(t,1) \\
\qquad -(M-1)\beta(t)\bar{V}(t,2) + (M-1)\beta(t)c - \sum_{i=1}^{k_1} \lambda_i(t)p_i, \\
\cdots\cdots\cdots \\
\frac{\partial \bar{V}(t,M-1)}{\partial t} = -\sum_{i=1}^{k_{M-1}} \lambda_i(t)\bar{V}(t,M-2) + [\beta(t) + \sum_{i=1}^{k_{M-1}} \lambda_i(t)]\bar{V}(t,M-1) \\
\qquad -\beta(t)\bar{V}(t,M) + \beta(t)c - \sum_{i=1}^{k_{M-1}} \lambda_i(t)p_i, \\
\frac{\partial \bar{V}(t,M)}{\partial t} = -\sum_{i=1}^{k_M} \lambda_i(t)\bar{V}(t,M-1) + \sum_{i=1}^{k_M} \lambda_i(t)\bar{V}(t,M) - \sum_{i=1}^{k_M} \lambda_i(t)p_i.
\end{array}\right. \tag{13.5}$$

A primary concern of a seat control policy at any time is whether to accept or reject booking requests from each of fare classes. In view of Corollary 13.3, this decision is made with reference to the value of MAF. Hence, our main interest is to solve MAF. Let $f(n)$ be a function of integer and D be a one-step difference operator such that $Df(n) = f(n) - f(n-1)$. By definition, $DV(t,n) = V(t,n) - V(t,n-1)$ for $n = 1, \cdots, M$. Assuming $V(t,n) = \bar{V}(t,n)$ where $\bar{V}(t,n)$ is given in Lemma 13.1, we derive through algebraic manipulation that (13.5) is equivalent to

$$\left\{\begin{array}{l}
\frac{\partial DV(t,1)}{\partial t} = [M\beta(t) + \sum_{i=1}^{k_1} \lambda_i(t)]DV(t,1) - (M-1)\beta(t)DV(t,2) - \beta(t)c - \sum_{i=1}^{k_1} \lambda_i(t)p_i, \\
\frac{\partial DV(t,2)}{\partial t} = -\sum_{i=1}^{k_1} \lambda_i(t)DV(t,1) + [(M-1)\beta(t) + \sum_{i=1}^{k_2} \lambda_i(t)]DV(t,2) \\
\qquad -(M-2)\beta(t)DV(t,3) - \beta(t)c - \sum_{i=k_1+1}^{k_2} \lambda_i(t)p_i, \\
\cdots\cdots\cdots \\
\frac{\partial DV(t,M-1)}{\partial t} = -\sum_{i=1}^{k_{M-2}} \lambda_i(t)DV(t,M-2) + [2\beta(t) + \sum_{i=1}^{k_{M-1}} \lambda_i(t)]DV(t,M-1) \\
\qquad -\beta(t)DV(t,M) - \beta(t)c - \sum_{i=k_{M-2}+1}^{k_{M-1}} \lambda_i(t)p_i, \\
\frac{\partial DV(t,M)}{\partial t} = -\sum_{i=1}^{k_{M-1}} \lambda_i(t)DV(t,M-1) + [\beta(t) + \sum_{i=1}^{k_M} \lambda_i(t)]DV(t,M) \\
\qquad -\beta(t)c - \sum_{i=k_{M-1}+1}^{k_M} \lambda_i(t)p_i.
\end{array}\right.$$

Similarly, the above system can be written in matrix form such as follows:

$$\frac{d[\mathbf{d}(t)]}{dt} = \mathbf{B}(t)\mathbf{d}(t) - \mathbf{b}(t), \tag{13.6}$$

where vectors

$$\mathbf{d}(t) = \begin{pmatrix} DV(t,1) \\ DV(t,2) \\ \cdots \\ DV(t,M-1) \\ DV(t,M) \end{pmatrix}, \qquad \mathbf{b}(t) = \begin{pmatrix} \beta(t)c + \sum_{i=1}^{k_1} \lambda_i(t)p_i \\ \beta(t)c + \sum_{i=k_1+1}^{k_2} \lambda_i(t)p_i \\ \cdots \\ \beta(t)c + \sum_{i=k_{M-2}+1}^{k_{M-1}} \lambda_i(t)p_i \\ \cdots \\ \beta(t)c + \sum_{i=k_{M-1}+1}^{k_M} \lambda_i(t)p_i \end{pmatrix},$$

and matrix

$$\mathbf{B}(t) = \mathbf{B}_1(t) + \mathbf{B}_2(t),$$

where

$$\mathbf{B}_1(t) = \begin{pmatrix} M & -(M-1) & 0 & \cdots & 0 & 0 & 0 \\ 0 & (M-1) & -(M-2) & \cdots & 0 & 0 & 0 \\ \cdots & \cdots & \cdots & \cdots & \cdots & \cdots & \cdots \\ 0 & 0 & 0 & \cdots & 0 & 0 & 1 \end{pmatrix} \cdot \beta(t),$$

and

$$\mathbf{B}_2(t) = \begin{pmatrix} \sum_{i=1}^{k_1} \lambda_i(t) & 0 & 0 & \cdots & 0 & 0 & 0 \\ -\sum_{i=1}^{k_1} \lambda_i(t) & \sum_{i=1}^{k_2} \lambda_i(t) & 0 & \cdots & 0 & 0 & 0 \\ \cdots & \cdots & \cdots & \cdots & \cdots & \cdots & \cdots \\ 0 & 0 & 0 & \cdots & 0 & -\sum_{i=1}^{k_{M-1}} \lambda_i(t) & \sum_{i=1}^{k_M} \lambda_i(t) \end{pmatrix}.$$

Note that in system (13.6) we have assumed $k_n \le k_{n+1}$. If $k_n > k_{n+1}$, then $\sum_{k_n+1}^{k_{n+1}}$ should be understood as $-\sum_{k_{n+1}}^{k_n+1}$.

13.2.2 Properties of MAF

Equation (13.6) indicates that $\mathbf{d}(t)$ can be solved if at each seat inventory level j, $j = 1, \cdots, M$, the index of the lowest active fare k_j is known. In this section we show that k_j is an increasing function of j because of the monotone property of MAF. This fundamental property states that MAF decreases with respect to the remaining seats. While the property is known to hold for various airline seat control models, it is not yet clear whether it remains true in the continuous-time booking control models when cancellations are permitted.

Lemma 13.4 If $k_1(t) \le k_2(t) \le \cdots \le k_M(t)$ holds for any t, $0 \le t \le T$, then $DV(t,n)$ is decreasing in n, $1 \le n \le M$.

Proof. The proof of Lemma 13.4 needs two additional lemmas. We begin with Lemma 13.5.

Lemma 13.5 Suppose that at time $0 \le t \le T$, $k_1(t) \le k_2(t) \le \cdots \le k_M(t)$, then for $n \ge 2$,

$$\begin{aligned}
\frac{\partial[DV(t,n) - DV(t,n-1)]}{\partial t} &= -\sum_{i=1}^{k_{n-2}(t)} \lambda_i(t)[DV(t,n-1) - DV(t,n-2)] \\
&- (M-n)\beta(t)[DV(t,n+1) - DV(t,n)] \\
&+ \left[(M-n+2)\beta(t) + \sum_{i=1}^{k_n(t)} \lambda_i(t)\right][DV(t,n) - DV(t,n-1)] + g(t,n),
\end{aligned}$$

where

$$\begin{aligned}
g(t,n) &= \sum_{i=k_{n-2}(t)+1}^{k_{n-1}(t)} \lambda_i(t)[p_i - DV(t,n-1)] \\
&+ \sum_{i=k_{n-1}(t)+1}^{k_n(t)} \lambda_i(t)[DV(t,n-1) - p_i] \ge 0.
\end{aligned}$$

According to computing procedure for $V(t,n)$, $k_n(t)$ is piecewise constant. Hence, for sake of simplicity, we use k_n to replace $k_n(t)$ in the following proof. For $3 \le n \le M$,

$$\begin{aligned}
&\frac{\partial[DV(t,n) - DV(t,n-1)]}{\partial t} \\
&= \sum_{i=1}^{k_{n-2}} \lambda_i(t) DV(t,n-2) - [(M-n+2)\beta(t) \\
&+ 2\sum_{i=1}^{k_{n-1}} \lambda_i(t)] DV(t,n-1) \\
&+ \left[2(M-n+1)\beta(t) + \sum_{i=1}^{k_n} \lambda_i(t)\right] DV(t,n) \\
&- (M-n)\beta(t) DV(t,n+1) + \sum_{i=k_{n-2}+1}^{k_{n-1}} \lambda_i(t) p_i - \sum_{i=k_{n-1}+1}^{k_n} \lambda_i(t) p_i \\
&= -\sum_{i=1}^{k_{n-2}} \lambda_i(t)\,[DV(t,n-1) - DV(t,n-2)] \\
&- (M-n)\beta(t)[DV(t,n+1) - DV(t,n)] \\
&+ \left[(M-n+2)\beta(t) + \sum_{i=1}^{k_n} \lambda_i(t)\right][DV(t,n) - DV(t,n-1)]
\end{aligned}$$

$$
\begin{aligned}
&+ \sum_{i=k_{n-2}+1}^{k_{n-1}} \lambda_i(t)\,[p_i - DV(t,n-1)] + \sum_{i=k_{n-1}+1}^{k_n} \lambda_i(t)\,[DV(t,n-1) - p_i] \\
&= -\sum_{i=1}^{k_{n-2}} \lambda_i(t)\,[DV(t,n-1) - DV(t,n-2)] \\
&- (M-n)\beta(t)[DV(t,n+1) - DV(t,n)] + g(t,n) \\
&+ \left[(M-n+2)\beta(t) + \sum_{i=1}^{k_n} \lambda_i(t)\right][DV(t,n) - DV(t,n-1)],
\end{aligned}
$$

where $g(t,n) \geq 0$ holds because of Lemma 13.1.

Note that the preceding analysis is applicable to $n = 2$ as well. When $n = 2$, the desired relation is

$$
\begin{aligned}
\frac{\partial[DV(t,2) - DV(t,1)]}{\partial t} &= -\left[M\beta(t) + 2\sum_{i=1}^{k_1} \lambda_i(t)\right] DV(t,1) \\
&+ \left[2(M-1)\beta(t) + \sum_{i=1}^{k_2} \lambda_i(t)\right] DV(t,2) \\
&\quad -(M-2)\beta(t)DV(t,3) + \sum_{i=1}^{k_1} \lambda_i(t)p_i - \sum_{i=k_1+1}^{k_2} \lambda_i(t)p_i \\
&= \left[M\beta(t) + \sum_{i=1}^{k_2} \lambda_i(t)\right][DV(t,2) - DV(t,1)] \\
&\quad -(M-2)\beta(t)[DV(t,3) - DV(t,2)] \\
&\quad +\sum_{i=1}^{k_1} \lambda_i(t)[p_i - DV(t,1)] + \sum_{i=k_1+1}^{k_2} \lambda_i(t)[DV(t,1) - p_i] \\
&= [M\beta(t) + \sum_{i=1}^{k_2} \lambda_i(t)][DV(t,2) - DV(t,1)] \\
&\quad -(M-2)\beta(t)[DV(t,3) - DV(t,2)] + g(t,2).
\end{aligned}
$$

□

The next lemma requires some preliminaries in matrix exponential. Given any square matrix $\mathbf{W}$, the matrix exponential of $\mathbf{W}$ is defined to be

$$
e^{\mathbf{W}} = \sum_{k=0}^{\infty} \frac{(\mathbf{W})^k}{k!}.
$$

It is apparent that the matrix exponential is a square matrix of the same dimension. If $\mathbf{W}_1$ and $\mathbf{W}_2$ are square matrices of the same dimension, and are *commutative* in multiplication; i.e., $\mathbf{W}_1\mathbf{W}_2 = \mathbf{W}_2\mathbf{W}_1$, then

$$
e^{\mathbf{W}_1+\mathbf{W}_2} = e^{\mathbf{W}_1}e^{\mathbf{W}_2}.
$$

Because $\mathbf{W}$ is commutative with $-\mathbf{W}$, and $\mathbf{W} + (-\mathbf{W}) = \mathbf{W} - \mathbf{W} = \mathbf{0}$, we have

$$e^{\mathbf{W}} e^{-\mathbf{W}} = \mathbf{I},$$

where $\mathbf{I}$ is the identity matrix. In other words, $e^{\mathbf{W}}$ is invertible.

To show that $DV(t, n)$ is decreasing in $0 \le n \le M$, we define

$$\mathbf{h}(t) = \begin{pmatrix} D^2V(t,2) \\ D^2V(t,3) \\ \cdots \\ D^2V(t,M) \end{pmatrix}, \quad \mathbf{g}(t) = \begin{pmatrix} g(t,2) \\ g(t,3) \\ \cdots \\ g(t,M) \end{pmatrix}.$$

In addition, let $\mathbf{G}(t) = \mathbf{G}_1(t) + \mathbf{G}_2(t)$ where

$$\mathbf{G}_1(t) = \begin{pmatrix} M & -(M-2) & 0 & \cdots & 0 & 0 & 0 \\ 0 & (M-1) & -(M-3) & \cdots & 0 & 0 & 0 \\ \cdots & \cdots & \cdots & \cdots & \cdots & \cdots & \cdots \\ 0 & 0 & 0 & \cdots & 0 & 0 & 2 \end{pmatrix} \cdot \beta(t),$$

and $\mathbf{G}_2(t)$ equals

$$\begin{pmatrix} \sum_{i=1}^{k_2} \lambda_i(t) & 0 & 0 & \cdots & 0 & 0 & 0 \\ -\sum_{i=1}^{k_1} \lambda_i(t) & \sum_{i=1}^{k_3} \lambda_i(t) & 0 & \cdots & 0 & 0 & 0 \\ \cdots & \cdots & \cdots & \cdots & \cdots & \cdots & \cdots \\ 0 & 0 & 0 & \cdots & 0 & -\sum_{i=1}^{k_{M-2}} \lambda_i(t) & \sum_{i=1}^{k_M} \lambda_i(t) \end{pmatrix}.$$

With these notations, the result of Lemma 13.5 can be expressed as

$$\frac{d\mathbf{h}(t)}{dt} = \mathbf{G}(t)\mathbf{h}(t) + \mathbf{g}(t). \tag{13.7}$$

It is now suffices to show $\mathbf{h}(t) \le \mathbf{0}$. For $\mathbf{G}(t) = (g_{ij}(t))$, the integral of $\mathbf{G}(t)$ in the interval (a, b) is defined to be

$$\int_a^b \mathbf{G}(s)ds = \left(\int_a^b g_{ij}(s)ds \right),$$

which is a matrix of the same dimension. We note that all the diagonal elements of $\int_a^b \mathbf{G}(s)ds$ are positive and all the off-diagonal elements are negative. Moreover, for each i, $1 \le i \le M-1$, if $k_1 \le k_2 \le \cdots \le k_M$, then

$$\int_a^b g_{ii}(s)ds + \sum_{j \ne i} \int_a^b g_{ij}(s)ds > 0.$$

Hence, matrix $\int_a^b \mathbf{G}(s)ds$ is non-singular whose inverse is of non-negative entries.

To derive $\mathbf{h}(t)$ from (13.7), we consider an auxiliary system

$$\frac{d\bar{\mathbf{h}}(t)}{dt} = \bar{\mathbf{G}}(t)\bar{\mathbf{h}}(t) + \mathbf{g}(t). \tag{13.8}$$

where $\bar{\mathbf{h}}(t)$ satisfies the terminal condition $\bar{\mathbf{h}}(T) = 0$, and $\bar{\mathbf{G}}(t)$ is a piecewise constant matrix whose elements will be defined. We examine the auxiliary system because its solution will be shown approximate to that of system (13.7).

We will first show that $\bar{\mathbf{h}}(t) \leq \mathbf{0}$, and from this we will then argue that $\mathbf{h}(t) \leq \mathbf{0}$. Divide $[0, T]$ into N subintervals $[t_i, t_{i+1}]$, $i = 0, 1, \cdots, N-1$, with an equal length $h = T/N$. For $t \in (t_i, t_{i+1}]$, $i = 1, \cdots, N-1$ or $t \in [0, t_1]$, let $\bar{t}_i = \frac{t_i + t_{i+1}}{2}$ and $\bar{\mathbf{G}}(t) = \mathbf{G}(\bar{t}_i)$. Given any norm for vectors and matrices, which is denoted $||\cdot||$, it is easy to see

$$||\mathbf{G}(t) - \bar{\mathbf{G}}(t)|| = O||h||. \tag{13.9}$$

Since $\bar{\mathbf{G}}(t)$ is a constant matrix over the interval $(t_i, t_{i+1}]$, we have

$$\int_t^{t_{i+1}} \bar{\mathbf{G}}(u)du = \mathbf{G}(\bar{t}_i)(t_{i+1} - t)$$

for $t \in (t_i, t_{i+1}]$. Moreover, both the matrix exponential $e^{\int_t^{t_{i+1}} \bar{\mathbf{G}}(u)du}$ and its inverse $e^{-\int_t^{t_{i+1}} \bar{\mathbf{G}}(u)du}$ are well defined.

Lemma 13.6 Suppose $k_1 \leq k_2 \leq \cdots \leq k_M$ at t, $0 \leq t \leq T$, then for $s \geq t$, $e^{-\mathbf{G}(\bar{t}_i)(s-t)}$ is a non-negative matrix.

Proof. Let

$$\mathbf{G}(\bar{t}_i)(s-t) = \alpha\mathbf{I} - \mathbf{E}(s,t),$$

where $\mathbf{I}$ is the identity matrix and $\alpha > 0$ is an upper bound of all diagonal entries of $\mathbf{G}(\bar{t}_i)(s-t)$. It is clear that $\alpha\mathbf{I}$ is *commutative* with $\mathbf{E}(s,t)$ in matrix multiplication; i.e.,

$$(\alpha\mathbf{I})\mathbf{E}(s,t) = \mathbf{E}(s,t)(\alpha\mathbf{I}).$$

In addition, since $\mathbf{G}(\bar{t}_i)$ is diagonally dominant, $\mathbf{E}(s,t)$ is of non-negative entries. As a result,

$$e^{-\mathbf{G}(\bar{t}_i)(s-t)} = e^{-\alpha\mathbf{I}+\mathbf{E}(s,t)} = e^{-\alpha}e^{\mathbf{E}(s,t)}$$

is of the non-negative entries. □

With Lemmas 13.5 and 13.6 on hand, we complete the proof of Lemma 13.4. Multiplying both sides of equation (13.8) by $e^{\int_t^{t_{i+1}}\bar{\mathbf{G}}(s)ds} = e^{\mathbf{G}(\bar{t}_i)(t_{i+1}-t)}$, we have

$$e^{\mathbf{G}(\bar{t}_i)(t_{i+1}-t)}\frac{d\bar{\mathbf{h}}(t)}{dt} = e^{\mathbf{G}(\bar{t}_i)(t_{i+1}-t)}\mathbf{G}(t)\bar{\mathbf{h}}(t) + e^{\mathbf{G}(\bar{t}_i)(t_{i+1}-t)}\mathbf{g}(t),$$

which is equivalent to

$$\frac{\partial\left[e^{\int_t^{t_{i+1}}\bar{\mathbf{G}}(u)du}\bar{\mathbf{h}}(t)\right]}{\partial t} = e^{\int_t^{t_{i+1}}\bar{\mathbf{G}}(u)du}\mathbf{g}(t). \tag{13.10}$$

Notice $\mathbf{G}(\bar{t}_i) = \bar{\mathbf{G}}(t)$ for $t_i \le t \le t_{i+1}$ and $\int_t^{t_{i+1}}\bar{\mathbf{G}}(u)du = \mathbf{G}(\bar{t}_i)(t_{i+1}-t)$. Integrating (13.10) from t to t_{i+1} leads to

$$\bar{\mathbf{h}}(t_{i+1}) - e^{\mathbf{G}(\bar{t}_i)(t_{i+1}-t)}\bar{\mathbf{h}}(t) = \int_t^{t_{i+1}} e^{\mathbf{G}(\bar{t}_i)(t_{i+1}-s)}\mathbf{g}(s)ds,$$

which implies

$$\begin{aligned}\bar{\mathbf{h}}(t) &= e^{-\mathbf{G}(\bar{t}_i)(t_{i+1}-t)}\bar{\mathbf{h}}(t_{i+1}) - \int_t^{t_{i+1}} e^{-\mathbf{G}(\bar{t}_i)(s-t)}\mathbf{g}(s)ds \\ &\le e^{-\mathbf{G}(\bar{t}_i)(t_{i+1}-t)}\bar{\mathbf{h}}(t_{i+1}).\end{aligned} \tag{13.11}$$

The last inequality holds because $e^{-\mathbf{G}(\bar{t}_i)(s-t)}$ is a non-negative matrix and $\mathbf{g}(s) \ge \mathbf{0}$.

Applying inequality (13.11) repeatedly results in

$$\begin{aligned}\bar{\mathbf{h}}(t) &\le e^{-\mathbf{G}(\bar{t}_i)(t_{i+1}-t)}\bar{\mathbf{h}}(t_{i+1}) \\ &\le e^{-\mathbf{G}(\bar{t}_i)(t_{i+1}-t)}e^{-\bar{\mathbf{G}}(\bar{t}_{i+1})(t_{i+2}-t_{i+1})}\bar{\mathbf{h}}(t_{i+2}) \\ &\le e^{-\mathbf{G}(\bar{t}_i)(t_{i+1}-t)}e^{-\bar{\mathbf{G}}(\bar{t}_{i+1})h}\cdots e^{-\bar{\mathbf{G}}(\bar{t}_{N-1})h}\bar{\mathbf{h}}(T) \\ &= \mathbf{0}.\end{aligned}$$

Hence, $\bar{\mathbf{h}}(t) \le 0$ for all $t \in [0, T]$.

To show the proximity of $\bar{\mathbf{h}}(t)$ to $\mathbf{h}(t)$, consider $\mathbf{e}(t) = \bar{\mathbf{h}}(t) - \mathbf{h}(t)$. Subtracting (13.7) from (13.8) leads to

$$\frac{d\mathbf{e}(t)}{dt} = \bar{\mathbf{G}}(t)\mathbf{e}(t) + \bar{\mathbf{g}}(t), \tag{13.12}$$

where $\bar{\mathbf{g}}(t) = [\bar{\mathbf{G}}(t) - \mathbf{G}(t)]\mathbf{h}(t) = \mathbf{O}(h)$ from (13.9). Similar to the way (13.11) is derived, we establish

$$\begin{aligned}\mathbf{e}(t) &= e^{-\mathbf{G}(\bar{t}_i)(t_{i+1}-t)}\mathbf{e}(t_{i+1}) - \int_t^{t_{i+1}} e^{-\bar{\mathbf{G}}(\bar{t}_i)(s-t)}\bar{\mathbf{g}}(s)\,ds \\ &= e^{-\mathbf{G}(\bar{t}_i)(t_{i+1}-t)}\mathbf{e}(t_{i+1}) + \mathbf{O}(h^2),\end{aligned} \tag{13.13}$$

for $t \in [t_i, t_{i+1}]$. Note that $e^{-\bar{\mathbf{G}}(\bar{t}_i)(s-t)}$ is bounded above by a constant matrix in any subinterval within $[0, T]$. Hence,

$$\mathbf{e}(t_i) = e^{-\mathbf{G}(\bar{t}_i)h}\mathbf{e}(t_{i+1}) + \mathbf{O}(h^2),\ i = 0, 1, \cdots, N-1.$$

Because $\mathbf{e}(t_N) = \mathbf{e}(T) = 0$ and $(e^{-\mathbf{G}(\bar{t}_i)h})^i = e^{-\mathbf{G}(\bar{t}_i)ih}$ is bounded above by some constant matrix, we claim

$$||\mathbf{e}(t_i)||_2 \leq CNh^2 \leq Ch,\ i = 0, 1, \cdots, N,$$

where C is a constant and independent of h, N and t. Furthermore, since matrix $e^{-\mathbf{G}(\bar{t}_i)(t_{i+1}-t)}$ is bounded above and (13.13) holds for all $t \in [0, T]$, we have

$$||\bar{\mathbf{h}}(t) - \mathbf{h}(t)|| \leq Ch. \tag{13.14}$$

It is easy to induce $\mathbf{h}(t) \leq 0$ from (13.14). Suppose to the contrary, the inequality does not hold for some t, $0 \leq t \leq T$. Let $\mathbf{h}(t) = (\gamma_i(t))$ and $\bar{\mathbf{h}}(t) = (\bar{\gamma}_i(t))$, then there exists a t' and an index i' such that $\gamma_{i'}(t') > \delta > 0$. However, when h is sufficiently small and satisfies $h < \delta/C$, (13.14) will ensure $\gamma_{i'}(t') \leq Ch + \bar{\gamma}_{i'}(t') < \delta$. Since $\bar{\mathbf{h}}(t) \leq 0$, it leads to a contradiction. □

Lemma 13.4 enables us to establish a critical result in the following theorem.

Theorem 13.7 For any given t, $0 \leq t \leq T$, $DV(t, n)$ decreases in n, $1 \leq n \leq M$.

Proof. Since $V(T, n) = H(m-n)$ for all $1 \leq n \leq m$, and $H(x)$ is concave, $V(T, n)$ is convex and hence, there exists $\bar{k}_n$ for $n = 1, \cdots, M$, such that

$$V(T, n) - V(T, n-1) = H(m-n) - H(m-(n-1)) \leq p_k, \quad k = 1, \cdots, \bar{k}_n,$$

and $1 \leq \bar{k}_1 \leq \bar{k}_2 \leq \cdots \leq \bar{k}_m \leq K$. Notice that $\bar{k}_n$ is defined as the maximum $1 \leq k \leq K$ such that $H(m-n) - H(m-(n-1)) \leq p_k$. Because $DV(t, n)$ is continuous near T, there exists a neighborhood where $DV(t, n) \leq p_k$, for $k = 1, \cdots, \bar{k}_n$ as well. In view of Lemma 13.4, within the neighborhood,

$$DV(t, 1) \geq DV(t, 2) \geq \cdots \geq DV(t, M).$$

The optimal fare composite will be changed at t_ϵ such that

$$DV(t_\epsilon, n) = p_{\bar{k}_n},$$

for some $n = 1, \cdots, m$. But for $l \neq n$, $DV(t_\epsilon, l) < p_{\bar{k}_l}$. Since actually

$$DV(t_\epsilon, 1) > DV(t_\epsilon, 2) > \cdots > DV(t_\epsilon, m),$$

this does not change the fact

$$k_1(t_\epsilon) = \bar{k}_1 \leq \cdots \leq k_n(t_\epsilon) = \bar{k}_n - 1 \leq k_{n+1}(t_\epsilon) = \bar{k}_{n+1} \leq \cdots \leq \bar{k}_m = k_m(t_\epsilon).$$

Thus when $[0, T]$ is divided into a number of subintervals such as

$$0 < t_1 < \cdots < t_{a-1} < t_a < T,$$

all $k_n, n = 1, \cdots, m$ are invariant in each of these subintervals, we will have $t_a = t_\epsilon$ and in the interval $[t_{a-1}, t_a)$, $k_n = \bar{k}_n - 1$ and $k_l = \bar{k}_l, l \neq n$. In words, fare $p_{\bar{k}_n}$ is no longer active and removed in $[t_{a-1}, t_a)$. Employing the same argument warrants that in the interval $[t_{a-1}, t_a)$

$$DV(t, 1) > DV(t, 2) > \cdots > DV(t, m),$$

and $k_1 \leq k_2 \leq \cdots \leq k_m$ in the interval $[t_{a-1}, t_a)$. The theorem is proved by repeating the above procedure. □

For most continuous-time seat control models, the $DV(t, n)$ curve has also been shown to be decreasing in elapsed time. When cancellations are permitted, the curve, however, may display some peculiar behavior and may not decline over time. Despite that, the time-monotone property remains true under certain conditions, as the next theorem shows.

Theorem 13.8 *When $m = M$, namely when the model is basic, suppose that $\lambda_i(t), i = 1, \cdots, K$, is a decreasing function of t, $0 \leq t \leq T$, and $\beta(t) = \beta$ is a constant. Then $DV(t, n)$ is a decreasing function of t.*

Proof. First of all, we assume that the sales season is divided into $L \geq 1$ subintervals by $0 < T_1 < T_2 < \cdots < T_L < T = T_{L+1}$ such that in the interval $(T_i, T_{i+1}]$, $\lambda_i(t) = \lambda_i(T_{i+1})$. Assume, in particular, that $\lambda_i(0) = \lambda_i(T_1)$. Hence, in the interval $(T_i, T_{i+1}]$, Equation (13.6) is specified as

$$\frac{d[\mathbf{d}(t)]}{dt} = \mathbf{B}(T_{i+1})\mathbf{d}(t) - \mathbf{b}(T_{i+1}), \tag{13.15}$$

where vectors

$$\mathbf{d}(t) = \begin{pmatrix} DV(t,1) \\ DV(t,2) \\ \cdots \\ DV(t, M-1) \\ DV(t, M) \end{pmatrix}, \qquad \mathbf{b}(T_{i+1}) = \begin{pmatrix} \beta c + \sum_{i=1}^{k_1} \lambda_i(T_{i+1}) p_i \\ \beta c + \sum_{i=k_1+1}^{k_2} \lambda_i(T_{i+1}) p_i \\ \cdots \\ \beta c + \sum_{i=k_{M-2}+1}^{k_{M-1}} \lambda_i(T_{i+1}) p_i \\ \cdots \\ \beta c + \sum_{i=k_{M-1}+1}^{k_M} \lambda_i(T_{i+1}) p_i \end{pmatrix},$$

and matrix $\mathbf{B}(T_{i+1}) = \mathbf{B}_1(T_{i+1}) + \mathbf{B}_2(T_{i+1})$ where

$$\mathbf{B}_1(T_{i+1}) = \begin{pmatrix} M & -(M-1) & 0 & \cdots & 0 & 0 & 0 \\ 0 & (M-1) & -(M-2) & \cdots & 0 & 0 & 0 \\ \cdots & \cdots & \cdots & \cdots & \cdots & \cdots & \cdots \\ 0 & 0 & 0 & \cdots & 0 & 0 & 1 \end{pmatrix} \cdot \beta,$$

and $\mathbf{B}_2(T_{i+1})$ equals

$$\begin{pmatrix} \sum_{i=1}^{k_1} \lambda_i(T_{i+1}) & 0 & \cdots & 0 & 0 \\ -\sum_{i=1}^{k_1} \lambda_i(T_{i+1}) & \sum_{i=1}^{k_2} \lambda_i(T_{i+1}) & \cdots & 0 & 0 \\ \cdots & \cdots & \cdots & \cdots & \cdots \\ 0 & 0 & \cdots & -\sum_{i=1}^{k_{M-1}} \lambda_i(T_{i+1}) & \sum_{i=1}^{k_M} \lambda_i(T_{i+1}) \end{pmatrix}.$$

System (13.15) yields a continuous solution $\mathbf{d}(t)$ within the interval $(T_i, T_{i+1}]$. In addition, $\mathbf{d}(t)$ is continuous at T_{i+1} where the booking curve changes its value. However, the solution is not differentiable at T_{i+1} as its left and right derivatives are not equal. In fact, for any $1 \leq n \leq M$, and $t \in (T_i, T_{i+1}]$,

$$\begin{aligned}
\frac{\partial DV(t,n)}{\partial t} &= -\sum_{i=1}^{k_{n-1}} \lambda_i(T_{i+1}) DV(t, n-1) \\
&+[(M-n+1)\beta + \sum_{i=1}^{k_n} \lambda_i(T_{i+1})] DV(t,n) \\
&-(M-n)\beta DV(t,n+1) - \beta c - \sum_{i=k_{n-1}+1}^{k_n} \lambda_i(T_{i+1}) p_i \\
= & -\sum_{i=1}^{k_{n-1}} \lambda_i(T_{i+1})[DV(t,n-1) - DV(t,n)] + \beta[DV(t,n+1) - c] \\
&+(M-n)\beta[DV(t,n) - DV(t,n+1)] \\
&+ \sum_{i=k_{n-1}+1}^{k_n} \lambda_i(T_{i+1})[DV(t,n) - p_i] \\
\leq & -\sum_{i=1}^{k_{n-1}} \lambda_i(T_{i+2})[DV(t,n-1) - DV(t,n)] \\
&+ \sum_{i=k_{n-1}+1}^{k_n} \lambda_i(T_{i+2})[DV(t,n) - p_i] \\
&+(M-n)\beta[DV(t,n) - DV(t,n+1)] + \beta[DV(t,n+1) - c].
\end{aligned}$$

Letting $t = T_{i+1}$, we have

$$\begin{aligned}\lim_{t<T_{i+1},t\to T_{i+1}} \frac{\partial DV(t,n)}{\partial t} &= \left.\frac{\partial DV(t,n)}{\partial t}\right|_{t=T_{i+1}-} \\ \leq \left.\frac{\partial DV(t,n)}{\partial t}\right|_{t=T_{i+1}+} &= \lim_{t>T_{i+1},t\to T_{i+1}} \frac{\partial DV(t,n)}{\partial t}.\end{aligned}$$

The above inequality indicates that when passing over different subintervals, MAF drops as demand intensity decreases.

Differentiating both sides of (13.15) leads to

$$\frac{d^2[\mathbf{d}(t)]}{dt^2} = \mathbf{B}(T_{i+1})\frac{d[\mathbf{d}(t)]}{dt}. \tag{13.16}$$

The solution to (13.16) is yielded by

$$\frac{d[\mathbf{d}(t)]}{dt} = e^{-\mathbf{B}(T_{i+1})(T_{i+1}-t)}\left.\frac{d[\mathbf{d}(t)]}{dt}\right|_{T_{i+1}-} \leq e^{-\mathbf{B}(T_{i+1})(T_{i+1}-t)}\left.\frac{d[\mathbf{d}(t)]}{dt}\right|_{T_{i+1}+}$$

since matrix $e^{-\mathbf{B}(T_{i+1})(T_{i+1}-t)}$ is non-negative. Because when $i = L$,

$$\begin{aligned}\left.\frac{d[\mathbf{d}(t)]}{dt}\right|_{T-} &= \frac{d[\mathbf{d}(T)]}{dT} \\ &= -(\beta c + \sum_{k=1}^{K} \lambda_k(T)p_k, \cdots, \beta c + \sum_{k=1}^{K} \lambda_k(T)p_k)',\end{aligned}$$

it follows that for any $t \in (T_i, T_{i+1}]$,

$$\frac{d[\mathbf{d}(t)]}{dt} \leq \frac{d[\mathbf{d}(T)]}{dT} < \mathbf{0}.$$

Hence, $DV(t,n)$ decreases in t.

When $\lambda_i(t)$ changes continuously, one can use an approximation method through discretizing the sales season into L equal subintervals $(T_i, T_{i+1}]$ in which all system parameters are constant. The conclusion can be proved by letting $L \to \infty$. □

Theorem 13.8 provides sufficient conditions for $DV(t,n)$ decreasing over time. However, it must be noted that the condition that $\lambda_i(t)$ decreases in t is fairly strong and may not be obeyed in reality. For example, the arrival rates of business travelers often increase near the time of departure. We will show in the numerical experiments that the decreasing property of $DV(t,n)$ with respect to t does not hold in general.

13.2.3 Threshold Control and Computation of the Value Function

Computing value function $V(t,n)$ and MAF $DV(t,n)$ is an iterative process. First, we solve system (13.6) in the neighborhood of T. Since $k_n = K$ for $n = 1, \cdots, m$, and $DV(T,n) = H(m-n)$, all $DV(t,n)$ can be determined. As time moves backwards, $DV(t,n)$ is constantly compared with two threshold fares: the lowest fare p_{k_n} in the active set and the highest fare p_{k_n+1} in the non-active set. This is because $DV(t,n)$ may exceed p_{k_n} or fall below p_{k_n+1}. According to Lemma 13.1, either of these events requires adjustment of k_n. The procedure continues until time 0 is encountered.

Two characteristics make the algorithm effective pronouncedly. One of them is recursively updating optimal fare mix. The other is incorporating effective numerical methodologies to the ordinary differential equations.

When computing $DV(t,n)$, the optimal active fares are altered at some time points. Those points, which are called *time thresholds*, and the by-product of the computation process, play an important role in implementing optimal booking control. Since for each remaining inventory level n, $DV(t,n)$ is continuous for $0 \le t \le T$, it must be uniformly continuous in this closed sales interval. Since for a $0 \le t \le T$, there exists a $\delta(t) > 0$ such that when $|s-t| < \delta(t)$, $|DV(t,n) - DV(s,n)| < \min_{k=1,\cdots,K-1}(p_k - p_{k+1})$. In terms of the theory of the finite coverings in the calculus, and the uniform continuity of $DV(t,n)$, there exists a $\delta(n)$ such that for any $0 \le t_1 < t_2 \le T$ and $t_2 - t_1 \le \delta$, the absolute value of $DV(t_1,n) - DV(t_2,n)$ is not greater than $\min_{k=1,\cdots,K-1}(p_k - p_{k+1})$. As a result, there exist a finite number of these thresholds dividing the sales horizon $[0,T]$ into a number of subintervals, which are denoted by U_n^k, for $1 \le k \le K$. Within a U_n^k, the active set of fares $\{p_1, \cdots, p_k\}$ is invariant. These intervals are crucial indicators while conducting optimal booking control. For instance, if $t \in U_n^k$ and $n(t) = n$, then all active fares are known. If no reservations are received during the remaining period of U_n^k, fare p_{k+1} will be added to the active set when the interval U_n^{k+1} is crossed over. On the other hand, suppose a seat is booked at t, then the minimum acceptable fare (MAF) is upgraded from $DV(t,n)$ to $DV(t,n-1)$. If t falls into U_{n-1}^k, no active fares are deleted at t. Otherwise, one (p_k) or more fares may need to be removed from the list of active fares after the transaction. Finally, if a booked seat is cancelled at t where $t \in U_n^k$ and $n(t) = n$, then the increased inventory and decreased MAF may result in the inclusion of p_j if $t \in U_n^j$ for some $j \ge k$. Hence, while selling a seat may lift the threshold fare, cancellations often do the opposite. Because seat control can be simply implemented based on these thresholds, we call the optimal seat control policy *threshold control.*

The thresholds are crucial also for computing value function itself. To understand this point, we need to know how the algorithm proceeds and how the thresholds are yielded. The algorithm will start within the close neighborhood of $t = T$ where since $V(T,n) = H(m-n)$ all the k_n should be determined by the maximum $1 \le k \le K$ such that $H(m-n) - H(m-(n-1)) \le p_k$. We then will use equation (13.3) to resolve $V(t,n)$ for all $1 \le n \le m$ at $t < T$ within this neighborhood. The optimal fare composition will be updated for $t < t_0 < T$ when $DV(t_0, n) = \bar{p}_{k_n}$. This t_0 is the latest threshold at level n and at $t < t_0$ price $\bar{p}_{k_n}$ is removed as a result of updating. The algorithm is then alternating to update optimal fare composition and compute value function until terminating at $t = 0$.

The numerical solution to equation (13.3) is based on the *Runge-Kutta* method. It divides the time horizon $[0, T]$ into N equal intervals with step length $h = T/N$. It starts at $t_N = T$ and moves backwards to $t_0 = 0$. At each cutoff point t_i, $i = 0, 1, \cdots, N$, since $V(t_{i+1}, n), n = 0, 1, \cdots, m$, have been obtained, we can derive $V(t_i, n), n = 0, 1, \cdots, m$ using the Runge-Kutta method. Suppose $\frac{\partial V(t,n)}{\partial t} = f_n(t, V(t, n+1), V(t,n), V(t, n-1))$ where $f_n(\cdot)$ is a known function. The procedure of computing an approximate value to $V(t_i, n)$ is as follows. Let

$$
\begin{aligned}
\hat{v}(t_i, n+1) &= \hat{v}(t_{i+1}, n+1) \\
&+ h f_{n+1}(t_{i+1}, \hat{v}(t_{i+1}, n+2), \hat{v}(t_{i+1}, n+1), \hat{v}(t_{i+1}, n)), \\
\hat{v}(t_i, n) &= \hat{v}(t_{i+1}, n) \\
&+ h f_n(t_{i+1}, \hat{v}(t_{i+1}, n+1), \hat{v}(t_{i+1}, n), \hat{v}(t_{i+1}, n-1)), \\
\hat{v}(t_i, n-1) &= \hat{v}(t_{i+1}, n-1) \\
&+ h f_{n-1}(t_{i+1}, \hat{v}(t_{i+1}, n), \hat{v}(t_{i+1}, n-1), \hat{v}(t_{i+1}, n-2)).
\end{aligned}
$$

The approximation of $V(t_i, n)$, denoted by $v(t_i, n)$, is defined to be

$$
\begin{aligned}
v(t_i, n) &= v(t_{i+1}, n) + (h/2) f_n(t_i, \hat{v}(t_i, n+1), \hat{v}(t_i, n), \hat{v}(t_i, n-1)) \\
&\quad + (h/2) f_n(t_{i+1}, v(t_{i+1}, n+1), v(t_{i+1}, n), v(t_{i+1}, n-1)).
\end{aligned}
$$

Note that in order to compute $v(t_i, n)$, $\hat{v}(t_i, n)$ must be evaluated in advance. This can be done since the recursive process has produced $\hat{v}(t_{i+1}, k)$ for all k. This second-order Runge-Kutta method ensures $v(t_i, n) = V(t_i, n) + O(h^2), n = 0, 1, \cdots, m$ uniformly. Since each move in the time axis takes a multiple of $m + 1$ elementary operations, the total number of elementary operations are a multiple of $N \times m$.

The algorithm performs better than the dynamic programming model because the computation accuracy achieved by the latter is in the order of $O(h)$. Stemming from the continuous-time model, our model allows to choose various time integration schemes for higher order accuracy. The computation storage is $O(m)$ since the algorithm upgrades the value function at all capacity levels while moving backward in the time direction.

In summary, we already have proposed a computing procedure for value function $V(t, n)$. The significance of this procedure is to confirm the existence of $\bar{V}(t, n)$ Lemma 13.2 supposed.

Lemma 13.9 Suppose that $m > M$ and $\lambda_k(t), k = 1, \cdots, K$ is piece-wise continuous. Then there exists a $\bar{V}(t,n)$ which satisfies the conditions of Lemma 13.2.

Proof. The procedure proposed to compute value function in this section exactly follows the conditions $\bar{V}(t,n)$ satisfies in Lemma 13.2. Hence, this lemma holds obviously. □

13.3 Extensions of the Basic Model

In this section we extend the basic model by relaxing some assumptions to fit more practical situations.

13.3.1 Time-Dependent Air Fares

The model discussed in the preceding sections assumes that fares in each passenger class are constant throughout the sales season. In reality, however, airlines often adjust fares as the season proceeds. For example, discount fares are usually suspended three weeks prior to departure. Hence, it is more realistic to consider fares as functions of time.

Suppose that for any fare class k, $k = 1, \cdots, K$, fares and their demand intensities are both time dependent. We denote them by $p_k(t)$ and $\lambda_k(t)$, respectively. It is easy to see that when $p_k(t)$ replaces p_k for $k = 1, \cdots, K$, the previous analysis can be duplicated. All the results remain valid. In particular, optimal booking control is of monotone threshold.

13.3.2 Fare-Dependent Partial Refunds

Most of airlines refund cancelled seats based on fares. Hence, a more general model should allow fare-dependent refunds for cancellations and no-shows, and even fare-dependent oversold penalties although the latter is rarely observed in practice. We show that under certain circumstances these variations can be converted into the basic model.

For each fare class k, $k = 1, \cdots, K$, we assume that the cancellation and no-show refunds are c_k and $\bar{c}_k$, respectively. The oversale penalty $\bar{p}(x)$ is a convex function of x, satisfying $\bar{p}(x) = 0$ for $x \leq 0$. The likelihood that passengers in class k will show-up at departure is ξ_k, $0 < \xi_k < 1$. For simplicity, the cancellation rate $\beta(t)$ is assumed to be independent of fares. Finally, the aggregate overbooking upper bound $m \geq M$ is given.

Because of fare-dependent refunds, one needs to keep track of the number of seats booked within each fare class. Let $(t, r_1, \cdots, r_K)$ be the state at time t, $0 \leq t \leq T$, where r_k, $k = 1, \cdots K$, is the number of seats booked in fare class k up to t. Accordingly, the value function corresponding to the state is denoted by $V(t, r_1, \cdots, r_K)$. Let $r = \sum_{k=1}^{K} r_k$, then the unfilled capacity is equal to $m - r$.

Because of fare-dependent refunds, the expected revenue at T is denoted by $h(r_1, \cdots, r_K)$. For $j = 1, \cdots, r_k$, let

$$Y_j^k = \begin{cases} 1 & \text{w.p. } \xi_k, \\ 0 & \text{w.p. } 1 - \xi_k \end{cases}$$

be the status of the jth seat booked in fare class k. Since Y_j^k is a Bernoulli variable and independent of one another, the number of boarding passengers in fare class k is $\sum_{j=1}^{r_k} Y_j^k$ which is a binomial variable. Hence, the expected revenue at departure is expressed as

$$h(r_1, \cdots, r_K) = -\mathsf{E}\left[\bar{p}\Big(\sum_{k=1}^{K}\sum_{j=1}^{r_k} Y_j^k - M\Big) + \sum_{k=1}^{K} \bar{c}_k \sum_{j=1}^{r_k}(1 - Y_j^k)\right]$$

$$= -\mathsf{E}\left[\bar{p}\Big(\sum_{k=1}^{K}\sum_{j=1}^{r_k} Y_j^k - M\Big)\right] - \sum_{k=1}^{K} r_k \bar{c}_k (1 - \xi_k).$$

With this modification, a sufficient condition for the optimal solution is stated in Lemma 13.10.

Lemma 13.10 Suppose that (i) $\bar{V}^m(t, r_1, \cdots, r_K)$ is continuous and piecewise continuously differentiable over the interval $[0, T]$ for any fixed $0 \le r_k$, where $k = 1, \cdots, K$, and $m \ge r = \sum_{k=1}^{K} r_k$; (ii) $\bar{V}^m(T, r_1, \cdots, r_K) = h(r_1, \cdots, r_K)$; (iii) for $0 \le r < m$,

$$\begin{aligned}
&\sum_{i=1}^{K} \lambda_i(t)[\bar{V}^m(t, r_1, \cdots, r_{i-1}, r_i + 1, r_{i+1}, \cdots, r_K) - \bar{V}^m(t, r_1, \cdots, r_K) + p_i]^+ \\
&+ \sum_{k=1}^{K} r_k \beta(t)[\bar{V}^m(t, r_1, \cdots, r_{k-1}, r_k - 1, r_{k+1}, \cdots, r_K) - \bar{V}^m(t, r_1, \cdots, r_K) - c_k] \\
&+ \frac{\partial \bar{V}^m(t, r_1, \cdots, r_K)}{\partial t} = 0, \qquad (13.17)
\end{aligned}$$

and (iv) for $r = m$,

$$\begin{aligned}
0 = &\sum_{k=1}^{K} r_k \beta(t)[\bar{V}^m(t, r_1, \cdots, r_k - 1, \cdots, r_K) - \bar{V}^m(t, r_1, \cdots, r_K) - c_k] \\
&+ \frac{\partial \bar{V}(t, r_1, \cdots, r_K)}{\partial t}. \qquad (13.18)
\end{aligned}$$

Then $\bar{V}^m(t, r_1, \cdots, r_K) = V^m(t, r_1, \cdots, r_K)$.

In view of Lemma 13.10, solving $V^m(t, r_1, \cdots, r_K)$ seems much involved because of the multi-dimension control states. Fortunately, since this multi-dimension situation can be converted into a one-dimension model, it is still tractable by applying the analysis for the basic model. The conversion follows the approach proposed by Subramanian, Stidham, and Lautenbacher (1999) in their discrete time Markovian process model.

To illustrate the conversion, we define $W_k(t, r_1, \cdots, r_K)$ to be a solution to the following system:

$$\begin{aligned} 0 &= \sum_{k=1}^{K} r_k\beta(t)[W(t, r_1, \cdots, r_{k-1}, r_k - 1, r_{k+1}, \cdots, r_K) - W(t, r_1, \cdots, r_K) - c_k] \\ &+ \frac{\partial W(t, r_1, \cdots, r_K)}{\partial t}, \end{aligned} \tag{13.19}$$

where $W(T, r_1, \cdots, r_K) = -\sum_{k=1}^{K} r_k\bar{c}_k(1 - \xi_k)$ and $W(t, 0, \cdots, 0) = 0$. Let $W_k(t, r_k)$ be the solution to the single variable system

$$\frac{\partial W_k(t, r_k)}{\partial t} + r_k\beta(t)[W_k(t, r_k - 1) - W_k(t, r_k) - c_k] = 0, \tag{13.20}$$

subject to $W_k(T, r_k) = -r_k\bar{c}_k(1 - \xi_k)$ and $W_k(t, 0) = 0$. It can be shown that

$$W(t, r_1, \cdots, r_K) = \sum_{k=1}^{K} W_k(t, r_k).$$

In other words, $W_k(t, r_1, \cdots, r_K)$ can be decomposed into K functions, each of them is solved from (13.20).

The solution to (13.20) is remarkably simple. This is because $DW_k(t, r_k) = W_k(t, r_k) - W_k(t, r_k - 1)$ is independent of r_k when $r_k \geq 1$.

Lemma 13.11 Given $k = 1, \cdots, K$, $DW_k(t, r_k) = W_k(t, 1)$ for $r_k \geq 2$.

Proof. The k-th equality holds recursively

$$\begin{aligned} & DW_k(t, r_k) - DW_k(t, r_k - 1) \\ = & -\int_t^T e^{\int_t^s r_k\beta(u)du}(r_k - 2)[DW_k(s, r_k - 1) - DW_k(s, r_k - 2)]\beta(s)ds. \end{aligned}$$

When $r_k = 2$, it can be solved from

$$0 = \frac{\partial[DW_k(t, 2) - DW_k(t, 1)]}{\partial t} - 2\beta(t)[DW_k(t, 2) - DW_k(t, 1)],$$

with the terminal condition $DW_k(T, 2) = DW_k(T, 1)$. As a result,

$$DW_k(t, 2) - DW_k(t, 1) = e^{2\int_t^T \beta(s)ds}[DW_k(T, 2) - DW_k(T, 1)] = 0.$$

Furthermore, the general recursion ensures that $DW_k(t, r_k) = DW_k(t, r_k + 1) = \cdots = DW_k(t, 1) = W_k(t, 1)$ for $r_k \geq 1$. □

Let $\hat{V}^m(t, r_1, \cdots, r_K) = \bar{V}^m(t, r_1, \cdots, r_K) - W(t, r_1, \cdots, r_K)$, it then follows

$$\begin{aligned} & \bar{V}^m(t, r_1, \cdots, r_K) - \bar{V}^m(t, r_1, \cdots, r_{k-1}, r_k + 1, r_{k+1}, \cdots, r_K) \\ = \;& \hat{V}^m(t, r_1, \cdots, r_K) - \hat{V}^m(t, r_1, \cdots, r_{k-1}, r_k + 1, r_{k+1}, \cdots, r_K) \\ & -W_k(t, 1). \end{aligned}$$

Hence, (13.17) can be used to solve $\hat{V}^m(t, r_1, \cdots, r_K)$ with a substitution $\hat{p}_k(t) = p_k + DW_k(t, 1)$ for p_k in (13.17). That is,

$$\begin{aligned} & \sum_{i=1}^{K} \lambda_i(t)[\hat{V}^m(t, r_1, \cdots, r_{i-1}, r_i + 1, r_{i+1}, \cdots, r_K) - \hat{V}^m(t, r_1, \cdots, r_K) + \hat{p}_i(t)]^+ \\ & + \sum_{k=1}^{K} r_k \beta(t)[\hat{V}^m(t, r_1, \cdots, r_{k-1}, r_k - 1, r_{k+1}, \cdots, r_K) - \hat{V}^m(t, r_1, \cdots, r_K)] \\ & + \frac{\partial \hat{V}^m(t, r_1, \cdots, r_K)}{\partial t} = 0, \end{aligned}$$

where

$$\hat{V}^m(T, r_1, \cdots, r_K) = -E[\bar{p}(\sum_{k=1}^{K} \sum_{j=1}^{r_k} Y_j^k - M)].$$

As a result, acquiring MAF of fare p_k is tantamount to obtaining

$$\hat{V}(t, r_1, \cdots, r_K) - \hat{V}(t, r_1, \cdots, r_{k-1}, r_k + 1, r_{k+1}, \cdots, r_K).$$

If $\xi_k = \xi$ for $k = 1, \cdots, K$, then $\hat{V}^m(t, r_1, \cdots, r_K)$ can be solved in the way similar to the basic model. To see this, we denote $\hat{V}^m(T, r_1, \cdots, r_K) = -E[\bar{p}(\sum_{k=1}^{K} \sum_{j=1}^{r_k})$.

Lemma 13.12 Assume $\xi_k = \xi$ for $k = 1, \cdots, K$. Let $\hat{V}^m(t, r_1, \cdots, r_K) = v(t, m - r)$ where $r = \sum_{k=1}^{K} r_k$. Then $v(t, m - r)$, $r = 0, 1, \cdots, m$, satisfies (13.17) in which $c = \bar{c} = 0$ and p_k is replaced by $\hat{p}_k(t) = p_k + W_k(t, 1)$.

Proof. It is straightforward to note that $v(t, m - r)$ satisfies (13.21) and the terminal condition

$$v(T, m - r) = \hat{V}^m(T, r_1, \cdots, r_K) = -\mathsf{E}[\bar{p}(\sum_{j=1}^{r_k} Y_j^k - M)].$$

Since the solution to (13.21) is unique, $\hat{V}^m(t, r_1, \cdots, r_K) = v(t, m - r)$. □

Since $D_k V^m(t, r_1, \cdots, r_K) = Dv(t, m - r) - W_k(t, 1)$, it is easy to see that the optimal booking control policy is essentially one-dimensional threshold control. The value of MAF,

$$\begin{aligned} D_k V^m(t, r_1, \cdots, r_K) \;&=\; V^m(t, r_1, \cdots, r_K) \\ &-\; V^m(t, r_1, \cdots, r_{k-1}, r_k + 1, r_{k+1}, \cdots, r_K), \end{aligned}$$

is monotone, as will be shown in the ensuing theorem.

Theorem 13.13 (i) For $k = 1, \cdots, K$,

$$D_k V^m(t, r_1, \cdots, r_K) = D_k V^m(t, r'_1, \cdots, r'_K)$$

where $\sum_{k=1}^K r_k = \sum_{k=1}^K r'_k$; and (ii) for $i = 1, \cdots, K$,

$$D_k V^m(t, r_1, \cdots, r_K) \le D_k V^m(t, r_1, \cdots, r_{i-1}, r_i + 1, r_{i+1}, \cdots, r_K).$$

(iii) For $1 \le k_1, k_2 \le K$, if $c_{k_1} > c_{k_2}$, then

$$D_{k_1} V^m(t, r_1, \cdots, r_K) \ge D_{k_2} V^m(t, r_1, \cdots, r_K).$$

Proof. Assertions (i) and (ii) hold following Theorem 13.7 and

$$D_k V^m(t, r_1, \cdots, r_K) = Dv(t, m - r) - W_k(t, 1).$$

Assertion (iii) is valid if and only if $W_{k_1}(t, 1) \le W_{k_2}(t, 1)$. To show the last inequality, consider

$$\frac{\partial [W_{k_1}(t,1) - W_{k_2}(t,1)]}{\partial t} - \beta(t)[W_{k_1}(t,1) - W_{k_2}(t,1)] = \beta(t)(c_{k_1} - c_{k_2}) > 0,$$

which leads to the desired relation by integrating both hand sides from t to T. □

Assertion (iii) states that fares with higher cancellation refunds have higher MAF values and should stay open longer. Normally airlines give partial or no refunds to the economic classes while fully refund cancelled business seats. This is because business classes ordinarily have higher MAF values.

13.4 Numerical Experiments

We present numerical experiments in this section which are run on the same Compaq personal computer. Those $V(t, n)$ and $DV(t, n)$ in the experiments were computed by the algorithm described in Section 3.3. The computer is equipped with the Pentium II processor and 384 MB RAM and operated by Linux system. The total CPU time for solving Example 1 below, for instance, given time step $h = 0.001$, is 3.37 seconds.

As shown by these examples, $DV(t, n)$ decreases in n with no exception. It also decreases over time if the conditions in Theorem 13.8 are satisfied. However, in the absence of these conditions, the monotone property of MAF with respect to time may fail to hold. In general, $DV(t, n)$ does not equal a given price at most once in time so that the optimal active fares may not expand over time.

Example 13.14 A single-leg flight with 400 seats is scheduled to depart 2 months later. It targets at three classes of passengers whose average prices are $p_1 = \$500, p_2 = \$400$ and $p_3 = \$300$, respectively. Booking requests in each fare class form a non-homogeneous Poisson process. It is estimated that demand intensities may fall into one of the following patterns. (1) $\lambda_1(t) = 110 - 50t, \lambda_2(t) = 95 - 40t$ and $\lambda_3(t) = 80 - 30t$; and (2) $\lambda_1(t) = 10 + 50t, \lambda_2(t) = 15 + 40t$ and $\lambda_3(t) = 20 + 30t$. The booked seats are allowed to cancel any time prior to departure and the refund for a cancelled seat is \$200. A reserved seat will be cancelled at rate $\beta(t)$ which is equal to 0.05.

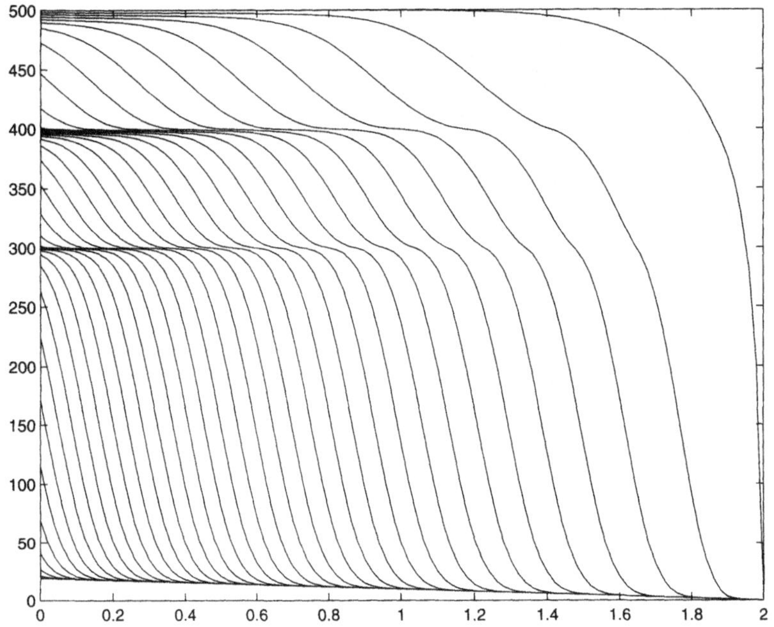

FIGURE 13.1. $DV(t, n)$ curves corresponding to case (1) of Example 13.14

Figure 13.1 exhibits the MAF ($DV(t, n)$) curves corresponding to various levels of seat inventory. The horizon axis represents the elapsed time and the vertical axis represents the values of $DV(t, n)$. In the first scenario demand intensities decrease over time. As a result, the $DV(t, n)$ curves are decreasing over time. In scenario (2), demand intensities grow as the season proceeds. It is interesting to see that the $DV(t, n)$ curves do not decrease over time, as Figure 13.2 shows. Because $DV(t, n)$ curves do not hold the monotone decreasing property, the set of active fares does not expand over the season.

Example 13.15 A single-leg flight with 400 seats is scheduled two months prior to its departure. There are three fare classes with $p_1 = \$200, p_2 = \$150$

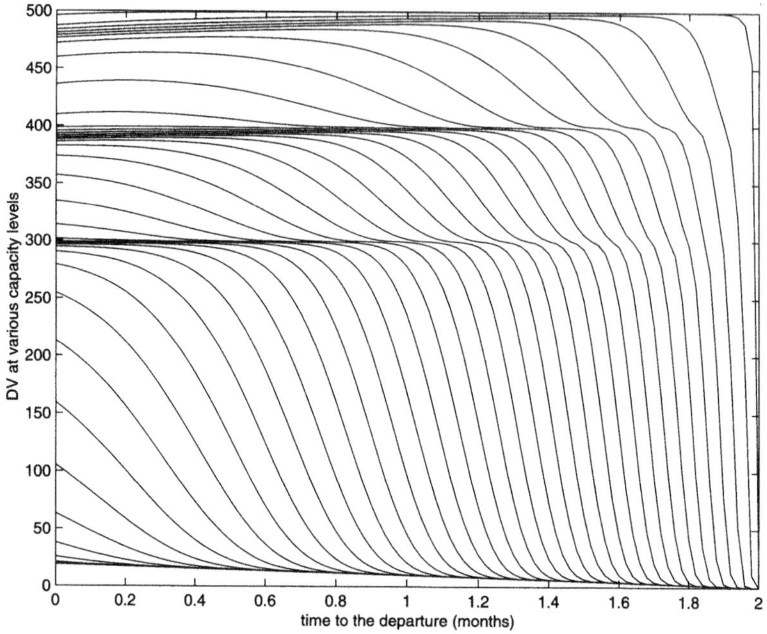

FIGURE 13.2. $DV(t,n)$ curves corresponding to case (2) of Example 13.14

and $p_3 = \$100$. Booking curves are given by respective functions $\lambda_1(t) = 50e^{-10(2-t)}, \lambda_2(t) = 30e^{-20(2-t)}$ and $\lambda_3(t) = 60e^{-40(2-t)}$. The airline gives a partial refund $c = \$50$ to a cancelled seat which arrives at the rate $\beta(t) = 4 - t^2$.

Figure 13.3 shows that $DV(t,n)$ declines slowly in time throughout most part of the sales season. Moreover, regardless of the levels of unbooked seats, the $DV(t,n)$ curves uniformly move down below the refund ($c = \$50$) line before they surge high near the end. This is because the joint effect of sales and cancellation is dominated by the elapsed time shortly after starting sales. However, towards the end of the season, booking curve rises sharply and cancellation rates approach zero. The combined impact of demand and cancellations tops the time effect. As a result, the $DV(t,n)$ curves show a steep increase. Generally, the effect of cancellation on revenues is complicated, and dynamic changes of both demand and cancellations often result in uncanny behavior of MAF.

Example 13.16 Following on Example 13.14, we assume that passengers come to boarding with a probability .85. Suppose that the refund for a no-show seat is $\bar{c} = c = 100$, and the oversold penalty function is $p(x) = 100x + 2x^2$ where x is the number of passengers who give up their boarding.

The optimal overbooking upper bound is determined to be $m^* = 610$. Figure 13.4 displays the MAF curves corresponding to this limit.

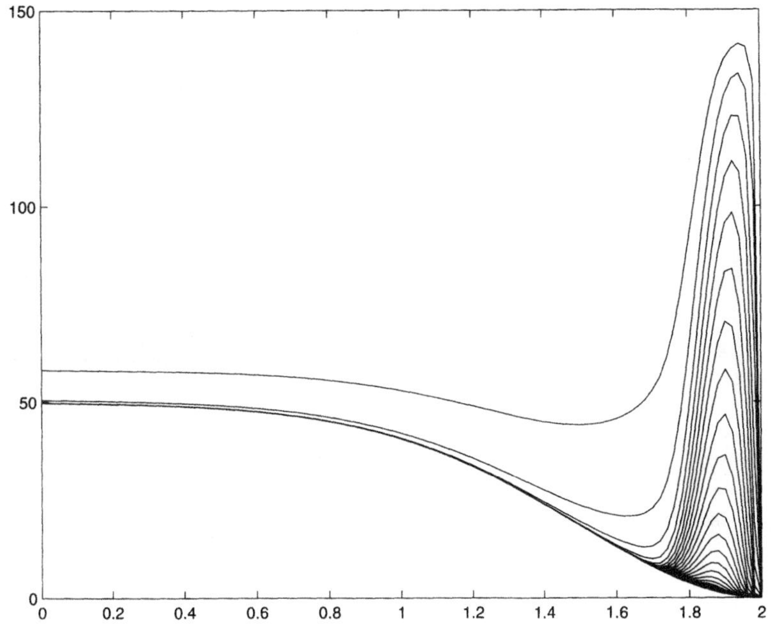

FIGURE 13.3. Illustration of opposite effects of cancellation and booking

Example 13.17 Reconsider Example 13.14. We further assume that the no-show probability is 0.15, which is independent of fare classes. However, refunds for no-show seats are fare-dependent and are equal to $\bar{c}_1 = c_1 = 400$, $\bar{c}_2 = c_2 = 300$ and $\bar{c}_3 = c_3 = 100$, respectively. The oversold penalty function is $p(x) = 100x + 2x^2$ where x is the number of no-shows.

The optimal overbooking upper bound is $m^* = 594$, which is lower than 610 in the previous example. This is because $p_1(T) = p_1 - \bar{c}_1(1-\xi) = 500 - 15 = 485$. Figure 13.5 draws the price curves for three fare classes. These price curves follow the discussion on the class-dependent refunds. For detail of this analysis, we refer the reader to Section 4. In view of both curves, it becomes easy to understand why $m^* = 594$. The overbooking limit control is therefore optimal threshold control associated with this limit.

13.5 Notes

Over the last three decades, research on airline seat inventory control has advanced rapidly. Rothstein (1971) first proposes a mathematical model to analyze the overbooking policy. Littlewood (1972) considers a two-class, single-leg seat control problem and develops the *marginal seat revenue* principle. The principle requires that a booking request at the low fare be denied

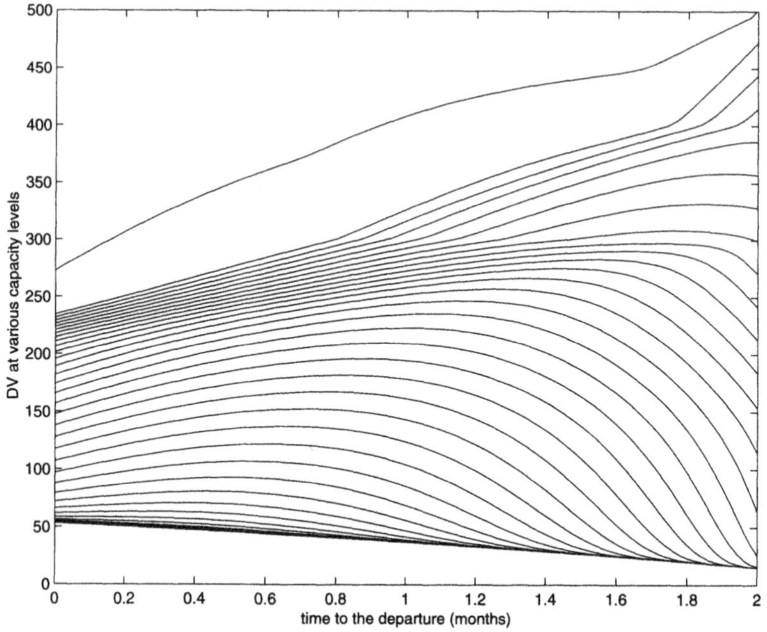

FIGURE 13.4. MAF curves for Example 13.16

if the seat saved can generate more revenue from the high-fare class. Bhatia and Parekh (1973), and Richter (1982) employ the marginal seat revenue principle in their multiple-fare models and find an optimal booking limit in the *nested*-fare inventory systems. Belobaba (1987,1989) extends Littlewood's model and proposes the *expected marginal seat revenue* (EMSR) rule in his probabilistic seat allocation model. The EMSR rule constantly revises the static estimation of optimal booking limits as information of the booking process grows. Belobaba improves on Littlewood's approach by incorporating the remaining time into his analysis. Wollmer (1992), Brumell and McGill (1993), and Robinson (1995) study the single-leg problem under the assumption that demand from different fare classes arrive sequentially. They show that Belobaba's heuristic is sub-optimal. They independently identify the first-order optimality condition and use it in finding the optimal booking policy. The optimal booking limits obtained relate the probability of stocking out with the ratio of the current fare to the highest fare. Curry (1990) further examines booking limits for a network of flights. Weatherford, Bodiy and Pfeifer (1993) investigate dynamic booking limits for two classes of passengers whose booking requests arrive concurrently, assuming that the distribution of remaining demand for each fare class is known. Lee and Hersh (1993) consider multiple fare classes and multi-seat booking requests under the same assumption. They divide the booking season into sufficiently small intervals and use a discrete dynamic programming technique in deriving optimal booking limits.

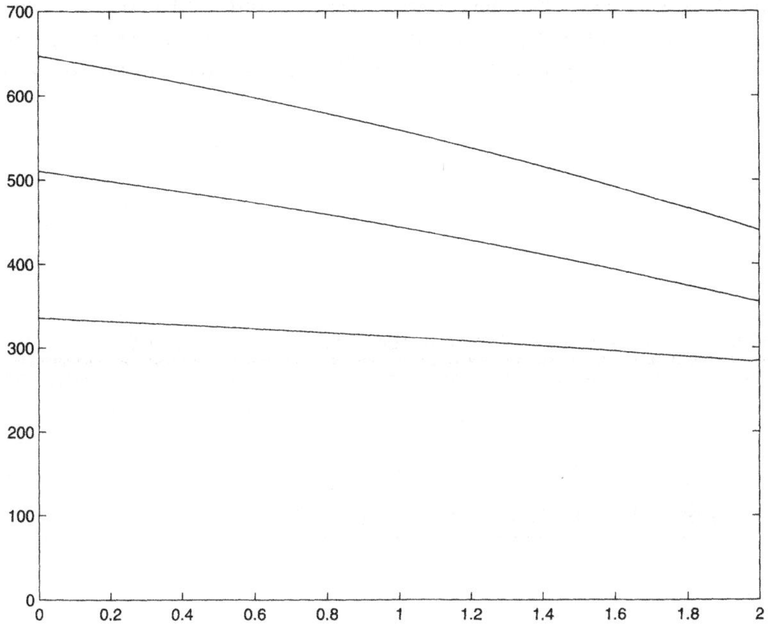

FIGURE 13.5. The curves of prices for Example 13.17

In pursuit of the optimal seat inventory control for single-leg flights, Feng and Xiao (1996,1997), and Liang (1999) study continuous-time seat allocation models and independently derive the optimal booking policy which considers the time and capacity factors simultaneously. Although tackling the problem with different approaches, they obtain very similar control rules, which are characterized as threshold control. Talluri and van Ryzin (1999) use opportunity costs, a standard term in economics, in their dynamic network model and show that the optimal seat control policy is determined through the comparison between the the requesting fares and the opportunity costs.

Another primary tool that airline managers rely on is the price policy. Since most passengers are either price-sensitive (vacationers) or time-sensitive (business travelers), price discrimination has been a very effective option to boost revenues. Airlines sell the same seats at distinct fares based on demand, seat inventory, and time remaining for the season. Although researchers generally agree that the remaining time and seat inventory are most critical factors in determining optimal prices, it is difficult to incorporate the two factors. The most eminent successes in integrating them have been seen in the continuous-time revenue management models. Gallego and van Ryzin (1994) present a comprehensive pricing management model which allows unlimited price changes for selling a finite stock of perishable products. They derive an exact solution in closed form when the demand function is exponential. For the two-price model with Poisson demand processes and a single price change, Feng and Gallego (1995) develop an optimal pricing policy. Feng and Gallego (1996) further extend their analysis to Markovian prices. Feng and Xiao (1999, 2000) generalize the two-price model by considering a risk factor and multiple prices. They show that the optimal control policy maintains the same threshold structure. Zhao and Zheng (2000) proved that a similar threshold control policy is optimal for a more general airline seat allocation model that allows diversion/upgrade and no-shows. They also showed that under certain conditions, the optimal threshold may not be monotone. Chatwin (2000) considers a continuous-time revenue management model in which price is dynamically adjusted. His results show that the optimal price declines as the time elapses and the inventory level goes up. In addition, these results hold when the prices and corresponding demand intensities are functions of the time-to-go, but not so when the demand intensities are functions of the inventory level. For more detailed and maybe latest literature review, we refer the reader to McGill and van Ryzin (1999).

Cancellations and no-shows have been studied in some recent publications. Subramanian, Stidham, and Lautenbacher (SSL,1999) formulate the airline seat allocation problem on a single-leg flight into a discrete-time Markov decision process. Chatwin (1998) analyzes a multi-period airline overbooking problem with non-stationary fares. A continuous-time version of his model (Chatwin 1996,1999) treats reservations and cancellations as a birth-and-death process.

Our research differs from SSL's work (1999) which deals with a discrete-time model. Our numerical experiments show that computation effort for continuous-time models is fairly mild, and does not grow noticeably as the inventory goes up or as the remaining season stays longer. Compared to Chatwin (1999), our model assumes multiple-fare classes and time-dependent demand intensities. In particular, we deal with two stochastic processes directly rather than to consolidate them. It is shown in Feng and Xiao (1997) that without cancellations, the optimal booking policy stems from the principle of MAF, which can be calculated through a recursive solution procedure. However, the same recursive approach fails to work in the presence of cancellations because of the interdependence between different inventory levels. Hence, technically our model is more challenging as it requires to solve a set of simultaneous differential equations, which was considered impractical in the past. In addition, numerical experiments show that the model is tractable and computationally effective.

13.6 References

[1] Ascher, U. and P. Lin., Sequential Regularization Methods for Higher Index DAEs with Constraint Singularities: Linear Index-2 Case. *SIAM J. Numer. Anal.*, **33** (1996), 1921-1940.

[2] BELOBABA, P. P., Airline Yield Management: An Overview of Seat Inventory Control. *Transportation Science*, **21** (1987), 63-73.

[3] BELOBABA, P. P., Application of a Probabilistic Decision Model to Airline Seat Inventory Control. *Operations Research*, **37** (1989), 183-197.

[4] BHATIA, A.V., AND S.C. PAREKH., Optimal Allocation of Seats by Fare. *Presentation by Trans World Airlines to AGIFORS Reservations Study Group*, (1973).

[5] BRÉMAUD, P., *Point Processes and Queues, Martingale Dynamics*, Springer-Verlag, New York, 1981.

[6] BRUMELLE, S.L. AND J.I. MCGILL., Airline Seat Allocation with Multiple Nested Fare Classes. *Operations Research*, **41** (1990), 127-137.

[7] BRUMELLE, S.L. AND J.I. MCGILL., Allocation of Airline Seats Between Stochastically Dependent Demand. *Transportation Science*, **24** (1993), 183-192.

[8] CHATWIN, R., Optimal Control of Continuous-Time Terminal-Value Birth-and-Death Processes and Airline Overbooking. *Naval Research Logistics*, **43** (1996), 159-168.

[9] CHATWIN, R. E., Continuous-Time Airline Overbooking with Time Dependent Fares and Refunds. *Transportation Science*, **33** (1999), 182-191.

[10] CHATWIN, R. E., Optimal Dynamic Pricing of Perishable Products with Stochastic Demand and a Finite Set of Prices. *European Journal of Operational Research*, **125** (2000), 149-174.

[11] CURRY, R., Optimal Seat Allocation with Fare Classes Nested by Origins and Destinations. *Trans. Science*, **24** (1990), 193-204.

[12] FENG, Y. AND G. GALLEGO., Optimal Stopping Times for End of Season Sales and Optimal Starting Times for Promotional Fares. *Management Science*, **41** (1995), 1371-1391.

[13] FENG, Y. AND G. GALLEGO., Perishable Asset Revenue Management with Markovian Time Dependent Demand Intensities. *Management Science*, **46** (2000), 941-956.

[14] FENG, Y. AND B. XIAO., A Dynamic Airline Seat Inventory Control Model and its Optimal Policy. *Operations Research*, **49** (2001), 938-949.

[15] FENG, Y. AND B. XIAO., Optimal Overbooking Policies for Single-Leg Flights with Time-Dependent Demands. Working paper. Department of Systems Engineering and Engineering Management, Chinese University of Hong Kong, Hong Kong, PRC.

[16] FENG, Y. AND B. XIAO., Maximizing Revenue of Perishable Assets with Risk Analysis. *Operations Research*, **47** (1999), 337-341.

[17] FENG, Y. AND B. XIAO., Optimal Policies of Yield Management with Multiple Predetermined Prices. *Operations Research*, **48** (2000), 332-343.

[18] GALLEGO, G. AND G. J. VAN RYZIN., A Multi-Product Dynamic Pricing Problem and its Applications to Network Yield Management. *Operations Research*, **45** (1997), 24-41.

[19] GLOVER, F., R. GLOVER, J. LORENZO, AND C. MCMILLAN., The Passenger-Mix Problem in the Scheduled Airlines. *Interfaces*, **12** (1982), 73-79.

[20] KLEYWEGT, A. J. AND J. D. PAPASTAVROU., The Dynamic and Stochastic Knapsack Problem. *Operational Research*, **46** (1998), 17-35.

[21] LEE, T.C. AND M. HERSH., A Model for Dynamic Airline Seat Inventory Control With Multiple Seat Bookings. *Transportation Science*, **27** (1993), 252-265.

[22] LIANG, Y., Solution to the Continuous Time Dynamic Yield Management Model. *Transportation Science*, **33** (1999), 117-123.

[23] LIBERMAN, V. AND U. YECHIALI., On the Hotel Overbooking Problem- An Inventory System with Stochastic Cancellations. *Management Science*, **24** (1978), 1117-1126.

[24] LITTLEWOOD, K., Forecasting and Control of Passengers. *12th AGIFORS symposium Proceedings*, (1972), 95-128.

[25] MCGILL, J. I. AND G.J. VAN RYZIN., Revenue Management: Research Overview and Prospects. *Transportation Science*, **33** (1999), 233-256.

[26] RICHTER, H., The Differential Revenue Method to Determine Optimal Seat Allocations by Fare Type. AGIFORS Symp. Proc. (1982), 339-362.

[27] ROBINSON, L. W., Optimal and Approximate Control Policies for Airline Booking with Sequential Fare Classes. *Operations Research*, **43** (1995), 252-263.

[28] ROTHSTEIN, M., An Airline Overbooking Model. *Transportation Science*, **5** (1971), 180-192.

[29] SMITH, B., J. LEIMKUHLER, R. DARROW, AND J. SAMUELS., Yield Management at American Airlines. *Interfaces*, **1** (1992), 8-31.

[30] SUBRAMANIAN, J., S. STIDHAM, AND A. C. LAUTENBACHER., Airline Yield Management with Overbooking, Cancellations and No-Shows. *Transportation Science*, **33** (1999), 147-167.

[31] SUBRAMANIAN, J., Airline Yield Management and Computer Memory Management via Dynamic Programming. Ph.D. Dissertation, University of North Carolina at Chapel Hill, 1995.

[32] TALLURI, K. AND G. VAN RYZIN., An Analysis of Bid-Price Controls for Network Revenue Management. *Management Science*, **44** (1999), 1577-1593.

[33] VAN SLYKE, R. AND Y. YOUNG., Finite Horizon Stochastic Knapsacks with Applications to Yield Management. *Operations Research*, **48** (2000), 155-172.

[34] WEATHERFORD, L.R. AND S.E. BODILY., A Taxonomy and Research Overview of Perishable-Asset Revenue Management: Yield Management, Overbooking and Pricing. *Operations Research*, **40** (1992), 831-844.

[35] WEATHERFORD, L.R. AND S.E. BODILY, AND P.E. PHEIFER., Modeling the Customer Arrival Process and Comparing Decision Rules in Perishable Asset Revenue Management Situations. *Transportation Science*, **27** (1993), 239-251.

[36] WOLLMER, R.D., An Airline Seat Management Model for a Single Leg Route when Lower Fare Classes Book First. *Operations Research*, **40** (1992), 26-37.

[37] ZHAO, W. AND Y.S. ZHENG., Optimal Dynamic Pricing for Perishable Assets with Nonhomogeneous Demand. *Management Science*, **46** (2000), 375-388.

14

Information Revision and Decision Making in Supply Chain Management

Houmin Yan and Hanqin Zhang

Facing a fast growing global economy and advances in the emerging e-commerce, companies must deal with customers who are well informed and have diverse needs. Many companies have taken a lead in obtaining advance information about customer demands. On the other hand, it is common that for the same sets of goods, companies provide their customers with choices of different lead times and delivery methods. These differences in lead times and delivery alternatives may result in different charging schemes, as faster delivery modes typically cost more. The models investigated in this chapter consider both advance demand information and multiple supply sources. We show that a state-dependent base-stock policy is optimal.

The chapter is organized as follows. In Section 14.1, we provide two real-life examples to motivate our study. In Sections 14.2 and 14.3, we solve, respectively, the multiple-period, two information revisions problem, and the single-period, multiple information revisions problem. For these two problems, we are able to construct the form of optimal supply chain decisions with information revisions, and provide structure properties. In Section 14.4, we demonstrate how our results can be applied to the two examples in Section 14.1. We conclude our study in Section 14.5 with an overview of the related literature.

14.1 Industrial Examples

In this section, we present two real-life examples to highlight the opportunities in this area. The first example is the procurement process of microcontrollers in a major security device and equipment company. With data obtained from the company, we demonstrate the moving forecast windows and the urgent need for decision support. The second example is from the deregulated energy industry which reveals the dynamics of information revision and how they impact the decision process.

14.1.1 The Procurement of Micro-Controller

A major security system manufacturing company produces and distributes security systems for military, residential, commercial and industrial applications. It has a design center in California, a manufacturing center in Asia, and three regional distribution centers in San Francisco, Amsterdam, and Singapore. The company sources components and sub-assemblies around the world. The management objectives are to improve the response time to meet market demand, to reduce inventory, and to shorten lead-time (including the time for manufacturing and distribution). In the security system market, customers expect to have the required device or the system within one month. Therefore, given the long lead-time in procurement and production, the manufacturing operation largely relies on forecasts.

From experience, forecasts are never accurate, but the company updates its demand forecasts until the exact demand is known. When too little raw material is ordered, the company has to either pay a higher price to secure them, or to use air shipment to expedite them if these options are feasible. When too much raw material and sub-assemblies are ordered, the company has to keep them in inventory. These materials often become obsolete. These changes in forecasting also make it difficult to allocate its production capacity efficiently.

In particular, a key component in security systems is the micro-controller which makes up 30–40% of the total materials cost. A micro-controller is a CPU chip with a built-in memory and interface circuits. The read-only memory (ROM) contains permanent data (program code). See Spasov [25] for related concepts and technology on micro-controllers. The company can order micro-controllers with user supplied data requirements. If user supplied data is provided, the semiconductor manufacturing includes a process known as custom photo-masking in the wafer fabrication process. Alternatively, the company can purchase micro-controllers with a programmable ROM (such as One Time Programmable (OTP) Read-Only Memory or Erasable Programmable Read-Only Memory (EPROM)). The company inputs the data into these programmable micro-controllers after the chips are received. To order custom-masked chips, the users are required to provide the data (program code) prior to manufacturing, and a significant lead-time is required. On the other hand, since programmable ROMs are generic, these micro-controllers can be produced in considerably shorter lead-time. However, the OTP chips are about twice as expensive as custom-masked chips and EPROM chips are more expensive. The company must decide how to order both custom-masked and OTP chips.

The company uses a half-year rolling window for demand forecasting. These forecasts are made and updated monthly by the regional offices. The headquarters coordinates the forecasts and passes them to its logistics and manufacturing functions. Procurement decisions are made based on the demand forecast and the lead-time required by its vendors. The company divides the raw materials into two classes: critical and regular. The components with fewer sources and a higher value content and requiring a longer lead-time are classified into the critical materials. Micro-controllers are a typical example. In what follows, we first analyze the demand forecast data, and then we study how an updated forecast affects the manufacturer's decision-making in the procurement process.

14.1.2 Analysis of Demand Forecast Data

For the seven products we investigated, the logistics and manufacturing functions receive a monthly demand update. In a six-month rolling horizon, the first five updates are forecasts, the last one is the real demand. We obtained data from February 1996 to September 1997. Our purpose is to investigate whether there is a trend of diminishing variance of forecast error, and what the possible managerial decisions and implications are.

We depict the demand and forecast in the scatter charts shown in Figures 14.1 and 14.2, where the x-axis is the actual demand, and y-axis is the forecasted demand. In both figures, a star ($\star$) represents the demand; therefore, the demand data are scattered on the 45° line. In Figure 14.1, a circle ($\circ$) represents a forecast made five months earlier. When the circle is above the 45° line, it indicates that the forecast is higher than the real demand; on the other hand, when it is below the 45° line, it indicates that the forecast is lower than the real demand. Similarly, in Figure 14.2, a dot ($\bullet$) represents the relationship between the forecast made one month earlier and the demand.

It can be easily observed that the later forecasts are much closer to the actual demand. We studied the forecast error and its variance which are shown in Table 14.1. Note that the numbers listed are scaled from the real data for the sake of confidentiality. In the table, errors and standard deviations (STDEV) of the five-month, three-month and one-month forecasts with respect to the real demand are tabulated. We also calculate the percentage of the standard deviation changes in percentage with respect to the standard deviations of the five-month forecast. As we have discussed earlier, the variance (standard deviation) is used to measure the accuracy of the demand forecast. From Table 14.1, comparing the five month and the one month forecasts, the latter improves remarkably for most of the products.

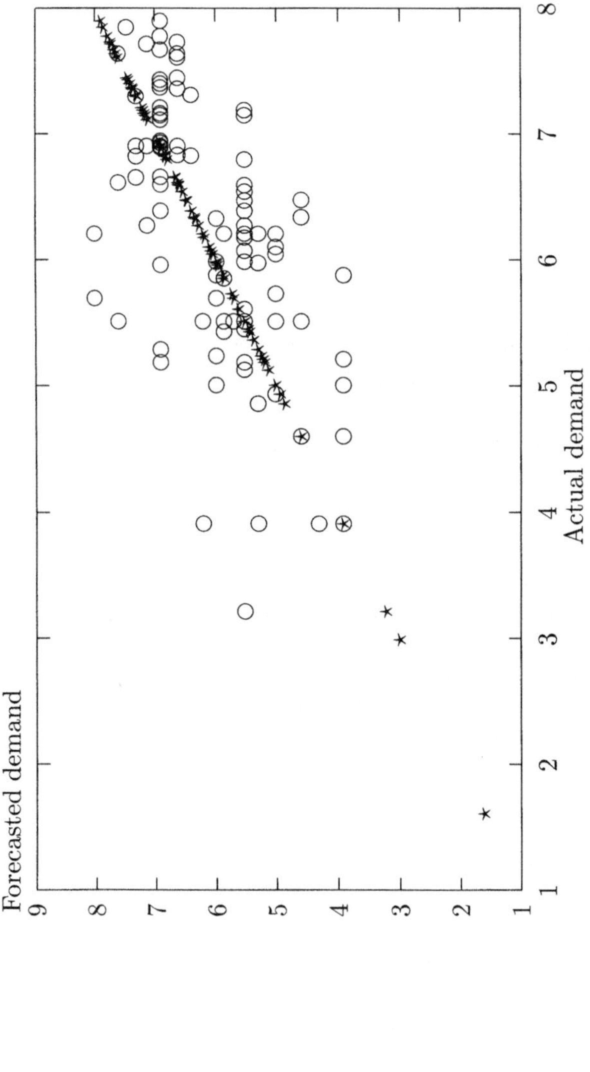

FIGURE 14.1. Scatter chart of five month forecast and the demand

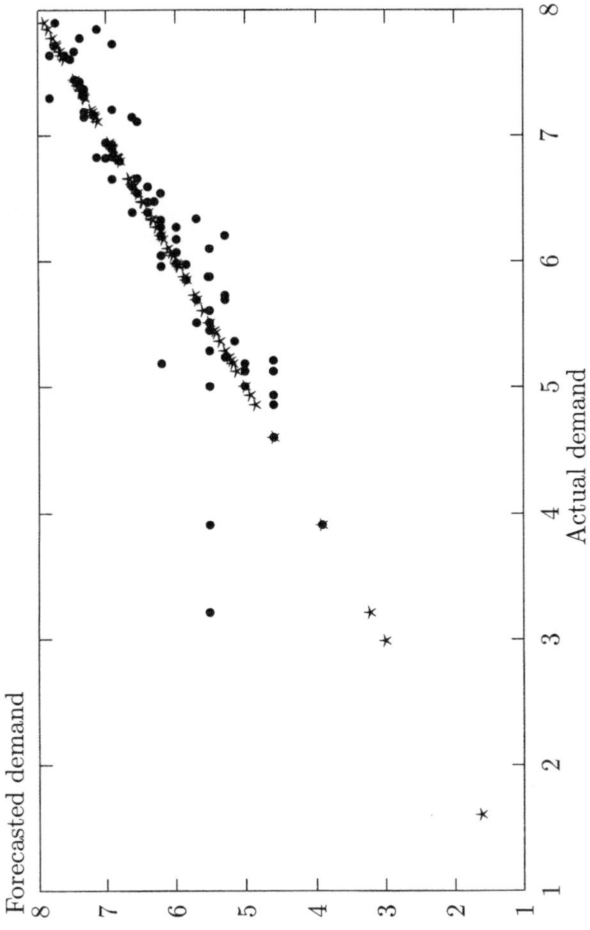

FIGURE 14.2. Scatter chart of one month forecast and the demand

Products	Five Month Forecast		Three Month Forecast			One Month forecast		
	Average	STDEV	Average	STDEV	Changes %	Average	STDEV	Changes %
Product 1	-21.5	64.7	-20.3	62.1	4.0	-30.2	60.3	6.8
Product 2	-52.0	208.6	22.5	250.9	20.3	-28.8	78.0	62.0
Product 3	33.0	1211.0	5.0	819.0	32.4	-139.8	366.3	69.8
Product 4	30.0	275.8	42.5	266.9	3.2	12.5	72.9	73.6
Product 5	-53.5	234.9	-73.5	139.9	40.4	-50.8	70.6	69.9
Product 6	-162.0	487.0	-90.0	528.0	8.4	-42.0	201.4	58.6
Product 7	86.0	962.0	-24.0	737.0	23.4	-47.5	443.9	53.9

TABLE 14.1. The errors of forecast and their improvement

14.1.3 The Deregulated Energy Markets

California Power Exchange (CalPX) operates the monthly forward, day-ahead and day-of markets. The trade processes in these markets are similar but differ in time frames. The monthly forward and day-ahead markets trade a month and a day ahead of the electricity delivery hour, and in the day-of market, trading occurs close to the delivery hour. For all these markets, CalPX aggregates electronic bids from participants to determine the Market Clearing Price (MCP) for buying and selling of electricity. The MCP is established at the intersection of aggregated energy supply and demand curves for each auction hour. A sample of aggregated energy supply and demand curves, and the market clearing price is depicted in Figure 14.3. We refer the reader to the annual report of California Power Exchange [6] for a detailed description and examples of the electricity auction markets.

The existence of these forward, day-ahead and day-of markets presents a challenging problem to energy buyers, who must determine how many Megawatts/hour (MWhs) to buy from each market, or how to allocate their demand into these time sequential markets. Note that the uncertain energy price in these markets is the major concern in demand allocation in addition to demand uncertainty. The price in each market is determined by the demand and supply bidding process. As a result of price elasticity, the energy buy cost is convex. For example, if a piece-wise linear bidding strategy is used in the bidding process, which is a common practice in the industry, the energy cost is in a quadratic form. Since these markets open at different times, the energy price in future markets is not known. In addition to the market price, the actual demand is not known before it is realized. Therefore, the energy price and the demand are the major uncertain factors in this demand allocation process.

14.2 A Multi-Period, Two-Decision Model

Sethi, Yan and Zhang [23] consider a discrete time, single product, periodic review inventory system. The dynamics of the system contain two parts: the material flows and the information flows. The in-bound material flows come from two supply sources, fast and slow, and the out-bound material flows to customers. After they are ordered at the beginning of a period, materials from the fast and slow sources arrive at the end of the current period and at the end of the next period, respectively. The information flows include the initial demand forecast, regular forecast updates, and the realized customer demand. At the beginning of each period, the forecast of the demand, which will materialize at the end of the period, is updated. When the realized customer demand occurs at the end of the period, the customer is satisfied if there is sufficient available inventory, and the excess is carried over to the next period. Otherwise, the customer demand is partially satisfied and

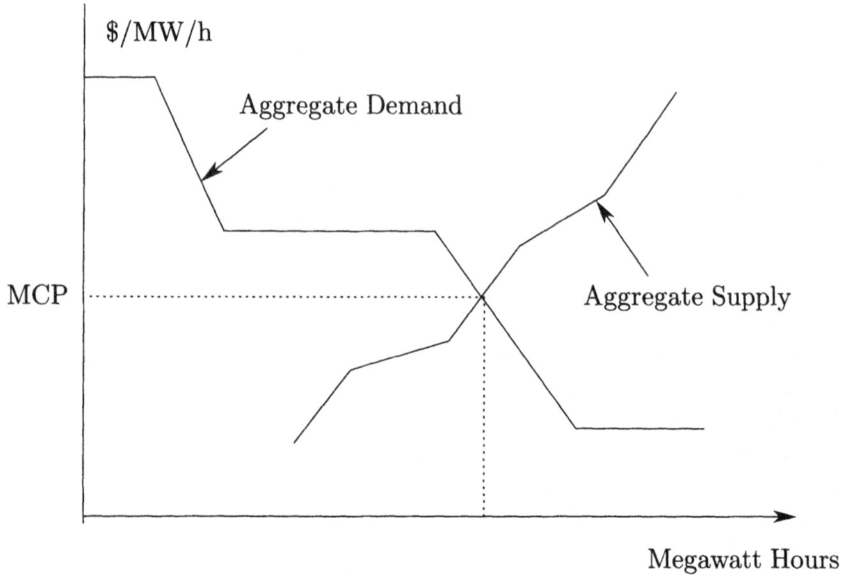

FIGURE 14.3. Aggregate energy demand and supply curves, and MCP

the unsatisfied demand is fully backlogged.

The decision variables are the quantities ordered from the fast and the slow sources at the beginning of each period. The decisions are made based on the current inventory position and the current (updated) demand information, where the inventory position at the beginning of the period is on-hand inventory plus the amount already on order to be delivered at the end of the period. A time-line of the system dynamics and the ordering decisions is illustrated in Figure 14.4.

Denote:

$\langle 1, N \rangle$: $\{1, 2, ..., N\}$, the time horizon;

$(\Omega, \mathcal{F}, P)$: the probability space;

F_k : the nonnegative fast order quantity in period k, $1 \le k \le N$;

S_k : the nonnegative slow order quantity in period k, $1 \le k \le N-1$;

$C_k^f(u)$: the cost of fast order $u \ge 0$ units in period k;

$C_k^s(u)$: the cost of slow order $u \ge 0$ units in period k;

I_k^1 : the first determinant (a random variable) of the demand in period k observed at the beginning of period k;

I_k^2 : the second determinant (a random variable) of the demand in period k observed at the end of period k;

v_k : the third determinant (a constant) of the demand in period k;

D_k : the demand in period k modeled as a function $g_k(I_k^1, I_k^2, v_k)$;

X_k : the inventory level at the beginning of period k;

Y_k : $X_k + S_{k-1}$ = inventory position at the beginning of period k;

X_{N+1} : the inventory level at the end of the last period N;

$H_k(x)$: the inventory holding/backlog cost when $X_k = x$;

$H_{N+1}(x)$: the holding cost if $X_{N+1} = x \geq 0$ or penalty cost if $X_{N+1} = x < 0$;

$\mathsf{E}[\cdot]$: the expectation operator defined on probability space $(\Omega, \mathcal{F}, P)$;

$\mathsf{E}_{(I_k^2, I_{k+1}^1)}[\cdot]$: $\mathsf{E}[\cdot | I_k^2, I_{k+1}^1]$ = the conditional expectation with respect to I_k^2 and I_{k+1}^1.

In addition, assume:

$$\{I_k^1\}_{k=2}^N \text{ and } \{I_k^2\}_{k=1}^N \text{ are independent sequences of independent random variables.} \tag{14.1}$$

For notational convenience, let $I_1^1 = i_1^1$ be a deterministic constant. Let $\mathcal{F}_{k+1}$, $k \geq 1$, be a σ-field defined as follows:

$$\mathcal{F}_{k+1} = \sigma\{(I_\ell^1, I_\ell^2), 1 \leq \ell \leq k, I_{k+1}^1\}, \quad 1 \leq k \leq N-1. \tag{14.2}$$

Let $\mathcal{F}_0 = \mathcal{F}_1 = \{\emptyset, \Omega\}$ and $\mathcal{F}_{N+1} = \mathcal{F}$. It is clear that the demand D_k is a $\mathcal{F}_{k+1}$-measurable random variable. Assume that

$$\mathsf{E}D_k = \mathsf{E}[g_k(I_k^1, I_k^2, v_k)] < \infty, \quad 1 \leq k \leq N. \tag{14.3}$$

Also assume that for each k,

$$C_k^f(u) \text{ and } C_k^s(u) \text{ are convex, } C_k^f(u) \geq c_k u \text{ and } C_k^s(u) \geq c_k u, \tag{14.4}$$

for some $c_k > 0$; and that the inventory cost function $H_k(x)$ satisfies:

$$H_k(x) \text{ is convex and } H_k(x) \leq c_H(1 + |x|), \quad 1 \leq k \leq N+1, \tag{14.5}$$

for some $c_H > 0$.

The inventory balance equations are as follows:

$$X_{k+1} = X_k + F_k + S_{k-1} - g_k(I_k^1, I_k^2, v_k), \ 1 \leq k \leq N, \tag{14.6}$$

where $S_0 = 0$ and

$$X_1 = x_1, \text{ initial inventory level.} \tag{14.7}$$

Furthermore, the decision F_k is adapted to $\mathcal{F}_k$ and the decision S_{k-1} is adapted to $\mathcal{F}_{k-1}$. From (14.6) therefore, we can see that X_k is an $\mathcal{F}_k$-measurable random variable and X_{k+1} is an $\mathcal{F}_{k+1}$-measurable random variable, since $D_k = g_k(I_k^1, I_k^2, v_k)$ is $\mathcal{F}_{k+1}$-measurable.

The objective is to choose a sequence of orders from the fast and the slow sources over time so as to minimize the total expected value of all the costs incurred during the interval $\langle 1, N\rangle$. Thus, the objective function is

$$J_1(x_1, s_0, i_1^1, (F,S)) = H_1(x_1) + \mathsf{E}\Big[\sum_{\ell=1}^{N-1}\Big(C_\ell^f(F_\ell) + C_\ell^s(S_\ell) + H_{\ell+1}(X_{\ell+1})\Big) + C_N^f(F_N) + H_{N+1}(X_{N+1})\Big], \tag{14.8}$$

where $(F,S) = ((F_1, ..., F_N), (S_1, ..., S_{N-1}))$ is a sequence of *nonanticipative admissible decisions*, that is, (F_k, S_k) is a nonnegative real-valued function of the demand information up to the beginning of period k, which is given by $\{I_\ell^1, I_\ell^2\}_{\ell=0}^{k-1}$ and I_k^1, F_N is a nonnegative real-valued function of the demand information up to the beginning of period N, which is given by $\{I_\ell^1, I_\ell^2\}_{\ell=1}^{N-1}$ and I_N^1, and s_0 is an outstanding slow order to be delivered at the end of period 1. Without loss of generality, we may assume $s_0 = 0$.

Finally, we define the value function for the problem over $\langle 1, N\rangle$ with the initial inventory level x_1 to be

$$V_1(x_1, s_0, i_1^1) = \inf_{(F,S)\in\mathcal{A}_1} J_1(x_1, s_0, i_1^1, (F,S)), \tag{14.9}$$

where $\mathcal{A}_1$ denotes the class of all nonanticipative admissible decisions.

Note that we use a general form $g_k(I_k^1, I_k^2, v_k)$ to represent the process of demand information updates and demand realization. This representation covers some specific models in the literature. For example, if we let I_k^1 be a deterministic constant, our system reduces to the Scheller-Wolf and Tayur [20] model of two sources of deliveries without demand information updates. If there is only one source for ordering, our system reduces to the advance demand information model of Gallego and Özlap Özer [11].

In addition, we would like to point out the role of v_k when $g_k(I_k^1, I_k^2, v_k)$ takes some special forms. If $g_k(I_k^1, I_k^2, v_k)$ is a multiplicative form, *i.e.*, $g_k(I_k^1, I_k^2, v_k) = v_k I_k^1 I_k^2$, where I_k^1 and I_k^2 are random variables with values in interval $[0,1]$, then v_k is the upper bound of the demand in period k. On the other hand, if $g_k(I_k^1, I_k^2, v_k) = v_k + I_k^1 + I_k^2$, where I_k^1 and I_k^2 are nonnegative random variables, then v_k is the lower bound of the period-k demand and I_k^1 and I_k^2 are components of demands observed one period apart. In particular, v_k could represent a contracted periodic demand and I_k^1 could represent the firm orders received at the beginning of period k to be filled at the end of period k. Finally I_k^2 represents the demand that arrives at the end of period k to be immediately filled (subject of course to a backlog situation if the inventory position $X_k + S_{k-1}$ is not sufficient to cover $v_k + I_k^1 + I_k^2$).

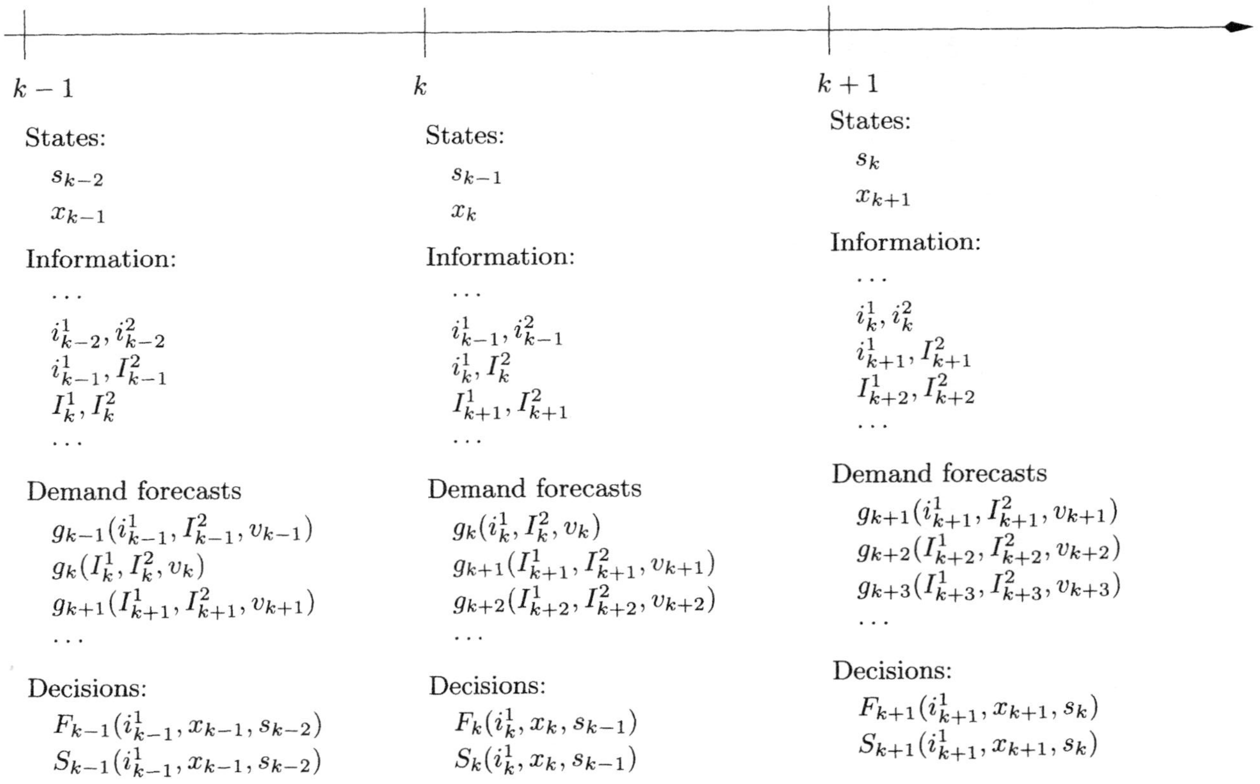

FIGURE 14.4. A time line of the system dynamics and decisions

In what follows, we use dynamic programming to study the problem. Let

$$\begin{aligned} & J_n(x_n, s_{n-1}, i_n^1, (F,S)) \\ &= H_n(x_n) + \mathsf{E}\Bigg[\sum_{\ell=n}^{N-1}\Big(C_\ell^f(F_\ell) + C_\ell^s(S_\ell) + H_{\ell+1}(X_{\ell+1})\Big) \\ &\qquad + C_N^f(F_N) + H_{N+1}(X_{N+1})\Bigg], \end{aligned} \tag{14.10}$$

where

$$(F,S) = ((F_n, ..., F_N), (S_n, ..., S_{N-1}))$$

is a *nonanticipative admissible decision* for the problem defined over periods $\langle n, N\rangle$. That is, given x_n, s_{n-1}, and i_n^1 as constants, (F_n, S_n) is a vector of nonnegative constants, (F_k, S_k) $(n < k < N)$ are nonnegative real-valued functions of the demand information from period n to the beginning of period k, given by $\{I_{\ell+1}^1, I_\ell^2\}_{\ell=n}^{k-1}$ and I_k^1, $k = n+1, ..., N-1$, and F_N is a nonnegative real-valued function of the demand information from period n to the beginning of period N, given by $\{I_{\ell+1}^1, I_\ell^2\}_{\ell=n}^{N-1}$ and I_N^1. Define the value function associated with the problem over periods $\langle n, N\rangle$ as follows:

$$V_n(x_n, s_{n-1}, i_n^1) = \inf_{(F,S)\in\mathcal{A}_n} J_n(x_n, s_{n-1}, i_n^1, (F,S)), \tag{14.11}$$

where $\mathcal{A}_n$ denotes the class of all nonanticipative admissible decisions for the problem over $\langle n, N\rangle$. The proof of the the following theorem can be found in Sethi, Yan and Zhang [23].

Theorem 14.1 The value function $V_1(x_1, s_0, i_1^1)$ is convex in x_1 on $(-\infty, \infty)$ and the value functions $V_n(x_n, s_{n-1}, i_n^1)$ $(2 \le n \le N)$ are convex in x_n and s_{n-1} on $(-\infty,\infty) \times [0,\infty)$.

In classical inventory models with convex cost, the value function is convex in the initial inventory level; see Bensoussan, Crouhy, and Proth [3]. Theorem 14.1 states that this classical result remains valid for the problem with dual delivery modes and forecast updates.

In view of (14.10), we can write the dynamic programming equation corresponding to the problem as follows:

$$\begin{aligned} U_\ell(x_\ell, s_{\ell-1}, i_\ell^1) &= H_\ell(x_\ell) + \inf_{\phi\ge 0, \sigma\ge 0}\{C_\ell^f(\phi) + C_\ell^s(\sigma) \\ &+ \mathsf{E}[U_{\ell+1}(x_\ell + \phi + s_{\ell-1} - g_\ell(i_\ell^1, I_\ell^2, v_\ell), \sigma, I_{\ell+1}^1)]\}, \\ &\qquad \ell = 1, ..., N-1, \end{aligned} \tag{14.12}$$

$$\begin{aligned} U_N(x_N, s_{N-1}, i_N^1) &= H_N(x_N) + \inf_{\phi\ge 0}\Big\{C_N^f(\phi) \\ &+ \mathsf{E}[H_{N+1}(x_N + s_{N-1} + \phi - g_N(i_N^1, I_N^2, v_N))]\Big\}. \end{aligned} \tag{14.13}$$

By the well-known selection theorem on Page 503 of Bensoussan, Crouhy, and Proth [3], there exists a Borel measurable function $f_N(x_N, i_N^1, s_{N-1})$ such that

$$\begin{aligned}
&C_N^f(f_N(x_N, i_N^1, s_{N-1})) \\
&+\mathsf{E}[H_{N+1}(x_N + s_{N-1} + f_N(x_N, i_N^1, s_{N-1}) - g_N(i_N^1, I_N^2, v_N))] \\
&= \inf_{0\le\phi<\infty} \left\{ C_N^f(\phi) + \mathsf{E}[H_{N+1}(x_N + s_{N-1} + \phi - g_N(i_N^1, I_N^2, v_N))] \right\},
\end{aligned}$$

and there exist Borel measurable functions

$$(f_\ell(x_\ell, i_\ell^1, s_{\ell-1}), s_\ell(x_\ell, i_\ell^1, s_{\ell-1})), 1 \le \ell \le N-1,$$

such that

$$\begin{aligned}
&C_\ell^f(f_\ell(x_\ell, i_\ell^1, s_{\ell-1})) + C_\ell^s(s_\ell(x_\ell, i_\ell^1, s_{\ell-1})) \\
&+\mathsf{E}\left[U_{\ell+1}(x_\ell + f_\ell(x_\ell, i_\ell^1, s_{\ell-1}) \right. \\
&\qquad \left. + s_{\ell-1} - g_\ell(i_\ell^1, I_\ell^2, v_\ell), s_\ell(x_\ell, i_\ell^1, s_{\ell-1}), I_{\ell+1}^1)\right] \\
&= \inf_{0\le\phi,\sigma<\infty} \left\{ C_\ell^f(\phi) + C_\ell^s(\sigma) \right. \\
&\qquad \left. + \mathsf{E}[U_{\ell+1}(x_\ell + \phi + s_{\ell-1} - g_\ell(i_\ell^1, I_\ell^2, v_\ell), \sigma, I_{\ell+1}^1)] \right\}.
\end{aligned}$$

Define:

$$\begin{aligned}
\hat{X}_1 &= x_1, \\
\hat{F}_1 &= f_1(x_1, s_0, i_1^1), \\
\hat{S}_1 &= s_1(x_1, s_0, i_1^1),
\end{aligned}$$

and

$$\begin{aligned}
\hat{X}_\ell &= \hat{X}_{\ell-1} + \hat{F}_{\ell-1} + \hat{S}_{\ell-2} - g_{\ell-1}(I_{\ell-1}^1, I_{\ell-1}^2, v_{\ell-1}), \\
\hat{F}_\ell &= f_\ell(\hat{X}_\ell, \hat{S}_{\ell-1}, I_\ell^1), \\
\hat{S}_\ell &= s_\ell(\hat{X}_\ell, \hat{S}_{\ell-1}, I_\ell^1)
\end{aligned}$$

for $2 \le \ell \le N-1$, where $\hat{S}_0 = s_0$ and

$$\begin{aligned}
\hat{X}_N &= \hat{X}_{N-1} + \hat{F}_{N-1} + \hat{S}_{N-2} - g_{N-1}(I_{N-1}^1, I_{N-1}^2, v_{N-1}), \\
\hat{F}_N &= f_N(\hat{X}_N, \hat{S}_{N-1}, I_N^1).
\end{aligned}$$

Theorem 14.2 $((\hat{F}_1, \dots, \hat{F}_{N-1}, \hat{F}_N), (\hat{S}_1, \dots, \hat{S}_{N-1}))$ specified above is an optimal solution to the problem. That is,

$$\begin{aligned}
H_1(x_1) + \mathsf{E}\Bigg[&\sum_{\ell=1}^{N-1} \left(C_\ell^f(\hat{F}_\ell) + C_\ell^s(\hat{S}_\ell) + H_{\ell+1}(\hat{X}_{\ell+1}) \right) \\
&+ C_N^f(\hat{F}_N) + H_{N+1}(\hat{X}_{N+1}) \Bigg] = V_1(x_1, s_0, i_1^1).
\end{aligned}$$

Theorem 14.2 establishes the existence of an optimal nonanticipative admissible policy, *i.e.*, there exists a policy in the class of all nonanticipative admissible policies, whose objective function value equals the value function defined in (14.9), and there exists a nonanticipative admissible policy characterized in Theorem 14.2 that achieves the same objective value. Under some mild conditions on the ordering costs, we can get the optimality of a base-stock policy. Formally, assume that

$$\begin{aligned} C_k^f(F) &= c_k^f \cdot F, \quad k = 1, \ldots, N, \\ C_i^s(S) &= c_i^s \cdot S, \quad i = 1, \ldots, N-1 \end{aligned}$$

with $c_i^f > c_i^s$. Let y_N^* be the value of ϕ that minimizes

$$c_N^f \phi + E[H_{N+1}(\phi - g_N(i_N^1, I_N^2, v_N)],$$

and let y_k^* be the value of ϕ that minimizes

$$(c_k^f - c_k^s)\phi + E[H_{k+1}(\phi - g_k(i_k^1, I_k^2, v_k)].$$

Let z_k^* be the value off σ that minimizes

$$\begin{aligned} &\Big\{(c_k^f - c_k^s)\sigma + E[H_{k+1}(\sigma - g_k(i_k^1, I_k^2, v_k)] \\ &- \Big((c_k^f - c_k^s)y_k^* + E[H_{k+1}(y_k^* - g_k(i_k^1, I_k^2, v_k)]\Big)\Big\}\, \delta(y_k^* - t) \\ &c_k^s t + E[U_{k+1}(t - g_k(i_k^1, I_k^2, v_k)], \end{aligned}$$

where

$$\delta(x) = \begin{cases} 1, & \text{if } x > 0 \\ 0, & \text{otherwise.} \end{cases}$$

Theorem 14.3 Assume the material purchase cost is linear and the cost of a fast order is more expensive than a slow order. If the initial inventory level at the beginning period k is x_k, the slow order quantity in period $(k-1)$ is denoted by s_{k-1}, and the observed value of I_k^1 is i_k^1 in period $(k-1)$, then the optimal fast order quantity f_k and the optimal slow order quantity s_k in period k are given by the following expressions:

(i) when $y_k^* < z_k^*$,

$$(f_k, s_k) = \begin{cases} (y_k^* - x_k - s_{k-1}, z_k^* - y_k^*), & \text{if} \quad x_k + s_{k-1} \le y_k^* \\ (0, z_k^* - x_k - s_{k-1}), & \text{if} \quad y_k^* < x_k + s_{k-1} \le z_k^* \\ (0, 0), & \text{if} \quad z_k^* \le x_k + s_{k-1}; \end{cases}$$

(ii) when $y_k^* \geq z_k^*$,

$$(f_k, s_k) = \begin{cases} (z_k^* - x_k - s_{k-1}, 0), & \text{if } \; x_k + s_{k-1} \leq z_k^* \\ (0, 0), & \text{if } \; z_k^* < x_k + s_{k-1} \leq y_k^* \\ (0, 0), & \text{if } \; y_k^* \leq x_k + s_{k-1}. \end{cases}$$

Here y_k^* and z_k^* do not depend on the inventory position $(x_k + s_{k-1})$. The theorem says that in case (i), there are two base-stock levels—fast and slow—with the fast base-stock level y_k^* being smaller than the slow base-stock level z_k^*. Moreover, when the inventory position is too low, i.e., smaller than y_k^*, then we order up to y_k^* via the fast mode and order an additional amount from y_k^* up to z_k^* via the slow mode. On the other hand, when the inventory position is too high, i.e., larger than z_k^*, then we order nothing. Finally, if the inventory position is neither too low nor too high, i.e., when it is between the level y_k^* and z_k^*, then simply order up to z_k^* via the slow mode. In case (ii), there is only one base-stock level z_k^*, and if the inventory position is too low, i.e., smaller than z_k^*, then we order up to z_k^* via the fast mode and order nothing via the slow mode. On the other hand, if the inventory position is too high, i.e., larger than z_k^*, then we order nothing.

Under the same condition, it is shown that the optimality of a base-stock policy still holds for α-discounted infinite-horizon problems [23].

14.3 A One-Period, Multi-Information Revision Model

Yan, Yan and Zhang [28] and Yan and Zhang [29] consider a problem of information revisions and decision making, such that there are N distinct time instance where the buyer has an opportunity to react to the revised information. At each time instance, the buyer revises his forecast about the future. The forecast includes customer demand, market price and others. Without loss of generality, we treat customer demand and product price as the information required for buyer. In addition, we assume that the supply chain decision is to decide how many units of product to purchase at each time instance. Note that other information and decisions can be modeled in a similar fashion.

Let I_k (a random variable, $1 \leq k \leq N$) be the information which can be observed between time instances k and $k+1$. We denote the product price for time instance k as a function $C_k(\cdot, I_0, I_1, ..., I_{k-1})$ $(1 \leq k \leq N)$ with the convenience

$$I_0 = i_0 \equiv 0.$$

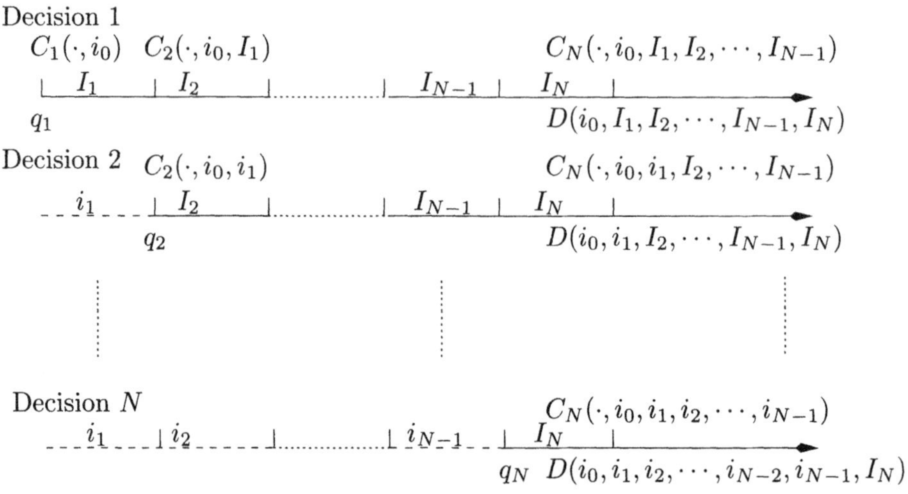

FIGURE 14.5. A time-line demonstration of decision and information dynamics

In addition, the demand forecast is also conditional on information. We model it as a function $D(i_0, I_1, ..., I_N)$. A time-line of the information dynamics and purchasing decisions is illustrated in Figure 14.5.

In Figure 14.5, each line represents a time instance, where the status of information and required supply chain decision are illustrated. When decision has to be made for time instance 1, product price $C_1(\cdot, i_0)$ for time instance 1 is available, while product prices for time instance k $(2 \leq k \leq N)$ are conditional on information I_1 that will be observed between instance 1 and instance 2. In particular, product price of instance 2 is conditional on information about what other players do at instance 1. Based on $C_k(\cdot, i_0, I_1, ..., I_{k-1})$ $(1 \leq k \leq N)$ and $D(i_0, I_1, ..., I_N)$, the decision is to allocate a part of the demand (known as q_1) at time instance 1. In general, for time instance k $(2 \leq k \leq N)$, the information $I_1, \ldots, I_{k-1}$ have been observed as $i_1, \ldots, i_{k-1}$; product price at time instance k is updated as $C_k(\cdot, i_0, i_1, ..., i_{k-1})$, and product prices from time instance $(k+1)$ to N are updated as $C_\ell(\cdot, i_0, i_1, ..., i_{k-1}, I_k, ..., I_{\ell-1})$ $(k+1 \leq \ell \leq N)$; The consumer demand is updated as $D(i_0, i_1, ..., i_{k-1}, I_k, ..., I_N)$. With revised information $C_k(\cdot, i_0, i_1, ..., i_{k-1})$, $C_\ell(\cdot, i_0, i_1, ..., i_{k-1}, I_k, ..., I_{\ell-1})$ $(k+1 \leq \ell \leq N)$ and $D(i_0, i_1, ...i_{k-1}, I_k, ..., I_N)$, the purchase decision is q_k for time instance k with respect to product having been purchased from previous $(k-1)$ time instances, $\sum_{\ell=1}^{k-1} q_\ell$. After these products are ordered at each time instance, all products arrive at time instance N to satisfy the customer demand. When the realized customer demand occurs at time instance N, the customer is satisfied if there is sufficient available inventory, and the excess is penalized by h for each unit. Otherwise, the customer demand is partially satisfied and the unsatisfied demand is penalized by p for each unit. The information revision is the demand and purchasing price forecast. As usual, we assume that the shortage cost p is larger than the holding cost

h, that is,

$$p > h. \tag{14.14}$$

For $i = 1, ..., N$, let

q_0 : the amount of product purchased at the beginning;
q_i : the amount of product purchased at the ith time instance.

For $k = 1, ..., N-1$, let

$\Phi_k(\cdot)$ and $\phi_k(\cdot)$: the distribution and density of I_k (the information will be observed in instance k) , respectively;
$\Psi_k(\cdot|(i_1, ..., i_k))$: the conditional distribution of $D(i_0, I_1, ..., I_N)$ given $(I_1, ..., I_k) = (i_1, ..., i_k)$;
$\psi_k(\cdot|(i_1, ..., i_k))$: the conditional density of $D(i_0, I_1, ..., I_N)$ given $(I_1, ..., I_k) = (i_1, ..., i_k)$;
$F(\cdot)$ and $f(\cdot)$: the distribution and density of $D(i_0, I_1, ..., I_N)$, respectively.

In addition, we make the following assumption through this chapter.

$$\begin{cases} C_1(x, i_0) & \text{is convex and continuously differentiable in } x, \\ C_k(x, i_0, i_1, ..., i_{k-1}) & \text{is convex and continuously differentiable in } x \\ & \text{for any observed value } (i_1, ..., i_{k-1}). \end{cases}$$

Furthermore, we assume that

$$I_1, ..., I_{N-1}, I_N \quad \text{are independent.} \tag{14.15}$$

In order to describe the cost function, we introduce the following notation:

$$\begin{aligned} & J_{N-1}(q, q_{N-1}, q_N, (i_1, ..., i_{N-2})) \\ = \; & \mathsf{E}\{C_{N-1}(q_{N-1}, i_0, i_1, ..., i_{N-2}) \\ & + J_N(q + q_{N-1}, q_N, (i_1, ..., i_{N-2}, I_{N-1}))\}, \end{aligned} \tag{14.16}$$

and

$$\begin{aligned} & J_N(q, q_N, (i_1, ..., i_{N-1})) \\ = \; & C_N(q_N, i_0, i_1, ..., i_{N-1}) \\ & + H_{N-1}(q + q_N, i_1, ..., i_{N-1}), \end{aligned} \tag{14.17}$$

where

$$
\begin{aligned}
&H_{N-1}(q+q_N, i_1, ..., i_{N-1}) \\
&= h\int_0^{q+q_N} [(q+q_N)-t]\psi_{N-1}(t|(i_1,...,i_{N-1}))dt \\
&+p\int_{q+q_N}^{\infty} [t-(q+q_N)]\psi_{N-1}(t|(i_1,...,i_{N-1}))dt. \qquad (14.18)
\end{aligned}
$$

$J_N(q, q_N, (i_1, ..., i_{N-1}))$ is the cost if the purchased quantity is q_N from the Nth time instance with q units having been purchased from first $(N-1)$ time instances and the observed information $(i_1, ..., i_{N-1})$. Similarly, with q units having been purchased from the first $(N-2)$ time instances and the observed information $(i_1, ..., i_{N-2})$, $J_{N-1}(q, q_{N-1}, q_N, (i_1, ..., i_{N-2}))$ is the cost if the unit purchased is q_{N-1} and q_N from the $(N-1)$th and Nth time instances, respectively. In the same way, for $k = N-2, ..., 2$, let

$$
\begin{aligned}
&J_k(q, q_k, ..., q_N, (i_1, ..., i_{k-1})) \\
&= \mathsf{E}\{C_k(q_k, i_0, i_1, ..., i_{k-1}) \\
&\quad +J_{k+1}(q+q_k, q_{k+1}, ..., q_N, (i_1, ..., i_{k-1}, I_k))\}. \qquad (14.19)
\end{aligned}
$$

Here $J_k(q, q_k, ..., q_N, (i_1, ..., i_{k-1}))$ has the same explanation as $J_{N-1}(q, q_{N-1}, q_N, (i_1, ..., i_{N-2}))$. With these notations, we have the following cost with the initial product purchased q_0, if $(q_1, ..., q_N)$ are purchased from these N sequential instances:

$$
V(q_0, q_1, ..., q_N) = C_1(q_1, i_0) + \mathsf{E}J_2(q_0 + q_1, q_2, ..., q_N, I_1). \qquad (14.20)
$$

Furthermore, we have the following property on the cost $V(q_0, q_1, ..., q_N)$.

Proposition 14.4 The cost function V in (14.20) is convex in $(q_1, ..., q_N)$.

Proof. This follows from the convexity of $C_1(q, I_0)$ and $C_k(q, I_0, i_1, ..., i_{k-1})$ $(2 \le k \le N)$, and the definition of $V(q_0, q_1, ..., q_N)$. □

Our objective is to determine the purchasing quantity q_k $(1 \le k \le N)$ at each instance to minimize the cost $V(q_0, q_1, ..., q_N)$. Let $V(q_0)$ be the minimum cost under the initial quantity q_0. Clearly, from the dynamic behavior of the system, we know that the optimal purchasing decision,

$$
(q_1^*, ..., q_N^*)
$$

should be a *nonanticipative admissible decision*, that is, q_k^* (the purchasing quantity from the kth instance, $1 \le k \le N$) is a nonnegative real-valued function of the market information observed from the first instance to the $(k-1)$th instance.

In view of (14.17)-(14.20), the dynamic programming equation corresponding to our problem is

$$\begin{cases} V_N(x,(i_1,...,i_{N-1})) &= \inf_{y\geq 0}\{C_N(y,i_0,i_1,...,i_{N-1}) \\ & \qquad +H_{N-1}(x+y,i_1,...,i_{N-1})\} \\ V_k(x,(i_1,...,i_{k-1})) &= \inf_{y\geq 0}\{C_k(y,i_0,i_1,...,i_{k-1}) \\ & \qquad +\mathsf{E}V_{k+1}(x+y,(i_1,...,i_{k-1},I_k))\}, \\ & \quad k=N-1,...,2, \\ V_1(x) &= \inf_{y\geq 0}\{C_1(y,i_0)+\mathsf{E}V_2(x+y,I_1)\}, \end{cases}$$

where $H_{N-1}(x+y,i_1,...,i_{N-1})$ is defined in (14.18). Just as Section 14.2, we know that there exists a Borel measurable function of $(x,(i_1,...,i_{N-1}))$, written as $v_N(x,(i_1,...,i_{N-1}))$, such that

$$\begin{aligned} &\inf_{y\geq 0}\{C_N(y,i_0,i_1,...,i_{N-1})+H_{N-1}(x+y,i_1,...,i_{N-1})\} \\ &= C_N(v_N(x,(i_1,...,i_{N-1})),i_0,i_1,...,i_{N-1}) \\ &\quad +H_{N-1}(x+v_N(x,(i_1,...,i_{N-1})),i_1,...,i_{N-1}). \end{aligned} \tag{14.21}$$

Similarly, there are Borel measurable functions of $(x,(i_1,...,i_{k-1}))$,

$$v_k(x,(i_1,...,i_{k-1})), \qquad k=2,...,N-1,$$

such that

$$\begin{aligned} &\inf_{y\geq 0}\{C_k(y,i_0,i_1,...,i_{k-1})+\mathsf{E}V_{k+1}(x+y,(i_1,...,i_{k-1},I_k))\} \\ &= C_k(v_k(x,(i_1,...,i_{k-1})),I_0,i_1,...,i_{k-1}) \\ &\quad +\mathsf{E}V_{k+1}(x+v_k(x,(i_1,...,i_{k-1})),(i_1,...,i_{k-1},I_k)), \end{aligned} \tag{14.22}$$

and a Borel measurable function of x, $v_1(x)$, such that

$$\begin{aligned} &\inf_{y\geq 0}\{C_1(y,i_0)+\mathsf{E}V_2(x+y,I_1)\} \\ &= C_1(v_1(x),i_0)+\mathsf{E}V_2(x+v_1(x),I_1). \end{aligned} \tag{14.23}$$

Now let

$$\begin{cases} q_1^*(q_0, i_0) = v_1(q_0) \\ q_k^*(q_0, i_0, I_1, ..., I_{k-1}) \\ = v_k\left(q_0 + \sum_{\ell=1}^{k-1} q_\ell^*(q_0, i_0, ..., I_{\ell-1}), (I_1, ..., I_{k-1})\right), \\ k = 2, ..., N. \end{cases} \tag{14.24}$$

Clearly, $q_1^*(q_0, i_0)$ is a deterministic constant only depending on the initial quantity q_0, and each $q_k^*(q_0, i_0, I_1, ..., I_{k-1})$ $(2 \le k \le N)$ is a random variable determined by $(I_1, ..., I_{k-1})$. Therefore,

$$(q_1^*(q_0, i_0), q_2^*(q_0, i_0, I_1), ..., q_N^*(q_0, i_0, I_1, ..., I_{N-1})) \tag{14.25}$$

is a nonanticipative purchasing policy. Furthermore, we have the following theorem. For notational simplicity, we shall write (14.25) as $(q_1^*, ..., q_N^*)$.

Theorem 14.5 $(q_1^*, ..., q_N^*)$ in (14.24) is an optimal nonanticipative purchasing policy from the N sequential time instances with the initial quantity q_0.

By Theorem 14.5, we know that the solution to the dynamic programming equation is an optimal nonanticipative purchasing policy. Furthermore, the selection theorem provides the existence of an optimal purchasing policy.

The next theorem reveals an interesting aspect of the optimal nonanticipative purchasing policy $(q_1^*, ..., q_N^*)$, i.e, the marginal expected purchasing costs are equal for all remaining sequential instances.

Theorem 14.6 For any j and k, if

$$P\{q_j^* \neq 0\} = P\{q_k^* \neq 0\} = 1, \tag{14.26}$$

then

$$\mathsf{E}\left(\left.\frac{\partial C_j(q_j, i_0, I_1, ..., I_{j-1})}{\partial q_j}\right|_{q_j = q_j^*}\right) = \mathsf{E}\left(\left.\frac{\partial C_k(q_k, i_0, I_1, ..., I_{k-1})}{\partial q_k}\right|_{q_k = q_k^*}\right), \tag{14.27}$$

where q_j^* and q_k^* are given in (14.24).

F_1	F_2	Demand	Alternatives	q_1	q_2	Inventory	Shortage	Sales	Cost ($)
75	50	15	Company Decision	150	0	135	0	15	585
			Our Model	115	0	100	0	15	445
50	100	110	Company Decision	100	100	90	0	110	990
			Our Model	95	54	39	0	110	648

TABLE 14.2. Comparisons of Proposed Model and Company Practice (without a major improvement in forecasting)

F_1	F_2	Demand	Alternatives	q_1	q_2	Inventory	Shortage	Sales	Cost ($)
400	250	150	Company Decision	800	0	650	0	150	3,050
			Our Model	390	0	240	0	150	1,410
150	500	500	Company Decision	300	400	200	0	500	3,500
			Our Model	140	350	0	10	490	2,660

TABLE 14.3. Comparisons of Proposed Model and Company Practice (with a major improvement in forecasting)

14.4 Applications

In this section, we provide two examples of application of our results developed in the previous section. The first example is the material purchase problem presented in Section 14.1, which is a multiple period problem with one information revision and two decisions in each period; the second example is a single period problem in which there are 2 information revisions and 2 decisions.

14.4.1 Decision-Making with Two Procurement Alternatives

As mentioned earlier, there exist multiple alternatives in implementing program codes in the ROM of micro-controllers. Moreover, the price of micro-controllers heavily depends on the technology. In addition, for the seven products we investigated, the micro-controller was the most expensive among all required components.

The company uses the custom-masked and one time programmable (OTP) chip as their choices in implementing program codes. The company chooses the custom-masked chip for its price; chooses the OTP chip for its relative flexibility (and its price compared with EPROM or EEPROM chips). Other than demand uncertainty, the company also faces the problem of the short product cycle time and frequent engineering change orders (ECO). When a product is phased out, or an ECO is made, the custom-masked chips may be obsolete. In addition, when a product is phased out, the OTP chips may also be obsolete. Due to the lack of a scientific method to determine order quantities for masking and OTP chips, the company purchases masking chips proportional to the first forecast. Very often, the order size is twice as much as the forecast. The OTP chips are only used when the company has sufficient evidence of a shortage in micro-controller chips. To demonstrate the approach developed in this chapter, we choose some sample data to evaluate the order quantity and the related costs. We pick Product 1 and Product 5. From Table 14.1, the forecast for Product 1 does not show a major improvement while the forecast for Product 5 improves significantly. For each product, two scenarios are investigated: one with a decreasing forecast and the other with an increasing forecast. The cost parameters are $c_1 = 3, c_2 = 6, h = 1$ and $p = 14$. Tables 14.2 and 14.3 provide the results of sample calculations. Using these sample calculations, it is not difficult to find that our approach has the advantage of maintaining a lower level of inventory while at the same time providing a similar level of service.

14.4.2 The Application to Deregulated Energy Markets

It is worth noting that the total purchased energy in PX markets are compared with the realized demand at the ISO market. The energy buyers are then forced to buy/sell the difference.

Taking the day-ahead and day-of markets as an example, in Figure 14.6, we illustrate trade-offs of demand allocation graphically. For the sake of concreteness, we fix a sample point $\omega \in \Omega$. Suppose that the energy buyers faces a demand $D(i_0, I_1(\omega), I_2(\omega))$, it is possible for the energy buyers to allocate its demand according to any one of the following:

(a) in the day-ahead market with a cost of $C_1(D(i_0, I_1(\omega), I_2(\omega)), i_0)$;

(b) in the day-of market with a cost of

$$C_2(D(i_0, I_1(\omega), I_2(\omega)), i_0, I_1(\omega));$$

(c) in both the day-ahead and the day-of market with a cost of

$$C_1(q_1, i_0) + C_2(q_2, i_0, I_1(\omega))$$

such that $q_1 + q_2 = D(i_0, I_1(\omega), I_2(\omega))$.

In Figure 14.6, energy purchasing costs from the day-ahead and day-of markets are depicted as solid and dash lines, respectively.

Note that demand and day-of market energy price are all uncertain at the day-ahead market. Suppose that the demand is $D(i_0, I_1(\omega), I_2(\omega))$. It will cost the energy buyer

$$C_1(D(i_0, I_1(\omega), I_2(\omega)), i_0) \quad (\ C_2(D(i_0, I_1(\omega), I_2(\omega)), i_0, I_1(\omega))\ )$$

to buy D units of energy from the day-ahead (day-of market). However, if demand $D(i_0, I_1(\omega), I_2(\omega))$ is allocated in both the day-ahead market q_1 and day-of market q_2, the cost is $C_1(q_1, i_0) + C_2(q_2, i_0, I_1(\omega))$. The problem is to choose q_1 and q_2 to minimize the energy purchasing cost, and can be solved by Theorem 14.5.

This result can be intuitively explained as follows. Suppose that we want to allocate D units of demand in the day-ahead market and the day-of market, respectively. The allocation process can be divided into two steps. The allocation should start from the market with the lower purchasing cost until the purchasing cost is the same for both markets. In the second step, the demand allocation should be proportional to the rate of increase of purchasing costs, i.e., the expected marginal cost of the optimal purchasing decisions should be the same. By Theorem 14.6, Figure 14.7 illustrates this fact graphically, i.e., the slopes of $C_1(q_1^*, i_0)$ and $C_2(q_2^*, i_0, I_1)$ must be the same.

14.5 Notes

Effective use of advance demand information has been a major concern in many industries, such as the apparel industry (Fisher, et al. [10] and Iyer and Bergen [17]), the toy industry (Barnes-Schuster, Bassok, and Anupindi [2]), and the computer and electronics industry (Tsay and Lovejoy [26],

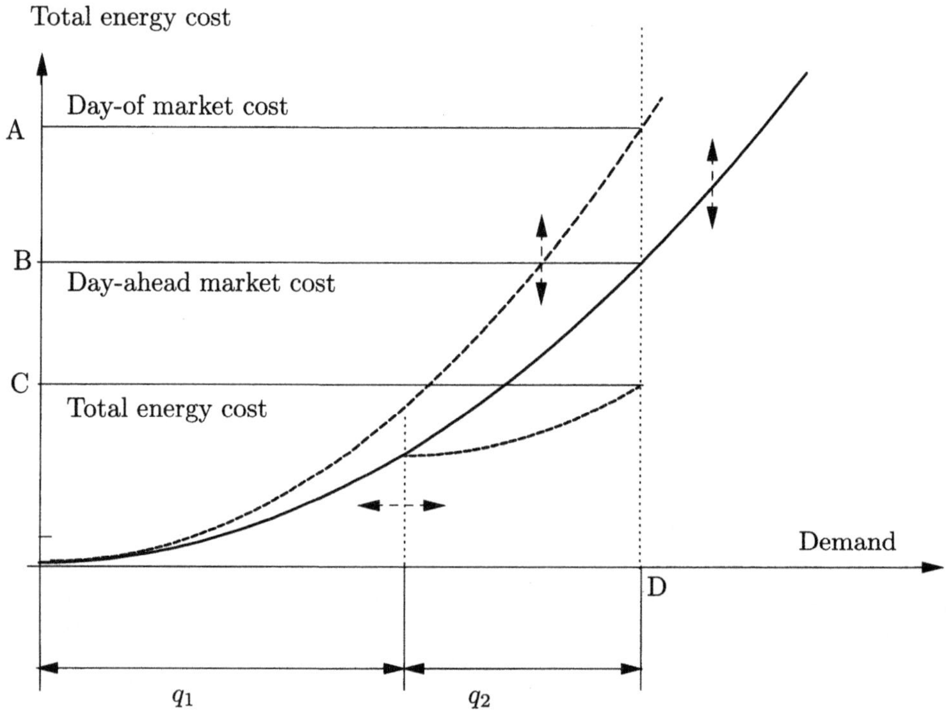

FIGURE 14.6. An Illustration of the Optimal Demand Allocation Condition

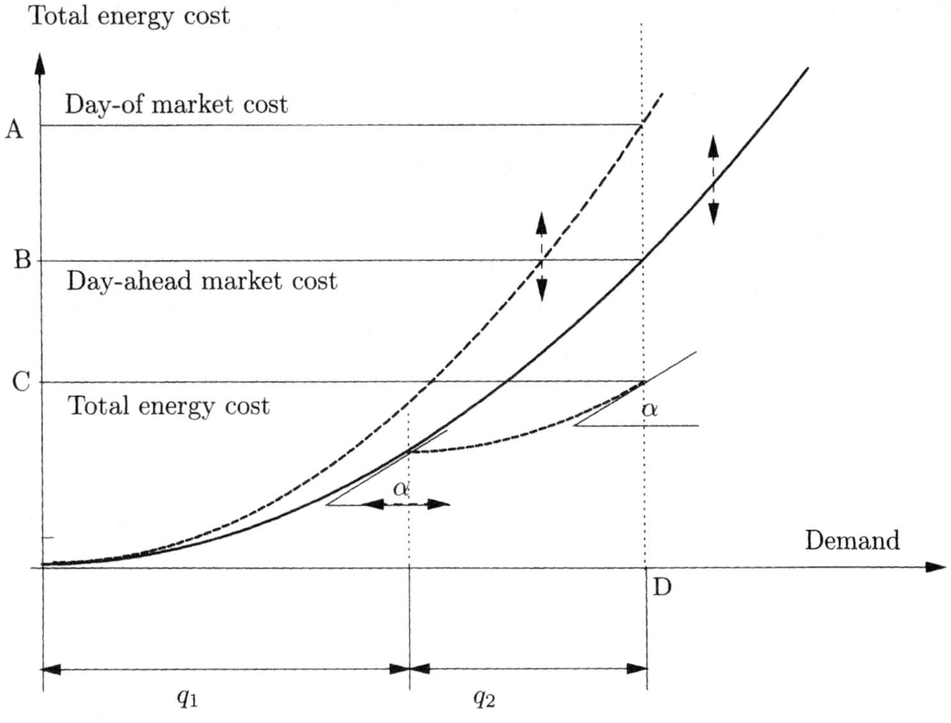

FIGURE 14.7. An Illustration of the Optimal Demand Allocation Condition

Brown and Lee [5], and Yan, Liu and Hsu [27]). Bayesian analysis and time-series models are the commonly used techniques in updating demand distribution functions and exploring inter-temporal relationships (see Dvoretzky, Keifer, and Wolfowitz [8], Eppen and Iyer [9], Johnson and Thompson [18], and Lovejoy [19]).

In contrast to Bayesian analysis and time-series models, this chapter is concerned with forecast revisions. Related studies include the following. Hausmann [13] models the evolution of the forecast as a quasi-Markovian process and provides optimal decision rules for sequential decision problems. Sethi and Sorger [22] formulate a quite general model, which allows unrestricted forecast updates. Heath and Jackson [15] use a martingale to model the forecast evolution, and analyze economic safety stock levels for a multi-product, multi-facility production system. Yan, Liu, and Hsu [27] obtain the optimal order quantity in a single-period, two-stage model with dual supply modes and demand information updates. Donohue [7] considers a risk-sharing supply contract between a buyer and a supplier, and discusses pricing issues when the demand information update is perfect. For a bivariate normal demand, Gurnani and Tang [12] provide an explicit solution in the cases of worthless and perfect information updates. Barnes-Schuster, Bassok, and Anupindi [2] consider a single-period, two-stage model with updating information arriving at the beginning of the second stage. Huang, Yan, and Sethi [16] consider a single-period, two-stage supply contract model with both fixed and variable costs and demand information updates. Gallego and Özlap Özer [11] model the forecast evolution as a supermartingale, and prove the optimality of a state-dependent (s, S) policy.

Bensoussan, Crouhy, and Proth [3] consider an inventory model with two supply modes, one instantaneous and the other with a one-period lead time, and obtain an optimal policy that generalization of the well-known (s, S) policy. Hausmann, Lee, and Zhang [14] study an inventory system with two supply modes (fast and slow) under the assumption of stationary demand, and develop an explicit formula for the optimal ordering decisions. Zhang [30] extends their work to allow three supply modes. Scheller-Wolf and Tayur [20] study a Markovian dual-source production inventory model but without the consideration of advance demand information, and prove the optimality of a state-dependent base-stock policy. For other studies on state-dependent policies, refer to Song and Zipkin [24], Sethi and Cheng [21], Beyer, Sethi, and Taksar [4].

Acknowledgments

The work was supported in part by RGC (Hong Kong) Competitive Earmarked Research Grants, a Distinguished Young Investigator Grant from the National Natural Sciences Foundation of China, and a grant from the Hundred Talents Program of the Chinese Academy of Sciences.

14.6 References

[1] Bassok, Y. and Anupindi, R., Analysis of Supply Contracts with Total Minimum Commitment, *IIE Transactions*, **29** (1997), 373-381.

[2] Barnes-Schuster, D., Bassok, Y., and Anupindi, R., Coordination and Flexibility in Supply Contracts and Options, Working Paper, Graduate School of Business, University of Chicago, IL, 1999.

[3] Bensoussan, A., Crouhy, M., and Proth, J.M., *Mathematical Theory of Production Planning*, North-Holland, New York, NY, 1983.

[4] Beyer, D., Sethi, S. and Taksar M., Inventory Models with Markovian Demands and Cost Functions of Polynomial Growth, *Journal of Optimization Theory and Applications*, **98** (1998), 281-323.

[5] Brown, A.O., and Lee, H.L., Optimal Pay-to-delay Capacity Reservation with Application to the Semiconductor Industry, Working Paper, Department of Industrial Engineering and Engineering Management, Stanford University, Stanford, CA, 1997.

[6] California Power Exchange, *1998-1999 Market Year Report to Californians*, 1999.

[7] Donohue, K.L., Efficient Supply Contracts for Fashion Goods with Forecast Updating and Two Production Modes, *Management Science*, **46** (2000). 1387-1411.

[8] Dvoretzky, A., Keifer, J., and Wolfowitz, J., The Inventory Problem: II. Case of Unknown Distribution Demand, *Econometrica*, **20** (1956), 450-466.

[9] Eppen, G.D., and Iyer, A.V., Improved Fashion Buying with Bayesian Updates, *Operations Research*, **45** (1997), 805-819.

[10] Fisher, M., Hammond, J.M., Obermeyer, W.R., and Raman, A., Making Supply Meet Demand in an Uncertain World, *Harvard Business Review*, May-June, 1994, 83-93.

[11] Gallego, G., and Özlap Özer, A., Integrating Replenishment Decisions with Advance Demand Information, *Management Science*, **47** (2001), 1344-1360.

[12] Gurnani, H. and Tang, C.S., Optimal Ordering Decisions with Uncertain Cost and Demand Forecast Updating, *Management Science*, **45** (1999), 1456-1462.

[13] Hausmann, W.H., Sequential Decision Problems: a Model to Exploit Existing Forecast, *Management Science*, **16** (1969), B93-B111.

[14] HAUSMANN, W.H., LEE, H.L., and ZHANG, V.L., Optimal Ordering for an Inventory System with Dual Lead Times, Working Paper, Department of Industrial Engineering and Engineering Management, Stanford University, Stanford, CA, 1993.

[15] HEATH, D., and JACKSON, P., Modeling the Evolution of Demand Forecast with Application to Safety Stock Analysis in Production/Distribution Systems, *IIE Transactions*, **26** (1994), 17-30.

[16] HUANG, H.Y., YAN, H., and SETHI, S., Purchase Contract Management with Demand Information Update, Working Paper, Dept. Systems Engineering and Engineering Management, The Chinese University of Hong Kong, Hong, Kong, 2000.

[17] IYER, A.V., and BERGEN, M.E., Quick Response in Manufacturing Retail Channels, *Management Science*, **43** (1997), 559-570.

[18] JOHNSON, O., and THOMPSON, H., Optimality of Myopic Inventory Policies for Certain Depended Processes, *Management Science*, **21** (1975), 1303-1307.

[19] LOVEJOY, W.S., Myopic Policies for Some Inventory Models with Uncertain Demand Distributions, *Management Science*, **36** (1990), 724-738.

[20] SCHELLER-WOLF, A. and TAYUR, S., A Markovian Dual-source Production-inventory Model with Order Bands, GSIA Working Paper, #1998-E200, Carnegie Mellon University, 1998.

[21] SETHI, S., and CHENG, F., Optimality of (s, S) Policies in Inventory Models with Markovian Demand, *Operations Research*, **45** (1997), 931-939.

[22] SETHI, S., and SORGER, G., A Theory of Rolling Horizon Decision Making, *Annals of Operations Research*, **29** (1991), 387-416.

[23] SETHI, S., YAN, H., and ZHANG, H., Peeling Layers of an Onion: a Periodic Review Inventory Model with Multiple Delivery Modes and Forecast Updates, *Journal of Optimization Theory and Applications*, **108** (2001), 253-281.

[24] SONG, J., and ZIPKIN, P., Inventory Control in a Fluctuating Demand Environment, *Operations Research*, **41** (1993), 351-370.

[25] SPASOV, P., *Microcontroller Technology*, Prentice Hall, Englewood Cliffs, New Jersey, 1993.

[26] TSAY, A., and LOVEJOY, W.S., Quantity Flexibility Contracts and Supply Chain Performance, Working Paper, Santa Clara University, Santa Clara, CA, 1998.

[27] YAN, H., LIU, K., and HSU, A., Order Quantity in Dual Supply Model with Updated Forecast, to appear in *Production and Operations Management*, 2002.

[28] YAN, H., YAN, H.J., and ZHANG, H., Demand Allocation Analysis for Energy Purchasers in Deregulated Energy Markets, *Proceedings of International Conference on Electric Utility Deregulation and Restructuring, and Power Technologies*, London, U.K., April, 2000, 344-348.

[29] YAN, H. and ZHANG, H., The Analysis of a Supply Contract with Limited Flexibility, Uncertain cost and Demand Forecast Updates, Working Paper, The Chinese University of Hong Kong, 2000.

[30] ZHANG, V.L., Ordering Policies for an Inventory System with Three Supply Modes, *Naval Research Logistics*, **43** (1996), 691-708.

About the Contributors

Xi-Ren Cao (Hong Kong University of Science and Technology) received the M.S. and Ph.D. degrees from Harvard University, in 1981 and 1984, respectively; and was a research fellow there from 1984 to 1986. He then worked as a principal and consultant engineer/engineering manager at Digital Equipment Corporation, U.S.A., until October 1993. Since then, he has been a Professor at the Hong Kong University of Science and Technology, where he is the Director of the Center for Networking. He held visiting positions at Harvard University, University of Massachusetts at Amherst, AT&T Labs, University of Maryland at College Park, University of Notre Dame, Shanghai Jiaotong University, Nankai University, Tsinghua University, and University of Science and Technology of China.

Professor Cao owns two patents in data communications, and has published two books: *Realization Probabilities - the Dynamics of Queuing Systems,* Springer-Verlag, 1994; and *Perturbation Analysis of Discrete-Event Dynamic Systems,* Kluwer Academic Publishers, 1991 (co-authored with Y.C. Ho). He received the Outstanding Transactions Paper Award from the IEEE Control System Society in 1987, and the Outstanding Publication Award from the Institution of Management Science in 1990. He is a Fellow of IEEE, Associate Editor at Large of *IEEE Transactions of Automatic Control*, and is/was on the Board of Governors of IEEE Control Systems Society, associate editor of a number of international journals, and chairman of a few technical committees of international professional societies. His current research areas include discrete event dynamic systems, communication systems, signal processing, stochastic processes, and system optimization.

Cheng-Shang Chang (National Tsing Hua University, Taiwan) received the B.S. degree from the National Taiwan University in 1983, and the M.S. and Ph.D. degrees from Columbia University in 1986 and 1989, respectively, all in Electrical Engineering. From 1989 to 1993 he was a Research Staff Member at the IBM Thomas J. Watson Research Center, Yorktown Heights, N.Y. In 1993, he joined the Department of Electrical Engineering at National Tsing Hua University, Taiwan, where he is a Professor. His current research interests are concerned with queueing theory, stochastic scheduling and performance evaluation of telecommunication networks and parallel processing systems. Professor Chang received an IBM Outstanding Innovation Award in 1992, an IBM Faculty Partnership Award in 2001; and Outstanding Research Awards from the National Science Council, Taiwan, in 1999 and 2001, respectively. He is the author of the book, *Performance Guarantees in Communication Networks,* Springer-Verlag, 1999. He served as an editor for Operations Research from 1992 to 1999.

Xiuli Chao (North Carolina State University) received his doctoral degree in Operations Research from Columbia University in 1989. He is currently a professor of Industrial Engineering and the co-director of Interdisciplinary Operations Research Programs at North Carolina State University. Before joining NCSU, he was a professor in the Industrial and Manufacturing Engineering Department at New Jersey Institute of Technology. He also held visiting positions in the Graduate School of Management at University of California at Irvine, Department of Information Sciences at Science University of Tokyo, and the Department of Industrial Engineering and Operations Research at Columbia University. Since 2001 he has been an adjunct professor in the Institute of Systems Science, Chinese Academy of Sciences.

His research interest includes queueing theory and queueing networks, scheduling, optimal control, financial engineering, inventory theory, systems development, and more recently, supply chain management. He is a co-author of two books: *Operations Scheduling with Applications in Manufacturing and Services* (Irwin/McGraw-Hill, 1998), and *Queueing Networks: Customers, Signals, and Product Solutions* (John Wiley & Sons, 1999). He received the 1998 Erlang prize from the Applied Probability Society of INFORMS. He is a department editor for *IIE Transactions*; an associate editor for *Operations Research, Naval Research Logistics,* and *Queueing Systems: Theory and Applications*; and also serves on the editorial board of *Probability in the Engineering and Informational Sciences, Journal of Systems Science and Complexity,* and *International Journal of Information Technology and Decision Making.*

Han-Fu Chen (Chinese Academy of Sciences) joined the Institute of Mathematics, Chinese Academy of Sciences in 1961. Since 1979 he has been with the Institute of Systems Science, Chinese Academy of Sciences. He is now a Professor of the Laboratory of Systems and Control of the Institute. His research interests include stochastic systems, system identification, adaptive control, and stochastic approximation and its applications to other areas. He has authored or co-authored more than 150 journal papers.

He was elected to the Chinese Academy of Sciences in 1993. He currently serves as the President of the Chinese Association of Automation, and on the Technical Board of IFAC. He is the Editor of *Systems Science and Mathematical Sciences*, and *Control Theory and Applications.*

Hong Chen (University of British Columbia) received his Ph.D. degree in Engineering-Economic Systems from Stanford University (1987), and was Postdoctoral Fellow at Harvard University (1987-1988). He has been a professor at the Faculty of Commerce and Business Administration of the University of British Columbia since 1998, where he also served as an associate professor (1993-1998) and an assistant professor (1989-1993).

He is a UBC Faculty Killam Research Fellow (1993-1994), and a recipient of the Commerce Research Excellence Award from the UBC Faculty of Commerce and Business Administration (1993). He received the Engineering Initiation Award from the (U.S.) National Science Foundation (1989-1991), and the Overseas Chinese Young Investigator Award in 2000 from the National Science Foundation of China.

Professor Chen's research interests are primarily in the analysis, design and control of stochastic systems, such as manufacturing systems, service processing networks, supply chain, and economic and financial systems. He has published in journals such as *Annals of Probability*, *Annals of Applied Probability*, *IEEE Transactions on Automatic Control*, *Mathematics of Operations Research*, *Operations Research*, and *Queueing Systems.* He is a co-author of a book (with David D. Yao), *Fundamentals of Stochastic Networks: Performance, Asymptotics and Optimization* (Springer-Verlag, 2001). He is a principal investigator of numerous grants from government and industrial sources. He was the Chair of the INFORMS (Institute for Operations Research and Management Sciences) Applied Probability Technical Section (now the INFORMS Applied Probability Society) (1995/96). He is an associate editor for *Operations Research* and *Queueing Systems.*

Jim Dai (Georgia Institute of Technology) received the B.A. and M.S. degrees in mathematics from Nanjing University, in 1982 and 1985, respectively, and the Ph.D. degree in mathematics from Stanford University in 1990. He joined Georgia Institute of Technology in 1990, and since 1998 has been a professor of Industrial & Systems Engineering and of Mathematics. He was a visiting professor first at Aarhus University and later at Stanford University during his sabbatical year in 1998. He was also a visiting professor at National University of Singapore in summer 2001. Since 1998 he has been a concurrent professor at Nanjing University.

Professor Dai has been an active researcher in applied probability, where he has employed various analytical tools including fluid models, diffusion models, and partial different equations. His primary research interest is in the performance analysis and control of stochastic networks, with applications in telecommunications and semiconductor manufacturing. In recent years, he has also been interested in supply chain management.

He is a member of Institute for Operations Research and the Management Sciences (INFORMS). He is also a member of Institute of Mathematical Statistics. He served as an associate editor for *Mathematics of Operations Research*, and is currently an associate editor for *Operations Research*, *Management of Science*, and *Queueing Systems.*

Among the honors he received are NSF Young Investigator Award in 1994 by the National Science Foundation, Bergmann Memorial Research Grant in 1995 by US-Israel Binational Science Foundation, the Best Paper Award in 1996 by Sigma Xi, the Best Publication Award in 1997 and the Erlang Prize in 1998 both by the Applied Probability Society of INFORMS, and the Overseas Chinese Young Investigator Award in 2000 from the National Science Foundation of China.

Jin-Chuan Duan (University of Toronto) has a Ph.D. in finance from the University of Wisconsin-Madison. He currently holds the Manulife Chair in Financial Services at the Joseph L. Rotman School of Management, University of Toronto. He has previously taught at McGill University and the Hong Kong University of Science and Technology.

His research interests cover pricing of derivative securities, time-series econometrics and banking. Currently his research focuses on developing pricing theory and devising numerical and statistical techniques for time-series based pricing models.

Youyi Feng (Chinese University of Hong Kong) received his M.Phil. and Ph.D. from the Department of Industrial Engineering and Operations Research, Columbia University, in 1991 and 1995, respectively. In addition, he obtained B.S. in 1982 and M.S. in 1985 from the Department of Mathematics, Nanjing University. From 1994 to 1995, he worked as a financial analyst for Baring Security and Lasser Marshall, both in New York. He started his academic carrier in 1995 when he joined the National University of Singapore as an assistant professor in the Department of Information System, School of Computing. In the new millennium he changed career to take up a manager position in the research group of Enron Corp., where he conducted a revenue management project for natural gas pipelines networks. In 2001 he returned to Asia and joined the Chinese University of Hong Kong, where he is currently an associate professor in the Department of Systems Engineering and Engineering Management.

His research interests include revenue management for airlines, hotel and energy industries; stochastic inventory and production systems; supply chain management and e-commerce. His research works have been published in *Management Science, Operations Research, IIE Trans., IEEE Trans. on Automatic Control, Journal of Applied Probability, European Journal of Operational Research,* and *Ann. of Operations Research.*

Peilin Fu (Chinese University of Hong Kong) received the B.Eng. degree in automatic control from the Qingdao University of Science and Technology (previously Qingdao Institute of Chemical Technology) in 1996 and the M.Eng. degree in electronic engineering from the Ocean University of Qingdao in 1999. She is currently a Ph.D. candidate in the Department of Systems Engineering and Engineering Management at the Chinese University of Hong Kong. Her research interests include stochastic control theory, operations research and biological sequence analysis.

Geneviéve Gauthier (l'École des Hautes Études Commerciales, Montréal) holds a Ph.D. degree in mathematics from Carleton University (Ottawa). Her research interests cover financial modeling with applications in derivative securities and risk management. She is currently an Associate Professor at l'École des Hautes Études Commerciales, Montréal.

Otis Brian Jennings (Massachusetts Institute of Technology) completed his graduate studies in Industrial Engineering at the Georgia Institute of Technology in May 2000 under the aegis of Jim Dai. Subsequently, he completed an 18-month post-doctoral stint at Stanford's Graduate School of Business. He received a bachelor of science degree from Princeton University in 1994, where his focus was Operations Research. He is currently a visiting lecturer in the Operations Management group of MIT's Sloan School of Management. He has a joint appointment with the Operations Research Center at MIT. His research interests are the stability, control and performance analysis of stochastic networks.

Duan Li (Chinese University of Hong Kong) graduated from Fudan University, received his M.E. degree in Automatic Control from Shanghai Jiaotong University, and his Ph.D. degree in Systems Engineering from Case Western Reserve University.

From 1987 to 1994 he was a faculty member at University of Virginia, where he was an Associate Professor in the Department of Systems Engineering and Associate Director of the Center for Risk Management of Engineering Systems. In December 1994, he joined the Chinese University of Hong Kong where he is currently a Professor in the Department of Systems Engineering and Engineering Management.

Professor Li's research interests include optimization, optimal control, decision-making methodologies, and financial engineering. He has authored and coauthored over 100 technical papers in these areas. He is a Senior Member of IEEE. He was a Member of the Editorial Board of *Control-Theory and Advanced Technology* from 1992 to 1995 and an Associate Editor of *Information and Decision Technologies* from 1992 to 1994. He served as a guest co-editor for several special issues of *Journal of Global Optimization, Journal of Optimization and Engineering, Control-Theory and Advanced Technology*, and *Reliability Engineering and System Safety.* He was a co-chair of the International Program Committee of the 5th International Conference on Optimization: Techniques and Applications, December, 2001, Hong Kong.

Ping Lin (National University of Singapore) received his Ph.D. in Applied Mathematics in December 1995 from the University of British Columbia, Canada. He was then a postdoctoral fellow at the Program of Scientific Computing and Computational Mathematics in Stanford University from 1996 to 1998 and a research associate at the Department of Computer Science in Rensselaer Polytechnic Institute in 1998. Afterwards he joined the National University of Singapore as an assistant professor in the Department of Mathematics. The main emphasis in his research is in the areas of scientific computing, numerical analysis, computational mechanics, and mathematical modeling including revenue management for airlines, hotel and energy industries. His research works have been published in *SIAM J. Scientific Computing, SIAM J. Numerical Analysis, Mathematics of Computation, Studies in Applied Mathematics, Physica D, International J. of solids and Structures, Computing, Applied Mathematics and Mechanics,* and other journals.

Liming Liu (Hong Kong University of Science and Technology) received his B.S. and MSc. from Huazhong University of Science and Technology and his Ph.D. in Industrial Engineering from University of Toronto. He is currently an associate professor of Industrial Engineering and Engineering Management at Hong Kong University of Science and Technology. His research interests include inventory control, supply chain, service and manufacturing systems modeling and optimization.

Fucai Qian (Xi'an University of Technology) received the B.S. degree in mathematics from Shaanxi Normal University in 1984, the M.S. degree in mathematics from Northwest University in 1988, and the Ph.D. degree in systems engineering from Xi'an Jiaotong University in 1998. He was a postdoctoral fellow at Chinese University of Hong Kong in 1999.

From 1988 to 1998 he was a lecturer at Xi'an Petroleum Institute. He was a visiting scholar at Chinese University of Hong Kong in 2001. Since 1999 he has been an Associate Professor at School of Automation and Information Engineering, Xi'an University of Technology. His current research interests include optimal control theory, dynamic programming, large-scale systems, and stochastic estimation, identification and control.

Jean-Guy Simonato (HEC Montréal) holds a Ph.D. in finance from McGill University, and has taught at Université du Québec á Montréal. He is currently Caisse de Retraite Hydro-Québec Professor of finance at HEC, Montréal. His research interests focus on derivative securities and econometric methods applied to financial problems.

Baichun Xiao (Long Island University) received his BA in Mathematics from Nanjing University, an MBA from the Catholic University of Leuven, Belgium, and a Ph.D. in Operations Research /Operations Management from the Wharton School, University of Pennsylvania. Currently he is a Professor of Operations Management at College of Management, Long Island University, C.W. Post. He has published in *Operations Research, Management Science, Mathematical Programming, European Journal of Operational Research, Advances in Applied Probability* and other leading journals. Currently he serves as a board member and Secretary of Revenue Management Subdivision of INFORMS.

Houmin Yan (University of Texas at Dallas) received his B.S. and M.S. degrees from the Department of Automation, Tsinghua University, in 1982 and 1985, respectively, and his Ph.D. degree from the Faculty of Management, University of Toronto in 1993. Since 1994, he has been with the Department of Systems Engineering and Engineering Management, the Chinese University of Hong Kong, as an Assistance, Associate and Full Professor. Currently, he alos holds a tenured position in the School of Management, The University of Texas at Dallas.

His main research areas are operations management, stochastic models, simulation, and supply chain management. He has published in journals such as *Operations Research, Production and Operations Management, Discrete Dynamic Event Systems: Theory and Applications, Journal of Classification, Journal of Optimization Theory and Applications, International Journal of Production Economics*, and *IEEE Trans. on Automatic Control.* He is a member of INFORMS.

David D. Yao (Columbia University) received his Ph.D. degree from the University of Toronto in 1983, and started his academic career at Columbia University, where he became a full professor in 1988, and a holder of Columbia's Thomas Alva Edison Chair in 1992. He has held a part-time appointment at IBM's T.J. Watson Research Center since 1990. His other academic appointments include Harvard (1986-88), Yale (1991/92), the Chinese University of Hong Kong, and Tsinghua University (Beijing).

He is an IEEE Fellow, and a recipient of numerous honors and awards, including the Guggenheim Fellowship (1991/92); the Presidential Young Investigator Award (1987-92) from the National Science Foundation (U.S.); the Franz Edelman Award (1999) from the Institute for Operations Research and Management Sciences; and from IBM Research, the Outstanding Technical Achievement Award (1999), Research Division Award (1996), and several Invention Achievement Awards.

Author/co-author of over 150 refereed publications, three books and four edited volumes, he has served on the editorial board of several leading journals, including *Discrete Event Dynamic Systems, IEEE Transactions on Automatic Control, Management Science, Operations Research*, and *Queueing Systems.* A principal investigator of over two dozen research grants and contracts, he has done extensive research and consulting work in various aspects of semiconductor manufacturing, computer systems scheduling, Internet and web-server performance optimization, and supply chain management. He is a holder of four U.S. patents in manufacturing operations and supply-chain logistics.

Gang George Yin (Wayne State University) received the B.S. degree in Mathematics (magna cum laude) from the University of Delaware in 1983, the M.S. degree in Electrical Engineering from Brown University in 1987, and the Ph.D. degree in Applied Mathematics from Brown University in 1987. Subsequently, he joined Wayne State University, where he is currently a Professor of mathematics. He has visited the Institute for Mathematics and Its Applications at the University of Minnesota, North Carolina State University, the University of Georgia, Exxon Research and Engineering Company, the Chinese University of Hong Kong, the University of Dortmund (Germany), the University of Melbourne (Australia), and the University of Toronto (Canada).

His research interests include applied probability, stochastic processes and their applications. He has authored or co-authored numerous journal papers and conference proceedings, and two research monographs, *Stochastic Approximation Algorithms and Applications* (with Harold J. Kushner), Springer-Verlag, 1997; and *Continuous-Time Markov Chains and Applications: A Singular Perturbation Approach* (with Qing Zhang), Springer-Verlag, 1998. He was also the co-editor of three other research volumes, and a co-author (with A. Antomoshkin, H.-P. Schwefel, A. Torn, and A. Zilinskas) of the textbook, *Introduction to Systems Analysis,* published by the Russian Academy of Sciences and Engineering Siberia Branch in 1993.

He was elected an IEEE Fellow, "for contributions to approximation, optimization, and control of stochastic systems" in 2002. He received the Career Development Chair Award in 1993, the Board of Governors Faculty Recognition Award in 1999, the Charles H. Gershenson Distinguished Faculty Fellowships in 2001, and the College of Science Excellence in Teaching Award in 2001, all from Wayne State University.

He is a member of AMS (American Mathematical Society) and SIAM (Society for Industrial and Applied Mathematics). He served on the editorial board of *Stochastic Optimization and Design*, a journal based in Russia; was the editor of SIAM Activity Group on Control and Systems Theory Newsletter, and was an associate editor of *IEEE Transactions on Automatic Control.* He has also served on the program committees of numerous international conferences.

Tim Zajic (Lockheed Martin) received a Ph.D. in Operations Research from Stanford University in 1994. From 1994 to 1995 he was a visitor at the Centro de Investigación en Matemáticas in Guanajuato, México. From 1995 to 1997, he was a National Science Foundation Postdoctoral Fellow at the Department of Industrial Engineering and Operations Research at Columbia University and the IBM T.J. Watson Research Center. He was a visiting assistant professor in the School of Mathematics at the University of Minnesota from 1997 to 1998.

Dr. Zajic is currently a principal engineer at Lockheed Martin. His research interests include nonlinear filtering, classification and data fusion.

Hanqin Zhang (Chinese Academy of Sciences) received his PhD in 1991 from the Institute of Applied Mathematics, Chinese Academy of Sciences, and is currently a full professor there.

He has visited and worked at Marburg University (Germany), University of British Columbia (Canada), University of Toronto (Canada), University of Texas (USA), the Chinese University of Hong Kong, and other institutions. He has been on the editorial board of *Acta Mathematicae Applicatae Sinica* (English Series and Chinese Series), *Systems Science and Mathematical Sciences, Journal of Systems Science and Complexity,* and *IEEE Transactions on Automatic Controls.* His areas of research interests include stochastic manufacturing systems, stochastic optimal control, queueing networks and supply chain management.

Qing Zhang (University of Georgia) received his B.Sc. in Control Theory from Nankai University in 1983, and M.Sc. and Ph.D. from Brown University in 1985 and 1988, respectively. He is currently a Professor of Mathematics at University of Georgia.

His research interests include stochastic systems and control, nonlinear filtering, and applications in manufacturing and finance. He has published more than 70 journal papers and two monographs, *Hierarchical Decision Making in Stochastic Manufacturing Systems,* Birkhauser Boston, 1994, co-authored with S.P. Sethi; and *Continuous-Time Markov Chains and Applications,* Springer-Verlag, 1998 co-authored with G. Yin. He received a Creative Research Medal in 200 from The University of Georgia for outstanding accomplishments in research and creativity.

He co-edited three research volumes, was an associate editor of *IEEE Transactions on Automatic control,* and is currently an associate editor of *SIAM Journal on Control and Optimization.* He also served on several organizing committees of international conferences, including a Co-Chair of the 1996 AMS-SIAM Summer Seminar in Applied Mathematics.

Shuzhong Zhang (Chinese University of Hong Kong) graduated from Department of Mathematics, Fudan University, in 1984, and received his Ph.D. degree in Operations Research at Tinbergen Institute, Erasmus University Rotterdam, The Netherlands, in 1991.

He was a faculty member from 1991 to 1993 with the Department of Econometrics, University of Groningen; and from 1993 to 1999 with the Econometric Institute, Erasmus University Rotterdam. He joined the Department of Systems Engineering & Engineering Management at The Chinese University of Hong Kong in 1999, and is currently a Professor. His research interests include the theory and practices of optimization. He received the Research Prize of Erasmus University (one recipient per year) in 1999, and was ranked number 6 of the top 40 Dutch economists in the same year. He was the recipient of Vice Chancellor's Exemplary Teaching Award for 2001, The Chinese University of Hong Kong.

Shaohui Zheng (Hong Kong University of Science and Technology) received the M.S. degree in Operations Research/Optimal Control from the Academy of Sciences of China, Beijing in 1987, and the Ph.D. degree in Operations Research from Columbia University, New York in 1994.

He joined the School of Business and Management at the Hong Kong University of Science and Technology in 1994 as an assistant professor, and is currently an associate professor there. His research interests include production and quality control, queueing systems, and supply chain management. He is co-author of a book (with David D. Yao), *Dynamic Control of Quality in Production-Inventory Systems: Coordination and Optimization*, Springer-Verlag, 2002. He was a guest editor for *Annals of Operations Research*, Volume 87 (1999), and is currently an associate editor of *Operations Research*. He is also a holder of a patent in manufacturing management.

Xun Yu Zhou (Chinese University of Hong Kong) received his Ph.D. in Applied Mathematics in 1989 from Fudan University. During 1989-1991 and 1991-1993, he was doing postdoctoral research at Kobe University and University of Toronto, respectively. He was awarded Alexander von Humboldt Fellowship of Germany in 1991 and DAAD Research Fellowship of Germany in 1998. He has been with the Chinese University of Hong Kong since 1993 and is currently a Professor.

As a principal investigator of numerous research grants, he has done research in optimal stochastic control, mathematical finance/insurance, and discrete-event manufacturing systems. He has served as Organizer and Program Committee Member for numerous international conferences, and has been invited to visit and/or give seminars at universities around the globe. He has published about 60 papers in refereed journals, and his book entitled *Stochastic Controls: Hamiltonian Systems and HJB Equations* (co-authored with J. Yong) was published by Springer-Verlag in 1999. He is a senior member of IEEE and a member of SIAM, and is on the editorial board of leading journals including *Operations Research, IEEE Transactions on Automatic Control,* and *Mathematical Finance.*

GPSR Compliance

The European Union's (EU) General Product Safety Regulation (GPSR) is a set of rules that requires consumer products to be safe and our obligations to ensure this.

If you have any concerns about our products, you can contact us on ProductSafety@springernature.com

In case Publisher is established outside the EU, the EU authorized representative is:

Springer Nature Customer Service Center GmbH
Europaplatz 3
69115 Heidelberg, Germany

Batch number: 10682572

Printed by Printforce, the Netherlands